REAL ANALYSIS

A FIRST COURSE
WITH
FOUNDATIONS

MALCOLM W. POWNALL
Colgate University

Messages from the Past
by
GERTRUDE D. POWNALL

WCB Wm. C. Brown Publishers
Dubuque, Iowa • Melbourne, Australia • Oxford, England

Book Team

Editor *Paula-Christy Heighton*
Developmental Editor *Jane Parrigin*
Publishing Services Coordinator *Julie Avery Kennedy*

Wm. C. Brown Publishers
A Division of Wm. C. Brown Communications, Inc.

Vice President and General Manager *Beverly Kolz*
Vice President, Publisher *Earl McPeek*
Vice President, Director of Sales and Marketing *Virginia S. Moffat*
National Sales Manager *Douglas J. DiNardo*
Marketing Manager *Julie Joyce Keck*
Advertising Manager *Janelle Keeffer*
Director of Production *Colleen A. Yonda*
Publishing Services Manager *Karen J. Slaght*
Permissions/Records Manager *Connie Allendorf*

Wm. C. Brown Communications, Inc.

President and Chief Executive Officer *G. Franklin Lewis*
Corporate Senior Vice President, President of WCB Manufacturing *Roger Meyer*
Corporate Senior Vice President and Chief Financial Officer *Robert Chesterman*

Copyediting and production by Custom Editorial Production, Inc.

A Times Mirror Company

Library of Congress Catalog Card Number: 93–7212

ISBN 0–697–12908–X

Printed in the United States of America by Wm. C. Brown Communications, Inc., 2460 Kerper Boulevard, Dubuque, IA 52001

10 9 8 7 6 5 4 3 2 1

For Kasey, Sammy, . . .

CONTENTS

REAL ANALYSIS

PREFACE

Unlike many upper level mathematics courses, a first course in real analysis can be described at the outset in terms the students can understand: this course will develop the theory of elementary calculus. One of its major goals can be specified clearly: to achieve a proof of the Fundamental Theorem of Calculus.

The unstarred parts of this book from Section 1.1 through Section 4.8 cover the "core" concepts leading to the Fudamental Theorem (completeness, sequences, limits, continuity, derivatives, and integrals). Many of these concepts will be familiar on an intuitive level from elementary calculus. Of course in real analysis they will be treated in a more formal manner.

An important feature of this text is its use of sequences as a unifying concept. Through simple properties of null sequences, the limit theorems for sequential limits are readily obtained, and the latter are used to define limits and continuity for real functions. Although proofs and examples are frequently carried out by means of sequences, the epsilon-delta and neighborhood formulations are not neglected. It is advantageous for students to have several ways to approach a proof or problem, and it is often instructive to compare different methods of proving the same theorem.

Most students learn theoretical mathematics by first seeing a number of examples and then working a variety of exercises. This text contains a good supply of both. Some exercises are hard enough to challenge even the most capable students. Others are routine—the kind of exercises that simply ask the students to think through the meaning of a definition or to investigate the application (or nonapplication) of a particular theorem. It is important to have some easy exercises because they help to build confidence. A number of exercises ask for proofs of theorems that parallel other theorems whose proofs are given in the text. These enhance the students' understanding by encouraging them to restudy the text. I especially favor "prove or disprove" exercises because they help to develop independence and analytical thinking.

My objective has been to present the concepts of analysis at an elementary, but honest, mathematical level, and so I have tried to provide explanations in considerable detail. I hope that students will find the results clear, even if challenging. I also hope that instructors will find enough attention to detail to spare them from lecturing on every aspect of every topic, thus allowing time for those digressions that are so welcome.

Chapter 1 introduces the completeness property of the real number system (R) and presents some major consequences of this property.

Chapter 2 contains the basic theory of real sequences, including the Bolzano-Weierstrass Theorem and the Cauchy Criterion. (The last two sections are devoted to infinite series and can be postponed or omitted if it is desired to reach the Fundamental Theorem quickly. However, these sections obviously will be needed if Chapter 5 is to be covered, and there are some pedagogical advantages to separating series of constants from series of functions.)

Chapter 3 is devoted to the theory of limits and continuity for real functions. Special emphasis is placed on the consequences of continuity on a closed, bounded interval. The chapter concludes with three sections that provide an introduction to the topology of the real line, and lead to a characterization of continuity in terms of open sets.

Chapter 4 is the centerpiece of the text and it is here that differential and integral calculus are brought to a fortuitous union in the Fundamental Theorem (Section 4.8). A number of applications are then discussed, including analytic definitions of several of the transcendental functions by means of integrals. Chapter 4 concludes with Taylor's Theorem, and sets the stage for series of functions.

The major theoretical topic in Chapter 5 is uniform convergence and power series is its main application. The chapter concludes with an example of an everywhere continuous, nowhere differentiable function.

Chapter 6 is intended to give some glimpse into where our earlier studies lead: metric spaces, topology, and functional analysis. The concluding unit on l^2 is especially good for bringing many interesting ideas under one roof.

The Appendix is devoted to the Isomorphism Theorem for complete ordered fields. For students who have studied abstract algebra, the proof of this theorem is a wonderful opportunity to see the interaction of algebraic and analytic ideas.

The Foundations Chapters

During the ten-year period in which this book has been developed, there have been several changes in the background of undergraduates entering the first real analysis course. One observation is that their calculus experience now tends to be more informal and intuitive than before. Also, the sophomore linear algebra course has virtually disappeared, and so in some cases, the students' experience with the language of formal mathematics and proofs does not extend much beyond the high school level. On the other hand, in resonse to these trends, there is a growing tendency to fill the vacancy left in the sophomore program with a "transitional" or "foundations" course.

The foundations chapters had their origin about five years ago in optional supplementary classes for some of my students who wished to strengthen their skills in proof-writing. Traditionally, teachers of mathematics have assumed that students learn to write proofs by an unconscious process of imitation;

only those who imitated successfully could succeed as mathematics majors. There is abundant evidence in the newer elementary analysis texts (and other texts as well) to indicate a growing trend toward a conscious, explicit effort to "teach proofs". Chapter L includes my attempt to do this. I found it desirable to begin with a very small amount of informal logic, such as truth tables, schemes of inference, and quantifiers, and then apply these tools to proving propositions about divisibility in the set of positive integers. In this very simple setting, there is much to discuss concerning proofs by cases, proofs by contradition, indirect proofs, existence proofs, and counter-examples. The mathematics is simple enough that it does not distract from the proof process; yet in most cases, the mathematics is interesting and significant. Chapter L is a natural place to include sets, relations, and functions. It concludes with an optional section discussing the use of equivalence relations and partitions to "construct" new number systems from old ones.

Chapter R summarizes the algebra and order properties of the real number system, with which most students will be quite familiar. Although the concept of *ordered field* is introduced, emphasis is on R. It is nevertheless important for mathematics majors to know that R is but one of many ordered fields, an idea that will be revisited in Chapter 1 when completeness is introduced. Chapter R seems a suitable place in which to locate material involving notation and terminology concerning real functions, and such properties as boundedness and periodicity. Section R.3 contains this mostly familiar material and can be used as a reference.

Chapter P focuses on the set of positive integers and shows how well-orderedness leads to the idea of recursive definition and the method of proof by induction.

Chapter C deals with infinite sets, making the distinction between countable and uncountable sets. Although this distinction is essential at only a few places in the text prior to Chapter 6, it is something that mathematics majors should find interesting.

In order to provide for selectivity in the use of the foundations chapters, I have tried to make them as independent of one another as possible. Naturally the function concept occurs in several places: in Section L.4, a general definition of *function* is given; in Section R.3, real functions are introduced in such a way that reference to L.4 is hardly necessary; and in Section C.1, when cardinal equivalence is defined, descriptions of *one-to-one* and *onto* mappings are repeated briefly in the text, with a footnote indicating that more detail will be found in Section L.4. Similarly in Chapter 1, the definitions relating to bounds for sets are repeated, in case the instructor has chosen to omit Chapter R.

Starred Sections and Exercises

Certain sections, subsections, and exercises are starred. A starred section is one that can be omitted or postponed without harm. For example, Section L.5 deals with construction of number systems and is a prerequisite only for certain exercises and projects of that type. Exercises that are dependent upon earlier starred sections or exercises are also starred, usually with a specific reference.

Sections 2.8 and 2.9 on infinite series are starred so that instructors may take up these sections in connection with series of functions (Chapter 5). Sections 3.10, 3.11, and 3.12 deal with the elementary topology of the line and are prerequisite only for Chapter 6. Sections 3.7 and 4.9 through 4.15 on the transcendental functions are starred in case the instructor wishes to accept their content without theoretical development.

Projects

A number of projects are suggested in order to extend and amplify the content of the course. These are usually presented in outline form and range in length from a long exercise to a term paper. Projects should encourage independence in pursuit of mathematical knowledge.

"Messages from the Past"

These portraits and brief biographical sketches of thirteen famous mathematicians are included as a reminder that mathematics today is the result of the cumulative efforts of many great minds over many centuries. They should help to assure students that insight and success in mathematics often do not come instantly, but after patience, perseverence, and learning from the work of others.

Use of This Text

This book contains more than enough material for a one-semester course, and is arranged so that instructors can use it selectively. Here are some possibilities for a one-semester course.

(1) It should be possible in one semester (13-15 weeks) to cover the four foundations chapters and the core material leading to the Fundamental Theorem. (Note that this excludes the starred sections.)

(2) For students with more background (say a foundations course or some other upper level theory course in mathematics), the foundations chapters can be omitted and designated as a reference. Then a course culminating in uniform convergence and infinite series can be built upon the following: Chapter 1; Chapter 2; the unstarred parts of Chapters 3 and 4; and Chapter 5. (Since the analytic definitions of transcendental functions are omitted, the properties of these functions will have to be taken without proof.)

(3) Another possibility for students with good background is to aim for Chapter 6; in this case, Chapter C should be included, together with all of Chapters 1 and 2; all of Chapter 3 except Section 3.7; the unstarred parts of Chapter 4; and Chapter 6. (It will be necessary in this arrangement to omit a few references to uniform convergence in the examples and exercises of Chapter 6.)

An ideal arrangement would be a two-quarter course, in which most of the book could be covered. Among the candidates for omission would be the starred parts of Chapters L, 4, and 5.

Acknowledgements

My sincere thanks go to many people who helped me throughout the development of this book: To the staff at William C. Brown and their reviewers for consistent encouragement and helpful advice. To the many students at Colgate University who bravely made their way through various early versions. To Cathy McFadden, who expertly typed and retyped the manuscript several times. To my colleagues David Lantz and David Yuen who taught from preliminary versions and made valuable suggestions that I have incorporated in the text. To my colleague William E. Mastrocola who has taught from the manuscript almost as many times as I have, and whose good friendship I have perhaps abused by continually asking for advice. Last but not least, to my wife Gertrude who has been a constant source of support and who lovingly prepared the "Messages from the Past".

MESSAGES FROM THE PAST

Mathematics, as presented in textbooks, seems an impressive structure, clean and strong, each level resting securely on a supporting level, each part dovetailed perfectly into the whole.

It should be recognized, however, that the concepts of mathematics were developed over many years, not by a process of evolution but by the creative efforts and inspirations of great minds, one mind building on the achievements of others. These human contributions come to us as legacies to be cherished: studied, used, and perhaps improved, before being passed on to others. They are messages from the past and it is only natural to seek their sources: individuals of talent who lived and struggled in the framework of other times and places. A sampling of such background is included in this book with very brief biographical sketches of some of the mathematicians whose messages are pertinent to our studies.

Readers who wish to learn more about these mathematicians and about the history of mathematics will find a selection of appropriate references at the end of this text.

A WORD TO THE READER

Real analysis deals with several concepts that you have already met in the study of calculus. There you learned to find derivatives and integrals of functions and to use those powerful tools to solve a variety of problems. Our approach will differ substantially from that in calculus. Here we will be interested specifically in the theory that lies behind that important subject. We will define concepts carefully, and then look into the reasons why they behave as they do. We will ask "why?" more often than "how?" and "how?" more often than "how much?"

Except for the properties of the number system, we will not take much for granted. Since derivatives and integrals are based upon limit concepts, Chapters 1, 2, and 3 are devoted to limits of various kinds and other closely related ideas. These concepts were many centuries in the making and so it is important to take a careful look at them. In Chapter 4, we will develop the theory of derivatives and integrals, including the important relationship between them known as the Fundamental Theorem of Calculus. The field of analysis does not end with calculus—far from it; limits, derivatives, and integrals are only the beginning. We take a glimpse at some of these things in Chapters 5 and 6.

In studying a "theory" course in mathematics, it is of the utmost importance to know and understand definitions. When you come to a definition in the text, you will usually find it followed by some examples and perhaps some nonexamples. Think these through carefully because they will help you understand the definition. With the meanings of terms firmly in mind you should be in a good position to tackle the reading and writing of proofs.

In analysis, it frequently happens that the meaning of a theorem can be grasped by making a graph or a sketch of some sort, which then may also help in approaching the proof. Thus, while you are reading the text, you should always have pencil and paper handy.

Mathematics is not to be read as if it were a novel. It may take several readings to absorb the contents of a single paragraph. Perhaps the best plan is to read over the material of a section quickly in order to form a general impression; then go back for more careful readings until the details fall into place.

Doing mathematics is quite different from reading mathematics but there are some parallels. Although you will be able to work many exercises on the first attempt, there will be some that require persistence (a quality that every professional mathematician must have). You may be surprised that a hard problem yields after being attacked several times at well-spaced time intervals.

Caution: In this book you will encounter a number of statements you have seen before; for example, "The limit of a sum is the sum of the limits." There is a temptation to regard this statement as "obvious". Please be careful to distinguish what is *obvious* from what is *familiar*. After a good definition of limit is formulated, we can and will prove that the limit of a sum is the sum of the limits. The proof is neither difficult nor obvious; you will find that it follows in a natural way from our definitions.

Theorems in analysis frequently come in pairs. Here are two theorems you will recall from calculus.

Theorem A *A function having a positive derivative on an interval is increasing there.*

Theorem B *A function having a negative derivative on an interval is decreasing there.*

These statements are very similar in form, and we can obtain one from the other if we just replace certain ideas by "dual" ideas. In this case "positive" and "negative" are dual ideas, as are "increasing" and "decreasing". Some other pairs of dual ideas in analysis are "greater than" and "less than", and "maximum" and "minimum".

We will often compress two statements like Theorems A and B into one by the use of parentheses:

A function having a positive (negative) derivative on an interval is increasing (decreasing) there.

A statement presented this way is read first without the parenthesized words; then an alternate reading substitutes the parenthesized words for those preceding them. This procedure not only saves space and avoids a certain amount of repetition, it also serves to emphasize the parallelness of ideas in the two statements.

As another example, let's recall the First Derivative Test, which you used in calculus to test critical points for being maxima or minima. It appears on page 272 as Theorem 4.3.5.

Theorem 4.3.5 **First Derivative Test.** *Let f be a real function continuous on a set that contains an open interval (a, b) and suppose that $a < c < b$. If $f'(x)$ exists and is positive (negative) for $x \in (a, c)$, and if $f'(x)$ exists and is negative (positive) for $x \in (c, b)$, then f assumes a relative maximum (minimum) at c.*

Take a moment to write out the two statements that are compressed into the concluding sentence of the theorem. And while you have your pencil out, why not make sketches illustrating these two statements. Do you see how Theorems A and B lie behind the First Derivative Test?

As your course in real analysis begins, I hope you will look through this text to get a general idea of what it contains that can assist you with your studies. Your instructor may not specifically cover all of the foundations chapters (Chapters L, R, P, and C)—in most cases because you are likely to be familiar with some of the material. Be aware of what is in those chapters in case you need to refer to them. (For example, you may have studied the Binomial Theorem in an earlier course; notice its presence in Section P.2, and if at some time later you need to refresh your memory, you know where to find the statement of the theorem and some examples. Similarly, algebraic properties of the real number system, as well as inequalities and absolute values, are located in Chapter R.) At the back of the book, you will find a list of special symbols, with reference to the page where each such symbol is introduced. And there is a list of references in case you wish to look elsewhere for help.

Keep in mind that mathematics has prospered, and continues to prosper, in a *community* of scholars, teachers, and students. By all means, ask questions, suggest your ideas to others. Talking about problems, examples, and theorems with others is just as vital a part of your mathematical experience as reading or writing mathematics. Good luck!

MWP

FOUNDATIONS

CHAPTER L

THE LANGUAGE OF MATHEMATICS

This chapter is intended especially for those of you who are taking a theoretical mathematics course for the first time. We review basic concepts, such as equality, sets and operations with sets; discuss the special way mathematics employs certain logical expressions; and illustrate the mathematical process known as proof by several examples. We also review the terminology and notation concerning relations and functions.

L.1 *Logic, Sets, Quantifiers*

Equality

The statement "a equals b", written $a = b$, means that the symbols "a" and "b" denote the same object. If a and b are not equal, we write $a \neq b$.

The following *properties of equality* hold for any objects a, b, and c whatsoever:

Reflexive Law:	$a = a$
Symmetric Law:	If $a = b$, then $b = a$.
Transitive Law:	If $a = b$ and $b = c$, then $a = c$.
Replacement Law:	If $a = b$, then in any statement in which the symbol "a" occurs we may replace "a" by "b" without changing the meaning of the statement.

As an instance of the replacement law, the statement "2 + 3 is less than 7" has the same meaning as "1 + 4 is less than 7" because "2 + 3" and "1 + 4" both denote the same number, 5.

We often need to add two equations. If $a = b$ and $c = d$, we can conclude that $a + c = b + d$. To see why, start with the reflexive law

$$a + c = a + c$$

On the right, replace "a" by "b" and "c" by "d". Similarly, if $a = b$ and $c = d$, then $ac = bd$. Of course, similar statements may be made with respect to other operations.

Hereafter the properties of equality are used without specific reference.

Logic

Logical Connectives

Mathematics uses certain words and phrases in a very special way. If you are studying theoretical aspects of mathematics for the first time, it is important to become aware of this special usage.

First, let us look at what is meant by a "statement" in mathematics. Grammatically, a statement must be a sentence—it must have a subject and a predicate; furthermore, a statement must be either true or false but, of course, not both! For example, the statement "3 is positive" is true; the statement "3 is even" is false.

Suppose we have several statements, denoted by p, q, r, ... We form new "compound" statements when we connect the given statements by means of such words as *or*, *and*, *if and only if*, which are called *logical connectives*. Examples are "p or q", "q and r", "p if and only if (q or r)". For the present we are not concerned with the specific content of the statements p, q, r, etc., nor with whether they are true or false; rather, we are concerned with how their truth or falsity determines the truth or falsity of the compound statements built up from them. Given the "truth value" (T for true, F for false) of each of p, q, r, . . ., we want to determine the truth value of "p or q", "q and r", "p if and only if (q or r)", etc.

Disjunction. The logical connective *or* is denoted by the symbol $\vee$; "p or q" is written "$p \vee q$" and is called the *disjunction* of p and q. The connective $\vee$ is used in mathematics in an inclusive sense, so that $p \vee q$ is true whenever *at least one* of p, q is true; it is false only when *both* p and q are false. Using T for true and F for false, we summarize disjunction in the following truth table:

p	q	$p \vee q$
T	T	T
T	F	T
F	T	T
F	F	F

Note: In ordinary conversation the word *or* is often used in an exclusive sense; that is, the possibility that both p and q are true is excluded. Keep in mind that mathematical usage is different.

Example 1: The statement

3 is positive or 3 is odd

is true; also

3 is positive or 3 is even

is true.

Conjunction. The logical connective *and* is denoted by the symbol $\wedge$; "p and q" is written "$p \wedge q$" and called the *conjunction* of p and q. The compound statement $p \wedge q$ is true when *both* p and q are true, otherwise it is false.

p	q	$p \wedge q$
T	T	T
T	F	F
F	T	F
F	F	F

Example 2: The statement

3 is positive and 3 is even

is false.

Conditional. A statement of the form "if p then q" is called a *conditional* statement and is written "$p \Rightarrow q$".* We refer to p as the *antecedent* and q as the *consequent* of $p \Rightarrow q$. In logic and mathematics the conditional statement $p \Rightarrow q$ is taken to be true whenever p is false or q is true. That is,

p	q	$p \Rightarrow q$
T	*T*	*T*
T	*F*	*F*
F	*T*	*T*
F	*F*	*T*

Note: The important mathematical convention that $p \Rightarrow q$ is true whenever p is false may seem strange at first. The following illustration from everyday life may be helpful. John promises Mary, "If it doesn't rain, I'll take you on a picnic." Now if it does rain, then no matter whether John does or does not take Mary on a picnic, he has not broken his promise.

Example 3: The statement

If 3 is positive, then 3 is even

is false; however,

If 3 is even, then 3 is positive

is true.

Biconditional. A statement of the form "p if and only if q" is called a *biconditional* statement and is written "$p \Leftrightarrow q$". It is true when p and q have the same truth values, otherwise it is false:

p	q	$p \Leftrightarrow q$
T	*T*	*T*
T	*F*	*F*
F	*T*	*F*
F	*F*	*T*

Example 4: The statement

3 is positive if and only if 3 is even

is false. The statement

3 is greater than 4 if and only if 3 is even

is true.

*Sometimes the statement $p \Rightarrow q$ is called an *implication* and $p \Rightarrow q$ is read "*p implies q*"; however, in this section we reserve these words for another purpose.

Statements involving several connectives can be analyzed by truth tables.

Example 5: The truth table for $p \Rightarrow (p \wedge q)$ is obtained by listing the four possibilities for the truth values of p and q and then "computing" in each case the truth value of $p \wedge q$ and finally the truth value of $p \Rightarrow (p \wedge q)$:

p	q	$p \wedge q$	$p \Rightarrow (p \wedge q)$
T	*T*	*T*	*T*
T	*F*	*F*	*F*
F	*T*	*F*	*T*
F	*F*	*F*	*T*

Example 6: Consider $p \Leftrightarrow (q \vee r)$. The eight possibilities for the truth values of p, q, and r are listed on the left side of the table below. In each case, the truth values of $q \vee r$ and $p \Leftrightarrow (q \vee r)$ are determined by using the tables for $\vee$ and $\Leftrightarrow$.

p	q	r	$q \vee r$	$p \Leftrightarrow (q \vee r)$
T	*T*	*T*	*T*	*T*
T	*T*	*F*	*T*	*T*
T	*F*	*T*	*T*	*T*
T	*F*	*F*	*F*	*F*
F	*T*	*T*	*T*	*F*
F	*T*	*F*	*T*	*F*
F	*F*	*T*	*T*	*F*
F	*F*	*F*	*F*	*T*

Negation. The statement "not p" is called the *negation* of p or the *denial* of p and is denoted "$\sim p$". The statement $\sim p$ has a truth value opposite to that of p.

p	$\sim p$
T	*F*
F	*T*

Logical Equivalence

Two compound statements involving p, q, r, etc., are *logically equivalent* provided they have the same truth values in every case. To test for the logical equivalence of two statements, we simply write out their truth tables and compare, case by case.

Example 7: Show that the statement $(\sim p) \vee q$ is logically equivalent to $p \Rightarrow q$.

Solution: The truth table for $(\sim p) \vee q$ is

p	q	$\sim p$	$(\sim p) \vee q$
T	T	F	T
T	F	F	F
F	T	T	T
F	F	T	T

Clearly it agrees in each case with the table for $p \Rightarrow q$ above. Because of the logical equivalence of $p \Rightarrow q$ and $(\sim p) \vee q$, the statements

$$\text{If } a \neq 2, \text{ then } a = 3$$

and

$$a = 2 \text{ or } a = 3$$

have the same mathematical meaning.

Example 8: Show that $\sim (p \wedge q)$ is logically equivalent to $(\sim p) \vee (\sim q)$.

Solution: The truth tables of the two statements are combined below:

p	q	$p \wedge q$	$\sim (p \wedge q)$	$\sim p$	$\sim q$	$(\sim p) \vee (\sim q)$
T	T	T	F	F	F	F
T	F	F	T	F	T	T
F	T	F	T	T	F	T
F	F	F	T	T	T	T

The columns for $\sim (p \wedge q)$ and $(\sim p) \vee (\sim q)$ are identical, and so the two statements are logically equivalent. Thus, to deny $p \wedge q$ is to assert not p or not q. Accordingly the statements

$$\text{It is not the case that } a = 0 \text{ and } b = 0$$

and

$$a \neq 0 \text{ or } b \neq 0$$

have the same meaning.

Example 9: In Exercise L.1.2*a*, you are asked to show that $\sim(p \vee q)$ is logically equivalent to $(\sim p) \wedge (\sim q)$. Thus to deny $p \vee q$ is to assert neither p nor q.

Example 10: In Exercise L.1.2*b*, you are asked to show that the negation of $p \Rightarrow q$ is logically equivalent to $p \wedge (\sim q)$. To deny $p \Rightarrow q$ is to assert $p \wedge \sim q$.

Example 11: $p \Leftrightarrow q$ is logically equivalent to $(p \Rightarrow q) \wedge (q \Rightarrow p)$. (See Exercise L.1.2*c*.)

Converse, Contrapositive

Many statements of mathematics have the form of a conditional.

Example 12: Recall the theorem from geometry:

If triangles A and B are congruent, then A and B are similar.

This theorem can be expressed logically as $p \Rightarrow q$, where p is "triangles A and B are congruent" and q is "triangles A and B are similar".

When a theorem has the form $p \Rightarrow q$, we say that p is a *sufficient condition* for q, and that q is a *necessary condition* for p. In Example 12, being congruent is a sufficient condition for two triangles to be similar; being similar is a necessary condition for them to be congruent.

Related to the conditional $p \Rightarrow q$ is its *converse*, $q \Rightarrow p$. The converse of a conditional is another conditional, in which the antecedent and consequent are interchanged. It is important to note that a conditional and its converse are not equivalent (work Exercise L.1.3a); the converse of a theorem is not necessarily a theorem itself. The converse of the theorem in Example 12 is

If triangles A and B are similar, then they are congruent.

It is *not* true.

The *contrapositive* of $p \Rightarrow q$ is the conditional $(\sim q) \Rightarrow (\sim p)$. It is easy to verify that $p \Rightarrow q$ and $(\sim q) \Rightarrow (\sim p)$ are logically equivalent (work Exercise L.1.3b). The contrapositive of the theorem in Example 12 is

If triangles A and B are not similar, then they are not congruent.

Example 13: Consider the statement

If a and b are nonzero real numbers, then their product is nonzero.

Let p be "$a \neq 0$", q be "$b \neq 0$" and r be "$ab \neq 0$". The given statement can be expressed as

$$(p \wedge q) \Rightarrow r$$

Its contrapositive,

$$\sim r \Rightarrow \sim (p \wedge q)$$

is equivalent by Example 8 to

$$\sim r \Rightarrow (\sim p) \vee (\sim q)$$

In words,

If $ab = 0$, then $a = 0$ or $b = 0$.

Thus, the given statement can be rephrased as follows:

> If the product of two numbers is zero, then at least one of the two numbers is zero.

When a theorem has the form $p \Leftrightarrow q$, p is both necessary and sufficient for q, and vice versa. A familiar example follows.

Example 14: Consider the Pythagorean Theorem:

Theorem: A triangle is a right triangle if and only if the sum of the squares of two of its sides is equal to the square of the third side.

Let p be "triangle T is a right triangle", and let q be "the sum of the squares of two sides of T is equal to the square of the third side". The theorem states that p is necessary and sufficient for q. Because $q \Rightarrow p$, we know that any triangle whose sides can be shown to satisfy the condition q must be a right triangle. Because $\sim q \Rightarrow \sim p$ (this is the contrapositive form of $p \Rightarrow q$), we can conclude that a triangle whose sides do not satisfy condition q is not a right triangle.

Some Logical Equivalences

Statement	Logically Equivalent Statement
$p \Rightarrow q$	$(\sim p) \vee q$
$\sim (p \wedge q)$	$(\sim p) \vee (\sim q)$
$\sim (p \vee q)$	$(\sim p) \wedge (\sim q)$
$\sim (p \Rightarrow q)$	$p \wedge (\sim q)$
$p \Rightarrow q$	$\sim q \Rightarrow \sim p$
$p \Leftrightarrow q$	$(p \Rightarrow q) \wedge (q \Rightarrow p)$

Logically True Statements

A statement whose truth value in every case is T is said to be *logically true*.

Example 15: By referring to its truth table, we see that $p \Rightarrow (p \vee q)$ is logically true:

p	q	$p \vee q$	$p \Rightarrow (p \vee q)$
T	T	T	T
T	F	T	T
F	T	T	T
F	F	F	T

A logically true statement is true because of its form rather than its content. Such a statement is also called a *tautology*. Some other tautologies are: $(p \wedge q) \Rightarrow p$, $p \Rightarrow p$, and $\sim(\sim p) \Leftrightarrow p$.

Implication

Consider two compound statements P and Q built up from simpler statements such as p, q, r, etc. by means of logical connectives. The statement P is said to *imply* Q provided that the statement $P \Rightarrow Q$ is a tautology.

Example 16: Let P be the statement $\sim(p \vee q)$ and Q be the statement $\sim p$. The truth table of $\sim(p \vee q) \Rightarrow \sim p$ is

p	q	$\sim p$	$\sim(p \vee q)$	$\sim(p \vee q) \Rightarrow \sim p$
T	T	F	F	T
T	F	F	F	T
F	T	T	F	T
F	F	T	T	T

and so P implies Q.

Schemes of Inference

A scheme of inference is a pattern of reasoning by which we deduce a "conclusion" from given "premises".

Example 17: Here is a simple scheme of inference: Our premises consist of two statements, $p \Rightarrow q$ and p. We deduce q.

We represent a scheme of inference by placing the premises above a horizontal bar and the conclusion below it, following a $\therefore$ sign. Thus, for the scheme in Example 17, we have

$$\begin{array}{l} p \Rightarrow q \\ \underline{p \qquad\;\;} \\ \therefore q \end{array}$$

A scheme of inference is considered to be valid if the conjunction of the premises implies the conclusion. We can test the validity of such a scheme by truth tables.

Example 18: We test the validity of the scheme in Example 17. The conjunction of the premises is $(p \Rightarrow q) \wedge p$; the conclusion is q. To show that $(p \Rightarrow q) \wedge p$ implies q, we check the truth table of

$$[(p \Rightarrow q) \wedge p] \Rightarrow q$$

p	q	$p \Rightarrow q$	$(p \Rightarrow q) \wedge p$	$[(p \Rightarrow q) \wedge p] \Rightarrow q$
T	T	T	T	T
T	F	F	F	T
F	T	T	F	T
F	F	T	F	T

Since the final statement is true in every case, the implication holds, and the scheme is valid.

There are various other schemes of inference commonly used in mathematics, some of which follow:

Some Valid Schemes of Inference

			$p_1 \Rightarrow q$
$p \Rightarrow q$	$p \Rightarrow q$	$p \Rightarrow q$	$p_2 \Rightarrow q$
p	$q \Rightarrow r$	$\sim q$	$p \Rightarrow (p_1 \vee p_2)$
$\therefore q$	$\therefore p \Rightarrow r$	$\therefore \sim p$	$\therefore p \Rightarrow q$

You are asked to show the validity of these schemes in Exercise L.1.5.

Sets

The concept of set is so fundamental in all mathematics that it is ordinarily left undefined. A set is formed whenever certain objects are collectively regarded as a single entity.

The objects from which a set is formed are called its *elements* or *members* and are said to *belong* to the set. The fact that an object x belongs to a set S is symbolized as $x \in S$. If x does not belong to S, we write $x \notin S$.

If all the elements of a set A are also elements of a set B, then we say that A is a *subset* of B and write $A \subseteq B$; symbolically, $x \in A \Rightarrow x \in B$. The relation $\subseteq$ is called *inclusion*. Clearly the following laws apply to all sets A, B, C:

Reflexive Law: $A \subseteq A$ This law holds because $x \in A \Rightarrow x \in A$ is logically true.

Transitive Law: If $A \subseteq B$ and $B \subseteq C$, then $A \subseteq C$. This law holds because

$$[(x \in A \Rightarrow x \in B) \wedge (x \in B \Rightarrow x \in C)] \Rightarrow (x \in A \Rightarrow x \in C)$$

is logically true.

The real numbers constitute a familiar set that will be denoted here by the letter R. The following are some important subsets of R.

(1) *Integers.* These are the whole numbers $0, \pm 1, \pm 2, \ldots$.* They include the positive integers, their negatives, and the number 0. We denote the set of all integers by Z.

(2) *Positive integers.* The positive integers 1, 2, 3, . . . are sometimes called the *natural numbers* because of the basic character of the counting process for which these numbers were invented. We denote the set of positive integers by Z^+. In Section P.1, we make some explicit assumptions about Z^+ and its relation to R, and explore some consequences of these assumptions.

(3) *Rational numbers.* A real number is *rational* provided it can be expressed as a ratio, or quotient, p/q, where $p \in Z$, $q \in Z$, and $q \neq 0$. Examples of rational numbers are $\frac{2}{3}, \frac{-5}{8}, \frac{22}{7}$. The set of all rational numbers is denoted by Q.

(4) *Irrational numbers.* A real number that is not rational is called *irrational*. The existence of irrational real numbers was discovered by the Greek mathematicians, perhaps as early as 500 B.C. The Greeks had believed devoutly that all "numbers" were rational, and so this discovery came as a profound shock. Its influence on the subsequent development of mathematics is difficult to exaggerate. Indeed, irrational numbers were not satisfactorily understood until the nineteenth century. In the next section we learn how the Greeks discovered the irrationality of $\sqrt{2}$. Other familiar irrationals are the number π from geometry and the base of natural logarithms, e.

Equality of Sets

Equality of two sets occurs if and only if each set contains the other: that is, $A = B$ if and only if $A \subseteq B$ and $B \subseteq A$.† If $A \subseteq B$ but $A \neq B$, then A is said to be a *proper* subset of B, meaning that every element of A is an element of B and that there is at least one element of B that is not an element of A. For

*The three dots can be read as "and so on".

†In set theory this statement is taken as an axiom, known as the *axiom of extension*.

instance, Z^+ is a proper subset of Z, which is a proper subset of Q, which in turn is a proper subset of R.

One way to define a particular set is to list its elements within a pair of braces. For instance, $A = \{1,2,3,4\}$. Another way to define a set is to state a criterion for membership, as in the next examples.

Example 19: The set $\{1,2,3,4\}$ can be indicated by the notation $\{x : x \in Z^+$ and $x < 5\}$, where the braces are read "set of all" and the colon is read "such that": thus, "the set of all x such that x is a positive integer and $x < 5$".

Example 20: $\{x : x \in Z^+ \text{ and } x^2 > 10\}$ denotes the set of positive integers whose squares are greater than 10. Another way of describing that set is $\{x : x \in Z^+ \text{ and } x > 3\}$.

Union, Intersection, and Difference

If A and B are sets, the *union* of A and B, denoted $A \cup B$, is the set that consists of the elements of A together with the elements of B (including any elements that belong to A and B). Using symbols of logic,

$$A \cup B = \{x : x \in A \vee x \in B\}$$

The *intersection* of A and B, denoted $A \cap B$, is the set that consists of those elements that belong to both A and B:

$$A \cap B = \{x : x \in A \wedge x \in B\}$$

The *difference* A minus B, denoted $A \dashv B$,* is the set that consists of those elements of A that are not elements of B:

$$A \dashv B = \{x : x \in A \wedge x \notin B\}$$

Example 21: Let $A = \{0,1,2\}$ and $B = \{1,2,3\}$. Then

$$\begin{aligned} A \cup B &= \{0,1,2,3\} \\ A \cap B &= \{1,2\} \\ A \dashv B &= \{0\} \\ B \dashv A &= \{3\} \end{aligned}$$

Example 22: $R \dashv Q$ is the set of irrational numbers.

The set having no elements at all is called the *empty* set. It is standard practice to denote the empty set by $\varnothing$. Because $\varnothing$ has no elements, the statement $x \in \varnothing$ is false for every x. Consequently (see Note on page 4), "$x \in \varnothing \Rightarrow x \in A$" is true for every x, and so $\varnothing \subseteq A$ for every set A.

*The symbol $\dashv$ for minus has been described as a broom that sweeps out of A any elements that also belong to B.

Definition L.1.1 The sets A and B are ***disjoint*** provided $A \cap B = \emptyset$.

Example 23: Let E be the set of even positive integers, O the set of odd positive integers. Then E and O are disjoint since there are no integers that are both even and odd. Also,

$$E \cup O = Z^+$$
$$E \dashv O = E$$
$$O \dashv E = O$$
$$E \dashv Z^+ = \emptyset.$$

Algebra of Unions and Intersections

The following inclusions hold for all sets A, B:

$$A \subseteq A \cup B; \qquad B \subseteq A \cup B$$
$$A \cap B \subseteq A; \qquad A \cap B \subseteq B$$

For example, $A \subseteq A \cup B$ because the defining statement

$$x \in A \Rightarrow (x \in A \vee x \in B)$$

is of the form $p \Rightarrow (p \vee q)$ and is logically true (see Example 15). You should have no trouble identifying the tautologies on which the other three inclusions are based.

We have said that $A = B$ if and only if $A \subseteq B$ and $B \subseteq A$. The following two equivalent statements express this fact in symbols:

$$(x \in A \Rightarrow x \in B) \wedge (x \in B \Rightarrow x \in A)$$
$$x \in A \Leftrightarrow x \in B$$

The following laws involving unions and intersections are part of the "algebra" of sets; they hold for all sets A, B, and C.

Commutative Laws: $A \cup B = B \cup A$
$A \cap B = B \cap A$

Associative Laws: $A \cup (B \cup C) = (A \cup B) \cup C$
$A \cap (B \cap C) = (A \cap B) \cap C$

Distributive Laws: $A \cap (B \cup C) = (A \cap B) \cup (A \cap C)$
$A \cup (B \cap C) = (A \cup B) \cap (A \cup C)$

The commutative laws follow from the tautologies

$$p \vee q \Leftrightarrow q \vee \mathrm{p}$$
$$p \wedge q \Leftrightarrow q \wedge p$$

The associative laws follow from the tautologies

$$p \vee (q \vee r) \Leftrightarrow (p \vee q) \vee r$$
$$p \wedge (q \wedge r) \Leftrightarrow (p \wedge q) \wedge r$$

In Exercise L.1.10 you are asked to identify the tautologies that establish the distributive laws.

Singletons, Pairs, Ordered Pairs

A set with one element is called a *singleton*. For instance, the set of all solutions of the equation $2x + 3 = 1$ is the singleton $\{-1\}$. The singleton $\{-1\}$ is different from the number -1 itself.

A set with exactly two elements is called a *pair*. For example, $\{2,3\}$ is a pair. It is the same as the pair $\{3,2\}$. A set with exactly three elements is called a *triple*, and so on.

There are times when it is necessary to use *ordered* pairs, triples, etc., in which the elements are assigned a definite order. A good illustration of the use of ordered pairs occurs in coordinate geometry. Each point in a plane is located by means of two numbers, its x coordinate and its y coordinate. In order to locate the point P from the given numbers, it is essential to know which of them is the x coordinate and which the y coordinate.

The ordered pair whose first element is a and whose second element is b is denoted (a, b) to distinguish it from the ordinary pair $\{a, b\}$. The ordered pairs (a, b) and (c, d) are equal if and only if $a = c$ and $b = d$ (Exercise L.1.11). In particular, ordered pairs, such as $(2, 3)$ and $(3, 2)$, which have the same elements but in reverse order, are different.

René Descartes

b. March 31, 1596; d. February 1, 1650

The most famous French mathematician of the seventeenth century was primarily a philosopher who lived his productive years in Holland. Having determined his existence ("Je pense, donc je suis"), he set forth to establish truth based purely on reasoning. His exposition of analytic geometry and his dealings with the nature of equations and their solutions appeared as an appendix in his publication, *Discours de la Méthode*. He had few friends and corresponded with those of his contemporaries who could not challenge him. He had a lifelong habit of meditating and writing while remaining in bed as late in the morning as suited him.

Cartesian Products

If A and B are sets, the *Cartesian product* $A \times B$ is the set of all ordered pairs (a, b) where $a \in A$ and $b \in B$; that is,

$$A \times B = \{(a, b) : a \in A, b \in B\}$$

The name is a tribute to Decartes, who is credited with the introduction of coordinates into geometry. The plane, equipped with a system of rectangular coordinates, is just $R \times R$.

Other examples of Cartesian products follow.

Example 24: If $A = \{x : 0 \le x \le 1\}$ and $B = \{y : 0 \le y \le 2\}$, then the points in $A \times B$ form the rectangular region shaded in Figure L.1.

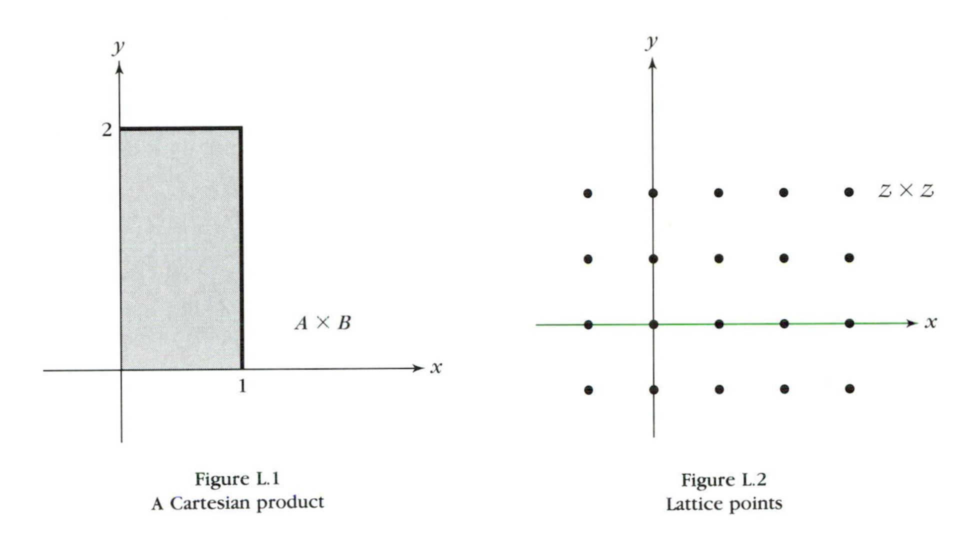

Figure L.1
A Cartesian product

Figure L.2
Lattice points

Example 25: The product $Z \times Z$ consists of the points in the plane with integer coordinates; these are known as *lattice points* (see Figure L.2).

Quantifiers

Consider the equation

$$x^2 = 25$$

in which x is an unspecified real number. If we replace x by 5 or by -5, the resulting equation is a true statement; any other replacement leads to a false statement. The equation itself cannot be said to be true or false until a value of x is specified. It is an example of an open sentence in x.

An *open sentence* in x is a sentence $P(x)$ that makes an assertion about an unspecified x belonging to a given set S. Because of the unspecified nature of x, it does not make sense to try to determine the truth or falsity of $P(x)$. However, it does make sense to consider whether $P(x)$ is true for *some* $x \in S$. The phrase "for some $x \in S$" is called an *existential quantifier* and is abbreviated "$\exists x \in S$". The statement

$$(\exists x \in S)P(x)$$

is read "for some $x \in S$, $P(x)$" or "there exists an $x \in S$ such that $P(x)$". (In mathematics, *some* means "at least one".) To show that $(\exists x \in S)\, P(x)$ is true it is enough to find a specific x belonging to S for which $P(x)$ is true.

Example 26: Let $P(x)$ be the equation

$$x^2 = 25$$

where $x \in R$. Then

$$(\exists x \in R)(x^2 = 25)$$

is true because $x = 5$ and $x = -5$ are values of x for which $P(x)$ is true.

It also makes sense to consider whether an open sentence in x is true for *every* $x \in S$. The phrase "for every $x \in S$" is called a ***universal quantifier*** and is abbreviated "$\forall x \in S$". The statement

$$(\forall x \in S)P(x)$$

is read "for every $x \in S$, $P(x)$". To show that this statement is true it is necessary to show that $P(x)$ is true for an arbitrary (i.e., unspecified) x belonging to S.

Example 27: Let $P(n)$ be the open sentence

$$n^2 \geq n$$

where $n \in Z^+$. We wish to show that

$$(\forall n \in Z^+)P(n)$$

We let n be an arbitrary element of Z^+. Because the elements of Z^+ are all greater than or equal to 1, we have $n \geq 1$. We can multiply both sides of this inequality by the positive integer n, obtaining $n^2 \geq n$. Thus, $P(n)$ holds for an arbitrary element of Z^+, and we can conclude that

$$(\forall n \in Z^+)(n^2 \geq n)$$

There are more examples of proofs involving quantifiers in the next section.

Negation of Quantified Sentences

Consider the negation of $(\forall x \in S)P(x)$:

It is not the case that, for every $x \in S$, $P(x)$ holds.

The negation means that there is some $x \in S$ for which $P(x)$ fails to hold; that is,

$$(\exists x \in S)[\sim P(x)]$$

Example 28: Let $P(x)$ be the open sentence "x is positive", where x is a real number. The fact that not all real numbers are positive is expressed in quantifiers by

$$\sim(\forall x \in R)P(x)$$

or, equivalently,

$$(\exists x \in R)[\sim P(x)]$$

The latter says, "For some $x \in R$, x is not positive"; or, "There exist real numbers that are not positive".

Next consider the negation of $(\exists x \in S)P(x)$:

There does not exist an $x \in S$ such that $P(x)$ holds.

It means that for every $x \in S$, it is the case that $P(x)$ does not hold. Symbolically,

$$(\forall x \in S)[\sim P(x)]$$

Example 29: Let $P(x)$ be the open sentence "$x^2 = -1$", where $x \in R$. Then

$$\sim(\exists x \in R)P(x)$$

means that there does not exist a real number whose square is -1; in other words, for every real number x, x^2 is not equal to -1:

$$(\forall x \in R)(x^2 \neq -1)$$

The logical equivalences in the table below are known as De Morgan's Laws:

Statement	Logically Equivalent Statement
$\sim(\forall x \in S)P(x)$	$(\exists x \in S)[\sim P(x)]$
$\sim(\exists x \in S)P(x)$	$(\forall x \in S)[\sim P(x)]$

When an open sentence involves two or more variables, several quantifiers may be applied. In the following examples we use symbols such as $P(m, n)$, $Q(m, n)$, and $R(m, n)$ to denote open sentences in m and n; the symbols m and n in these examples stand for elements of Z^+.

Example 30: Let us express the fact that the sum of any two positive integers is a positive integer. Denote by $P(m, n)$ the following open sentence:

$$m + n \in Z^+$$

Then the desired statement is

$$(\forall m \in Z^+)(\forall n \in Z^+)P(m, n)$$

Example 31: The fact that the equation $m^2 + n^2 = 25$ has positive integer solutions may be expressed as

$$(\exists m \in Z^+)(\exists n \in Z^+)Q(m, n)$$

where $Q(m, n)$ is the open sentence $m^2 + n^2 = 25$.

Example 32: Let $R(m, n)$ be the open sentence $m < n$. Then the statement

$$(\forall m \in Z^+)(\exists n \in Z^+)R(m, n)$$

says that for every positive integer (m) there is a greater one (n).

Example 33: Let us form the negation of the statement in the previous example:

$$(\forall m \in Z^+)(\exists n \in Z^+)(m < n)$$

According to De Morgan's Laws, the negation is

$$\begin{aligned}&\sim[(\forall m \in Z^+)(\exists n \in Z^+)(m < n)]\\&(\exists m \in Z^+)[\sim(\exists n \in Z^+)(m < n)]\\&(\exists m \in Z^+)(\forall n \in Z^+)[\sim(m < n)]\end{aligned}$$

Now the negation of $m < n$ is $m \geq n$, so the negation of the given statement is

$$(\exists m \in Z^+)(\forall n \in Z^+)(m \geq n)$$

In words, "There is a greatest positive integer" (a false statement!).

Example 34: Every even positive integer n has the property that $n = 2m$ for some positive integer m; in symbols,

$$(\forall n \in E)(\exists m \in Z^+)(n = 2m)$$

If we interchange the order of the quantifiers in this true statement, we obtain

$$(\exists m \in Z^+)(\forall n \in E)(n = 2m)$$

The new statement says that there is a certain positive integer m with the property that *every* even integer is equal to $2m$; of course, this is false. Observe that interchanging the order of an existential and universal quantifier can drastically alter the meaning of a statement.

Note: The purpose of this section has been to focus on some of the words and phrases that are used in mathematics in a special way; for that purpose the logical symbols for connectives and quantifiers have been emphasized. In the future, the use of these symbols will be reserved for emphasis or for situations where the logical structure of a mathematical statement is complicated and calls for special clarification.

EXERCISES L.1

1. Construct a truth table for each of the following statements.

(a) $p \Rightarrow (p \wedge q)$
(b) $p \Rightarrow (p \vee q)$
(c) $p \wedge (p \Rightarrow \sim q)$
(d) $(p \Rightarrow q) \vee (p \Rightarrow \sim q)$
(e) $[(p \Rightarrow q) \wedge (q \Rightarrow r)] \Rightarrow (p \Rightarrow r)$

2. Use truth tables to show that the following pairs of statements are logically equivalent.

(a) $\sim(p \vee q)$; $(\sim p) \wedge (\sim q)$
(b) $\sim(p \Rightarrow q)$; $p \wedge (\sim q)$
(c) $p \Leftrightarrow q$; $(p \Rightarrow q) \wedge (q \Rightarrow p)$
(d) $\sim(p \Leftrightarrow q)$; $[p \wedge (\sim q)] \vee [q \wedge (\sim p)]$
(e) $p \Rightarrow (q \Rightarrow r)$; $(p \wedge q) \Rightarrow r$

3. Test the following pairs of statements for logical equivalence.

(a) $p \Rightarrow q$; $q \Rightarrow p$
(b) $p \Rightarrow q$; $\sim q \Rightarrow \sim p$
(c) $p \Rightarrow q$; $\sim p \Rightarrow \sim q$

4. Use truth tables to show that the given statements are logically true.

(a) $(p \wedge q) \Rightarrow p$
(b) $p \Rightarrow (p \vee q)$
(c) $[(p \Rightarrow q) \wedge (q \Rightarrow r)] \Rightarrow (p \Rightarrow r)$
(d) $(\sim p \vee \sim q) \Leftrightarrow (p \Rightarrow \sim q)$

5. Which of the following are valid schemes of inference? Explain.

(a)
$$\begin{array}{l} p \Rightarrow q \\ \underline{q \Rightarrow r} \\ \therefore p \Rightarrow r \end{array}$$

(b)
$$\begin{array}{l} p \Rightarrow q \\ \underline{\sim q} \\ \therefore \sim p \end{array}$$

(c)
$$\begin{array}{l} p_1 \Rightarrow q \\ p_2 \Rightarrow q \\ \underline{p \Rightarrow (p_1 \vee p_2)} \\ \therefore p \Rightarrow q \end{array}$$

(d)
$$\begin{array}{l} p \Rightarrow q \\ \underline{q} \\ \therefore p \end{array}$$

(e)
$$\begin{array}{l} p \Rightarrow q \\ \underline{r \Rightarrow q} \\ \therefore p \Rightarrow r \end{array}$$

6. There are 16 subsets of $A = \{1,2,3,4\}$. List them.

7. (a) List all the subsets of $\{1,2,3,4,5\}$ that have exactly two elements.
(b) How many subsets of $\{1,2,3,4,5\}$ have exactly three elements?

8. Let $A = \{1,2,3,4,5\}$ and $B = \{3,4,5,6,7\}$. Determine the following sets.

(a) $A \cup B$
(b) $A \cap B$
(c) $A \dashv B$
(d) $B \dashv A$
(e) $A \cup (B \dashv A)$
(f) $A \dashv (B \dashv A)$
(g) $(A \cup B) \dashv A$

9. Let A and B be sets. What can you conclude if you are given the following information?

(a) $A \cap B = A$
(b) $A \cup B = A$
(c) $A \cap B = \varnothing$
(d) $A \cup B = \varnothing$

10. Identify the tautologies that establish the two distributive laws and verify that they are tautologies.

11. A set theoretic definition of the ordered pair (a, b) is the following:

$$(a, b) = \{\{a\}, \{a, b\}\}$$

That is, (a, b) is the set whose elements are the sets $\{a\}$ and $\{a, b\}$. Explain why $(a, b) = (c, d)$ if and only if $a = c$ and $b = d$.

12. Sketch or describe the Cartesian product $A \times B$ in the coordinate plane for the following choices of A and B.

(a) $A = \{1,2,3\}$ $B = \{4,5\}$
(b) $A = \{4,5\}$ $B = \{1,2,3\}$
(c) $A = \{1\}$ $B = R$
(d) $A = \{1\}$ $B = \{0\}$
(e) $A = \{1\}$ $B = Q$
(f) $A = \{x : x < -1\} \cup \{x : x > 1\}$
$B = \{y : -1 \leq y \leq 1\}$

13. Express each of the following sets as a Cartesian product or as a union of Cartesian products.

(a) $\{(x, y) : x > 0 \text{ and } y > 0\}$
(b) $\{(x, y) : 0 \leq x \leq 1 \text{ and } y > 0\}$
(c) The set of points in the Cartesian plane with exactly one rational coordinate.

14. (a) Explain why $A \times \varnothing = \varnothing = \varnothing \times A$.
(b) What can you conclude if $A \times B = B \times A$?
(c) What can you conclude if $A \times B = A \times C$?

15. Let $P(x, y)$ be the open sentence "x likes y," where x and y belong to the set S of people in your class. Translate the following statements into smooth English. Then write the negation of each statement in symbolic form and translate.

(a) $(\forall x \in S)(\forall y \in S)P(x, y)$
(b) $(\forall x \in S)(\exists y \in S)P(x, y)$
(c) $(\exists x \in S)(\forall y \in S)P(x, y)$
(d) $(\exists x \in S)(\exists y \in S)P(x, y)$

16. The following statements refer to the operations of addition and multiplication in R. Translate them into smooth English. Decide which are true.

(a) $(\forall x \in R)(\forall y \in R)(x + y = y + x)$
(b) $(\forall x \in R)(\exists y \in R)(x + y = 0)$
(c) $(\exists y \in R)(\forall x \in R)(x + y = 0)$
(d) $(\forall m \in R)(\forall x \in R)(\exists y \in R)(x + y = m)$
(e) $(\forall m \in R)(\exists y \in R)(\forall x \in R)(x + y = m)$
(f) $(\exists y \in R)(\forall x \in R)(xy = x)$
(g) $(\forall x \in R)(\forall y \in R)(\forall z \in R)(xy = xz \Rightarrow y = z)$
(h) $(\forall x \in R)(\forall y \in R)(\forall z \in R)[(x \neq 0 \wedge xy = xz) \Rightarrow y = z]$

17. Use quantifiers to write the following statements about real numbers.

(a) There is a largest real number.
(b) There is no largest real number.
(c) There is a smallest positive real number.
(d) There is no smallest positive real number.

18. Let $F(p, t)$ denote the open sentence "You can fool person p at time t."

(a) Now use quantifiers and logical connectives to translate into symbols: "You can fool some of the people all the time, and all of the people some of the time, but you cannot fool all of the people all of the time."
(b) Form the negation of your result in (a) and translate it into smooth English.

L.2 *Proofs*

As a student of mathematics you will often be required to construct proofs. A proof is a special type of essay—a sequence of statements explaining how a particular conclusion follows logically from the stated hypothesis and/or other propositions already known or given. In high school geometry this sequence of statements sometimes takes the form of a list of "arguments and reasons" with arguments on the left and reasons on the right. In more advanced mathematics we are less formal and ordinarily write proofs in paragraph style. Nevertheless, when you write a proof you should be sure that it consists of accurately expressed statements and that your reasons are clearly explained.

Learning to devise proofs takes time, thought, and patience. Some proofs require ingenuity or even pure luck! In the following paragraphs we will look at several common types of proofs that illustrate a variety of approaches since there is no single plan of attack that is effective in all cases. If there were, mathematics would not be the challenging subject that it is!

From your experience with proofs in geometry you are aware that every discussion involving mathematical reasoning must start somewhere. Thus, in geometry, one begins by taking such terms as *point* and *line* as undefined and accepting several properties of points and lines without proof. Then additional properties (theorems) are proved, new terms defined, and so on.

In order to proceed with our plan to illustrate various kinds of proofs, we need a mathematical context in which to work. This will be your experience with the algebra of real numbers. In fact, most of the following examples depend only on simple facts about positive integers: for example, that both the sum and the product of two positive integers are also positive integers, that $a + b$ is always equal to $b + a$, and other facts equally basic and familiar.

Divisibility; Existence Proofs

Our first examples deal with the relation of *divisibility* among the positive integers. In discussing them we will encounter some very simple existence arguments in which the objective is to show the existence of an object having a certain property.

Definition L.2.1 Let n and d be positive integers. Then n is ***divisible*** by d (in symbols, $d|n$) provided that* there exists a positive integer q such that $n = qd$.

There are several other ways of stating the fact that $d|n$: for example, "n is a *multiple* of d"; "d is a *factor* of n"; "d is a *divisor* of n." All of these statements mean the same thing. The process of expressing a positive integer as a product of factors is called *factorization*.

In logical notation, Definition L.2.1 is expressed with an existential quantifier:

$$d|n \Leftrightarrow (\exists q \in Z^+)(n = qd)$$

Proving a statement of the form $d|n$ therefore involves an existence argument.

Example 1: The positive integer 6 is divisible by 2 because there is a positive integer q, namely $q = 3$, such that $6 = q \cdot 2$.

The existence argument in Example 1 is as simple as possible and consists of exhibiting a number having the required property.

Example 2: If n is a positive integer, then $n|n$ and $1|n$ because we have

$$n = n \cdot 1 = 1 \cdot n$$

Example 3: The proof of the following proposition involves an existence argument of a little more substance:

Theorem L.2.1 *Let d, m, and n be positive integers. If $d|m$ and $m|n$, then $d|n$.*

PROOF: We wish to find $q \in Z^+$ such that $n = qd$. However, the hypothesis that $d|m$ and $m|n$ guarantees the existence of positive integers r and s such that

$$n = rm \qquad \text{and} \qquad m = sd$$

*In a definition the phrase "provided that" is understood to mean "if and only if".

Upon substituting sd for m in the first equation, we find that

$$n = r(sd) = (rs)d$$

Now by taking $q = rs$ we have found the required positive integer q. ■

Remainders; Proof by Cases

When n is not divisible by d, the division process can be carried out with a remainder. For example, if 17 is divided by 3, the quotient is 5 with a remainder of 2:

$$17 = 5 \cdot 3 + 2$$

In general, for given positive integers n and d, there exist nonnegative integers q and r such that

$$n = qd + r \tag{1}$$

For the given n and d, values of q and r satisfying Equation (1) can be chosen in various ways, for example:

$$17 = 4 \cdot 3 + 5 \qquad \text{and} \qquad 17 = 0 \cdot 3 + 17$$

But by taking q as large as possible, a value of r that is less than d will be determined. In fact, there is just one value of r satisfying both Equation (1) and the condition $0 \le r < d$; this unique r is called the *remainder* when n is divided by d. Clearly, $r = 0$ if and only if n is divisible by d.

Division Algorithm If n and d are positive integers, then there exists a unique integer r satisfying the conditions

(i) $0 \le r < d$
(ii) $n = qd + r$ for some integer $q \ge 0$.

The proof of the Division Algorithm is based on the property of well orderedness, which is discussed in Section P.1 (see also the project following that section).

The Division Algorithm is an example of an "existence and uniqueness" theorem. The definition of remainder is meaningful because of the uniqueness aspect of the theorem. The following discussion relies on the existence aspect. When $d = 2$, there are just two possible remainders, 0 and 1. Thus, each positive integer n has one of the forms

$$n = 2q + 0 \qquad \text{or} \qquad n = 2q + 1$$

where q is a nonnegative integer. It is this fact that underlies the splitting of Z^+ into two sets, E (the evens, $r = 0$) and O (the odds, $r = 1$) (see Example 23 in Section L.1). The Division Algorithm assures us that every $n \in Z^+$ belongs either to E or to O and (because of the uniqueness of r) that E and O are disjoint.

The next example involves a "proof by cases." In this simple proof, we prove a proposition about positive integers n by showing that it holds when n is even and that it holds when n is odd. Since every n is either even or odd, we conclude that the proposition holds for all positive integers n.

Example 4: We prove that if n is any positive integer, then the sum of n and its square is even.

For $n \in Z^+$, there are two cases to consider, either (1) $n \in E$ or (2) $n \in O$. We examine these two cases separately.

Case 1: If $n \in E$, then there exists a positive integer q such that $n = 2q$. The sum of n and its square is given by

$$\begin{aligned} n + n^2 &= 2q + (2q)^2 \\ &= 2q + 4q^2 \\ &= 2(q + 2q^2) \end{aligned}$$

Therefore, $n + n^2 = 2q_1$, where $q_1 = q + q^2$, and so $n + n^2 \in E$.

Case 2: If $n \in O$ then there is a nonnegative integer q such that $n = 2q + 1$. Thus, we have

$$\begin{aligned} n + n^2 &= (2q + 1) + (2q + 1)^2 \\ &= (2q + 1) + (4q^2 + 4q + 1) \\ &= 4q^2 + 6q + 2 \\ &= 2(2q^2 + 3q + 1) \\ &= 2q_2 \end{aligned}$$

where $q_2 = 2q^2 + 3q + 1$. Again we see that $n + n^2 \in E$.

Since these two cases cover all possibilities, we conclude that $n + n^2$ is even for every $n \in Z^+$.

The logical structure of the proof in Example 4 follows the scheme of inference

$$\begin{array}{l} p_1 \Rightarrow q \\ p_2 \Rightarrow q \\ \underline{p \Rightarrow p_1 \vee p_2} \\ \therefore p \Rightarrow q \end{array}$$

(see Exercise L.1.5c).

The oddness or evenness of a positive integer depends on the remainder when the integer is divided by 2. When an integer is divided by 3, a three-way classification of Z^+ arises: The remainder can be 0, in which case $n = 3q$ for some $q \in Z^+$, and n is divisible by 3; the remainder can be 1, in which case $n = 3q + 1$ for some nonnegative integer q; or the remainder can be 2, in which case $n = 3q + 2$ for some nonnegative integer q. Each positive integer n belongs to just one of three classes:

$$C_1 = \{n : n = 3q \text{ for some } q \in Z^+\}$$

$$C_2 = \{n : n = 3q + 1 \text{ for some integer } q \geq 0\}$$
$$C_3 = \{n : n = 3q + 2 \text{ for some integer } q \geq 0\}$$

Example 5: We prove that if n is any positive integer, then $n^3 + 2n$ is divisible by 3.

Let $n \in Z^+$. There are three cases to consider, depending upon which of the above classes n belongs to. We show that $n^3 + 2n$ is divisible by 3 in each case.

Case 1: If $n \in C_1$, then for some $q \in Z^+$, we have

$$\begin{aligned} n^3 + 2n &= (3q)^3 + 2(3q) \\ &= 27q^3 + 6q \\ &= 3(9q^3 + 2q) \end{aligned}$$

Since $9q^3 + 2q \in Z^+$, we see by Definition L.2.1 that $n^3 + 2n$ is divisible by 3.

Case 2: If $n \in C_2$, then for some nonnegative integer q, we have

$$\begin{aligned} n^3 + 2n &= (3q + 1)^3 + 2(3q + 1) \\ &= (27q^3 + 27q^2 + 9q + 1) + (6q + 2) \\ &= 27q^3 + 27q^2 + 15q + 3 \\ &= 3(9q^3 + 9q^2 + 5q + 1) \end{aligned}$$

Since $9q^3 + 9q^2 + 5q + 1 \in Z^+$, we see that $n^3 + 2n$ is divisible by 3.

Case 3: If $n \in C_3$, then for some nonnegative integer q, we have

$$\begin{aligned} n^3 + 2n &= (3q + 2)^3 + 2(3q + 2) \\ &= (27q^3 + 54q^2 + 36q + 8) + (6q + 4) \\ &= 27q^3 + 54q^2 + 42q + 12 \\ &= 3(9q^3 + 18q^2 + 14q + 4) \end{aligned}$$

and again we see that $n^3 + 2n$ is divisible by 3.

Since each positive integer must belong to one of the three classes, and the conclusion of the theorem holds in each case, then it holds for all positive integers. (See Exercise L.2.8.)

A proof by cases may involve any number of cases; the essential point is that the cases must cover all possibilities. If the conclusion is shown to hold for each case, then it must hold in general.

"If and Only If" Proofs

As noted above, a definition is always understood to be in the form of a biconditional statement. Many important theorems are also of this form (for example, the Pythagorean Theorem; see Example 14 in Section L.1). The proof of an "if and only if" theorem often involves two parts, based on the scheme of inference

$$\begin{array}{l} p \Rightarrow q \\ \underline{q \Rightarrow p} \\ \therefore p \Leftrightarrow q \end{array}$$

The following definition and example serve to illustrate a proof of this type.

Definition L.2.2 Two positive integers have the same ***parity*** provided that they are either both even or both odd.

Example 6: 2 and 40 have the same parity; 17 and 25 have the same parity; 2 and 3 do not have the same parity.

Example 7: We prove that the sum of two positive integers is even if and only if they have the same parity.

Note: In this proof, let p be "the sum is even" and q be "have the same parity." The first paragraph below establishes $q \Rightarrow p$ in a proof involving two cases. The second paragraph establishes $p \Rightarrow q$ in its contrapositive form $\sim q \Rightarrow \sim p$.

(1) Suppose that two positive integers n_1 and n_2 are of the same parity. Two cases arise: n_1 and n_2 are both even; and n_1 and n_2 are both odd. In the first case, there are positive integers m_1 and m_2 such that $n_1 = 2m_1$ and $n_2 = 2m_1$; but then $n_1 + n_2 = 2m_1 + 2m_2 = 2(m_1 + m_2)$, and we see that the sum is even. In the second case, there are nonnegative integers m_1 and m_2 such that $n_1 = 2m_1 + 1$ and $n_2 = 2m_2 + 1$; thus, $n_1 + n_2 = 2m_1 + 2m_2 + 2 = 2(m_1 + m_2 + 1)$, and again the sum is even.

(2) Suppose that two integers n_1 and n_2 are *not* of the same parity; then one of them (say, n_1) is odd and the other (n_2) is even. Thus, for certain nonnegative integers m_1, m_2, we have

$$n_1 = 2m_1 + 1 \qquad \text{and} \qquad n_2 = 2m_2$$

Then their sum

$$n_1 + n_2 = 2(m_1 + m_2) + 1$$

is odd, *not* even. Therefore, if the sum is even, n_1 *and* n_2 must have the same parity.

Prime Numbers and Unique Factorization

For many centuries the set of prime numbers has been an object of fascination for professional mathematicians as well as for amateurs.

Definition L.2.3 A positive integer n greater than 1 is ***prime*** provided that it is divisible only by itself and by 1; it is ***composite*** provided it is not prime.

Some examples of primes are 2,3,5,7,11; the numbers 4,6,8,9,10 are composite. Notice that the definition excludes 1, so that 1 is neither prime nor composite. A fundamental theorem in the theory of numbers relates to the factorization of numbers into primes:

Theorem L.2.2 **Unique Factorization** *Every positive integer greater than 1 can be expressed as a product of primes; and except for the order in which the factors are written, this expression is unique.*

The proof of the Unique Factorization Theorem involves the property of well-orderedness of Z^+, discussed in Section P.1; an outline of that proof is given in a project following Section P.1.

To illustrate the theorem, consider the number 48. It can be factored in several different ways; for example, $6 \cdot 8$, $12 \cdot 4$, $3 \cdot 16$, $3 \cdot 8 \cdot 2$, and so on. However, if we continue factoring until we have only prime factors, we will inevitably arrive at a product consisting, in some order, of four factors of 2 and one of 3; that is, $48 = 2^4 \cdot 3$.

The Unique Factorization Theorem (UFT) was known to early Greek mathematicians; it is a tribute to their remarkable sophistication that they not only were able to prove this theorem but that they recognized the need to do so.*

Infinitude of Primes; Proof by Contradiction

It was known quite early that the list of primes has no end—that is, that there is no largest prime. This fact is known as Euclid's Theorem.

The proof of Euclid's Theorem is based upon the UFT and illustrates the method of "proof by contradiction". In a proof by contradiction, we assume that the proposition to be proved is false and then derive a contradiction; it then follows that the original proposition must be true. The relevant scheme of inference is

$$\begin{array}{l} \sim p \Rightarrow q \\ \underline{\sim q \qquad} \\ \therefore p \end{array}$$

Theorem L.2.3 **Euclid's Theorem** *There is no greatest prime.*

PROOF: Suppose the statement is false. Then there exists a greatest prime P. We form the product of all primes from 2 to P inclusive and to this product we add 1. The resulting number

$$N = 2 \cdot 3 \cdot 5 \cdot 7 \cdot 11 \cdot \cdots \cdot P + 1$$

is certainly greater than 1. By the UFT, N can be expressed as a product of primes and is therefore divisible by some prime. But N is not divisible by any

*A system of numbers that is similar in many ways to Z^+ but in which factorization into primes is not unique, is described in (e) of the project following Section P.1. It shows that uniqueness of factorization is not obvious.

EUCLID

(dates not available—about 300 B.C.)

Nearly 2300 years ago, this great Greek scientist organized the geometry dealing with measures of earthly parts and their relations to one another. While the information was of value, the *reasoning* demonstrated throughout became the model for logical thought. Along with Aristotle, Euclid is credited with the discovery of the procedure called *deductive logic*. His *Elements* sets forth definitions, postulates, and axioms from which theorems are deduced in a sequence of steps. The reasoning is clear and concise, proceeding from the simple to the complex. Euclid founded a school at Alexandria that became the scientific center of his world. His known contributions include mathematics, astronomy, optics, music, and mechanics.

one of the primes less than or equal to P. There must therefore exist a prime greater than P, contradicting our assumption at the outset. ■

Although we will not have occasion to use Theorem L.2.3 in this course (except here, in order to illustrate a method of proof), it would be a serious error to think that Euclid's Theorem is unrelated to the field of analysis. The fact that the list of primes continues without end is not at all obvious; in fact, empirical data suggest that the primes occur along the list of positive integers with diminishing frequency (see, for example, Exercise L.2.20). A search for some pattern describing the distribution of the primes has proved extremely difficult and has spurred the development of great parts of the literature of nineteenth and twentieth century analysis.

Irrationality of the Square Root of 2

As another example of proof by contradiction let's look at how the Greeks discovered the fact that there is no rational square root of 2. This fact is highly relevant for our work because it indicates that the rational number system is inadequate in certain aspects. One difficulty is that not all geometrical lengths can be measured by rational numbers; for example, the diagonal of a square of unit edge cannot be measured by the rational numbers. The need for a larger number system leads us to the very important concept of "completeness" in Chapter 1.

Theorem L.2.4 *There does not exist a rational number whose square is 2.*

PROOF: Again, we use the method of contradiction. Suppose that there were a rational number whose square is 2. It then would follow that there are integers m and n such that $n \neq 0$, and $(m/n)^2 = 2$. Thus, we would have

$$m^2 = 2n^2$$

for certain nonzero integers m and n, and since $(-x)^2 = x^2$ for any integer x, we can assume that m and n are positive integers. Now let us ask how often the prime factor 2 occurs on the right-hand side of the above equation. Since 2 occurs an even number of times, if at all, as a factor of n^2, it occurs an *odd* number of times in $2n^2$. On the left side of the equation, 2 occurs, if at all, an *even* number of times as a factor of m^2. We have contradicted the Unique Factorization Theorem because we have shown that the number m^2 has two essentially different factorizations into primes: one with an even number of 2s, another with an odd number of 2s. Because of this contradiction we conclude that there does not exist a rational number whose square is 2. ∎

The proofs of Theorems L.2.3 and L.2.4 are both based on the UFT—another existence and uniqueness theorem. Observe that Theorem L.2.3 rests on the *existence* of a factorization of N into primes; in that proof we needed to have a prime factor of N. In Theorem L.2.4, the contradiction we achieved was based on the *uniqueness* of the factorization of m^2 into primes.

More about Universal Quantifiers

To prove a universal statement of the form $(\forall x \in S)P(x)$, we let x be an arbitrary (unspecified) element of S and show that $P(x)$ holds. It is essential that there be no conditions imposed upon x other than that $x \in S$.

Example 8: In Example 2 we proved that if n is a positive integer, then $n|n$ and $1|n$. Since no conditions were imposed upon n except that $n \in Z^+$, we have

$$(\forall n \in Z^+)(n|n \wedge 1|n)$$

Universal statements can involve several variables; for example, the commutative law for addition of positive integers is expressed as follows:

$$(\forall m, n \in Z^+)(m + n = n + m)$$

Theorem L.2.1 can be stated as follows:

$$(\forall d, m, n \in Z^+)(d|m \wedge m|n \Rightarrow d|n)$$

Sometimes a universal statement can be proved by contradiction. Here is an example from elementary number theory:

Theorem L.2.5 *Every composite positive integer has a prime factor less than or equal to its square root.*

PROOF: Suppose the statement false. Then there exists a composite positive integer n such that all prime factors of n are greater than $\sqrt{n}$ (see Exercise L.2.17). The UFT tells us that n can be expressed as a product of primes, and since n is composite there are at least two prime factors, p_1 and p_2, in this product; thus,

$$n = p_1 p_2 q$$

where q is either 1 or the product of the remaining prime factors of n; in any case, $q \geq 1$. Now p_1 and p_2 are both greater than $\sqrt{n}$, and so we have

$$n = p_1 p_2 q > \sqrt{n}\sqrt{n}q = nq \geq n$$

This contradiction establishes the theorem. ■

Theorem L.2.5 is useful in testing numbers for being prime: We need only check the primes less than or equal to the square root of a number n; if none of these is a divisor of n, then n is prime.

Indirect Proof

Recall from Section L.1 that the conditional statement $p \Rightarrow q$ and its contrapositive $\sim q \Rightarrow \sim p$ are logically equivalent. A proof based on this equivalence is known as an *indirect* proof.

Example 9: Prove that if the average of a set of test scores is greater than 90, then at least one of the scores is greater than 90.

PROOF: Let the set of scores be $S = \{a_1, a_2, \ldots, a_n\}$. The proposition to be proved can be expressed symbolically by $p \Rightarrow q$, where p is

$$\frac{a_1 + a_2 + \cdots + a_n}{n} > 90$$

and q is

$$(\exists x \in S)(x > 90)$$

The contrapositive is $\sim q \Rightarrow \sim p$, and so we begin by assuming the negation of q:

$$(\forall x \in S)(x \leq 90)$$

Thus, if q is false, all scores in the set are ≤ 90, and by calculating the average we get

$$\frac{a_1 + a_2 + \cdots + a_n}{n} \leq \frac{90 + 90 + \cdots + 90}{n} = \frac{90n}{n} = 90$$

This inequality is the negation of p, so we have established the contrapositive $\sim q \Rightarrow \sim p$. We conclude that $p \Rightarrow q$. ■

Note: Example 9 shows that an existence theorem can be proved without actually exhibiting the object whose existence is asserted.

Existence and Uniqueness Proofs

Theorems that state the existence and uniqueness of certain things can be very useful, and so it is important we understand what is required to prove such statements. (The Division Algorithm and the UFT are examples of existence and uniqueness theorems whose proofs are deferred to the project following Section P.1 because they are based upon the property of the well-orderedness of Z^+. We will discuss some more elementary examples here.)

First, let us look carefully at the meaning of existence and uniqueness statements. To say that there exists a unique x in a set S having a certain property $P(x)$ means that

(1) There is an $x \in S$ such that $P(x)$
(2) No two distinct* elements of S both have this property

Condition (1) is simply a statement of existence: $(\exists x \in S)P(x)$. The meaning of (2) is that different elements of S cannot both have the property; in other words,

$$x \neq y \Rightarrow \sim [P(x) \wedge P(y)]$$

By writing this conditional statement in its contrapositive form, we have, for $x, y \in S$:

$$P(x) \wedge P(y) \Rightarrow x = y$$

Thus, condition 2 can be expressed logically as

$$(\forall x, y \in S)([P(x) \wedge P(y)] \Rightarrow x = y)$$

That is, if x and y both have the property, then x and y are the same element of S.

Example 10: A theorem of elementary algebra states that if $a, b \in R$ and $a \neq 0$, then the equation $ax = b$ has a unique solution; that is, there exists a unique real number x such that $ax = b$. In the proof below we make use of the fact that the nonzero real number a has a reciprocal a^{-1}:

(1) *Existence.* The real number $a^{-1}b$ is a solution because

$$a(a^{-1}b) = (aa^{-1})\,b = 1 \cdot b = b$$

(2) *Uniqueness.* If x and y are both solutions, so that $ax = b$ and $ay = b$, then

$$ax = ay$$

By multiplying both sides by a^{-1}, we get $x = y$.

*In mathematics, "distinct" means "different" or "unequal".

Note: The solution of the equation in Example 10 is often written simply as a one-step derivation:

$$ax = b$$
$$x = a^{-1} b$$

Strictly speaking, it has not been shown that the equation has a solution, but rather that *if* there is a solution, it is equal to $a^{-1} b$ and is therefore unique. The proof of existence requires substituting $a^{-1} b$ for x in the equation; as in step 1. Our point is perhaps driven home a little harder by the next example.

Example 11: Solve the equations below in R:

(i) $\sqrt{x} + 4 = 0$
(ii) $3 - x = 2\sqrt{x}$

Solution:

(i) Writing $\sqrt{x} = -4$ and squaring both sides gives $x = 16$; so *if* there is a solution, it must be 16. But substitution shows that 16 is not a solution, and so the equation has no solutions.

(ii) Squaring both sides of (ii) gives

$$9 - 6x + x^2 = 4x$$

or

$$9 - 10x + x^2 = 0$$
$$(x - 1)(x - 9) = 0$$

Therefore, *if* x is a solution of $3 - x = 2\sqrt{x}$, then $x = 1$ or $x = 9$. Direct substitution shows that $x = 1$ is a solution but $x = 9$ is not. Therefore, the only solution is $x = 1$.

Counterexamples

You will find a number of exercises in this text that ask you to prove or disprove a statement. Often the statement makes a general assertion of the form $(\forall x \in S)P(x)$. If you decide the general statement is true, your job is to prove it—to take an arbitrary $x \in S$ and show that $P(x)$ holds. If you decide the statement is false, then you have decided in favor of its negation, which says $(\exists x \in S)[\sim P(x)]$. Of course, an effective way of proving the negation is to find an $x \in S$ for which $P(x)$ is false; such an x is a *counterexample*.

If at first you are uncertain, you might begin by trying to prove the given assertion. If you succeed, you are done. If you encounter obstacles in your attempt, these obstacles may help you either to find a counterexample or to discover in some other way that the statement is false.

Example 12: Prove or disprove the following statements (in which d, m, n refer to arbitrary positive integers):

(i) If $d|m$ and $d|n$, then $d^2|mn$.
(ii) If $d|mn$, then either $d|m$ or $d|n$.
(iii) There exists a positive rational number whose square is 18.

Solution:

(i) We have a general statement here:

$$(\forall d, m, n \in Z^+)(d|m \wedge d|n \Rightarrow d^2|mn)$$

It is a true statement because by hypothesis there exist positive integers q_1 and q_2 such that $m = q_1 d$ and $n = q_2 d$. Therefore, $mn = (q_1 d)(q_2 d) = (q_1\, q_2)d^2$, so that $d^2|mn$.

(ii) The general statement is

$$(\forall d, m, n \in Z^+)(d|mn \Rightarrow d|m \vee d|n)$$

It would not be surprising if you should already have found a counterexample to this false statement. However, suppose we try to prove it. Given that $mn = qd$ for some $q \in Z^+$, we are to show that either m or n is divisible by d. Starting with the equation $mn = qd$, we might try solving it for m:

$$m = \frac{qd}{n} = \left(\frac{q}{n}\right)d$$

in hopes of expressing m as the product of d with some positive integer. But we encounter difficulty because n is perhaps not a factor of q. We might now become suspicious and try to find values of the variables on the right that give us a counterexample. We try $q = 5$, $n = 2$ (so that q/n is not a positive integer) and $d = 4$. Then $m = \frac{5}{2} \cdot 4 = 10$, and $mn = 20$. The number $d = 4$ is a divisor of mn but not of either m or n. This counterexample proves that the statement is false.

(iii) Let us try to find a rational number m/n ($m, n \in Z^+$) such that $(m/n)^2 = 18$. Then we are led to

$$m^2 = 18n^2$$

and taking the square root, we find

$$m = 3\sqrt{2}n$$

If this equation should hold for positive integers m,n, then $\sqrt{2}$ would be a quotient of positive integers; that is,

$$\sqrt{2} = \frac{m}{3n}$$

But $\sqrt{2}$ is irrational, and so we have reached a contradiction. The statement is false.

A Famous Counterexample and a Famous Unsolved Problem

The great mathematician Fermat conjectured that every number of the form

$$F_n = 2^{2^n} + 1 \qquad n \text{ a nonnegative integer}$$

is prime. Indeed, the numbers F_n are prime for $n = 0,1,2,3$, and 4:

$$F_0 = 3 \quad F_1 = 5 \quad F_2 = 17 \quad F_3 = 257 \quad F_4 = 65537$$

Many years after Fermat made his conjecture, however, Euler (see page 164) showed that the number $F_5 = 2^{32} + 1$ is divisible by the prime 641:

$$F_5 = 4,294,967,297 = 641 \cdot 6700417$$

Fermat's conjecture was thus proved false by Euler's counterexample.

In fairness to Fermat, we emphasize that he only *conjectured* that F_n is prime for every n; he never claimed to have proved this conjecture. On another occasion, Fermat claimed to have *proved* the following:

"Fermat's Last Theorem" For every positive integer n greater than 2, the equation

$$x^n + y^n = z^n$$

is not satisfied by positive integers x, y, and z.

Pierre de Fermat

(b. August 17, 1601 (?); d. December 12, 1665)

Fermat was a jurist who, in his thirties, made mathematics an avocation. Although he published very little, this amateur corresponded with many famous mathematicians, including Descartes. He independently developed what is now called analytic geometry in his work, *Introduction to Plane and Surface Loci*. His work on tangents to a curve and his method of maxima and minima led him to applications in optics. He shares credit with Pascal for the basics of probability theory. His greatest joy was in number theory, which he brought to the spotlight by indicating proofs while omitting "detail." A pure mathematician, he did not seek publicity; today his name is used in connection with Fermat numbers, Fermat's Little Theorem, and Fermat's "Last Theorem" (a conjecture long unsettled).

"For this," wrote Fermat in the margin of a book, "I have discovered a truly wonderful proof but the margin is too small to contain it." After many years of effort, the best mathematical minds have been unable to find a proof of Fermat's Last Theorem, nor have they been able to prove him wrong by finding a counterexample.*

EXERCISES L.2

1. Prove: If d, m, and n are positive integers such that $d|m$ and $d|n$, then $d|(m+n)$.

2. Prove: If d, e, m, and n are positive integers such that $d|m$ and $e|n$, then $de|mn$.

3. Prove: If n is a positive integer, then

$n^3 + 3n^2 + 2n$ is divisible by 3.

4. Prove: If n is a positive integer, then

$n^4 + 2n^3 + 3n^2 + 2n$ is divisible by 4.

5. (a) Prove: If n is a positive integer, and if p and q are different primes such that $p|n$ and $q|n$, then $pq|n$. (Use the Unique Factorization Theorem.)

(b) Prove: If n is a positive integer greater than 1, then $n^3 - n$ is divisible by 6.

6. Prove: If n is a positive integer greater than 1, then $n^5 - n$ is divisible by 5.

7. Prove: If n is a positive integer greater than 1, then $n^5 - n$ is divisible by 30.

8. What scheme of inference lies behind the proof in Example 5? Show that the scheme is valid.

9. Prove: If a and b are real numbers and $a < b$, then there is a real number c such that $a < c < b$.

10. Prove: There is no smallest positive rational number.

11. Prove: There is no greatest positive integer.

12. Use the Unique Factorization Theorem to show that the following numbers are irrational.

(a) $\sqrt{3}$

(b) $\log_{10} 2$ (Recall that $\log_b a$ is a number x such that $b^x = a$.)

13. Prove: If a, b, and c are real numbers such that $c = a + b$, then either $a \leq \frac{1}{2}c$ or $b \leq \frac{1}{2}c$.

14. Prove or disprove the following statements.

(a) The product of any two rational numbers is rational.

(b) The sum of any rational and any irrational number is irrational.

(c) The product of any two irrational numbers is irrational.

(d) The sum of any two irrational numbers is irrational.

15. Prove or disprove: Every prime greater than 2 is odd.

16. Prove or disprove: If m and n are positive integers and p is a prime such that $p|mn$, then $p|m$ or $p|n$.

17. Write Theorem L.2.5 in logical notation with quantifiers and then form its negation.

18. Prove or disprove: If p is a prime, then $2^p - 1$ is prime.

19. Another famous conjecture that has never been proved or disproved is due to Goldbach. It states:

Every positive even integer greater than 4 is a sum of two odd primes.

(a) Use quantifiers to express Goldbach's conjecture symbolically.

(b) Use quantifiers to write the negation of Goldbach's conjecture symbolically.

20. Show that for any positive integer $N > 1$, it is possible to find a list of N consecutive composite positive integers. [*Hint:* Start the list with $(N+1)! + 2$, where $(N+1)!$ denotes the product of all the positive integers from 1 to $N+1$.]

*As this book goes to press, it appears that "Fermat's Last Theorem" may have been proved. See the article, "A Curvy Path Leads to Fermat's Last Theorem", *Science News*, volume 144, pp. 5–6.

L.3 *Relations*

Relations and Their Inverses

The following examples illustrate the mathematical concept of "relation".

Example 1: The relation R that relates the real number x to the real number y if and only if $y = x + 1$.

Example 2: The relation D that relates the positive integer m to the positive integer n if and only if m is a divisor of n.

Example 3: The relation S that relates the real number x to the real number y if and only if $y = x^2$.

Example 4: The relation L that relates the real number u to the real number v if and only if $u \leq v$.

Example 5: The relation T that relates the point P in the coordinate plane to the nonnegative real number d if and only if d is the distance of P from the origin O.

In mathematics we identify a relation with the set of ordered pairs whose elements are related by it. If the first elements of these ordered pairs belong to a set A, and the second elements belong to B, then the relation is a subset of $A \times B$. We take this as our definition.

Definition L.3.1 A ***relation*** from A to B is a subset R of $A \times B$. The ***domain*** of R is the subset of A consisting of all first elements of the ordered pairs in R:

$$\text{Domain } R = \{x : (\exists y \in B)(x, y) \in R\}$$

The ***range*** of R is the subset of B consisting of all second elements of the ordered pairs in R:

$$\text{Range } R = \{y : (\exists x \in A)(x, y) \in R\}$$

Let us now express Examples 1 to 5 in ordered pair notation:

Example 1: $R = \{(x, y) : x, y \in \mathsf{R} \text{ and } y = x + 1\}$; the domain and range of R are both the set R of real numbers.

Example 2: $D = \{(m, n) : m, n \in \mathsf{Z}^+ \text{ and } m|n\}$. The domain and range of D are both Z^+.

Example 3: $S = \{(x, y) : x, y \in \mathsf{R} \text{ and } y = x^2\}$. The domain of S is R, and the range of S is the set of nonnegative reals.

Example 4: $L = \{(u, v) : u, v \in R \text{ and } u \leq v\}$. The domain and range are both R.

Example 5: $T = \{(P, d) : P \in R \times R, d \in R \text{ and } d = |OP|^*\}$. The domain of T is the plane $R \times R$, and the range is the set of all nonnegative real numbers.

The fact that the ordered pair (x, y) belongs to a relation R is often written xRy meaning "x is related by R to y". If we denote the relation L in Example 4 by the symbol $\leq$, then the statement $(x, y) \in \leq$ takes the form $x \leq y$.

If the domain and range of R are subsets of R (as in Examples 1 to 4), then we can construct a graph of R in the familiar way. (See Figure L.3.1.)

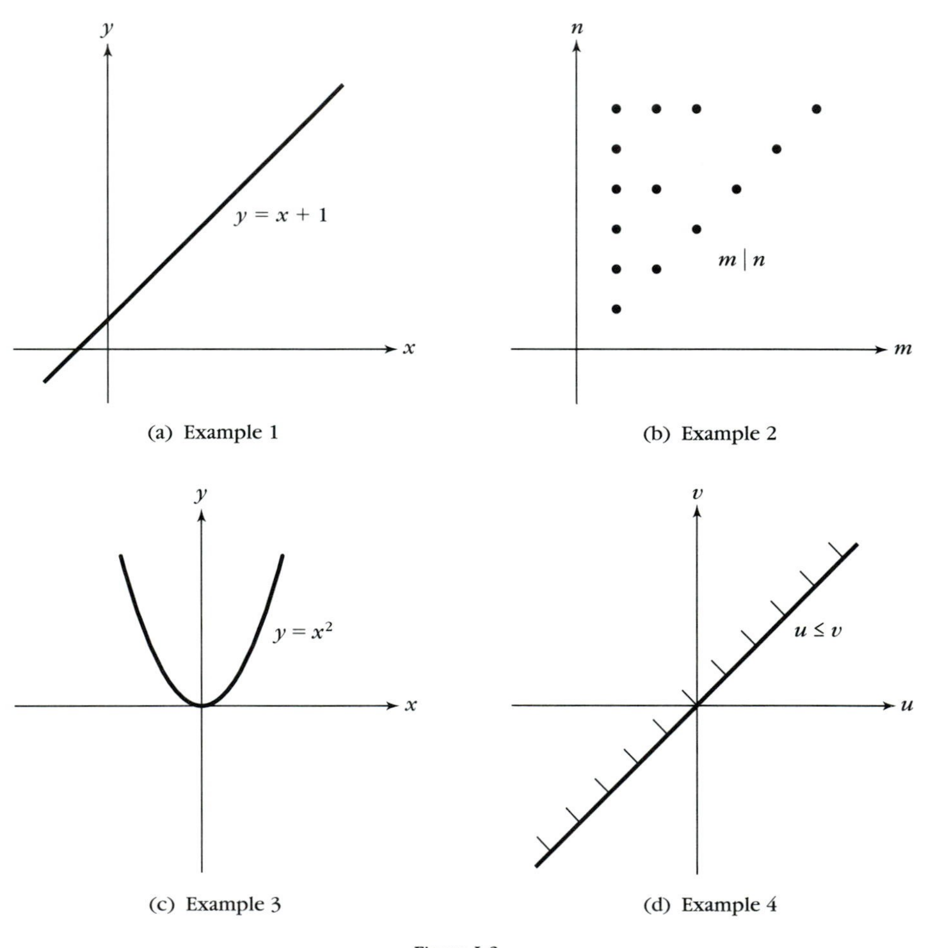

Figure L.3
Graphs of Examples 1–4

$^*|OP|$ means the distance from O to P.

The inverse of a relation is obtained by reversing its ordered pairs:

Definition L.3.2 The ***inverse*** of a relation R is the relation

$$R^{-1} = \{(x, y) : (y, x) \in R\}$$

The domain of R^{-1} is the range of R, the range of R^{-1} is the domain of R, and we see that $x\, R^{-1} y$ if and only if yRx.

Example 6: The inverse of the relation R in Example 1 is

$$\begin{aligned} R^{-1} &= \{(x, y) : x, y \in \mathbb{R} \text{ and } x = y + 1\} \\ &= \{(x, y) : x, y \in \mathbb{R} \text{ and } y = x - 1\} \end{aligned}$$

We see that while the original relation "adds 1," its inverse "subtracts 1".

Example 7: In Example 2, the inverse is

$$D^{-1} = \{(m, n) : m, n \in \mathbb{Z}^+ \text{ and } n \mid m\}$$

Thus, the inverse of "being a divisor of" is "being divisible by".

Example 8: In Example 3, the inverse is

$$\begin{aligned} S^{-1} &= \{(x, y) : x, y \in \mathbb{R} \text{ and } x = y^2\} \\ &= \{(x, y) : x, y \in \mathbb{R}, x \geq 0 \text{ and } y = \pm\sqrt{x}\} \end{aligned}$$

The restriction $x \geq 0$ occurs because only nonnegative real numbers have real square roots.

Example 9: The inverse of $\leq$ is $\geq$.

If the domain and range of R are subsets of $\mathbb{R}$, then the graph of the inverse can be obtained by reflecting the original graph through the line bisecting the first and third quadrants. (See Figure L.4 and Exercise L.3.3.)

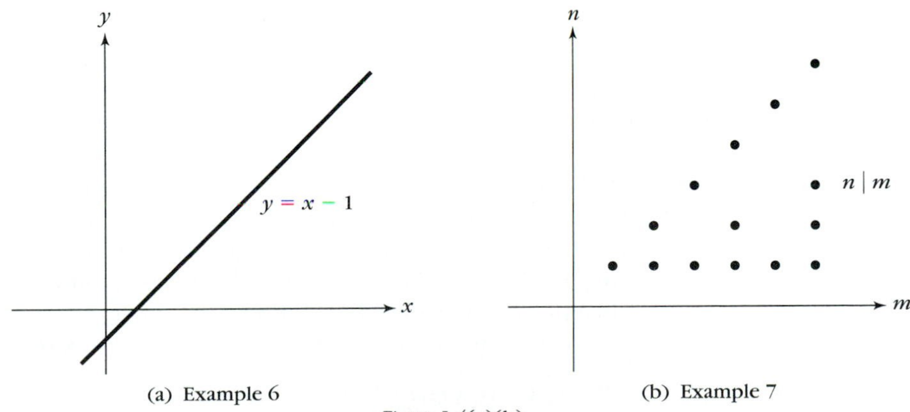

(a) Example 6 (b) Example 7

Figure L.4(a)(b)
Inverses of Relations

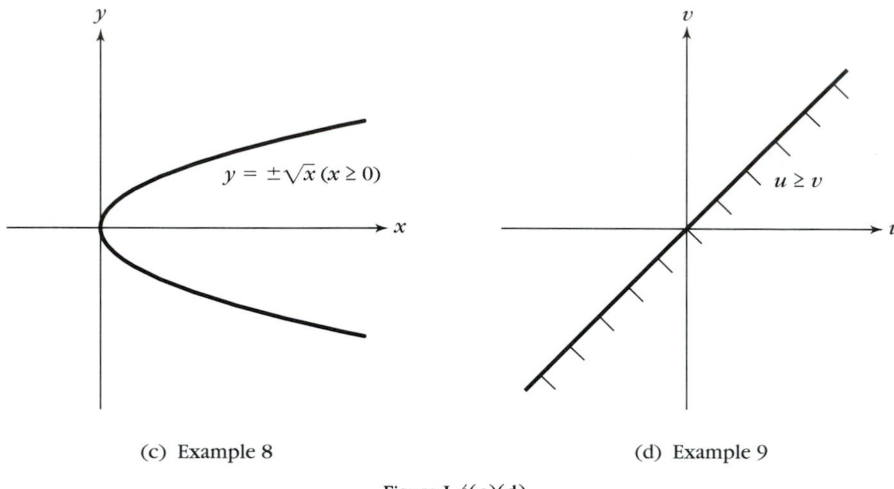

(c) Example 8

(d) Example 9

Figure L.4(c)(d)
Inverses of Relations

Equivalence Relations

A relation from A to A is simply called a "relation in A". The following terminology is used in connection with such relations:

Definition L.3.3 Let R be a relation in a set A.

(1) R is ***reflexive*** provided that xRx for every $x \in A$
(2) R is ***symmetric*** provided that, for every x and $y \in A$, if xRy, then yRx
(3) R is ***transitive*** provided that, for every x, y, and $z \in A$, if xRy and yRz, then xRz

An ***equivalence*** relation is one that is reflexive, symmetric, and transitive.

Of course, equality itself is an equivalence relation in any set; and from plane geometry we have familiar examples such as congruence and similarity of plane figures. Here is another geometric example.

Example 10: In a polar coordinate system, each ray emanating from the pole (origin) is the terminal side of an angle measured (in radians) counterclockwise from the polar axis to the ray itself. Now, by defining two angles to be related if they differ by an integral multiple of 2π, we obtain a relation R in which $\theta_1 R \theta_2$ if and only if $\theta_1 - \theta_2 = 2\pi k$ for some integer k. It is easy to verify that R is an equivalence relation:

(1) For any angle θ, $\theta - \theta = 2\pi \cdot 0$

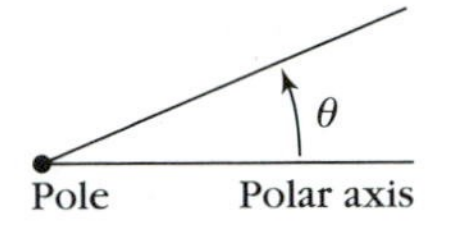

Figure L.5
Polar coordinate system

(2) If $\theta_1 - \theta_2 = 2\pi k$ for some integer k, then $\theta_2 - \theta_1 = 2\pi(-k)$
(3) If $\theta_1 - \theta_2 = 2\pi k_1$ and $\theta_2 - \theta_3 = 2\pi k_2$ for integers k_1 and k_2, then $\theta_1 - \theta_3 = 2\pi(k_1 + k_2)$.

Thus, R satisfies the reflexive, symmetric, and transitive laws.

Example 11: The relation L ($\leq$) in Example 4 is reflexive and transitive; however, it is not symmetric and so it is not an equivalence relation.

An equivalence relation can be thought of as a kind of "quasi-equality". Equivalent objects are indistinguishable in certain respects. In Example 10, equivalent angles are indistinguishable insofar as their terminal sides are concerned. Similar triangles are indistinguishable insofar as the ratios of corresponding parts are concerned.

Congruence modulo *m*

In order to discuss an interesting equivalence relation in Z, we need first to extend the concept of divisibility from Z^+ to all of Z:

Definition L.3.4 If n and d are integers, then d is a ***divisor*** of n provided that there is an integer q such that $n = qd$.

Example 12: The divisors of -6 are ± 1, ± 2, ± 3, and ± 6.

Definition L.3.5 If $a, b \in Z$ and $m \in Z^+$, then a is ***congruent*** to b ***modulo*** m, written $a \equiv_m b$, provided that m is a divisor of $a - b$.

Example 13: Taking $m = 5$, we have $8 \equiv_5 13$; $35 \equiv_5 0$; $4 \equiv_5 -1$. With $m = 19$, $118 \equiv_{19} 23$.

Theorem L.3.1 *Let $m \in Z^+$. Then the relation $\equiv_m$ is an equivalence relation in Z.*

PROOF:

(1) Since $a - a = 0$ is divisible by every $m \in Z^+$, we have $a \equiv_m a$ for every $a \in Z$, and the relation $\equiv_m$ is reflexive.
(2) Clearly $b - a$ is divisible by m whenever $a - b$ is, and it follows that $\equiv_m$ is symmetric.
(3) Suppose $a \equiv_m b$ and $b \equiv_m c$; then $b - a$ and $c - b$ are both divisible by m, say, $b - a = q_1 m$ and $c - b = q_2 m$. Then

$$\begin{aligned} c - a &= (c - b) + (b - a) \\ &= q_2 m + q_1 m = (q_2 + q_1)m, \end{aligned}$$

and we see that $c - a$ is divisible by m. Thus, $\equiv_m$ is transitive. ∎

Congruence modulo m has a "replacement" property similar to equality.

Theorem L.3.2 *Let $m \in Z^+$ and let a, b, c, and $d \in Z$. If $a \equiv_m b$ and $c \equiv_m d$, then $a + c \equiv_m b + d$ and $ac \equiv_m bd$.*

PROOF: Our hypothesis is that $b - a$ and $d - c$ are both divisible by m. Arguing as in the proof of part (3) of Theorem L.3.1, it follows from the equation

$$(b + d) - (a + c) = (b - a) + (d - c)$$

that $(b + d) - (a + c)$ is divisible by m, and so $b + d \equiv_m a + c$. To show that $bd \equiv_m ac$, we write

$$\begin{aligned} bd - ac &= bd - bc + bc - ac \\ &= b(d - c) + (b - a)c. \end{aligned}$$

In this form it is clear that $bd - ac$ is also divisible by m. ■

Equivalence Classes

An equivalence relation R in a set A imposes an interesting structure in A: it "partitions" A into subsets, called *cells,* in such a way that each element of A belongs to one and only one cell. As we shall see, each cell consists of elements of A that are related by R.

Definition L.3.6 A ***partition*** of a nonempty set A is a collection C of sets such that

(1) Every $S \in C$ is a nonempty subset of A
(2) For each $x \in A$, there is an $S \in C$ such that $x \in S$
(3) For S and $T \in C$, if $S \neq T$, then $S \cap T = \emptyset$

The sets belonging to C are called ***cells***.

Condition (3) in this definition states that two different cells in a partition cannot overlap; thus, no element of A belongs to more than one cell.

Example 14: Let O be the set of odd positive integers, E the set of even positive integers. The collection $C = \{O, E\}$ is a partition of Z^+ because (1) these sets are nonempty subsets of Z^+, (2) every positive integer belongs to O or E, and (3) $O \cap E = \emptyset$.

Theorem L.3.3 *Let R be an equivalence relation in a nonempty set A. For each $x \in A$, let*

$$S_x = \{y : y \in A \text{ and } xRy\}$$

Then the collection

$$C = \{S_x : x \in A\}$$

is a partition of A.

PROOF: By definition each S_x is a subset of A. Since R is reflexive, we know that xRx for every $x \in A$. It follows that $x \in S_x$ and $S_x \neq \varnothing$, thus establishing conditions (1) and (2) of the definition.

To prove condition (3), we need to show, for $x, y \in A$, that if the sets S_x and S_y are different, then they have no elements in common; that is, we must show that

$$S_x \neq S_y \Rightarrow S_x \cap S_y = \varnothing$$

We do this by establishing the contrapositive:

$$S_x \cap S_y \neq \varnothing \Rightarrow S_x = S_y$$

So suppose that S_x and S_y have an element z of A in common. Then xRz and yRz, and because R is both symmetric and transitive, we conclude that xRy. Now if u is any element of S_y, we have yRu; and by invoking the transitive property again, we obtain xRu, so that $u \in S_x$. Therefore, $S_y \subseteq S_x$. It can be shown similarly that $S_x \subseteq S_y$, and we conclude that $S_x = S_y$. We have shown that if S_x and S_y have an element in common, then $S_x = S_y$. ■

Definition L.3.7 Let R be an equivalence relation in a nonempty set A. The cells in the partition of A described in Theorem L.3.3 are called ***equivalence classes*** of A with respect to R.

Example 15: Let $m \in Z^+$. The set of integers is partitioned into equivalence classes by the relation $\equiv_m$; these classes are sometimes called *residue classes* of Z *modulo* m. By means of the Division Algorithm we can show that every integer is congruent modulo m to one and only one of the numbers $0, 1, \ldots, m - 1$. Thus, there are exactly m residue classes of Z modulo m. In the notation of Theorem L.3.3, S_0 consists of all the multiples of m; S_1 consists of those integers that are congruent to 1 modulo m, and so on. If, for example, $m = 5$, we have the following residue classes modulo 5:

$$\begin{aligned} S_0 &= \{0, \pm 5, \pm 10, \ldots\} \\ S_1 &= \{1, 1 \pm 5, 1 \pm 10, \ldots\} \\ S_2 &= \{2, 2 \pm 5, 2 \pm 10, \ldots\} \\ S_3 &= \{3, 3 \pm 5, 3 \pm 10, \ldots\} \\ S_4 &= \{4, 4 \pm 5, 4 \pm 10, \ldots\}. \end{aligned}$$

Example 16: Equivalence of Fractions It seems natural to define a *fraction* to be an ordered pair of integers (numerator, denominator) in which the denominator is not zero. Then the set of all fractions is

$$F = \{(m, n) : m, n \in Z \text{ and } n \neq 0\}$$

where we use the notation (m, n) instead of the customary m/n. In elementary arithmetic we are taught that $\frac{4}{6}$ "is the same as" $\frac{2}{3}$—as are $\frac{8}{12}$,

$-\frac{4}{6}$, and many other fractions. However, according to the fundamental property of ordered pairs, $(a, b) = (c, d)$ if and only if $a = c$ and $b = d$, so that $(4, 6) \neq (2, 3)$. The conflict can be resolved by introducing a relation $\sim$ in F by defining $(p, q) \sim (r, s)$ if and only if $ps = qr$. The relation $\sim$ is an equivalence relation in F, and it is easy to see that $(2, 3)$ is equivalent by $\sim$ to $(4, 6)$, $(8, 12)$, $(-4, 6)$, and so on (Exercise L.3.10). Thus, we are on safe ground if we interpret "is the same as" to mean $\sim$. Moreover, if we define a *rational number* to be an equivalence class in the partition of F relative to $\sim$, and denote by m/n the equivalence class to which the pair (m, n) belongs, then we really do have $\frac{2}{3} = \frac{4}{6}$ because the classes to which $(2, 3)$ and $(4, 6)$ belong are identical. (In Section L.5 we show how to define "sum" and "product" of these classes.)

Order Relations

Let A be a nonempty set. We wish to introduce the concept of "order."

Definition L.3.8 A relation L in A is ***antisymmetric*** provided that if x and y are elements of A such that xLy and yLx, then $y = x$.

Definition L.3.9 A relation L in A is an ***order*** relation provided that L is reflexive, antisymmetric, and transitive.

Example 17: The relation $\leq$ in R is antisymmetric because, for $x, y \in R$, if both $x \leq y$ and $y \leq x$, we know that $y = x$. We observed earlier that this relation is also reflexive and transitive, so $\leq$ is an order relation in R.

Example 18: Let S and T be sets of real numbers, and define S to be related to T provided that $S \subseteq T$. The relation $\subseteq$ so defined is an order relation in the set of all subsets of R (Exercise L.3.11).

Example 19: The relation $m|n$ is an order relation in Z^+ but not in Z (Exercise L.3.12).

An essential difference between the order relation in Example 17 and those in Examples 18 and 19 is brought out in the following definition.

Definition L.3.10 If L is an order relation in A, then the elements $x, y \in A$ are ***comparable*** with respect to L provided that either xLy or yLx. An order relation in A is ***linear*** provided that any two elements of A are comparable.

The order relation $\leq$ in R is linear because for any two real numbers x, y, it must be the case that either $x \leq y$ or $y \leq x$. However, neither of the order relations in Examples 18 and 19 is linear (work Exercises L.3.11 and L.3.12).

EXERCISES L.3

1. Identify the domain and the range of each of the following relations in R.

(a) $\{(x, y) : 0 \leq x + 2y \leq 2\}$
(b) $\{(x, y) : y = \frac{9}{5}x + 32 \text{ and } x \geq -273\}$
(c) $\{(x, y) : y^2 = x - 1\}$
(d) $\{(x, y) : y = 1$ if x is rational; $y = 0$ if x is irrational$\}$
(e) $\{(x, y) : x^2 + 4y^2 \leq 9\}$

2. Determine the inverse of each relation in Exercise L.3.1.

3. Use plane coordinate geometry to show that the line $y = x$ is the perpendicular bisector of the line connecting the points $P(a, b)$ and $Q(b, a)$. (This result means that P and Q are symmetric with respect to the line, and verifies the remark following Example 9.)

4. Determine which of the following indicated relations are (1) reflexive, (2) symmetric, or (3) transitive in the given set A. Justify your conclusions.

(a) $A = R$; xRy if and only if $x < y$
(b) $A = R$; xRy if and only if $|x - y| < 1$
(c) $A = R$; xRy if and only if $x^2 = y^2$
(d) $A = R$; xRy if and only if $x^2 + y^2 = 1$
(e) $A = Z$; mRn if and only if $m - n$ is an even integer
(f) $A = \{n : n \in Z^+, n \leq 5\}$; $R = \{(1, 1), (2, 2), (2, 4), (3, 3), (4, 2), (4, 4), (5, 5)\}$
(g) $A =$ the set of all subsets of R; SRT if and only if $S \subseteq T$
(h) $A =$ the set of nonvertical lines in the coordinate plane; $\ell_1 R \ell_2$ if and only if ℓ_1 and ℓ_2 have the same slope

5. Verify that a relation R in a set A is symmetric if and only if $R^{-1} = R$.

6. Give an example of a relation R in a set A having the following properties:

(a) R is reflexive and symmetric but not transitive.
(b) R is symmetric and transitive but not reflexive.
(c) R is transitive but neither reflexive nor symmetric.

7. (a) For $x, y \in R$, define a relation R by the condition xRy if and only if there exists an integer n such that

$$|x - n| < \tfrac{1}{2} \text{ and } |y - n| < \tfrac{1}{2}$$

Find the domain A of the relation R and show that R is an equivalence relation in A. Describe the equivalence classes determined by this relation.

(b) Show that if the conditions $|x - n| < \frac{1}{2}$ and $|y - n| < \frac{1}{2}$ are replaced by $|x - n| \leq \frac{1}{2}$ and $|y - n| \leq \frac{1}{2}$ in part (a), then the resulting relation R is not an equivalence relation.

8. For (x, y) and $(u, v) \in R \times R$, define $(x, y) R(u, v)$ provided that $x + y = u + v$. Show that R is an equivalence relation in $R \times R$. Describe the equivalence classes determined by this relation.

9. Let $A = \{n : n \in Z^+, n \leq 10\}$. Verify that

$$C = \{\{1\}, \{2, 3\}, \{4, 5, 6\}, \{7, 8, 9, 10\}\}$$

is a partition of A, and list the ordered pairs in the equivalence relation R that determines this partition of A.

10. Show that the relation $\sim$ defined in Example 16 is an equivalence relation in F.

11. Show that the relation $\subseteq$ in Example 18 is an order relation. Explain why it is not linear.

12. (a) Explain why the relation $m|n$ is an order relation in Z^+ but not in Z.
(b) Explain why the order relation $m|n$ in Z^+ is not linear. Find a subset of Z^+ in which this order relation is linear.

13. (a) Show that if L is an order relation in A, then so is L^{-1}.
(b) Show that if an order relation L is linear, then so is L^{-1}.

L.4 *Functions*

In calculus you were probably taught that a function is a rule that assigns to each element of a first set a unique element of a second set. The following definition expresses this idea in terms of relations.

Definition L.4.1 Let A and B be sets. A ***function from A to B*** is a relation f from A to B with the property that for each $x \in A$ there exists one and only one $y \in B$ such that $(x, y) \in f$.

Since a function from A to B is a special kind of relation, it has a domain and a range; the domain of f is just A, and the range of f is a subset of B.

Example 1: The set $f = \{(x, x^2) : x \in R\}$ is a function from R to R. The domain of f is R; its range is the set of nonnegative real numbers. The rule conveyed by this function is "square the real number x".

Example 2: The set $T = \{(P, d) : P \in R \times R, d = |OP|\}$ in Example 5 of Section L.3 is a function from $R \times R$ to R. Its domain is the coordinate plane, and its range is the set of nonnegative real numbers. The rule conveyed by this function is "take the distance of P from the origin".

Example 3: The set $H = \{(1, 3), (2, 4), (3, 5), (3, -1)\}$ fails to be a function because there are two values of y, namely, $y = 5$ and $y = -1$, for which $(3, y) \in H$.

Example 4: The set $\{(m, n) | m, n \in Z^+ \text{ and } m|n\}$ is not a function. In fact, it fails badly since for any given positive integer m, there are many positive integers n that are divisible by m.

Notational Convention

If f is a function, then for each x in the domain of f, the symbol $f(x)$ denotes the unique y in the range for which $(x, y) \in f$. We call $f(x)$ the *value of f at x* or the *image of x under f*; it is the result of applying the rule f to the element x.

Example 5: If f is the function of Example 1, then $f(x) = x^2$, $f(a + b) = a^2 + 2ab + b^2$, and so on.

The function concept appears in mathematics in a variety of situations, as the following synonyms for function suggest: *operator, transformation*, and *mapping*.

We often think of a function as acting on elements of its domain and producing elements of the range. The differentiation *operator* acts on a differentiable function by producing its derivative; a vector is *transformed* by rotating it through a right angle; a point on the surface of the earth is *mapped* onto a point on a globe.

Using the "mapping" idea, if we have a function f from A to B, then we say that *f maps A into B*, and we write $f : A \to B$. In Figure L.6,

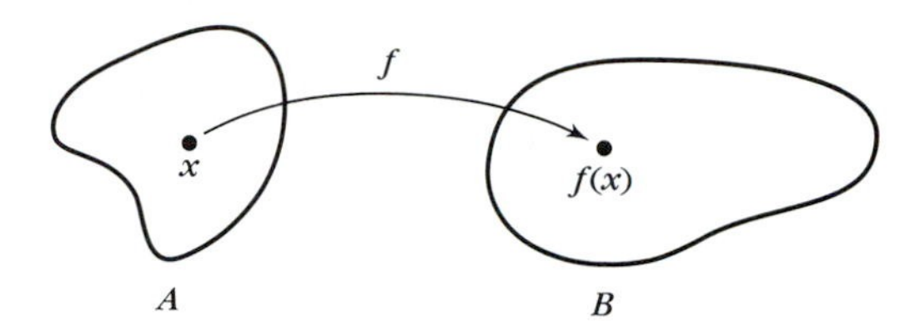

Figure L.6
f as a mapping

we show x being mapped to $f(x)$ by f. The arrow represents the function, and we distinguish f itself from its value $f(x)$.

If $f : A \to B$ and if S is a subset of A, then $f(S)$ denotes the set of all images of elements of A; that is,

$$f(S) = \{y : y = f(x) \text{ for some } x \in S\}$$

We call $f(S)$ the *image of S under f* (see Figure L.7). In case $S = A$, the set $f(A)$ is just the range of f.

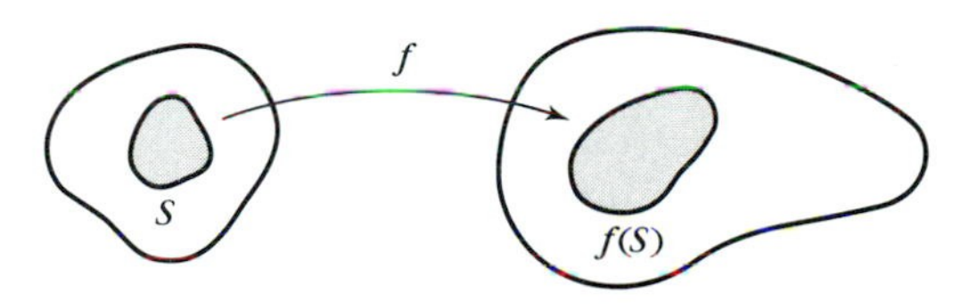

Figure L.7
Image of S under f

Example 6: Let $f = \{(x, y) : x, y \in R \text{ and } y = 2x + 1\}$. Then $f : R \to R$, and for each $x \in R$, the value of f at x is given by $f(x) = 2x + 1$. If $S = \{x : x \in R \text{ and } 0 < x < 2\}$, then the image of S under f is

$$f(S) = \{y : y \in R \text{ and } 1 < y < 5\}$$

The range of f is

$$f(R) = R$$

because each real number y is the value of f at a real number that can be found by solving the equation $y = 2x + 1$ for x—namely, $x = \frac{1}{2}(y - 1)$.

Composition

Definition L.4.2 If A, B, C are sets and if f, g are functions such that $f : A \to B$ and $g : B \to C$, then we can introduce a function $h : A \to C$ whose value is defined for $x \in A$ by

$$h(x) = g(f(x))$$

The function h is called the ***composition*** of g on f, denoted $g \circ f$. (Notice that the definition requires that the range of f be contained in the domain of g; see Figure L.8.)

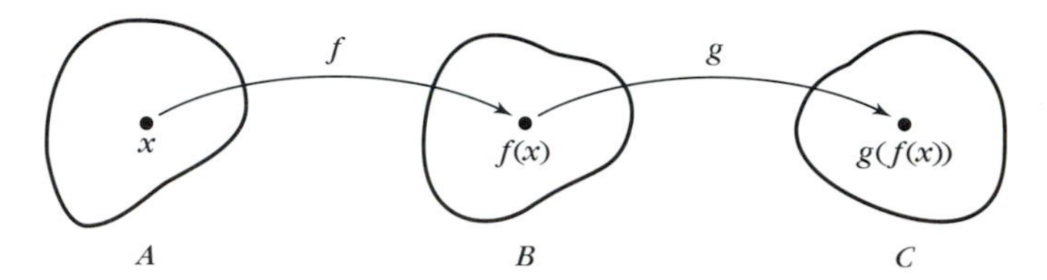

Figure L.8
Composition of functions

Example 7: Let $f = \{(x, y) : x, y \in R \text{ and } y = x^2 + 1\}$ and $g = \{(y, z) : y, z \in R \text{ and } z = 1/y\}$. Notice that the range of f is the set S of reals greater than or equal to 1; while the domain of g is the set of nonzero reals. Since the range of f is contained in the domain of g, we can form the composition $h = g \circ f$; its values are given for $x \in R$ by

$$h(x) = g(f(x)) = \frac{1}{x^2 + 1}$$

Example 8: If we change the last example slightly, making f equal to

$$\{(x, y) : x, y \in R \text{ and } y = x^2 - 4\}$$

with g as before, then the range of f includes the number 0, which is not in the domain of g. In order to form the composition, we need to remove from the domain of f the numbers 2 and -2 for which the values of f are 0. The resulting composite function is defined for all real numbers $x \neq \pm 2$ by

$$h(x) = \frac{1}{x^2 - 4}$$

The Inverse of a Function

Because a function is a relation, it has an inverse; however, the inverse may or may not be a function itself.

Example 9: Let $f = \{(x, y) : x, y \in R \text{ and } y = 2x + 1\}$. Then f is a function whose domain is R. The inverse of f is given by

$$\begin{aligned} f^{-1} &= \{(x, y) : (y, x) \in f\} \\ &= \{(x, y) : x, y \in R \text{ and } x = 2y + 1\} \\ &= \left\{(x, y) : x, y \in R \text{ and } y = \frac{x - 1}{2}\right\} \end{aligned}$$

It is clear that f^{-1} is also a function whose domain is R.

Example 10: Let $f = \{(x, y) : x, y \in R \text{ and } y = x^2\}$. Then

$$\begin{aligned} f^{-1} &= \{(x, y) : (y, x) \in f\} \\ &= \{(x, y) : x, y \in R \text{ and } x = y^2\}. \end{aligned}$$

Clearly f^{-1} is not a function because, for instance, the ordered pairs $(4, 2)$ and $(4, -2)$ both belong to f^{-1}.

Example 11: Let $g = \{(x, y) : x, y \in R, x \geq 0 \text{ and } y = x^2\}$, where we have restricted x to nonnegative values. Then

$$\begin{aligned} g^{-1} &= \{(x, y) : (y, x) \in g\} \\ &= \{(x, y) : x, y \in R, y \geq 0 \text{ and } x = y^2\}. \end{aligned}$$

Because $y \geq 0$, we have

$$g^{-1} = \{(x, y) : x, y \in R \text{ and } y = \sqrt{x}\}$$

where $\sqrt{\ }$ means "nonnegative square root". In this case, the inverse is a function.

Definition L.4.3 A function f is ***invertible*** provided that f^{-1} is a function.

Examples 10 and 11 show that it may be necessary to restrict the domain of a function in order to make it invertible.

One-to-One Mappings (Injections)

Let $f : A \to B$. Then f is invertible if and only if for each y in the range of f, there is only one $x \in A$ such that $(x, y) \in f$. This condition can be stated in either of two equivalent ways:

(1) For every $x_1, x_2 \in A$, if $f(x_1) = f(x_2)$, then $x_1 = x_2$; or (in the contrapositive form)

(2) For every $x_1, x_2 \in A$, if $x_1 \neq x_2$, then $f(x_1) \neq f(x_2)$.

Definition L.4.4 Let $f : A \to B$. Then f is said to map A ***one-to-one*** into B, provided that f satisfies either of the two conditions above (hence both of them). Also, under these circumstances, $f : A \to B$ is said to be a ***one-to-one*** mapping, or ***injection***.

The following theorem is an immediate consequence of the last definition and the remarks preceding it:

Theorem L.4.1 *A function f is invertible if and only if it is one-to-one.*

Example 12: Let $g = \{(x, y) : x, y \in R, x \geq 1, \text{ and } y = x^3 - 3x\}$. We prove that g is invertible by showing that condition (1) holds. Suppose that $g(x_1) = g(x_2)$ for certain real numbers x_1, x_2 that are both ≥ 1; then

$$x_1^3 - 3x_1 = x_2^3 - 3x_2$$

so that

$$x_1^3 - x_2^3 - 3(x_1 - x_2) = 0$$

With a little algebra, we find that

$$(x_1 - x_2)(x_1^2 + x_1 x_2 + x_2^2 - 3) = 0$$

For the product to be zero, we must have either $x_1 - x_2 = 0$ or $x_1^2 + x_1x_2 + x_2^2 = 3$. The first of these equations means that $x_1 = x_2$; and because neither x_1 nor x_2 can be less than 1, the second equation can hold only if both x_1 and x_2 are equal to 1. In any case, we have determined that if $g(x_1) = g(x_2)$, then $x_1 = x_2$, and so g is invertible.

Figure L.9 is a schematic diagram showing a one-to-one function f mapping A into B. Its inverse f^{-1} maps the range $f(A)$ of f into A:

$$f : A \to B$$
$$f^{-1} : f(A) \to A$$

From the definition of inverse it follows that, for $x \in A$ and $y \in f(A)$, the equations

$$y = f(x) \qquad \text{and} \qquad x = f^{-1}(y)$$

are equivalent.

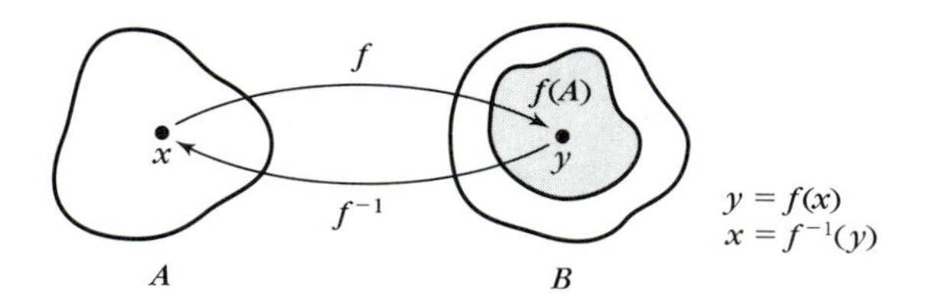

Figure L.9
A one-to-one function and its inverse

Bijections

Definition L.4.5 Let $f : A \to B$.

(1) If the range of f is all of B, then f is said to map A ***onto*** B, and the mapping $f : A \to B$ is called a ***surjection***.

(2) If f maps A one-to-one onto B, then $f : A \to B$ is called a ***bijection***.

Note: Obviously any function maps its domain *onto* its range.

Example 13: Straight Line Function Let m be a fixed real number, different from 0, and let $L = \{(x, y) : x, y \in R \text{ and } y = mx + b\}$. We claim that $L : R \to R$ is a bijection. First, if $L(x_1) = L(x_2)$, then

$$mx_1 + b = mx_2 + b$$

and it follows by elementary algebra that $x_1 = x_2$. Thus, L is one-to-one. To prove that the range of L is R, we must choose an arbitrary $y \in R$ and show that there exists an $x \in R$ such that $L(x) = y$. Such an x is easily obtained by solving the equation

$$y = mx + b$$

for x:

$$x = \frac{y - b}{m}$$

Thus, L maps R onto R, and $L : R \to R$ is a bijection.

Example 14: The function $T : R \times R \to R$ in Example 2 assigns to each point P of the coordinate plane $R \times R$, the distance of P from the origin. Since distances cannot be negative, this mapping is not surjective. However, T does map $R \times R$ onto $S = \{d : d \in R \text{ and } d \geq 0\}$, so we can say that $T : R \times R \to S$ is a surjection. Now at a given distance from the origin there are many points, and so the mapping $T : R \times R \to S$ is not one-to-one, hence not a bijection.

If $f : A \to B$ is a bijection, then we know that f is invertible because it is one-to-one. Moreover, the range of f is all of B, and so Figure L.9 takes the form of Figure L.10.

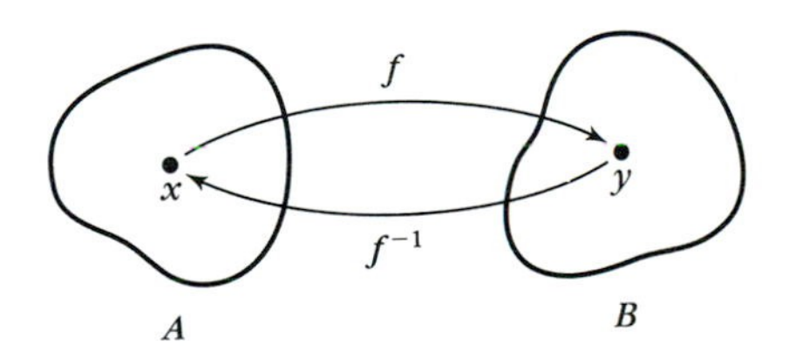

Figure L.10
Inverse of a bijection

It is easy to show that $f^{-1} : B \to A$ is also a bijection and that its inverse is just f.

The following theorem presents a useful observation about composition of bijections.

Theorem L.4.2 *If $f : A \to B$ and $g : B \to C$ are bijections, then $g \circ f : A \to C$ is a bijection, and $(g \circ f)^{-1} = f^{-1} \circ g^{-1}$.*

PROOF: First we show that $g \circ f$ maps A onto C. Let c be an arbitrary element of C; then, because g maps B onto C, there exists an element $b \in B$ such that $g(b) = c$. Furthermore, since f maps A onto B, there exists an $a \in A$ such that $f(a) = b$. Then

$$c = g(b) = g(f(a)) = (g \circ f)(a)$$

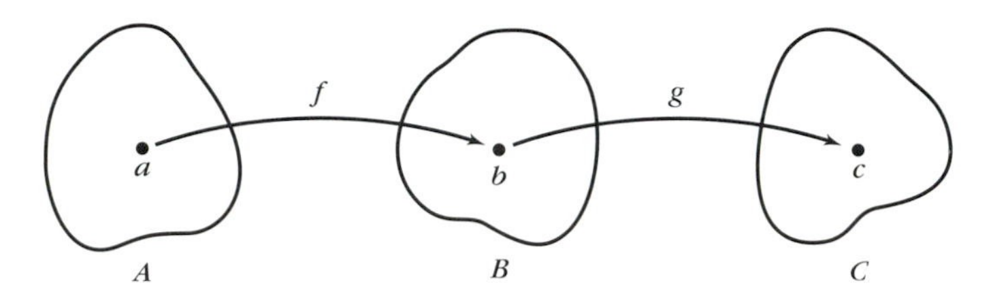

Figure L.11

and we see that c is in the range of $g \circ f$, so that $g \circ f$ maps A onto C (see Figure L.11).

Next we show that $g \circ f$ is one-to-one. If x_1 and x_2 are elements of A such that $x_1 \neq x_2$, then $f(x_1) \neq f(x_2)$ (because f is one-to-one) and $g(f(x_1)) \neq g(f(x_2))$ (because g is one-to-one). Thus, $(g \circ f)(x_1) \neq (g \circ f)(x_2)$, and so $g \circ f$ is one-to-one, completing the proof that $g \circ f : A \to C$ is a bijection.

Now since $h = g \circ f$ maps A one-to-one onto C, its inverse h^{-1} maps C one-to-one onto A. We let c be an arbitrary element of C, and let $a = h^{-1}(c)$. Then

$$\begin{aligned} c &= h(a) \\ &= g(f(a)) \\ &= g(b) \end{aligned}$$

where we have set $b = f(a)$. Then $a = f^{-1}(b)$ and $b = g^{-1}(c)$, so that $a = f^{-1}(g^{-1}(c)) = (f^{-1} \circ g^{-1})(c)$. Thus,

$$h^{-1}(c) = a = (f^{-1} \circ g^{-1})(c)$$

for every $c \in C$, and we conclude that $h^{-1} = f^{-1} \circ g^{-1}$. ∎

Theorem L.4.2 says that the inverse of the composition of bijections is the composition of their inverses *in reverse order*. (As an everyday illustration of this fact, consider inverting the following operation: put on your socks, then put on your shoes.)

Example 15: Let $f = \{(x, y) : x, y \in R \text{ and } y = 2x - 1\}$ and let $g = \{(y, z) : y, z \in R \text{ and } z = y + 1\}$. It is easy to determine that $(g \circ f)(x) = 2x$ for $x \in R$, so that $(g \circ f)^{-1}(z) = \frac{1}{2}z$ for $z \in R$. On the other hand, if we compute $(f^{-1} \circ g^{-1})(z)$, we get

$$\begin{aligned} f^{-1}(g^{-1}(z)) &= f^{-1}(z-1) \\ &= \frac{(z-1)+1}{2} = \frac{z}{2} \end{aligned}$$

Restrictions and Extensions

In Examples 10 and 11 we encountered the need to restrict the domain of a function in order to obtain an inverse.

Definition L.4.6

(1) If f is a function with domain D and if $S \subseteq D$, then the function g defined by $g(x) = f(x)$ $(x \in S)$ is the ***restriction of f to S***.
(2) If f is a function with domain D, if $S \supseteq D$ and if F is a function with the property that $F(x) = f(x)$ for all $x \in D$, then F is an ***extension of f to S***.

Example 16: The square function

$$f = \{(x, y) : x, y \in R \text{ and } y = x^2\}$$

is not invertible. However, we saw in Example 11 that its restriction to the set of nonnegative real numbers,

$$g = \{(x, y) : x, y \in R, x \geq 0 \text{ and } y = x^2\},$$

is invertible.

Example 17: The function

$$f = \left\{(x, y) : x, y \in R \text{ and } y = \frac{x^2 - 9}{x - 3}\right\}$$

is defined on the set of real numbers different from 3. The function

$$F(x) = \{(x, y) : x, y \in R \text{ and } y = x + 3\}$$

is defined on all of R, and, for all $x \neq 3$, it is true that $F(x) = f(x)$. Thus, F is an extension of f to R.

The restriction of a given function to a particular subset of its domain is unique; however, an extension to a larger set is not in general unique.

Operations

Addition and multiplication are familiar operations in the real numbers. Let us formulate a definition of "operation". The effect of an operation is that two elements of a set (not necessarily different elements) can be combined in some way to form a third element of that set. Thus, to each ordered pair of elements, the operation assigns a third element, and so an operation is really a function.

Definition L.4.7 An ***operation*** $\circ$ in a set S is a function mapping $S \times S$ into S, that is,

$$\circ : S \times S \to S$$

If $a, b, c \in S$ and if c is the image of (a, b) under $\circ$, then we write

$$c = a \circ b$$

Definition L.4.8 An operation $\circ$ in a set S is ***commutative*** provided $a \circ b = b \circ a$ for every $a, b \in S$. It is ***associative*** if $a \circ (b \circ c) = (a \circ b) \circ c$ for every $a, b, c \in S$.

Addition and multiplication in R are commutative and associative.

Example 18: If $a, b \in R$, we can define an operation $\circ$ by $a \circ b = (a + b)/2$, where $+$ means ordinary addition. The operation $\circ$ is commutative because $(a + b)/2 = (b + a)/2$. It is not associative, however.

Example 19: If $a, b \in Z$, we can define an operation $\circ$ by $a \circ b = a - b$. This operation (ordinary subtraction) is neither commutative nor associative.

Example 20: Division is not an operation in Q because we cannot divide by 0. However, it is an operation in $Q \dashv \{0\}$.

EXERCISES L.4

1. Determine which of the following relations are functions, and find the domain and range of each.

(a) $\{(x, y) : x, y \in R \text{ and } x + 2y = 2\}$
(b) $\{(x, y) : x, y \in R \text{ and } y^2 = 4x\}$
(c) $\{(x, y) : x \in R \text{ and } y = 1 \text{ if } x \text{ is rational}, y = 0 \text{ if } x \text{ is irrational}\}$
(d) $\{(u, v) : u, v \in R \text{ and } u \leq v\}$
(e) $\left\{(x, y) : x, y \in R \text{ and } y = \frac{1}{x}\right\}$
(f) $\left\{(x, y) : x, y \in R \text{ and } y = \frac{1}{\sqrt{x}}\right\}$

2. For each function f and set A, determine the set $f(A)$.

(a) $f = \{(x, y) : x, y \in R \text{ and } y = x^2\}$; $A = \{x : -1 < x \leq 2\}$
(b) $f = \left\{(x, y) : x, y \in R \text{ and } y = \frac{1}{x}\right\}$; $A = \{x : 0 < x < 1\}$
(c) $f = \{(u, v) : u, v \in R \text{ and } v = \sqrt{4 - u^2}\}$; $A = \{u : -1 < u < 1\}$

3. For each pair of functions, determine the composite function $h = g \circ f$ by stating its domain and giving a rule for finding $h(x)$.

(a) $\begin{cases} f = \{(x, y) : x, y \in R \text{ and } y = 3x + 4\} \\ g = \{(y, z) : y, z \in R \text{ and } z = y - 2\} \end{cases}$

(b) $\begin{cases} f = \{(x, y) : x, y \in R \text{ and } y = 1 + x^2\} \\ g = \{(y, z) : y, z \in R \text{ and } z = \sqrt{y + 1}\} \end{cases}$

(c) $\begin{cases} f = \{(x, y) : x, y \in R \text{ and } y = 1 + x^2\} \\ g = \{(y, z) : y, z \in R \text{ and } z = \sqrt{y - 1}\} \end{cases}$

(d) $\begin{cases} f = \{(x, y) : x, y \in R, x \geq 2 \\ \quad \text{and } y = 1 + x^2\} \\ g = \{(y, z) : y, z \in R \text{ and } z = \sqrt{y - 5}\} \end{cases}$

4. Let $f : A \to B$.

(a) Show that if S and T are subsets of A and if $S \subseteq T$, then $f(S) \subseteq f(T)$.

(b) Prove or disprove: If S and T are subsets of A such that $f(S) \subseteq f(T)$, then $S \subseteq T$.

(c) Prove or disprove: If S and T are subsets of A, then $f(S \cup T) = f(S) \cup f(T)$.

(d) Prove or disprove: If S and T are subsets of A, then $f(S \cap T) = f(S) \cap f(T)$.

5. Show that if $f : A \to B$ is a bijection, then f^{-1} is a bijection mapping B onto A and show that $(f^{-1})^{-1} = f$.

6. Let $f = \{(x, y) : x, y \in R \text{ and } y = x^2 - 2x\}$.

(a) Show that f is not invertible.

(b) If g is the restriction of f to the set of real numbers greater than or equal to 1, show that g is invertible and find an expression for its inverse.

7. Show that the following functions are invertible:

$$f = \{(x, y) : x, y \in R, x \geq 6 \text{ and } y = x - 3\}$$
$$g = \{(y, z) : y, z \in R, y \geq 3, z \geq 0, \text{ and } y^2 - z^2 = 9\}.$$

Find expressions for the values of f^{-1}, g^{-1}, $(g \circ f)^{-1}$, $f^{-1} \circ g^{-1}$, and verify that the conclusion of Theorem L.4.2 holds.

8. Which of the following are operations in the given set? For those that are operations, determine whether they are commutative or associative.

(a) Ordinary multiplication; Z^+

(b) Subtraction; Z^+

(c) Division; R

(d) Division; $R \dashv \{0\}$

(e) $m \circ n = (m + n)/2$; $m, n \in Z$

(f) $m \circ n = 2^{m+n}$; $m, n \in Z^+$

(g) $m \circ n = 2^{mn}$; $m, n \in Z^+$

(h) $x \circ y = \max(x, y) = \begin{cases} x & \text{if } x \geq y \\ y & \text{if } x < y \end{cases}$; $x, y \in R$

(i) $\rho(x, y) = x$; $x, y \in R$

(j) Composition of functions; F is the set of all functions mapping R into R.

★ L.5 Construction of Number Systems

In this section, we discuss, in two somewhat different ways, the construction of a "new" number system from a given "old" one. The old system is the familiar system Z of integers, with its basic operations and algebraic laws. We take these for granted and use the tools of set theory (sets, equivalence relations, functions, operations) to construct the new systems: Z_5 (the integers modulo 5) and Q (the rationals).

Algebraic Properties of the System Z of Integers

Here are the familiar facts about Z that we are taking for granted: Z has two basic operations, addition and multiplication. The result of adding two integers a and b is called their *sum* and is denoted $a + b$; the result of multiplying a and b is called their *product* and is denoted $a \cdot b$, or, more often, just ab. The following equations hold for all integers a, b, c and will be called here the "laws of arithmetic":

Commutative Laws: $a + b = b + a$ and $ab = ba$
Associative Laws: $a + (b + c) = (a + b) + c$ and $a(bc) = (ab)c$
Distributive Law: $a(b + c) = ab + ac$

The integer 0 has the property that $a + 0 = a$ for every integer a; the integer 1 has the property that $a \cdot 1 = a$ for every integer a. For each integer a there is a unique integer b such that $a + b = 0$; b is denoted $-a$ and is called the opposite (or negative) of a. Subtraction can then be defined as follows: $b - a = b + (-a)$.

The reciprocal of an integer is not in general another integer. However, the system of integers satisfies a "cancellation" law: if $ab = ac$ and $a \neq 0$, then $b = c$.

Construction of Z_5

The elements ("numbers") of Z_5 are the five residue classes of Z modulo 5, listed in Example 15 of Section L.3. Thus, $Z_5 = \{S_0, S_1, S_2, S_3, S_4\}$, where

$$S_0 = \{0, \pm 5, \pm 10, \ldots\}$$
$$S_1 = \{1, 1 \pm 5, 1 \pm 10, \ldots\}$$
$$S_2 = \{2, 2 \pm 5, 2 \pm 10, \ldots\}$$
$$S_3 = \{3, 3 \pm 5, 3 \pm 10, \ldots\}$$
$$S_4 = \{4, 4 \pm 5, 4 \pm 10, \ldots\}$$

We will make this set into a number system by defining the operations addition mod 5 and multiplication mod 5. First, we introduce notation that makes the elements of Z_5 look like numbers: We will denote S_0 by $\overline{0}$, S_1 by $\overline{1}$, S_2 by $\overline{2}$, S_3 by $\overline{3}$, and S_4 by $\overline{4}$. Furthermore, if a is *any* integer, we will denote by $\overline{a}$ the residue class modulo 5 to which a belongs, and we will call a a *representative* of that class. Thus, 3 is a representative of S_3, but so are 8, -2, and many other integers.

Now, let's define the operations in Z_5: We let $x, y \in Z_5$. Then x and y are residue classes mod 5, and we choose a to be an integer representing x, and b to be an integer representing y, so that $x = \overline{a}$ and $y = \overline{b}$. We define the *sum* $x \oplus y$ mod 5 to be the residue class to which the ordinary sum $a + b$ belongs. We define the *product* $x \odot y$ mod 5 to be the residue class to which the ordinary product belongs. Thus,

$$x \oplus y = \overline{a} \oplus \overline{b} = \overline{a + b}$$
$$x \odot y = \overline{a} \odot \overline{b} = \overline{ab}$$

Example 1:

$$\overline{3} \oplus \overline{4} = \overline{3+4} = \overline{7} = \overline{2}$$
$$\overline{3} \odot \overline{2} = \overline{3 \cdot 2} = \overline{6} = \overline{1}$$
$$\overline{3} \odot (\overline{2} \oplus \overline{4}) = \overline{3} \odot \overline{1} = \overline{3 \cdot 1} = \overline{3}$$

There is a problem, however: The definition of each operation seems to depend on the choice of the integers representing x and y. Actually, this apparent dependence is not real.

Example 2: With $x = \overline{3}$, $y = \overline{4}$, we calculated $x \oplus y = \overline{3} \oplus \overline{4} = \overline{3+4} = \overline{2}$. Suppose we had chosen different representatives from the classes x and y; for instance, suppose we had chosen 8 as a representative of x, 9 as a representative of y. We would have obtained the same sum because

$$\overline{8} \oplus \overline{9} = \overline{17} = \overline{2}$$

In this case, the matter of choice of representative caused no harm.

Let us look at the question in more generality. Suppose the residue class x is represented by the integer a and also by another integer c (so that $x = \overline{a} = \overline{c}$); and that a second class y is represented by the integers b and d (so that $y = \overline{b} = \overline{d}$). Then if we calculate their sum using $\overline{a} \oplus \overline{b} = \overline{a+b}$, we get the same result as from $\overline{c} \oplus \overline{d} = \overline{c+d}$. This result is an immediate consequence of Theorem L.3.2, which assures us that if $a \equiv_5 c$ and $b \equiv_5 d$, then $a + b \equiv_5 c + d$. Our argument shows that the definition of $\oplus$ makes sense; it does not depend on the choice of representatives in the two classes that are being added. The situation regarding multiplication is completely analogous.

Tables for the two operations just defined on Z_5 are:

$\oplus$	$\overline{0}$	$\overline{1}$	$\overline{2}$	$\overline{3}$	$\overline{4}$
$\overline{0}$	$\overline{0}$	$\overline{1}$	$\overline{2}$	$\overline{3}$	$\overline{4}$
$\overline{1}$	$\overline{1}$	$\overline{2}$	$\overline{3}$	$\overline{4}$	$\overline{0}$
$\overline{2}$	$\overline{2}$	$\overline{3}$	$\overline{4}$	$\overline{0}$	$\overline{1}$
$\overline{3}$	$\overline{3}$	$\overline{4}$	$\overline{0}$	$\overline{1}$	$\overline{2}$
$\overline{4}$	$\overline{4}$	$\overline{0}$	$\overline{1}$	$\overline{2}$	$\overline{3}$

$\odot$	$\overline{0}$	$\overline{1}$	$\overline{2}$	$\overline{3}$	$\overline{4}$
$\overline{0}$	$\overline{0}$	$\overline{0}$	$\overline{0}$	$\overline{0}$	$\overline{0}$
$\overline{1}$	$\overline{0}$	$\overline{1}$	$\overline{2}$	$\overline{3}$	$\overline{4}$
$\overline{2}$	$\overline{0}$	$\overline{2}$	$\overline{4}$	$\overline{1}$	$\overline{3}$
$\overline{3}$	$\overline{0}$	$\overline{3}$	$\overline{1}$	$\overline{4}$	$\overline{2}$
$\overline{4}$	$\overline{0}$	$\overline{4}$	$\overline{3}$	$\overline{2}$	$\overline{1}$

Algebraic properties of Z_5 strongly resemble those for the real numbers: addition and multiplication are commutative and associative, and a distributive law relates the two operations.

Theorem L.5.1 *For all $x, y, z \in Z_5$, the laws of arithmetic hold:*

Commutative Laws: $x \oplus y = y \oplus x$; $x \odot y = y \odot x$

Associative Laws: $x \oplus (y \oplus z) = (x \oplus y) \oplus z;$
$x \odot (y \odot z) = (x \odot y) \odot (z)$

Distributive Law: $x \odot (y \oplus z) = (x \odot y) \oplus (x \odot z)$

PROOF: The commutative property of addition is easily verified: Given x, $y \in Z_5$, we choose a, $b \in Z$ such that $x = \overline{a}$ and $y = \overline{b}$. Then $x \oplus y = \overline{a + b} = \overline{b + a} = y \oplus x$. To verify the distributive law, we let $z \in Z_5$ and choose $c \in Z$ with $z = \overline{c}$. Then

$$y \oplus z = \overline{b + c}$$

so that

$$x \odot (y \oplus z) = \overline{a(b + c)}$$

On the other hand,

$$\begin{aligned} (x \odot y) \oplus (x \odot z) &= \overline{ab} \oplus \overline{ac} \\ &= \overline{ab + ac} \end{aligned}$$

Now $a(b + c) = ab + ac$ because of the distributive law for Z. Therefore,

$$(x \odot y) \oplus (x \odot z) = x \odot (y \oplus z)$$

You should have no trouble proving the remaining laws. ∎

In addition to the laws of arithmetic, Z_5 has other properties resembling Z. It is easy to see just by looking at the operation tables that Z_5 possesses a zero, $\overline{0}$ ($x \oplus \overline{0} = x$ for every $x \in Z_5$), and a unity, $\overline{1}$ ($x \odot \overline{1} = x$ for every $x \in Z_5$). For each element $x \in Z_5$ there is a unique element y (the opposite of x) such that $x \oplus y = \overline{0}$ (every row in the addition table has exactly one $\overline{0}$).

Moreover, for each nonzero element $x \in Z_5$, there is a unique y (the reciprocal of x), such that $x \odot y = \overline{1}$. This last property is stronger than the cancellation law, for if $ab = ac$, $a \neq \overline{0}$, we can multiply both sides by the reciprocal of a and obtain $b = c$.

The System of Rational Numbers

Let us now look at one way of constructing the rational numbers from Z. Recall that in Example 16 of Section L.3 we introduced the set F of fractions:

$$F = \{(m, n) : m, n \in Z \text{ and } n \neq 0\}$$

and defined an equivalence relation $\sim$ as follows:

$$(p, q) \sim (r, s) \quad \text{if and only if} \quad ps = qr$$

In terms of this relation, the ordered pairs $(2, 3)$ and $(4, 6)$ are equivalent, even though they are not identical as ordered pairs. We define a *rational number* to be an equivalence class in the partition of F relative to $\sim$, and we denote by m/n the class to which the ordered pair (m, n) belongs. Thus, the set of all rational numbers is

$$Q = \{m/n: m, n \in Z, n \neq 0\}$$

and it is our objective to define two operations, addition and multiplication, in Q and to establish their fundamental properties.

Now suppose $x, y \in Q$. Then x and y are both equivalence classes relative to $\sim$; and these classes consist of ordered pairs of integers. Suppose x is the class containing (p, q) and y is the class containing (r, s). Then $x = p/q$, and $y = r/s$, and we define

$$x \oplus y = \frac{p}{q} \oplus \frac{r}{s} = \frac{ps + qr}{qs}$$
$$x \odot y = \frac{p}{q} \odot \frac{r}{s} = \frac{pr}{qs}$$

As in the construction of Z_5, it would appear that the definitions of these operations might depend on which elements of F are used to represent x and y. They do not. Let us look at the definition of multiplication and suppose that we had chosen (p', q') as a representative of the class x, (r', s') as a representative of y. Then $(p, q) \sim (p', q')$ and $(r, s) \sim (r', s')$. By using these representatives we would have obtained

$$x \odot y = \frac{p'r'}{q's'}$$

Thus, the question before us is whether the same class has been obtained for $x \odot y$. The answer is yes because, as the following argument shows, the pairs $(p'r', q's')$ and (pr, qs) are equivalent:

$$\begin{aligned} pq' &= p'q && \text{because } (p, q) \sim (p', q') \\ rs' &= r's && \text{because } (r, s) \sim (r', s') \\ (pq')(rs') &= (p'q)(r's) && \text{by multiplying} \\ (pr)(q's') &= (p'r')(qs) && \text{by rearranging factors} \\ (pr, qs) &\sim (p'r', q's') && \text{by the definition of } \sim. \end{aligned}$$

A similar argument shows that addition is well-defined.

The algebraic properties of Q under $\oplus$ and $\odot$ are similar to those of Z_5.

Theorem L.5.2 *For all $x, y, z \in Q$ the laws of arithmetic hold:*

Commutative Laws: $x \oplus y = y \oplus x$; $x \odot y = y \odot x$

Associative Laws: $x \oplus (y \oplus z) = (x \oplus y) \oplus z$; $x \odot (y \odot z) = (x \odot y) \odot z$

Distributive Law: $x \odot (y \oplus z) = (x \odot y) \oplus (x \odot z)$

PROOF: We prove the distributive law and leave the remaining parts for you to prove. We let $x = m/n$, $y = p/q$, and $z = r/s$. Then

$$x \odot (y \oplus z) = \frac{m}{n} \odot \frac{ps + qr}{qs} = \frac{m(ps + qr)}{n(qs)}$$

$$\begin{aligned}(x \odot y) \oplus (x \odot z) &= \frac{mp}{nq} \oplus \frac{mr}{ns} \\ &= \frac{mpns + nqmr}{(nq)(ns)} \\ &= \frac{mps + mqr}{nqs} = \frac{m(ps + qr)}{nqs}\end{aligned}$$

In the last step we used the fact that two fractions (k, l) and (nk, nl) are equivalent if $n \neq 0$. ■

If we identify each rational number of the form $n/1$ with the integer n, then we can think of Q as containing Z. The integers 0 and 1 appear as the rational numbers 0/1 and 1/1, respectively, and have the properties that $x \oplus 0 = x$ and $x \odot 1 = x$ for every $x \in Q$. It is easy to show that for each $x \in Q$, there is a unique $y \in Q$ such that $x \oplus y = 0$; y is the *opposite* of x, $-x$. Also, for each nonzero element x of Q, there is a unique element $z \in Q$ such that $x \odot z = 1$; z is the *reciprocal* of x, $1/x$.

There is no longer any need to distinguish the addition operation in Z from the one defined for Q because they give the same results. Consider, for example, the equation $2 + 3 = 5$; if we interpret the $+$ sign as rational $\oplus$, then this equation means $2/1 \oplus 3/1 = 5/1$, but that is equivalent to saying $2 + 3 = 5$ in the integers. Thus, we can dispense with two symbols for addition and replace the symbol $\oplus$ by $+$. The same applies to $\odot$.

* * *

The two number systems constructed from Z are very different but each is an example of an algebraic structure known as a *field*. In Section R.1, field is defined and the properties of fields are discussed.

EXERCISES L.5

1. For $x, y \in Z_5$, define the *difference* $x - y$ to be $x \oplus (-y)$, where $-y$ is the opposite of y; and define the *quotient* $x \div y$ $(y \neq \overline{0})$ to be $x \odot (1/y)$, where $1/y$ is the reciprocal of y. (Observe that $-$ is an operation on Z_5 and $\div$ is an operation on $Z_5 \dashv \{\overline{0}\}$.)

   (a) Calculate: $-\overline{3}$, $\overline{-3}$, $\overline{2} - \overline{3}$, $\overline{2} \odot \overline{-3}$, $\overline{2}/\overline{3}$

   (b) Solve the following equations in Z_5 for x. (Interpret x^2 as $x \odot x$.)

   (i) $\overline{2} \odot x = \overline{3}$
   (ii) $(\overline{2} \odot x) \oplus \overline{4} = \overline{3}$
   (iii) $(-\overline{2}) \odot x = \overline{4}$
   (iv) $x^2 \oplus x = \overline{2}$
   (v) $x^2 = \overline{2}$
   (vi) $x^2 = \overline{1}$
   (vii) $x^2 = -\overline{1}$

2. Follow the construction outlined for Z_5, but use the equivalence relation $\equiv_6$ to construct the system of integers modulo 6. The tables of addition and multiplication for Z_6 are given below (the bars over the numerals are omitted).

$\oplus$	0	1	2	3	4	5
0	0	1	2	3	4	5
1	1	2	3	4	5	0
2	2	3	4	5	0	1
3	3	4	5	0	1	2
4	4	5	0	1	2	3
5	5	0	1	2	3	4

$\odot$	0	1	2	3	4	5
0	0	0	0	0	0	0
1	0	1	2	3	4	5
2	0	2	4	0	2	4
3	0	3	0	3	0	3
4	0	4	2	0	4	2
5	0	5	4	3	2	1

(a) Solve: $3 \odot x = 0$
(b) Solve: $3 \odot x = 1$
(c) Solve: $5 \odot x = 2$
(d) Solve: $x^2 = 3$
(e) Solve: $x^2 = -1$

3. Verify the commutative and associative properties for addition and multiplication in Q.

CHAPTER R

REAL NUMBERS AND REAL FUNCTIONS

The objective of this book is the study of real functions; that is, functions whose domains and ranges are subsets of R. Section R.3 is devoted to a review of a number of functions of this type, many of which will be familiar from your earlier study of mathematics.

The first two sections of Chapter R are devoted to the properties of R itself, particularly those having to do with algebra and order. As we review these properties, we will find it instructive to compare and contrast the real number system with several other number systems having some of the same properties. Later (Chapter 1) we will introduce an additional property of R that serves to distinguish it from these other systems.

R.1 *Algebra and Order*

The set R of real numbers with its basic operations of addition and multiplication forms an algebraic structure known as a *field*. Furthermore, there is superimposed on this algebraic structure an order relation, so that R is an *ordered field*. The following paragraphs discuss these facts briefly.

Algebra

Definition R.1.1 Let F be a set whose elements may be combined by means of two binary operations, ***addition*** and ***multiplication***. If $a, b \in F$, then the result of adding a and b is their ***sum***, denoted $a + b$; the result of multiplying a and b is their ***product***, denoted $a \cdot b$ or frequently just ab. Both $a + b$ and ab belong to F. Then F forms a ***field*** with respect to these operations provided that the following seven postulates hold.

Postulate R.1.1 Commutative Laws

If $a, b \in F$, then $a + b = b + a$ and $ab = ba$.

Postulate R.1.1 assures us that the order in which two elements are added or multiplied does not affect their sum or their product.

Postulate R.1.2 Associative Laws

If $a, b, c \in F$, then $a + (b + c) = (a + b) + c$ and $a(bc) = (ab)c$.

Postulate R.1.2 states that the manner in which three elements are grouped by parentheses does not affect their sum or their product. An illustration is the equation

$$2 + (4 + 1) = (2 + 4) + 1$$

which is equivalent to

$$2 + 5 = 6 + 1$$

Because $a + (b + c)$ and $(a + b) + c$ are equal, we may denote each of them by $a + b + c$.

Postulate R.1.3 Zero Law

There exists an element $0 \in F$, such that the equations

$$a + 0 = 0 + a = a$$

hold for every $a \in F$. The element 0 is called the ***zero*** of F.

Postulate R.1.4 Unity Law

There exists an element $1 \in F$, different from 0, such that the equations $a \cdot 1 = 1 \cdot a = a$ hold for every $a \in F$. The element 1 is called the ***unity*** of F.

Postulates R.1.3 and R.1.4 assert the existence of elements in F having special properties. Addition of the element 0 to any other element a of F leaves that element unchanged; for this reason 0 is called the *identity for addition*. The element 1, which is different from 0, serves as an *identity for multiplication*. Exercise R.1.3a asks you to show that a field can have only one zero and only one unity.

Postulate R.1.5 Law of Opposites

For each $a \in F$ there exists an element $b \in F$, for which

$$a + b = b + a = 0$$

We call b the ***opposite*** of a and denote it by $-a$.

Exercise R.1.3b asks you to show that an element of a field can have only one opposite.

Postulate R.1.6 Law of Reciprocals

For each $a \in F$ with the exception of 0 there exists an element $c \in F$ such that $ac = ca = 1$. We call c the ***reciprocal*** of a and denote it by $1/a$ or a^{-1}.

You are asked in Exercise R.1.3c to show that a nonzero element of a field can have only one reciprocal.

Postulates R.1.5 and R.1.6 guarantee the existence of inverse elements. The addition of a to its opposite $-a$ always results in the identity element 0; if $a \neq 0$, the product of a and its reciprocal $1/a$ is always the identity element 1. As we will see, these elements are important for determining solutions of equations, as well as for defining the operations of subtraction and division.

Postulate R.1.7 Distributive Law

If $a, b, c \in F$, then $a(b + c) = ab + ac$.

Postulate R.1.7 states a relationship between the two basic operations in a field. A companion law

$$(a + b)c = ac + bc$$

also holds in every field, but it is not included here among the postulates because it can easily be proved from the others. (Exercise R.1.4a asks you to do this.)

It is our *assumption* that the set R of all real numbers forms a field with respect to ordinary addition and multiplication. Thus we take for granted that the seven postulates stated above apply to R.*

Before proceeding, note that there are many fields other than the reals; for example: Q, the system of rationals; C, the system of complex numbers; and the system Z_5 of integers modulo 5. (See Section L.5 for a definition of Z_5.)

In any field, subtraction may be introduced by defining $a - b$ as $a + (-b)$. Division may also be introduced by defining a/b as $a \cdot (b^{-1})$ provided $b \neq 0$. Thus, in any field, the processes of arithmetic are available, and the rules of algebra can be shown to hold (see Exercise R.1.4).

Order

Order can be introduced by means of "positiveness".

Definition R.1.2 A field F is an ***ordered*** field provided that there is a nonempty subset P of F such that the following two postulates hold:

Postulate R.1.8

For each $a \in F$ one and only one of the following conditions holds:

$$a \in P \qquad \text{or} \qquad a = 0 \qquad \text{or} \qquad -a \in P$$

Postulate R.1.9

If $a, b \in P$, then $a + b \in P$ and $ab \in P$.

In an ordered field F, the elements of the set P are called ***positive***. All other elements of F, except zero, are called ***negative***. If two elements of F are both positive, or both negative, then we say they have the ***same sign***; if one is positive and the other is negative, they have ***opposite signs***.

*Our approach is "descriptive" inasmuch as we are simply describing R by stating enough of its properties to enable us to work with the real numbers. In a "constructive" approach we would have to define what we mean by a real number (not an easy job) and then prove that the stated properties actually do hold. That approach involves a long excursion that we do not wish to make now, although some steps along the way are outlined. For example, in Section L.5 we have seen how to construct the rational numbers from the integers. We will outline a construction of the reals from the rationals in a project following Section 2.7. And in a project at the end of the present section, you will be asked to construct the complex numbers from the reals.

The real number system R satisfies Definition R.1.2 with the usual understanding of "positive," and so does the system Q of rationals. However, Z_5 and C do not, as we will presently see.

Inequalities can be introduced in any ordered field F. For $a, b \in F$ we make the following definitions:

(1) $a < b$ (read, "a is *less than* b") provided $b - a$ is positive
(2) $a > b$ (read, "a is *greater than* b") provided $b < a$
(3) $a \leq b$ (read, "a is *less than or equal to* b") provided $a < b$ or $a = b$
(4) $a \geq b$ (read, "a is *greater than or equal to* b") provided $a > b$ or $a = b$

It follows that an element $b \in F$ is positive if and only if $b > 0$; an element a is negative if and only if $-a$ is positive.

The relation $<$ has several important properties, stated in the theorems that follow.

Theorem R.1.1 **Law of Trichotomy** *If a and b are elements of an ordered field, then one and only one of the following conditions holds:*

$$a < b \quad \text{or} \quad a = b \quad \text{or} \quad b < a$$

PROOF: This result follows immediately from the definition of $<$ by applying Postulate R.1.8 to $b - a$. ∎

Theorem R.1.2 **Transitive Law** *If a, b, and c are elements of an ordered field such that if $a < b$ and $b < c$, then $a < c$.*

PROOF: By hypothesis, $b - a$ and $c - b$ are positive. By Postulate R.1.9, their sum belongs to P: $(c - b) + (b - a) \in P$. But $(c - b) + (b - a) = c - a$ (work Exercise R.1.4f), so $c - a$ is positive and $a < c$. ∎

Theorem R.1.3 **Addition Property of <** *If a, b, and c are elements of an ordered field, then $a < b$ if and only if $c + a < c + b$.*

PROOF: Work Exercise R.1.7. ∎

Corollary R.1.1 *If a, b, c, and d are elements of an ordered field such that*

$$a < b \quad \text{and} \quad c < d$$

then

$$a + c < b + d$$

PROOF: Since $c < d$, we have $a + c < a + d$, by Theorem R.1.3. Similarly, from $a < b$ we obtain $a + d < b + d$. By Theorem R.1.2, $a + c < b + d$ ∎

Theorem R.1.4 **Multiplication Property of <** *If a, b, and c are elements of an ordered field, then:*

(i) *For positive c,* $a < b$ *if and only if* $ca < cb$
(ii) *For negative c,* $a < b$ *if and only if* $ca > cb$

PROOF:

(i) You are asked to prove part (i) in Exercise R.1.8.
(ii) Now c is negative, so $-c$ is positive. To prove $a < b \Rightarrow ca > cb$, we assume $a < b$. Then $b - a$ is positive, and so $(-c)(b - a)$ is also positive. Now we can readily show that $(-c)(b - a) = ca - cb$, so that $ca - cb > 0$ and $ca > cb$. To prove that $ca > cb \Rightarrow a < b$, we establish the contrapositive: $a \geq b \Rightarrow ca \geq cb$. By assuming $a \geq b$, we get $a - b \geq 0$, $(-c)(a - b) \geq 0$, $cb - ca \geq 0$, and, finally, $ca \leq cb$. ∎

The preceding theorems enable us to solve certain inequalities in much the same way that we solve equations. We may add the same element to both members of an inequality, or we may multiply both members by a positive element, and in either case the resulting inequality is equivalent to the original. However, multiplication by a negative element reverses the inequality.

Example 1: To solve the inequality

$$2x - 4 < 3x + 9$$

in R we use Theorem R.1.3 to add $-3x$ to both sides; then we add 4 to both sides to obtain

$$-x < 13$$

We use part (ii) of Theorem R.1.4 to multiply both sides by -1, obtaining

$$x > -13$$

The set of solutions of the given inequality is simply the set of real numbers greater than -13.

Theorem R.1.5 **Laws of Sign** *If two elements of an ordered field have the same sign, then their product is positive; if they have opposite signs, then their product is negative.*

PROOF: The proof is left for you to work out in Exercise R.1.10. ∎

Corollary R.1.2

(1) *The square of any nonzero element of an ordered field is positive.*
(2) *The unity of an ordered field is positive; its opposite is negative.*
(3) *Any nonzero element of an ordered field has the same sign as its reciprocal.*

Example 2: In the field C of complex numbers, the number i has the property that $i^2 = -1$; since $i \neq 0$ and -1 is negative, we conclude that C cannot be an ordered field. (The project at the end of this section contains an outline showing how to construct the field C.)

Example 3: The system Z_5 of integers modulo 5 is a field but not an ordered field because the sum of positive elements $\overline{1} + \overline{1} + \overline{1} + \overline{1} + \overline{1}$ is not positive.

If F is an ordered field, the set of ordered pairs

$$\{(x, y) : x, y \in F \text{ and } x \leq y\}$$

is an order relation as defined in Section L.3. It is reflexive by definition ($x \leq x$ always holds). It is antisymmetric by the Law of Trichotomy (if $x \leq y$ and $y \leq x$, then y is not $< x$ and x is not $< y$, so we must have $x = y$). It is transitive by the Transitive Law for $<$.

EXERCISES R.1

1. Explain why each of the following number systems fails to be a field.

(a) The set Z^+ of positive integers
(b) The set Z of all integers
(c) The set of positive real numbers
(d) The set of rational numbers greater than 1
(e) The set of irrational numbers
(f) The system Z_6 of integers modulo 6 whose operations are shown in the tables.

$\oplus$	0	1	2	3	4	5
0	0	1	2	3	4	5
1	1	2	3	4	5	0
2	2	3	4	5	0	1
3	3	4	5	0	1	2
4	4	5	0	1	2	3
5	5	0	1	2	3	4

$\odot$	0	1	2	3	4	5
0	0	0	0	0	0	0
1	0	1	2	3	4	5
2	0	2	4	0	2	4
3	0	3	0	3	0	3
4	0	4	2	0	4	2
5	0	5	4	3	2	1

2. Let a, b be any fixed elements of a field F.

(a) Show that the equation $a + x = b$ has a unique solution. (That is, show that there exists one and only one $x \in F$ such that $a + x = b$.)
(b) Show that if $a \neq 0$, then the equation $ax = b$ has a unique solution.

3. (a) Show that the zero and the unity of a field are unique.
(b) Show that if a is an element of a field, then the opposite of a is unique.
(c) Show that if a is a nonzero element of a field, then its reciprocal is unique.

4. Show that the following equations hold for all choices of a, b, c, and d in a field.

(a) $(a + b)c = ac + bc$
(b) $-(-a) = a$
(c) $-(a + b) = (-a) + (-b)$
(d) If $a \neq 0$, then $\dfrac{1}{1/a} = a$
(e) If $a \neq 0$ and $b \neq 0$, then $\dfrac{1}{ab} = \dfrac{1}{a} \cdot \dfrac{1}{b}$
(f) $(c - b) + (b - a) = c - a$
(g) $(b + c) - (a + c) = b - a$
(h) $a \cdot 0 = 0$ [*Hint:* Show that $a \cdot 0 + a \cdot 0 = a(0 + 0) = a \cdot 0 = a \cdot 0 + 0$, so that $a \cdot 0$ and 0 are both solutions of the equation $a \cdot 0 + x = 0$. Use Exercise R.1.2a.]

(i) If $ab = 0$, then $a = 0$ or $b = 0$
(j) If a and c are nonzero, then $b/a = d/c$ if and only if $bc = ad$
(k) If a and c are nonzero, then

$$\frac{b}{a} + \frac{d}{c} = \frac{bc + ad}{ac} \quad \text{and} \quad \frac{b}{a} \cdot \frac{d}{c} = \frac{bd}{ac}$$

5. Use the result of Exercise R.1.4i to solve by factoring the following equations in the given fields.

(a) $x^2 = 4$ field Q
(b) $x^2 = 2$ field R
(c) $x^3 - 2x^2 + x = 0$ field R
(d) $x^2 = 1$ field Z_5

6. We define $Q[\sqrt{2}]$ to be the subset of R consisting of all real numbers of the form $a + b\sqrt{2}$, where $a, b \in Q$. Show that $Q[\sqrt{2}]$ is a field under the operations of R.

7. Use Exercise R.1.4g to prove Theorem R.1.3.

8. Prove part (i) of Theorem R.1.4.

9. Solve the following inequalities in R.

(a) $4 + 3x < 16$
(b) $3x - 17 > 22$
(c) $x^2 - 5x + 6 < x^2 + 5x - 14$
(d) $5 + x^2 < 4$
(e) $x - 3 \leq -x + 3$
(f) $3 - \frac{x}{2} \leq 2x + 1$

10. Prove Theorem R.1.5.

11. Discuss the following "solutions" of inequalities in R. Find correct solutions.

(a)
$$\begin{aligned} x(x+1) &> x(x+4) \\ x + 1 &> x + 4 \\ 1 &> 4 \end{aligned}$$

(b)
$$\begin{aligned} (x+2)(x+3) &\leq (x+2)(x-2) \\ x + 3 &\leq x - 2 \\ 3 &\leq -2 \end{aligned}$$

12. (a) Show that if $a, b \in R$, then $ab \leq (a^2 + b^2)/2$. (*Hint:* The square of $a - b$ is nonnegative.)

(b) Let u and v be positive real numbers. Assuming that $\sqrt{u}$ and $\sqrt{v}$ exist, show that the *geometric mean* $\sqrt{uv}$ of u and v is less than or equal to their *arithmetic mean* $\frac{1}{2}(u + v)$.

13. Show that if $a, b, c \in R$, then

$$a^2 + b^2 + c^2 \geq ab + ac + bc$$

14. Show that the sum of a positive real number and its reciprocal is greater than or equal to 2.

PROJECT: The System of Complex Numbers

The real numbers are part of a larger system of numbers called *complex numbers*. In this project you are asked to construct this number system, starting from the reals.

(a) Let $C = \{(a, b) : a \in R \text{ and } b \in R\}$. Let addition $\oplus$ and multiplication $\odot$ on C be defined as follows:

$$\begin{aligned} (a, b) \oplus (c, d) &= (a + c, b + d) \\ (a, b) \odot (c, d) &= (ac - bd, ad + bc) \end{aligned}$$

where the operations being defined are denoted $\oplus$ and $\odot$, and the ordinary operations of addition and multiplication of real numbers are denoted in the usual way. Show that C is a field under these operations, with zero $(0, 0)$ and unity $(1, 0)$.

(b) Show that if $a, b \in R$, then

$$(a, b) = (a, 0) \oplus [(b, 0) \odot (0, 1)]$$

This equation is frequently written

$$(a, b) = a + bi$$

in which i stands for $(0, 1)$ and each complex number $(a, 0)$ is identified with the real number a, and in which we have replaced the special operation symbols for C by their counterparts for R.

(c) Show that in the notation introduced in part (b), the equation

$$x^2 \oplus (1, 0) = (0, 0)$$

can be written

$$x^2 = -1$$

Show that this equation has two solutions in C.

(d) Explain how the result in part (c) shows that C is not an ordered field.

R.2 *Absolute Value, Intervals, Bounds*

In any ordered field, concepts such as absolute value, upper and lower bounds of sets, and betweenness can be introduced.

Absolute Value

Definition R.2.1 Let F be an ordered field and let $a \in F$. The ***absolute value** of* a is a if a is nonnegative and is $-a$ if a is negative. The absolute value of a is denoted $|a|$.

Example 1: In R we have: $|5| = 5$, $|-\frac{4}{3}| = \frac{4}{3}$, $|\sqrt{2} - 2| = 2 - \sqrt{2}$, etc.

Some basic properties of absolute values are stated in the following theorem, whose proof is left for you to work out in Exercises R.2.1 through R.2.3.

Theorem R.2.1 **Properties of Absolute Values** *If F is an ordered field and a, $b \in F$, then:*

(i) $|a| = |-a|$
(ii) $|a|$ *is nonnegative*
(iii) $|a| = 0$ *if and only if* $a = 0$
(iv) $|ab| = |a|\,|b|$
(v) $-|a| \leq a \leq |a|$
(vi) *If δ is positive, then $|a| \leq \delta$ if and only if $-\delta \leq a \leq \delta$*
(vii) *If δ is positive, then $|a| < \delta$ if and only if $-\delta < a < \delta$*
(viii) ***Triangle Inequality*** $|a + b| \leq |a| + |b|$
(ix) $\big||a| - |b|\big| \leq |a - b|$

Lower and Upper Bounds

Definition R.2.2 Let S be a subset of an ordered field F. Then:

(i) If there is an element a of F such that $a \leq x$ for every $x \in S$, then S is ***bounded below***, and a is a ***lower bound*** of S.
(ii) If there is an element b of F such that $x \leq b$ for every $x \in S$, then S is ***bounded above***, and b is an ***upper bound*** of S.
(iii) If S is both bounded above and bounded below, then S is ***bounded***.
(iv) If S is not bounded, then it is ***unbounded***.

Example 2: Consider the subset $S = \{x : 0 \leq x \leq 3\}$ of R. 0 and all numbers less than 0 are lower bounds of S; 3 and all numbers greater than 3 are upper bounds.

Example 2 shows that a lower (or upper) bound of a set is not unique.

Example 3: In any ordered field F, the set P of positive elements is not bounded above. We prove this by contradiction. Let us assume that some element b of F is an upper bound of P. Since the unity 1 is positive, we have $b \geq 1 > 0$, showing that $b \in P$. Furthermore, $b + 1 \in P$ by Postulate R.1.9, and since $b + 1 > b + 0 = b$, we see that $b + 1$ is an element of P greater than b, contradicting the assumption that b is an upper bound of P. Since P fails to be bounded above, it is unbounded.

Theorem R.2.2

(i) A subset S of an ordered field F is bounded if and only if there exists a positive element p of F such that $|x| \leq p$ for every $x \in S$.

(ii) A subset S of an ordered field F is unbounded if and only if for every positive element p of F there exists an $x \in S$ such that $|x| > p$.

PROOF:

(i) Assuming that S is bounded, there exist $a, b \in F$ such that $a \leq x \leq b$ for every $x \in S$. In case a and b are both 0, we can take p to be 1. Otherwise we let p be the larger of the two numbers $|a|$ and $|b|$; then we see that $|x| \leq p$ for every $x \in S$. Conversely, if there exists some $p \in P$ such that $|x| \leq p$ holds for all $x \in S$, then by part (vi) of Theorem R.2.1, it follows that $-p \leq x \leq p$ for $x \in S$. So S is bounded above and below.

(ii) The second part is an immediate logical consequence of part (i), because the negation of

$$(\exists p \in P)(\forall x \in S)(|x| \leq p)$$

is

$$(\forall p \in P)(\exists x \in S)(|x| > p)$$

Betweenness

Definition R.2.3

Let F be an ordered field and let a, b and $x \in F$. Then x is ***between*** a and b provided that $a < x < b$ or $b < x < a$.

If a and δ are elements of an ordered field F, and δ is positive, it follows from (vii) of Theorem R.2.1 and Theorem R.1.3 that x satisfies the inequality $|x - a| < \delta$ if and only if $a - \delta < x < a + \delta$; that is, if and only if x is between $a - \delta$ and $a + \delta$.

Example 4: In R, $|x - 2| < 3$ holds if and only if x is between -1 and 5.

Example 5: If a and b are different real numbers, then $(a + b)/2$ is between them. In case $a < b$, then by Theorem R.1.3 we have $a + a < a + b$ and $a + b < b + b$, so that $2a < a + b < 2b$; by multiplying by $\frac{1}{2}$ we get $a < (a + b)/2 < b$ (Theorem R.1.4). A similar argument applies when $b > a$.

Intervals

There are several types of sets that are known as "intervals".

Definition R.2.4 Let F be an ordered field and let $a, b \in F$ with $a \leq b$.

(i) The set $\{x : x \in F \text{ and } a < x < b\}$ is the ***open interval*** from a to b and is denoted (a, b).
(ii) The set $\{x : x \in F \text{ and } a \leq x \leq b\}$ is the ***closed interval*** from a to b and denoted $[a, b]$.
(iii) The sets $\{x : x \in F \text{ and } a \leq x < b\} = [a, b)$ and $\{x : x \in F \text{ and } a < x \leq b\} = (a, b]$ are ***half-open*** intervals.

The numbers a and b are the ***left*** and ***right endpoints***, respectively, of the intervals in (i), (ii), and (iii).

The open interval (a, b) contains just the elements between a and b, while the closed interval $[a, b]$ contains the endpoints a and b as well.

The familiar geometric model of the real number system is the number line. When we wish to represent intervals of real numbers on the line, we use the convention that a square bracket at an endpoint signifies that the endpoint is included in the interval while a round parenthesis signifies that it is not included (see Figure R.1).

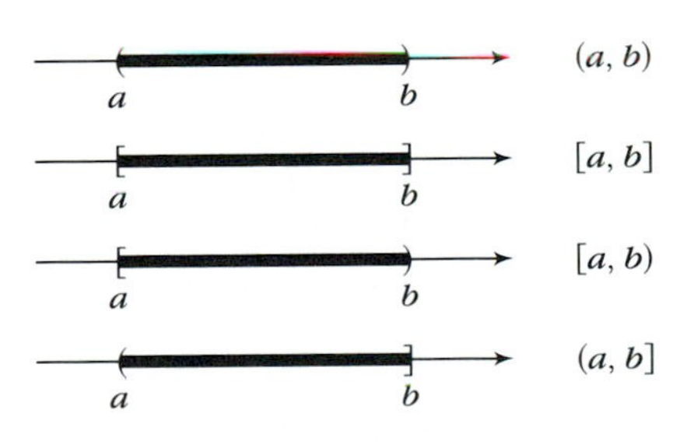

Figure R.1
Bounded intervals

The intervals defined in Definition R.2.4 are bounded subsets of R. There are also unbounded intervals.

Definition R.2.5 Let F be an ordered field and let $a \in F$. The following sets are ***unbounded intervals***:

(i) $\{x : x \in F \text{ and } x \geq a\}$
(ii) $\{x : x \in F \text{ and } x > a\}$
(iii) $\{x : x \in F \text{ and } x \leq a\}$
(iv) $\{x : x \in F \text{ and } x < a\}$
(v) The set F itself

The number a is the ***left endpoint*** of intervals (i) and (ii); it is the ***right endpoint*** of intervals (iii) and (iv). Intervals (i) and (iii) are ***closed***, while (ii) and (iv) are ***open***. Interval (v) is considered to be *both* closed and open.

In the ordered field of real numbers, we denote intervals (i) through (v) as follows: (i) $[a, \infty)$; (ii) (a, ∞); (iii) $(-\infty, a]$; (iv) $(-\infty, a)$; and (v) $(-\infty, \infty)$. Figure R.2 shows two unbounded intervals of real numbers.

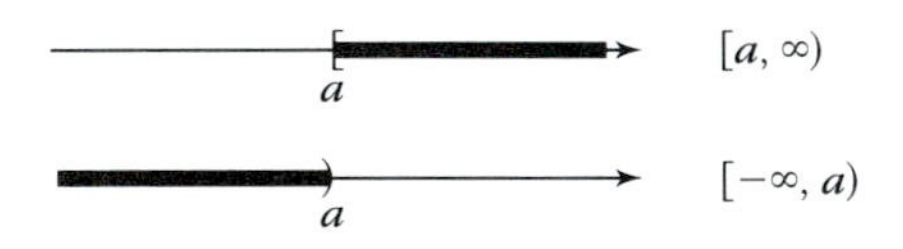

Figure R.2
Unbounded intervals

Henceforth the word *interval* will refer to any one of the types of intervals described in Definitions R.2.4 and R.2.5.

Note that the closed interval $[a, a]$ consists of the single element a, so a singleton set is an interval. Also note that the open interval (a, a) is empty because the postulates of order prohibit the condition $a < x < a$; thus, $\varnothing$ is an interval.

Example 6: In R, the solution set of the inequality $|x - 2| < 3$ is the open interval $(-1, 5)$ (see Example 4). The solution set of the inequality $|x| \leq 5$ is the closed interval $[-5, 5]$. The solution set of $0 < |x - 1| < 1$ is the union of $(0, 1)$ and $(1, 2)$.

Example 7: We solve the following inequalities in R and express the solution sets as intervals or unions of intervals.

(a) $2x^2 + 5x < 12$ (b) $4x + x^3 \geq 0$ (c) $\dfrac{x}{x+1} \geq 0$

(a) By collecting terms on the left and factoring, we get equivalent inequalities

$$2x^2 + 5x - 12 < 0 \quad \text{and} \quad (x + 4)(2x - 3) < 0$$

For a product to be negative it is necessary and sufficient that the factors be of opposite sign. Two cases arise: (i) If $x + 4 > 0$ and $2x - 3 < 0$, we obtain $-4 < x < \frac{3}{2}$. (ii) If $x + 4 < 0$ and $2x - 3 > 0$, we obtain the inconsistent inequalities $x < -4$, $x > \frac{3}{2}$. The solution set is the open interval $\left(-4, \frac{3}{2}\right)$.

(b) By factoring, we obtain

$$x(4 + x^2) \geq 0$$

Regardless of what value x may assume, $4 + x^2$ is positive, and we may multiply by its reciprocal. This gives $x \geq 0$, and so the solution set is the unbounded closed interval $[0, \infty)$.

(c) For $x \neq -1$, we may multiply both members by $(x + 1)^2$, which is positive. We obtain

$$x(x + 1) \geq 0$$

By analyzing the signs of the factors, we obtain $x \leq -1$ or $x \geq 0$. Since the number -1 must be excluded, the solution set of the given inequality is $(-\infty, -1) \cup [0, \infty)$.

Neighborhoods in R

Definition R.2.6 In the ordered field R, the set $\{x : |x - a| < \delta\}$, where a is a fixed real number and $\delta > 0$, is the ***δ-neighborhood of a***, denoted $N_\delta(a)$.

The condition $|x - a| < \delta$ is equivalent to $a - \delta < x < a + \delta$, and so $N_\delta(a)$ is simply the open interval $(a - \delta, a + \delta)$ (see Figure R.3). The number δ is the *half-width* or *radius* of the neighborhood.

Figure R.3
Neighborhood of a

Definition R.2.7 In R, the set $\{x : 0 < |x - a| < \delta\}$ where a is a fixed real number and $\delta > 0$, is the ***deleted δ-neighborhood of a***, denoted $N^*_\delta(a)$.

The deleted neighborhood of a is the union of two open intervals (see Figure R.4):

$$N^*_\delta(a) = (a - \delta, a) \cup (a, a + \delta)$$

$a - \delta$ a $a + \delta$

Figure R.4.
Deleted neighborhood of a

EXERCISES R.2

1. Prove parts (i) through (v) of Theorem R.2.1. [*Hint:* For part (iv), proceed by cases: If a is positive and b is negative, then $|a| = a$ and $|b| = -b$ so that $|a|\ |b| = a(-b) = -(ab) = |ab|$.]

2. Prove parts (vi) and (vii) of Theorem R.2.1.

3. Prove parts (viii) and (ix) of Theorem R.2.1. [*Hint:* For part (viii), add the inequalities $-|a| \leq a \leq |a|$ and $-|b| \leq b \leq |b|$. For part (ix), start with $|a| = |b + (a - b)| \leq |b| + |a - b|$, obtaining $|a| - |b| \leq |a - b|$. Then interchange a and b, and combine the two results.]

4. Determine whether the following sets are bounded above, bounded below, or bounded, in the ordered field R.

   (a) $\{r : r \in Q \text{ and } 0 < r < 1\}$
   (b) $\{m : m \in Z \text{ and } m^2 < 26\}$
   (c) $\{x : x \in R \text{ and } |x - 2| \geq 3\}$
   (d) $\{x : x \in R \text{ and } 0 < \frac{1}{x} < 1\}$

5. Express the solution set of each of the following inequalities in R as an interval or as a union of intervals.

   (a) $x^2 > 4x$
   (b) $x^2 \leq 5x - 6$
   (c) $4x^2 - 9x + 2 < 0$
   (d) $2x^2 - 1 > 7$
   (e) $5x + 4 \leq 6x^2$
   (f) $2x^2 + 6x + 3 \leq 0$
   (g) $9x + x^3 > 0$
   (h) $16 - x^4 \geq 0$
   (i) $x/(9 + x^2) \geq 0$
   (j) $x^3 - x \leq 0$
   (k) $x^3 + 2x \leq 3x^2$
   (l) $x|x - 3| > 0$
   (m) $x|x - 3| \geq 0$
   (n) $x > |x|$

6. Show that if a and $b \in R$, then $|a| < |b|$ if and only if $a^2 < b^2$.

7. Use the result of Exercise R.2.6 to solve the following inequalities in R. Express each solution set as an interval or as a union of intervals.

   (a) $|x + 1| < |x - 3|$
   (b) $|1 - 2x| > |3 + 2x|$
   (c) $|1 + 2x| < |1 + 3x|$
   (d) $|1 + x| \geq |3 + 2x|$

8. Express the following sets in R as intervals or as unions or intervals.

   (a) $N_1(3) \cap N_2(3)$
   (b) $N_1^*(3) \cap N_2^*(3)$
   (c) $N_1(3) \cup N_1(5)$
   (d) $N_1^*(3) \cup N_1^*(5)$
   (e) $N_1(3) \cap N_1(5)$

9. (a) The union of two neighborhoods of a is also a neighborhood of a. What is its half-width?
   (b) The intersection of two neighborhoods of a is also a neighborhood of a. What is its half-width?

10. Show that if $a, b \in R$ and $a \neq b$, then there exists a positive number δ such that $N_\delta(a) \cap N_\delta(b) = \varnothing$.

R.3 *Real Functions*

A function whose domain and range are subsets of R is called simply a "real function". Among the real functions are those that are studied in the usual first-year calculus course.

Traditional Terminology and Notation

Traditionally, a real function is defined by stating its domain D and giving a formula (or another rule of some kind) by means of which the value $f(x)$ can be determined for each $x \in D$. It is also traditional in calculus to omit naming the domain D when D is the set of all real numbers for which the formula for $f(x)$ is meaningful. We will adhere to these traditions.

Example 1: The square function

$$f = \{(x, y) : y = x^2\}$$

is indicated by the rule $f(x) = x^2$, and the domain is understood to be R.

Example 2: If a real function g is given by the formula

$$g(x) = \sqrt{1 - \frac{1}{x}}$$

with no mention of the domain, we determine the domain D by noting that x must be nonzero and that $1 - \frac{1}{x}$ must be nonnegative. It follows that $D = (-\infty, 0) \cup [1, \infty)$.

Example 3: For the real function

$$F = \{(x, y) : y = 2x + 3 \text{ and } x > 0\}$$

the traditional way to define F is

$$F(x) = 2x + 3 \qquad x > 0$$

Here the domain must be specified since it is not the set of all real numbers for which $2x + 3$ makes sense.

When a real function is given by an equation of the form $y = f(x)$ $(x \in D)$, it is traditional in calculus to say that "y is a function of x". Strictly speaking, this is not correct because we distinguish the function f (a set of ordered pairs) from its value $f(x)$ (which is a number). However, it is convenient to permit saying "the function $f(x) = x^2$" rather than "the function f defined for $x \in R$ by the equation $f(x) = x^2$"; this kind of statement is accepted as an abbreviated form of speech. Thus, when y is a function of x, there is a function f and a set $D \subseteq R$ such that $y = f(x)$ for $x \in D$. It is traditional also to say that x is the *independent variable* and y is the *dependent variable*, reflecting the fact that x can be any element of D while the value of y depends on the chosen value of x. Of course, our use of the variables x and y in these roles is arbitrary—any other variables can be (and frequently are) used, and sometimes the roles of x and y are reversed.

In Example 3, we had the function

$$y = F(x) = 2x + 3 \qquad x > 0$$

Here the independent variable is x, the dependent variable is y. Let us show that F is invertible and find its inverse: We let $x_1, x_2 > 0$ and suppose that $F(x_1) = F(x_2)$. Then $2x_1 + 3 = 2x_2 + 3$, and it follows that $x_1 = x_2$, so that F is one-to-one and thus invertible. Its inverse is obtained by interchanging the roles of the variables:

$$F^{-1} = \{(x, y) : x = 2y + 3 \text{ and } y > 0\}$$

where the "interchanged" equation $x = 2y + 3$ is equivalent to $y = (x - 3)/2$. Since $y > 0$, we must have $x > 3$. Thus, the inverse function is given by

$$y = F^{-1}(x) = \frac{x - 3}{2} \qquad x > 3$$

Note: A slightly different practice for inverting a function is to solve the original equation for x in terms of y (without first interchanging). In that case, the result of inverting Example 3 would appear as

$$F^{-1}(y) = \frac{y - 3}{2} \qquad y > 3$$

It should be clear that the same function has been obtained, the only difference being that y is now the independent variable.

* * *

We next recall several types of real functions that are familiar from calculus and review some of their basic properties.

Polynomial Functions

A *polynomial function* is one that is defined by an equation of the form

$$p(x) = a_n x^n + a_{n-1} x^{n-1} + \cdots + a_1 x + a_0$$

where n is a nonnegative integer and the *coefficients* $a_0, a_1, \ldots, a_n$ are real constants. If $a_n \neq 0$, then the *degree* of p is n, and a_n is its *leading coefficient*. The domain of a polynomial function is R. The expression on the right-hand side of the equation for $p(x)$ is called a *polynomial*.

Example 4: The equation $p(x) = 4x^3 - 2x^2 + 3x + 1$ defines a polynomial function of degree 3, with leading coefficient 4.

A polynomial function of degree 0 has the form $p(x) = a_0$, where $a_0 \neq 0$. Such a function associates with each $x \in R$ the constant value a_0; it is therefore called a *constant function*. The zero function, $Z(x) = 0$ for $x \in R$ is also a constant function; for technical reasons the zero function does not have a degree.

Algebra of Real Functions

Definition R.3.1: Let f and g be real functions with domains D_f and D_g, respectively. The ***sum*** of f and g is the function $f + g$ having domain $D = D_f \cap D_g$ whose value at x is given by

$$(f + g)(x) = f(x) + g(x)$$

The ***difference*** $f - g$ and ***product*** fg are similarly defined with domain D:

$$(f - g)(x) = f(x) - g(x)$$
$$(fg)(x) = f(x) \cdot g(x)$$

The ***quotient*** f/g is defined for those $x \in D$ for which $g(x) \neq 0$ by the equation

$$(f/g)(x) = \frac{f(x)}{g(x)}$$

Rational Functions

A *rational function* is one that can be expressed as a quotient of polynomial functions:

$$r = \frac{p}{q}$$

where p and q are polynomial functions and q is not the zero function. The domain of r is the set of all real numbers x for which the denominator $q(x)$ is not zero.

Example 5: The following rational functions have the indicated domains:

$$r(x) = \frac{4x + 1}{x^2 + 3x} \qquad x \neq 0, -3$$
$$s(x) = \frac{1}{x} \qquad x \neq 0$$
$$t(x) = \frac{x}{x^4 + 8} \qquad x \in R$$

Other Algebraic Functions: Square Roots, nth Roots

By a *square root* of a number a we mean a number b such that $b^2 = a$. The field postulates (Postulates R.1.1 through R.1.7 in Section R.1) guarantee neither the existence nor the uniqueness of square roots within a given field.

Example 6: In the field R of real numbers, no negative numbers have square roots. In fact, this property holds for all ordered fields by Corollary R.1.2.

Example 7: In Z_5, we have $\overline{0}^2 = \overline{0}$, $\overline{1}^2 = \overline{4}^2 = \overline{1}$, $\overline{2}^2 = \overline{3}^2 = \overline{4}$, and so the only squares in that field are $\overline{0}$, $\overline{1}$, and $\overline{4}$. The elements $\overline{2}$ and $\overline{3}$ do not have square roots at all, while $\overline{4}$ has two square roots, $\overline{2}$ and $\overline{3}$.

Example 8: In Section L.2, we saw that there is no rational number whose square is 2. Thus, there is no square root of 2 in Q.

It is true that each nonnegative element of R has a unique nonnegative square root, but we are not yet ready to prove this. By accepting this fact without proof for the present, we can define a function $f(x) = \sqrt{x}$ $(x \geq 0)$ where $\sqrt{x}$ means the unique nonnegative real number whose square is x. Some basic properties of square roots are left for you to work out in Exercise R.3.20.

By an nth root ($n \in Z^+$) of a number a, we mean a number b such that $b^n = a$. (The case $n = 2$ is just the square root.) In the field of real numbers, the following can be shown:

(1) For n odd, every real number has a unique nth root in R, denoted $\sqrt[n]{a}$; if $a \neq 0$, then $\sqrt[n]{a}$ has the same sign as a.

(2) For n even, every positive real number has two nth roots in R, one positive and one negative; the positive nth root of a is denoted $\sqrt[n]{a}$.

(3) The equation $\sqrt[n]{ab} = \sqrt[n]{a}\,\sqrt[n]{b}$ holds whenever the nth roots on the right both exist.

Only the *existence* aspects of (1) and (2) are difficult (we discuss these in Sections 3.6 and 3.7). In the meantime, keep in mind that the real function

$$f(x) = \sqrt[n]{x}$$

is defined for all x when n is odd, but only for $x \geq 0$ when n is even (in which case, $\sqrt[n]{x}$ stands for the positive nth root of x).

Example 9:

(1) The domain of the real function $f(x) = \sqrt{1 - x^2}$ is the closed interval $[-1, 1]$.

(2) The domain of $g(x) = \sqrt{1 + x^2}$ is R.

Absolute Value and Greatest Integer Functions

The next two functions are used repeatedly in later examples and exercises.

Example 10: The *absolute value function* A is defined for $x \in R$ by the conditions

$$A(x) = \begin{cases} x & \text{if } x \geq 0 \\ -x & \text{if } x < 0 \end{cases}$$

Its graph is shown in Figure R.5. It is customary to denote $A(x)$ by $|x|$. Many important properties of this function follow from Theorem R.2.1.

Example 11: The *greatest integer function* G is defined for $x \in R$ as follows: $G(x)$ is the greatest integer that is less than or equal to x (see Figure R.6). In Section 1.2, we show that such an integer always exists for a given x

and that it can be characterized as the unique integer m such that $x = m + d$ for some real number d, with $0 \leq d < 1$. The greatest integer less than or equal to x is denoted $[x]$.

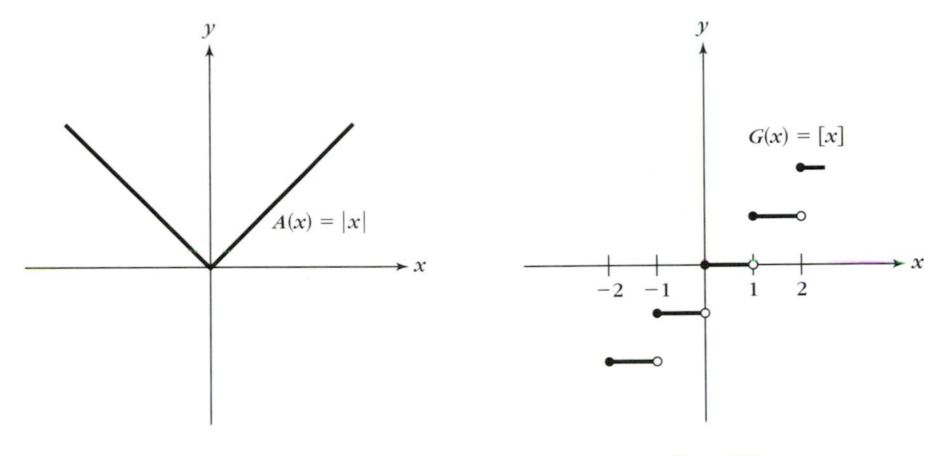

Figure R.5
Absolute value function

Figure R.6
Greatest integer function

Boundedness

Definition R.3.2 Let f be a real function with domain D and let $S \subseteq D$. Then:

(i) f is ***bounded above*** on S provided there is a real number b such that $f(x) \leq b$ for every $x \in S$
(ii) f is ***bounded below*** on S provided there is a real number a such that $a \leq f(x)$ for every $x \in S$
(iii) f is ***bounded*** on S provided it is bounded above and below on S
(iv) f is ***unbounded*** on S if it fails to be either bounded above or bounded below

Observe that f is bounded on S if and only if there exists a number $M > 0$ such that $|f(x)| \leq M$ for all $x \in S$.

Example 12: The polynomial function $f(x) = 3 + 2x$ is bounded on the interval $[-1, 1]$ because $|f(x)| \leq 5$ for $-1 \leq x \leq 1$. It is unbounded on the set of all positive real numbers because it fails to be bounded above there: For every choice of $M > 0$ there exist values of $x \in R$ with $3 + 2x > M$ [take any positive x such that $x > (M - 3)/2$].

Example 13: The rational function

$$f(x) = \frac{1}{x^2 + 1}$$

is bounded on the set of all real numbers: clearly for $x \in R$ we have $0 < f(x) \leq 1$.

Sine and Cosine

The definitions of polynomial and rational functions are straightforward inasmuch as the formulas that define them involve only the operations of arithmetic. Certain other functions studied in elementary calculus (the trigonometric, exponential, and logarithmic functions, for example) are not so easily defined, and we must await the development of certain analytic concepts before formally introducing them (see Section 4.12). Even though these functions play no role in our theoretical development for the time being, it is often convenient, for purposes of examples and exercises, to refer to the trigonometric functions *sine* and *cosine*. The basic properties of these functions are summarized in the following example.

Example 14: Let $x \geq 0$. On the unit circle C, whose equation in the u, v plane is $u^2 + v^2 = 1$, let P_x be the point such that the arc from $(1, 0)$ to P_x, measured counterclockwise, has length x* (see Figure R.7). Then the first coordinate of P_x is defined to be the *cosine* of x ($\cos x$), and its second coordinate is defined to be the *sine* of x ($\sin x$):

$$P_x = (\cos x, \sin x)$$

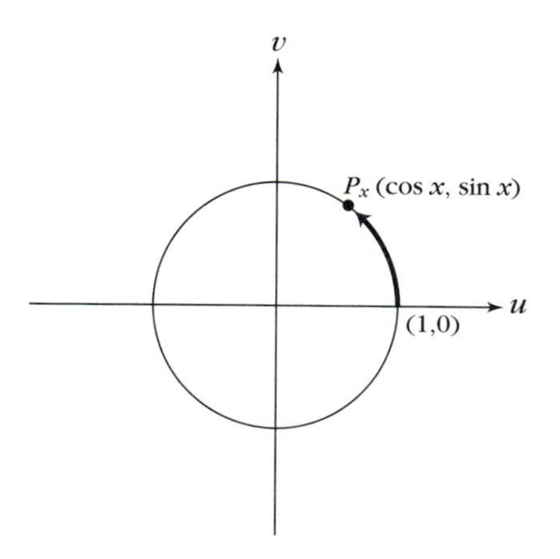

Figure R.7
Cosine and sine

*The difficulty in defining sine and cosine properly in this way lies in proving that the "length" of such an arc is meaningful.

For $x < 0$, the definitions of $\sin x$ and $\cos x$ are the same except that we determine P by measuring the arc length $|x|$ in the clockwise (negative) sense. Because the arc length of the whole circle is 2π, we see that $P_{x+2\pi} = P_x$ and so the cosine and sine satisfy the equations

$$\cos(x + 2\pi) = \cos x \qquad \text{and} \qquad \sin(x + 2\pi) = \sin x \qquad x \in R \quad (1)$$

From the unit circle equation we obtain the fundamental identity

$$\cos^2 x + \sin^2 x = 1 \qquad x \in R \qquad (2)$$

and it follows immediately from Equation (2) that these functions are bounded on R:

$$|\cos x| \leq 1 \qquad \text{and} \qquad |\sin x| \leq 1 \qquad x \in R$$

Periodicity

An important functional property, known as *periodicity*, is exhibited by the sine and cosine functions in Equations (1).

Definition R.3.3 A real function f with domain D is ***periodic*** provided there exists a positive real number p such that, for each $x \in D$,

$$x \pm p \in D \qquad \text{and} \qquad f(x \pm p) = f(x)$$

The number p is a ***period*** of f.

Example 15: The sine and cosine functions both have period 2π by Equations (1). It can be shown that 2π is the smallest real number that is a period of these functions.

The properties of sine and cosine discussed above can be visualized in their graphs (Figure R.8), which are constructed initially in the interval $[0, 2\pi]$ and are then extended as their periodicity requires.

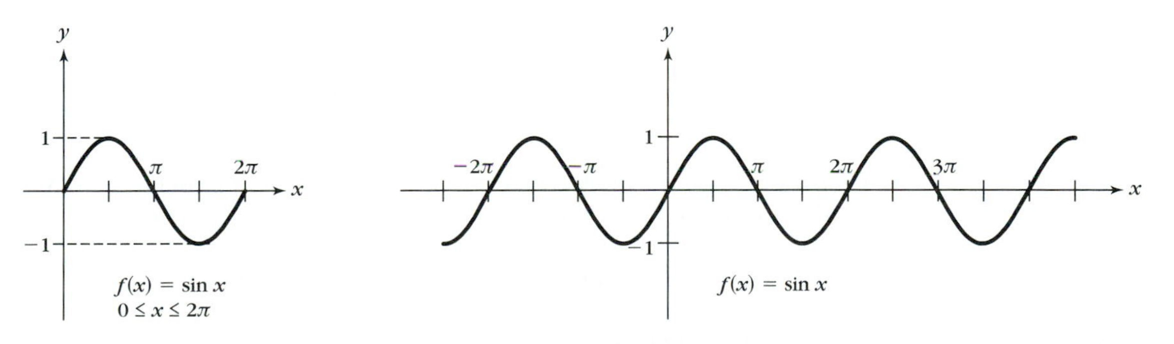

Figure R.8 (a)
Graphs of cosine and sine functions

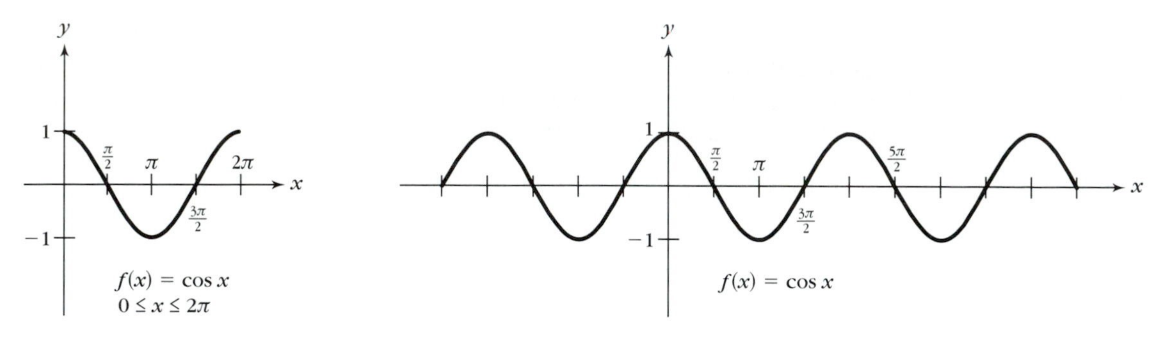

Figure R.8 (b)
Graphs of cosine and sine functions

Example 16: The *tangent* function is defined as the quotient of the sine by the cosine:

$$\tan x = \frac{\sin x}{\cos x}$$

Its domain is the set of real numbers for which the cosine is nonzero:

$$\{x : x \neq \frac{\pi}{2} + n\pi \text{ for } n \in Z\}$$

It can be shown that the tangent function is periodic with period π.

Even/Odd Functions

Definition R.3.4 Let f be a real function whose domain D has the property that $-x \in D$ whenever $x \in D$. Then

(i) f is ***even*** provided that $f(-x) = f(x)$ for every $x \in D$
(ii) f is ***odd*** provided that $f(-x) = -f(x)$ for every $x \in D$

It should be easy to remember this "even/odd" terminology by thinking of the function $f(x) = x^n$, where n is a fixed integer. This function is even when n is even, odd when n is odd.

Example 17: Figure R.9 illustrates the fact that the cosine function is even and the sine function odd. The tangent function is odd because

$$\tan(-x) = \frac{\sin(-x)}{\cos(-x)} = \frac{-\sin x}{\cos x} = -\tan x$$

for all x in its domain.

Caution: Not all functions are even or odd; for example, $\sin x + \cos x$ is neither even or odd.

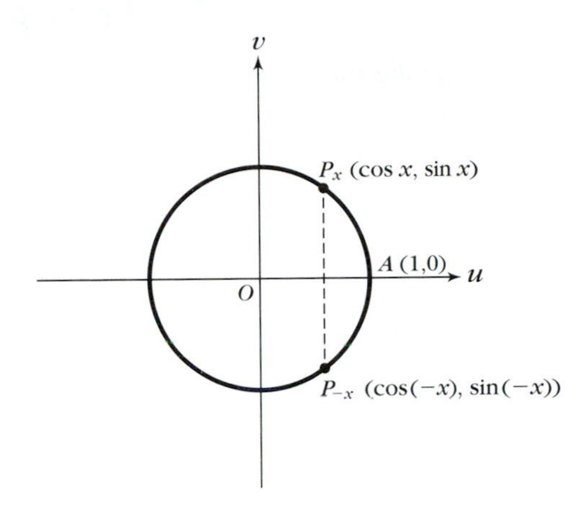

Figure R.9
Cosine is even, sine is odd

Composition of Real Functions

By applying one real function after another we can build up composite functions, as described in Section L.4 (see Definition L.4.2):

> If f and g are real functions such that the range of f is contained in the domain of g, then the ***composition*** of g on f is the function $h = g \circ f$ whose domain is the domain D of f and whose values are given by
>
> $$h(x) = g(f(x)) \qquad x \in D$$

Example 18: Let A be the absolute value function and f the sine function. Then $A \circ f$ is defined by

$$(A \circ f)(x) = |\sin x| \qquad x \in R$$

while $f \circ A$ is defined by

$$(f \circ A)(x) = \sin|x| \qquad x \in R$$

Notice that these composite functions are different, so that composition is not a commutative operation.

Functions Defined Piecewise

It is perfectly permissible to have a function defined by different formulas over different sets provided these sets do not overlap. The absolute value function (Example 10) is an example; here are some others.

Example 19: The function

$$f(x) = \begin{cases} x^2 & \text{if } -1 \leq x \leq 1 \\ 1 & \text{if } x > 1 \text{ or } x < -1 \end{cases}$$

is graphed in Figure R.10.

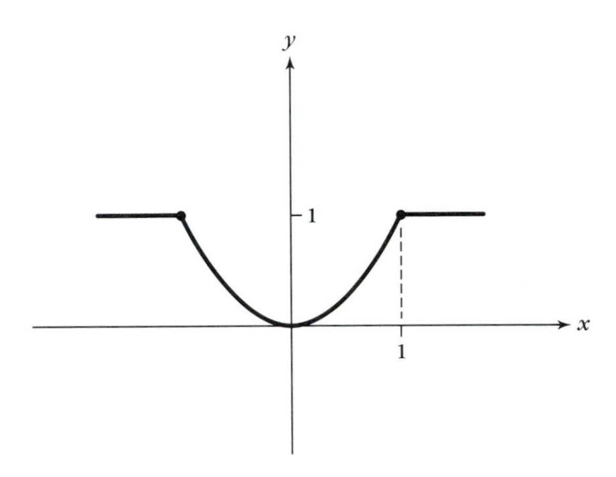

Figure R.10
Piecewise-defined function

Example 20: If S is a set of real numbers, then the function

$$\chi_S(x) = \begin{cases} 1 & \text{if } x \in S \\ 0 & \text{if } x \in R - S \end{cases}$$

is known as the *characteristic function* of S.

EXERCISES R.3

1. Given the real function $f(x) = (x^2 + 1)/x (x > 0)$, find the values $f(2)$ and $f(\frac{1}{2})$. Show that $f(t) = f(\frac{1}{t})$ for every $t > 0$.

2. Given the real function $f(x) = (1 - x)/(1 + x)$, find $f(0)$ and $f(1)$. Show that $(f \circ f)(x) = x$ for $x \neq -1$ and that $f^{-1} = f$.

3. Which of the following real functions have the property that $f(2x) = 2f(x)$ for $x \in R$?
   (a) $f(x) = ax$ (where a is a constant)
   (b) $f(x) = ax + b$ (where a and b are constants)
   (c) $f(x) = |x|$
   (d) $f(x) = [x]$ (where the brackets indicate the greatest integer function)
   (e) $f(x) = \begin{cases} 2x & \text{if } x > 1 \\ 3x & \text{if } x \leq 1 \end{cases}$

4. Determine the domains of the following real functions.
   (a) $f(x) = \sqrt{25 - x^2}$
   (b) $g(x) = \dfrac{x}{x^2 - 9}$
   (c) $h(x) = \dfrac{x}{\sqrt{x^2 - 9}}$

5. Determine in each of the following cases whether or not the real function f satisfies the condition $f(a + b) = f(a) + f(b)$ for a, b belonging to the domain.
   (a) $f(x) = 3x$
   (b) $f(x) = x^2$
   (c) $f(x) = 1/x$

(d) $f(x) = |x|$
(e) $f(x) = [x]$ (where the brackets indicate the greatest integer function)

6. Determine which of the functions in Exercise R.3.5 satisfy the condition $f(ab) = f(a)f(b)$ for a, b belonging to the domain.

7. Prove or disprove:

(a) The degree of the sum of two polynomials is always equal to the sum of their degrees.
(b) The degree of the product of two polynomials is always equal to the product of their degrees.
(c) The degree of the product of two polynomials is always equal to the sum of their degrees.

8. Let $f(x) = x + \frac{1}{x}$. On which of the following sets is f bounded? Justify your conclusions.

(a) $\{x : 0 < x < 1\}$
(b) $\{x : 1 < x < 2\}$
(c) $\{x : x > 1\}$

9. (a) Give a geometric argument to show that the tangent function has period π. (*Hint:* $\tan x$ is the slope of the line segment connecting the origin to P_{x}.)

(b) Let D be the set of real numbers x for which $\cos x \neq 0$; that is, $D = \{x : x \neq \frac{\pi}{2} + n\pi$ for $n \in Z\}$. Define the *secant* function to be the reciprocal of the cosine, that is,

$$\sec x = \frac{1}{\cos x} \qquad x \in D$$

Show that, for all $x \in D$,

$$\sec^2 x = 1 + \tan^2 x$$

(c) Show that the following functions have period π: $\sec^2 x$; $\cos^2 x$; $\sin^2 x$; $|\cos x|$; $|\sin x|$.

10. Let $d(x) = x - [x]$. Show that d is periodic with period 1.

11. (a) Show that the sum and the difference of two rational numbers are both rational.
(b) Show that the sum of an irrational number and a rational number must be irrational.
(c) Show that the characteristic function of the rationals is periodic but that it has no least period.

12. Let f be a real function whose domain D satisfies the condition that $-x \in D$ whenever $x \in D$. Show that f is the sum of an even function and an odd function. (*Hint:* $\dfrac{f(x) + f(-x)}{2}$ is even.)

13. Suppose you are given a real function f defined for $x \geq 0$.

(a) Indicate how to define an extension* of f to R that is even.
(b) Assuming that $f(0) = 0$, indicate how to define an extension of f to R that is odd.

14. Show that if a real function f is odd and invertible, then f^{-1} is odd.

15. Show that $f(x) = x^3 + 3x$ is one-to-one.

16. Show that the real function defined for $x > 1$ by $f(x) = x + \frac{1}{x}$ is one-to-one. Find its inverse and give a formula for $f^{-1}(x)$.

17. Show that $f(x) = x^2 + 4x + 5$ $(x \geq -2)$ is one-to-one. Use the quadratic formula to find the range of f, and obtain a formula for $f^{-1}(x)$.

18. Show that $f(x) = (x - 1)/(x + 1)(x \neq -1)$ is one-to-one, with range $\{y : y \neq 1\}$, and find f^{-1}.

19. Using logical connectives and quantifiers, write out the definitions of each of the following statements about a real function f on a set $S \subseteq R$.

(a) f is bounded above on S.
(b) f is bounded below on S.
(c) f is bounded on S.
(d) f is unbounded on S.

20. (a) Let F be a field and let $a \in F$. Show that if b is a square root of a belonging to F, then $-b$ is also a square root of a, and these are the only square roots of a in F.
(b) Let F be an ordered field. Show that no negative elements have square roots in F, but that if a positive element of F has square roots in F, then it has exactly two square roots in F, one positive and one negative.
(c) Show that if F is an ordered field and if a, b are elements of F that have square roots in F, then ab also has square roots in F and $\sqrt{ab} = \sqrt{a}\sqrt{b}$ (where $\sqrt{\ }$ denotes a nonnegative square root).

21. Show that if a and b have nth roots in R, then ab also has an nth root in R and $\sqrt[n]{ab} = \sqrt[n]{a}\sqrt[n]{b}$.

*Extensions of functions are discussed in Section L.4.

22. Show that if a, b are nonnegative elements of R, then $\sqrt{ab} \leq (a + b)/2$.

23. Show that if S and T are subsets of R, then $\chi_{S \cup T}(x) \geq \chi_S(x)$ and $\chi_{S \cap T} \leq \chi_S(x)$, where χ is the characteristic function discussed in Example 20.

24. Draw or describe the graph of the characteristic function of S:

(a) $S = [0, 1]$
(b) $S = (0, 1)$
(c) $S = Q$

CHAPTER P

POSITIVE INTEGERS AND MATHEMATICAL INDUCTION

Of all the subsystems of the real numbers, the system Z^+ of positive integers is in many ways the most basic. This chapter takes a close look at Z^+ and its relationship to R.

Although our description of Z^+ as "the set of positive whole numbers 1, 2, 3, and so on" may seem quite clear, it is imprecise because of the vagueness of the phrase "and so on." Yet we know Z^+ well enough to identify some of its striking characteristics. For one thing, Z^+ has a least element 1. And Z^+ is "discrete" in the sense that $n + 1$ is the "next" element after n (there are no elements between them). But we cannot prove these properties on the basis of our vague description of Z^+.

We need to make some formal assumptions about Z^+ and its relationship to R. These assumptions, together with the algebra and order properties of R, will form the basis on which we will derive properties of the number system, including those mentioned in the preceding paragraph, as well as an important technique of proof known as mathematical induction.

P.1 *Well-Orderedness of* Z^+

Least and Greatest Elements

Definition P.1.1 A ***least*** element of a set T of real numbers is an element $s \in T$ such that $s \leq x$ for every $x \in T$. A ***greatest*** element of T is an element $t \in T$ such that $t \geq x$ for every $x \in T$.

Example 1: The least element of the closed interval $[0, 3]$ is 0; its greatest element is 3.

Example 2: The open interval $(0, 3)$ has no least element. For suppose s is a least element; then $0 < s < 3$ and $0 < \frac{s}{2} < s < 3$ so that $\frac{s}{2}$ is an element of S less than s. This contradiction establishes our claim. You can easily show in the same way that $(0, 3)$ has no greatest element.

Although a set of real numbers does not necessarily have a least (or greatest) element, such an element is unique when it does exist. (Why?)

The Property of Well-Orderedness

Definition P.1.2 A subset S of R is ***well-ordered*** provided that every nonempty subset of S has a least element.

Example 3: Any finite subset of R is well-ordered.

Example 4: The set P of positive real numbers is not well-ordered. For, as we just saw, the set $(0, 3)$ has no least element and it is a nonempty subset of P. (Clearly P has many nonempty subsets that do not have least elements—including P itself.)

Example 5: Even if we adjoin the number 0 to P, the resulting set $\{x : x \geq 0\}$ is not well-ordered. It does have a least element, but there are nonempty subsets that do not.

To require a set to be well-ordered is a strong requirement indeed.

Assumptions about Z^+

We are now ready to state our assumptions about Z^+, one of which is well-orderedness:

(1) The elements of Z^+ are positive real numbers
(2) $1 \in Z^+$

(3) If a, $b \in Z^+$, then $a + b$ and $ab \in Z^+$; if a, $b \in Z^+$ and $a < b$, then $b - a \in Z^+$

(4) *Well-Orderedness:* Z^+ is well-ordered

The power of these assumptions, especially (4), will soon be evident. We begin by proving some of the observations made about Z^+ in the introductory paragraph.

Theorem P.1.1 *1 is the least element of* Z^+.

PROOF: By assumption (2), $1 \in Z^+$, and so it is sufficient to show that the set S of positive integers less than 1 is empty. If $S \neq \varnothing$ then, by the property of well-orderedness (4), S has a least element s, and since $s \in S$, $s < 1$. Now s is a positive integer and is therefore positive by assumption (1), so we have $0 < s < 1$. By multiplying the members of these inequalities by s, we find $0 < s^2 < s$. Now $s^2 \in Z^+$ because of assumption (3); moreover, since $s < 1$, the Transitive Law for $<$ shows that $s^2 < 1$. Thus, s^2 belongs to S. We have reached a contradiction because $s^2 < s$, and s was defined to be the *least* element of S. The assumption that S is nonempty has led to a contradiction, and we conclude that S must be empty. ■

Theorem P.1.2 *If $n \in Z^+$, there is no positive integer between n and $n + 1$.*

PROOF: Again we use a proof by contradiction. Suppose there is a positive integer m between n and $n + 1$: $n < m < n + 1$. Then, by subtracting n from each member of these inequalities, we obtain $0 < m - n < 1$. Since $m > n$, $m - n \in Z^+$ (by assumption 3). But then $m - n$ is a positive integer less than 1. This contradicts Theorem P.1.1 and establishes the theorem. ■

The set Z of all integers consists of the elements of Z^+ and their negatives as well as the number 0. It can be shown that addition, subtraction, and multiplication (but not division) can be carried out in Z. The following corollary is a consequence of Theorem P.1.2.

Corollary P.1.1 *If $m \in Z$ then there is no integer between m and $m + 1$.*

Rational Numbers in Lowest Terms

Example 6: Any rational number is expressible as a quotient of integers, but a given rational can be so expressed in various ways; for example, $\frac{2}{3} = \frac{4}{6} = \frac{(-6)}{(-9)}$, and so on. There are, however, occasions when we wish to speak of a rational number in "lowest terms." If $r \in Q$, consider the set of all ordered pairs (p, q) where $p \in Z$, $q \in Z^+$, and $p/q = r$. If (p_0, q_0) is the ordered pair in the set having *minimal* q, then we say that p_0/q_0 represents r in lowest terms. It is the well-orderedness of Z^+ that allows us to make this choice.

The Induction Theorem

An important method of proof, known as *mathematical induction*, is discussed in the next section. It is based on the property of well-orderedness and the concept of an inductive set.

Definition P.1.3 A subset S of Z^+ is ***inductive*** provided that $(k + 1) \in S$ whenever $k \in S$.

For an inductive set S, we see that if a positive integer k belongs to S, then $k + 1$ belongs to S, and so $k + 2 = (k + 1) + 1$ belongs to S; likewise, $k + 3 = (k + 2) + 1$ belongs to S; and so on.

Example 7: The set of all positive integers greater than 5 is inductive, for if k is a positive integer greater than 5, then $k + 1$ is also a positive integer greater than 5.

Example 8: The set $\{1, 2, 3, 4, 5\}$ is not inductive: Although 5 belongs to the set, $5 + 1 = 6$ does not belong to it.

Theorem P.1.3 **Induction Theorem** *If S is an inductive set and if $1 \in S$, then all positive integers belong to S.*

PROOF: Let S be an inductive set with $1 \in S$. Suppose S does not include all elements of Z^+. Then the set $Z^+ \dashv S$, consisting of those positive integers not in S, is nonempty and therefore has a least element s by assumption (4). Since $1 \in S$, $s \neq 1$. According to Theorem P.1.1 and the law of trichotomy, we must have $s > 1$. By assumption (3), $s - 1 \in Z^+$. Since s is the least element of $Z^+ \dashv S$, we must have $s - 1 \in S$; otherwise $s - 1$ would be an element of $Z^+ \dashv S$ less than its least element. But if $s - 1 \in S$, it follows from the inductive property of S that $s \in S$, which contradicts the definition of s. We therefore conclude that S contains all elements of Z^+. ∎

Note: The Induction Theorem is the fifth of five axioms for Z^+ (known as Peano's Postulates) that can be used to construct the real and complex number systems "from scratch," including the ordering and completeness property of R. Such an approach will be found in Edmund Landau's little book, *Foundations of Analysis.**

*Edmund Landau, *Foundations of Analysis*, 2nd ed., (Bronx, NY: Chelsea Publishing Co., 1960).

The set Z of all integers is clearly not well-ordered; however, it can be shown that any subset of Z that is bounded below must be well-ordered. The following theorem is a special case of that result.

Theorem P.1.4 *Every subset of Z that has a least element is well-ordered.*

PROOF: You are asked to prove this theorem in Exercise P.1.6. ■

EXERCISES P.1

1. Show that the set of positive real numbers has no least element and no greatest element.
2. Show that if a and b are real numbers, then the interval (a, b) has no least element and no greatest element.
3. Which of the following subsets of R are well-ordered? Which have a least element?
   (a) $\{1, 2, 3, 4\}$
   (b) The set of positive even integers $\{2, 4, 6, \ldots\}$
   (c) $\{n : n \in Z^+ \text{ and } n \geq 10\}$
   (d) The open interval $(5, 10)$
   (e) The closed interval $[5, 10]$
   (f) $Z^+ \cup [5, 10]$
   (g) The set of integers greater than -6
   (h) The set of odd integers
4. Which of the following sets are inductive?
   (a) The set of integers
   (b) The set of all real numbers
   (c) The set of positive even integers
   (d) The set of rational numbers
5. Show that every subset of a well-ordered set is also well-ordered.
6. Prove Theorem P.1.4.

⋆ PROJECT: A Proof of the Unique Factorization Theorem

Several important proofs in Section L.2 are based upon the Unique Factorization Theorem, one of the most fundamental theorems in mathematics. The proof of the Unique Factorization Theorem depends on the well-orderedness of Z^+, hence its postponement to the present section. Three major preliminary steps in the proof of the Unique Factorization Theorem are indicated in (a), (b), and (c) below; for each of them and for (d), the Unique Factorization Theorem itself, you are given an outline of the proof and are expected to fill in the details. Finally, in order to dispel any illusion that uniqueness of factorization is "obvious", we look in (e) at an example of a number system in which factorization into primes is not unique.

(a) **Division Algorithm:** *If n and d are positive integers, then there exists a unique integer r satisfying the conditions*

(i) $0 \leq r < d$
(ii) $n = qd + r$ *for some integer* $q \geq 0$

PROOF (outline): The set

$$T = \{n - qd : q \in Z, q \geq 0\}$$

has a least nonnegative element r; this r satisfies conditions (i) and (ii). ■

* * *

★ ★ ★

Definition The ***greatest common divisor*** of two integers a and b (not both 0) is an integer d with the properties

(i) $d > 0$
(ii) d is a divisor of a and also of b
(iii) If e is a divisor of a and b, then e is a divisor of d

Example: The greatest common divisor of -12 and 15 is 3.

(b) **Theorem:** *The greatest common divisor d of two integers a and b (not both 0) exists and is unique; moreover, there exist integers x and y such that*

$$d = ax + by$$

PROOF (outline): Let $S = \{am + bn : m, n \in Z\}$. Then S has a least positive element d, which satisfies conditions (i), (ii), and (iii). ■

(c) **Euclid's Lemma:** *If a prime p divides the product of two integers a and b, then either p divides a or it divides b.*

PROOF (outline): Suppose that $p|ab$ but that p does not divide a. Then the greatest common divisor of a and p is 1. So there exist integers x and y such that

$$ax + py = 1$$

Now multiply both sides by b. ■

★ ★ ★

(d) **Unique Factorization Theorem:** *Every positive integer greater than* 1 *can be expressed as a product of primes; and except for the order in which the factors are written, this expression is unique.*

Note: A single prime is regarded as a product of primes.

PROOF (outline):
Existence: Show that the set of all positive integers greater than 1 that are not expressible as products of primes is empty.
Uniqueness: Show that the set of all positive integers greater than 1 that have more than one factorization into primes is empty. ■

★ ★ ★

(e) **Example of a number system in which factorization into primes is not unique:** Let $S = \{n : n \in Z^+ \text{ and } n \equiv 1 (\text{mod } 4)\}$. Show that the product of any pair of elements of S is an element of S. For $m, n \in S$ define m to be a *factor* of n (in S) provided that there exists $q \in S$ such that $n = mq$, and define an element of s to be *prime* (in S) provided it has no factors (in S) except itself and 1. Show that the numbers 9, 21, 33, and 77 are prime (in S) and conclude that S does not have unique factorization into primes.

P.2 *Recursive Definition; Proofs by Induction*

Theorem P.1.3, the Induction Theorem, is useful in two ways: it provides a method of defining certain expressions *recursively* and it is the basis of the method of proof known as *mathematical induction.*

Recursive Definition

Exponentials and factorials are among the important algebraic expressions that can be defined recursively.

Example 1: Definition of a^n Let $a \in R$. Stipulate that

$$\text{(i) } a^1 = a$$

and that

$$\text{(ii) } a^{k+1} = a \cdot a^k \qquad \text{for every } k \in Z^+$$

Equation (i) tells us what is meant by a^1. Now, by setting $k = 1$ and using Equation (ii), we find that $a^2 = a \cdot a^1 = a \cdot a$. Having found a^2, we may find a^3 by using (ii) again: $a^3 = a \cdot a^2 = a \cdot (a \cdot a)$. Indeed, the set S of positive integers n for which we can determine a^n is inductive by Equation (ii), and $1 \in S$ by Equation (i). Hence, $S = Z^+$. We say that Equations (i) and (ii) define a^n *recursively.*

Note: If a is nonzero, we can define $a^0 = 1$; then a^n is defined for all nonnegative integers n and Equation (ii) holds for all nonnegative integers.

Example 2: Definition of $n!$ The symbol $n!$ (read "n factorial") is defined recursively for positive integers n by stipulating that

$$\text{(i) } 1! = 1$$

and that

$$\text{(ii) } (k + 1)! = (k + 1) \cdot k! \qquad \text{for every } k \in Z^+$$

For example,

$$2! = 2 \cdot 1! = 2 \cdot 1 = 2$$
$$3! = 3 \cdot 2! = 3 \cdot 2 \cdot 1 = 6$$

Again, the set S of positive integers n for which we can determine $n!$ is inductive by Equation (ii), and $1 \in S$ by Equation (i), so $S = Z^+$.

Note: It is useful to define $0! = 1$ as it is then easy to see that Equation (ii) holds for all nonnegative integers.

Example 3: Consider the sum s_n of the first n positive integers: $s_n = 1 + 2 + \cdots + n$. In agreement with mathematical custom, we stipulate that the "sum" of one term is just that term itself. To obtain s_{k+1} from s_k we simply add $k + 1$:

$$\text{(i) } s_1 = 1$$
$$\text{(ii) } s_{k+1} = s_k + (k + 1) \text{ for every } k \in Z^+$$

The steps (i) and (ii) define s_n recursively for $n \in Z^+$.

In the examples above, step (ii), which tells us how to obtain s_{k+1} from s_k, is called the "recursion relation". Step (i), which tells us the value of s_1, is called the "initializing step."

Proofs by Mathematical Induction

We now see how the Induction Theorem can be used to prove certain theorems about the positive integers.

Example 4: We claim that $2^n > n$ for $n \in Z^+$. It is natural to begin by testing the statement to be proved for some values of n, for example,

$$\begin{aligned} 2^1 &= 2 > 1 \\ 2^5 &= 32 > 5 \\ 2^{10} &= 1024 > 10 \end{aligned}$$

So indeed our claim is verified for $n = 1, 5$, and 10. However, convincing as it may seem, our claim is not proved for all $n \in Z^+$, and we cannot hope to prove it by just testing various values of n; there will always remain values that are not tested.

The trick here is to prove a *conditional* statement: *If* the claim is true for some particular value of n, say, $n = k$, *then* it must also be true for the next value of n, $n = k + 1$:

$$\text{(ii) } 2^k > k \Rightarrow 2^{k+1} > k + 1$$

Let us suppose, then, that $2^k > k$ and try to deduce that $2^{k+1} > k + 1$. Now

$$2^{k+1} = 2 \cdot 2^k = 2^k + 2^k$$

and by using our hypothesis that $2^k > k$, together with the addition property for inequalities, we get

$$2^{k+1} > k + k$$

But the second "k" on the right may be replaced by 1 since $k \geq 1$ whenever $k \in Z^+$; thus,

$$2^{k+1} > k + 1$$

We now know that whenever the inequality $2^n > n$ holds for one value of n, it must hold for the next. Since

$$\text{(i) } 2^1 > 1$$

it now follows that $2^2 > 2$; then that $2^3 > 3$; and so on. Now let S be the set of positive integers n for which $2^n > n$. By Equation (ii), S is inductive, and by Equation (i), $1 \in S$; so by the Induction Theorem, $S = Z^+$.

This two-step method is summarized in the following theorem.

Theorem P.2.1 **Principle of Mathematical Induction** *Let $P(n)$ be an open sentence* in n, where $n \in Z^+$. If it can be proved that*

(i) $P(n)$ *holds when* $n = 1$
(ii) *For every* $k \in Z^+$, $P(k+1)$ *holds whenever* $P(k)$ *holds;*

then $P(n)$ holds for all positive integers; that is,

$$(\forall n \in Z^+)P(n)$$

PROOF: Let S be the set of positive integers n for which $P(n)$ is true. Then S is inductive by (ii) and $1 \in S$ by (i). By the Induction Theorem, S contains all positive integers. ■

Example 5: Recall that a positive integer m is *divisible* by a positive integer d provided that there is a $q \in Z^+$ such that $m = dq$. In Example 5 of Section L.2, we showed that for every $n \in Z^+$, $n^3 + 2n$ is divisible by 3. We now use induction to give a new proof of that fact. Denote by $P(n)$ the open sentence "$n^3 + 2n$ is divisible by 3". Then $P(1)$ is clearly true, so (i) holds. To establish (ii), we assume that $P(k)$ holds and derive $P(k+1)$. Our assumption means that $k^3 + 2k$ is divisible by 3, so $k^3 + 2k = 3q$ for some positive integer q. Then

$$\begin{aligned}(k+1)^3 + 2(k+1) &= (k^3 + 3k^2 + 3k + 1) + (2k + 2)\\ &= (k^3 + 2k) + (3k^2 + 3k + 3)\\ &= 3q + (3k^2 + 3k + 3)\\ &= 3(q + k^2 + k + 1)\end{aligned}$$

It follows that $(k+1)^3 + 2(k+1)$ is also divisible by 3, and so we have proved the conditional statement

$$P(k) \Rightarrow P(k+1) \qquad \text{for } k \in Z^+$$

By mathematical induction, $P(n)$ holds for all $n \in Z^+$.

Example 6: In Example 3 we gave a recursive definition for the sum

$$s_n = 1 + 2 + \cdots + n$$

of the first n positive integers. Now let us prove by induction that, for $n \in Z^+$, $s_n = n(n+1)/2$. In this case the open sentence $P(n)$ is

$$s_n = \frac{n(n+1)}{2}$$

(i) That $P(1)$ holds follows immediately from the definition $s_1 = 1$ in Example 3.
(ii) Take as a hypothesis that $P(k)$ holds for a certain $k \in Z^+$; then, $s_k = k(k+1)/2$.

*Open sentences are discussed on page 15.

Now consider s_{k+1}. By Equation (ii) in Example 3, we have

$$s_{k+1} = s_k + (k + 1)$$

so that

$$\begin{aligned} s_{k+1} &= \frac{k(k+1)}{2} + (k+1) \\ &= \frac{(k+1)(k+2)}{2} \end{aligned}$$

Hence $P(k + 1)$ also holds, and we have established the conditional

$$P(k) \Rightarrow P(k + 1)$$

for $k \in Z^+$. By mathematical induction, $P(n)$ holds for all $n \in Z^+$, giving us the summation formula

$$1 + 2 + \cdots + n = \frac{n(n+1)}{2} \tag{1}$$

Example 7: The sum of the cubes of the first n positive integers, $t_n = 1^3 + 2^3 + \cdots + n^3$, is defined recursively by the equations $t_1 = 1$, $t_{k+1} = t_k + (k + 1)^3$. Let us prove the formula

$$t_n = \frac{n^2(n+1)^2}{4}$$

We begin with

$$\text{(i)}\ \frac{1 \cdot 2^2}{4} = 1 = t_1$$

We assume that

$$\text{(ii)}\quad t_k = 1^3 + 2^3 + \cdots + k^3 = \frac{k^2(k+1)^2}{4}$$

giving us

$$\begin{aligned} t_{k+1} = t_k + (k+1)^3 &= \frac{k^2(k+1)^2}{4} + (k+1)^3 \\ &= \frac{(k+1)^2(k^2 + 4k + 4)}{4} \end{aligned}$$

so that

$$t_{k+1} = \frac{(k+1)^2(k+2)^2}{4}$$

Therefore, the formula

$$1^3 + 2^3 + \cdots + n^3 = \frac{n^2(n+1)^2}{4} \tag{2}$$

holds for all $n \in Z^+$.

Modified Induction

Induction is sometimes used to prove theorems that apply to integers greater than or equal to a given positive integer m. The proofs are based on the following variation of the Induction Theorem (Theorem P.1.3):

Theorem P.2.2 *If S is an inductive set, and if the positive integer m belongs to S, then all positive integers greater than m also belong to S.*

You are asked to prove Theorem P.2.2 in Exercise P.2.11a. It follows that to prove by induction that $P(n)$ holds for all positive integers $n \geq m$, we should show that

(i) $P(n)$ holds when $n = m$
(ii) For every $k \geq m$ $P(k + 1)$ holds whenever $P(k)$ holds

Example 8: Let us now show that $n! > 2^n$ for every positive integer $n \geq 4$.

(i) Clearly the inequality holds when $n = 4$:

$$4! = 24 > 16 = 2^4$$

(ii) Let $k \in Z^+$, $k \geq 4$, and assume that $k! > 2^k$. Then

$$(k + 1)! = (k + 1)k! > (k + 1) \cdot 2^k$$

and since $k + 1 > 4 > 2$, we have

$$(k + 1)! > 2 \cdot 2^k = 2^{k+1}$$

Sigma Notation for Sums

Recall from the study of integrals in calculus that there is a convenient way of abbreviating extended sums, such as those on the left sides of Equations (1) and (2). Given terms $u_1, u_2, \ldots, u_n$, we can express the sum $u_1 + u_2 + \cdots + u_n$ in "summation" ("sigma") notation as follows:

$$u_1 + u_2 + \cdots + u_n = \sum_{k=1}^{n} u_k$$

The expression on the right is read "summation of u_k as k varies from 1 to n"; the Greek letter sigma $(\sum)$ stands for "sum."

Example 9: We may denote $1^3 + 2^3 + \cdots + n^3$ by $\sum_{k=1}^{n} k^3$, so that Equation (2) takes the form

$$\sum_{k=1}^{n} k^3 = \frac{n^2(n + 1)^2}{4}$$

Similarly, Equation (1) may be written

$$\sum_{k=1}^{n} k = \frac{n(n+1)}{2}$$

More generally, if α and β are integers ($\alpha \leq \beta$), and if u_k is defined for all integers from α and β, inclusive, then we can write

$$u_\alpha + u_{\alpha+1} + \cdots + u_\beta = \sum_{k=\alpha}^{\beta} u_k$$

The variable k is the *index* of summation. The index can be changed to another variable without changing the meaning; for instance, if we wish to use m as the index, we have

$$u_\alpha + u_{\alpha+1} + \cdots + u_\beta = \sum_{m=\alpha}^{\beta} u_m$$

so that

$$\sum_{k=\alpha}^{\beta} u_k = \sum_{m=\alpha}^{\beta} u_m$$

Sometimes it is useful to "shift" the index as follows:

$$\sum_{m=\alpha}^{\beta} u_m = \sum_{m=\alpha+1}^{\beta+1} u_{m-1}$$

On the right, the subscript on u has been decreased by 1, but the values of m have been increased by 1. It is easy to verify the correctness of this equation by simply writing out the two sums the long way.

A Factorization Formula

Example 10: If a and b are any real numbers, then, for every positive integer $n \geq 2$,

$$a^n - b^n = (a - b)(a^{n-1} + a^{n-2}b + \cdots + b^{n-1}) \qquad \textbf{(3)}$$

Special cases of this important formula are

$$a^2 - b^2 = (a - b)(a + b)$$
$$a^3 - b^3 = (a - b)(a^2 + ab + b^2)$$

To prove Formula (3) for $n \geq 2$, we use modified induction. We let $P(n)$ be the open sentence (3). Suppose k is a positive integer, $k \geq 2$, and that Formula (3) holds for $n = k$. Then we have

$$a^k - b^k = (a - b)(a^{k-1} + a^{k-2}b + \cdots + b^{k-1}) \qquad \textbf{(4)}$$

Now consider $a^{k+1} - b^{k+1}$; a little computation shows that

$$a^{k+1} - b^{k+1} = a(a^k - b^k) + b^k(a - b)$$

By using Equation (4) we get

$$\begin{aligned} a^{k+1} - b^{k+1} &= a[(a - b)(a^{k-1} + a^{k-2}b + \cdots + b^{k-1})] + b^k(a - b) \\ &= (a - b)[a(a^{k-1} + a^{k-2}b + \cdots + b^{k-1}) + b^k] \\ &= (a - b)(a^k + a^{k-1}b + \cdots + ab^{k-1} + b^k) \end{aligned}$$

This is exactly what we obtain from (3) for $n = k + 1$. So $P(k + 1)$ holds whenever $P(k)$ holds and $k \geq 2$. Since we have already verified $P(2)$, we may now claim that $P(n)$ holds for all positive integers $n \geq 2$.

It follows from Example 10 that, if $a \neq b$, we can express the sum

$$a^{n-1} + a^{n-2}b + \cdots + b^{n-1}$$

as a quotient:

$$a^{n-1} + a^{n-2}b + \cdots + b^{n-1} = \frac{a^n - b^n}{a - b} \qquad \textbf{(5)}$$

or in summation notation:

$$\sum_{k=1}^{n} a^{n-k}b^{k-1} = \frac{a^n - b^n}{a - b} \qquad a \neq b$$

When $a = 1$ and $b = r$, Formula (5) takes the form of an equation we will need later:

$$1 + r + \cdots + r^{n-1} = \frac{1 - r^n}{1 - r} \qquad r \neq 1 \qquad \textbf{(6)}$$

The Binomial Theorem

The "binomial coefficients" (often called combinatorial numbers) play a role in the statement of the Binomial Theorem as well as in the study of probability, statistics, and combinatorics.

Definition P.2.1 If n is a positive integer and k is a nonnegative integer $\leq n$, then we define the ***binomial coefficient*** $\binom{n}{k}$ by the equation

$$\binom{n}{k} = \frac{n!}{k!(n - k)!}$$

In the theory of counting (combinatorics), it is shown that $\binom{n}{k}$ is the number of ways of choosing a subset of k objects from a given set of n objects. For that reason the symbol $\binom{n}{k}$ is frequently read "n choose k".

Our reason for introducing the binomial coefficients now is to present the Binomial Theorem. First we need a lemma.

Lemma P.2.1 *If n is a positive integer and k is a positive integer $\leq n$, then*

$$\binom{n}{k} + \binom{n}{k-1} = \binom{n+1}{k}$$

PROOF: The proof merely requires combining the fractions that define the numbers on the left (see Exercise P.2.12). ■

Theorem P.2.3 **Binomial Theorem** *If $a, b \in R$ and $n \in Z^+$, then*

$$\begin{aligned}(a+b)^n &= a^n + \binom{n}{1}a^{n-1}b + \binom{n}{2}a^{n-2}b^2 + \cdots + b^n \\ &= \sum_{m=0}^{n} \binom{n}{m} a^{n-m} b^m \qquad (7)\end{aligned}$$

PROOF: Note first that the initial and final coefficients $\binom{n}{0}$ and $\binom{n}{n}$ are both equal to 1. Clearly Equation (7) holds when $n = 1$. Now we proceed by induction. Let k be a positive integer for which (7) holds:

$$\begin{aligned}(a+b)^k &= a^k + \binom{k}{1}a^{k-1}b + \binom{k}{2}a^{k-2}b^2 + \cdots + b^k \\ &= \sum_{m=0}^{k} \binom{k}{m} a^{k-m} b^m\end{aligned}$$

By multiplying both sides by $a + b$, we obtain

$$\begin{aligned}(a+b)^{k+1} &= (a+b)\sum_{m=0}^{k} \binom{k}{m} a^{k-m} b^m \\ &= a\sum_{m=0}^{k} \binom{k}{m} a^{k-m} b^m + b\sum_{m=0}^{k} \binom{k}{m} a^{k-m} b^m \\ (a+b)^{k+1} &= \sum_{m=0}^{k} \binom{k}{m} a^{k-m+1} b^m + \sum_{m=0}^{k} \binom{k}{m} a^{k-m} b^{m+1} \qquad (8)\end{aligned}$$

The second summation on the right of Equation (8) can be rewritten by shifting the index m by 1 and adjusting the limits of summation:

$$\begin{aligned}\sum_{m=0}^{k} \binom{k}{m} a^{k-m} b^{m+1} &= \sum_{m=1}^{k+1} \binom{k}{m-1} a^{k-m+1} b^m \\ &= \sum_{m=1}^{k} \binom{k}{m-1} a^{k-m+1} b^m + b^{k+1}\end{aligned}$$

By separating the initial term of the first sum on the right of Equation (8), we now have

$$\begin{aligned}(a+b)^{k+1} = a^{k+1} &+ \sum_{m=1}^{k} \binom{k}{m} a^{k-m+1} b^m \\ &+ \sum_{m=1}^{k} \binom{k}{m-1} a^{k-m+1} b^m + b^{k+1}\end{aligned}$$

And by combining corresponding terms in the summations and using Lemma P.2.1, we obtain

$$\begin{aligned}(a+b)^{k+1} &= a^{k+1} + \sum_{m=1}^{k}\left[\binom{k}{m} + \binom{k}{m-1}\right]a^{k-m+1}\,b^m + b^{k+1}\\ &= a^{k+1} + \sum_{m=1}^{k}\binom{k+1}{m}a^{k+1-m}\,b^m + b^{k+1}\\ &= \sum_{m=0}^{k+1}\binom{k+1}{m}a^{k+1-m}\,b^m\end{aligned}$$

Now this is exactly what Equation (7) says when $n = k + 1$, so $P(k + 1)$ holds whenever $P(k)$ does, and the theorem is proved. ■

Example 11: The following equations are illustrations of the Binomial Theorem:

$$(a+b)^4 = a^4 + 4a^3b + 6a^2b^2 + 4ab^3 + b^4$$

$$\left(x - \frac{2}{y}\right)^3 = x^3 - 6\frac{x^2}{y} + 12\frac{x}{y^2} - \frac{8}{y^3}$$

EXERCISES P.2

1. Show by induction that if a and b are real numbers and $n \in Z^+$, then $(ab)^n = a^nb^n$.

2. Show that if a and b are positive real numbers, and if $a > b$, then $a^n > b^n$ for every $n \in Z^+$.

3. A real number u_n is defined for each $n \in Z^+$ by stipulating that (i) $u_1 = \frac{1}{2}$ and (ii) $u_{k+1} = \frac{1}{2}(1 + u_k)$ for every $k \in Z^+$
   (a) Calculate u_2, u_3, u_4, and u_5.
   (b) Prove by induction that $u_n < 1$ for every $n \in Z^+$.
   (c) Guess a formula for u_n and prove it by induction.

4. A real number x_n is defined for each $n \in Z^+$ by stipulating that (i) $x_1 = 1$ and (ii) $x_{k+1} = \frac{1}{2}\left(x_k + \frac{2}{x_k}\right)$ for every $k \in Z^+$. Calculate x_2, x_3, and x_4.

5. In each of the following inequalities, determine the set of positive integers n for which the given inequality holds. Prove your result by induction (or modified induction).
   (a) $3^n > 5n$
   (b) $(1 + a)^n \geq 1 + na$, where $a \geq -1$
   (c) $2^n > n^2$
   (d) $2^n \geq n^2$
   (e) $2^n < n^3$

6. Write out the indicated sum in long form as illustrated in part (a).
   (a) $\displaystyle\sum_{k=5}^{25}\frac{1}{k} = \frac{1}{5} + \frac{1}{6} + \cdots + \frac{1}{25}$
   (b) $\displaystyle\sum_{k=3}^{8}(2k + 1)$
   (c) $\displaystyle\sum_{k=2}^{10}(k^2 - 2k)$
   (d) $\displaystyle\sum_{m=3}^{n}\frac{1}{m^2+1}$ (assume $n > 3$)
   (e) $\displaystyle\sum_{m=3}^{n}\frac{m}{n}$ (assume $n > 3$)
   (f) $\displaystyle\sum_{k=1}^{n}\left(1 + \frac{a}{k}\right)^k$ (assume $n > 1$)

7. Express the following sum in summation notation.
   (a) $1 + 4 + 9 + \cdots + 36$
   (b) $\frac{1}{2} + \frac{2}{3} + \frac{3}{4} + \cdots + \frac{7}{8}$
   (c) $1 - \frac{1}{2} + \frac{1}{3} - \frac{1}{4} + \frac{1}{5} - \frac{1}{6}$

8. By induction, show that each of the following formulas holds for all positive integers.
   (a) $1^2 + 2^2 + \cdots + n^2 = \dfrac{n(n+1)(2n+1)}{6}$
   (b) $\dfrac{1}{1\cdot 2} + \dfrac{1}{2\cdot 3} + \cdots + \dfrac{1}{n(n+1)} = \dfrac{n}{n+1}$

(c) $\sum_{k=1}^{n} 2k = n(n + 1)$

(d) $\sum_{k=1}^{n}(2k - 1) = n^2$

(e) $\sum_{k=1}^{n}(2k - 1)^2 = \dfrac{n(2n - 1)(2n + 1)}{3}$

9. Prove by induction: If $n \in Z^+$ and u_m, v_m are real numbers defined for $m = 1, 2, \ldots, n$; and if a and b are fixed real numbers, then

$$\sum_{m=1}^{n}(au_m + bv_m) = a\sum_{m=1}^{n} u_m + b\sum_{m=1}^{n} v_m$$

10. (a) Prove by induction that if $n \in Z^+$ and a real number u_k is defined for each k ($k = 0, 1, \ldots, n$), then

$$\sum_{k=1}^{n}(u_k - u_{k-1}) = u_n - u_0$$

(b) Obtain the result in part (a) by writing out the sum and simplifying.

11. (a) Prove Theorem P.2.2.

(b) Show that the sum of the interior angles of a convex polygon of n sides ($n \geq 3$) is $(n - 2) \cdot 180$ degrees. (*Note:* A polygon is *convex* provided that each interior angle is less than 180 degrees.)

12. Prove Lemma P.2.1.

13. Expand the following by the Binomial Theorem.

(a) $(a + b)^5$

(b) $(3x - 4y)^4$

(c) $\left(1 + \frac{1}{n}\right)^n$, where $n \in Z^+$

14. Use the Binomial Theorem to show that if $p > 0$ and $n \in Z^+$, then $(1 + p)^n \geq 1 + np$.

15. Show that the number of subsets of $\{1, 2, \ldots, n\}$ is 2^n.

16. Expand $(1 + 1)^n$ by the Binomial Theorem and conclude that

$$\sum_{m=0}^{n}\binom{n}{m} = 2^n \qquad \text{for every } n \in Z^+$$

17. (a) Show that $4^n + 2$ is divisible by 3 for every $n \in Z^+$.

(b) Show that $10^n + 3 \cdot 4^{n+2} + 5$ is divisible by 9 for every $n \in Z^+$.

18. Discuss the following "proof" that all billiard balls are of the same color. By mathematical induction we prove that, for every positive integer n, the following statement holds: *In any set of n billiard balls, all the balls have the same color.*

Let S be the set of positive integers n for which the statement holds. If $k \in S$, then any set having just k billiard balls has the property that all the balls in the set have the same color. Now consider any set T_{k+1} of $k + 1$ billiard balls. Choose a particular ball, a, and remove it from the set; the resulting set is $T_{k+1} \dashv \{a\}$, which is a set of just k billiard balls. By our assumption, all the balls in this set are of one and the same color, C. Now choose a different ball, $b \in T_{k+1} \dashv \{a\}$; remove b, and put a back, obtaining the set $T_{k+1} \dashv \{b\}$, which is another set of k billiard balls. The balls in this set (to which a belongs) must all be of the same color, and therefore a has color C, too. We have now proved that $(k + 1) \in S$, and therefore S is inductive. Now it is obvious that $1 \in S$, and so the statement made at the outset has been proved by induction for all $n \in Z^+$.

19. (a) Show that if x and y are two positive real numbers such that $xy = 1$, then $x + y \geq 2$, with equality holding only if $x = y = 1$.

(b) Show that if $x_1, x_2, \ldots, x_n$ are n positive real numbers such that $x_1 x_2 \cdots x_n = 1$, then $x_1 + x_2 + \cdots + x_n \geq n$ with equality holding only if $x_1 = x_2 = \cdots = x_n$. [*Note:* Part (a) is the case $n = 2$.]

PROJECT: The Fibonacci Numbers

A variation of mathematical induction uses the following theorem:
If S is a set such that

(i) $1 \in S$ and $2 \in S$; and
(ii) $k \in S$ and $k + 1 \in S \Rightarrow k + 2 \in S$ for every positive integer k,

then S contains all positive integers.

(a) Prove the theorem.

(b) Explain how the theorem applies in the following definition of the Fibonacci numbers* F_n:

$$\text{(i) } F_1 = 1 \quad \text{and} \quad F_2 = 2$$
$$\text{(ii) } F_{k+2} = F_k + F_{k+1}$$

Find F_3, F_4, F_5, F_6, F_7.

(c) By induction, prove the formula

$$F_n = \frac{1}{\sqrt{5}}\left[\left(\frac{1+\sqrt{5}}{2}\right)^{n+1} - \left(\frac{1-\sqrt{5}}{2}\right)^{n+1}\right]$$

for every positive integer n.

(d) Prove that, for $n \in Z^+$,

$$F_n^2 - F_{n-1}F_{n+1} = (-1)^n$$

*The earliest study of these famous numbers is attributed to Leonardo of Pisa (alias Fibonacci) early in the thirteenth century. They are associated with a variety of natural phenomena. See, for example, David Burton, *Elementary Number Theory*, 3rd ed. (Dubuque: Wm. C. Brown, 1994), chapter 13.

CHAPTER C

CARDINAL EQUIVALENCE

There is a sense in which it can be said that the real numbers greatly outnumber the rationals. Now both the rationals and the reals are infinite sets, and so we owe the reader some explanation of how one infinite set can have more elements than another. It is a remarkable fact that there are different "orders" of infinity. To pursue these matters we now introduce *cardinal equivalence.*

The concept of cardinality arises in a very natural way from primitive notions about counting, yet the resulting theory contains many surprises, some of which we will encounter here. Our discussion is only a brief survey; more thorough coverage belongs to a course in set theory.

C.1 *Finite and Denumerable Sets*

Let us begin by recalling some relevant terminology about mappings.* Let A and B be sets (not necessarily sets of real numbers). If f is a function whose domain is A and whose range is contained in B, then we write $f: A \to$ B. We say that f *maps A into B*, and we speak of the *mapping* $f: A \to B$. When the range of f is all of B, we say that f *maps A onto B*, and the mapping

*A fuller discussion of mappings is found in Section L.4

$f: A \to B$ is said to be *surjective*. A mapping $f: A \to B$ is *one-to-one* if it has the property that

$$f(x_1) = f(x_2) \Rightarrow x_1 = x_2$$

such a mapping is also called *injective*.

Example 1: The mapping $f: R \to R$ defined by $f(x) = x^2$ is not surjective because no negative real number is the image under f of any real number. However, if $R^+ = \{y : y \in R, y \geq 0\}$ then f maps R onto R^+ and $f: R \to R^+$ is surjective, but not injective (because $f(-1) = f(1)$).

A mapping $f: A \to B$ is a *bijection* (or is *bijective*) provided it is both surjective and injective. In that case, f^{-1} exists and $f^{-1}: B \to A$ is also a bijection.

Example 2: The mapping $f: R \to R$ defined by $f(x) = x^3$ is a bijection. It is surjective because every real number is the cube of some number (its cube root); it is injective because if $x_1 \neq x_2$, then $x_1^3 \neq x_2^3$. The inverse of f is defined by $f^{-1}(x) = \sqrt[3]{x}$.

A bijection $f: A \to B$ provides a one-to-one correspondence between the elements of A and those of B: each element of A corresponds under f to exactly one element of B; and for each element of B, there is exactly one element of A that corresponds to it. When such a correspondence exists, we regard A and B as being "equinumerous."

Definition C.1.1 A set A is ***cardinally equivalent*** to a set B provided that there exists a bijection $f: A \to B$. When A is cardinally equivalent to B, we write $A \sim B$.

Example 3: Let A consist of the first three letters of the alphabet, $A = \{a, b, c\}$, and let $B = \{1, 2, 3\}$. Then the mapping $f: A \to B$ defined by $f(a) = 1, f(b) = 2, f(c) = 3$ is a bijection, and so $A \sim B$.

Example 4: Let E denote the set of positive even integers. The mapping $f: E \to Z^+$ defined by $f(m) = \frac{m}{2}$ is a bijection. It is surjective because each $m \in Z^+$ is the image under f of $2m \in E$. It is injective because for $m_1, m_2 \in E$,

$$f(m_1) = f(m_2) \Rightarrow \frac{m_1}{2} = \frac{m_2}{2} \Rightarrow m_1 = m_2$$

We conclude that $E \sim Z^+$.

Example 5: Let U be the open interval $(0, 1)$ in R, and let I be the interval $(0, 2)$. Then the mapping $f: U \to I$ defined by $f(x) = 2x$ is a bijection.

It is surjective because if $y \in I$, then $0 < y < 2$, so that $\frac{y}{2} \in U$ and $f\left(\frac{y}{2}\right) = y$. It is injective because if $x_1, x_2 \in U$ and $2x_1 = 2x_2$, it follows that $x_1 = x_2$; therefore, $U \sim I$.

Before proceeding with more examples, note that cardinal equivalence has the three properties required for any relation described as an equivalence relation (see Section L.3). For all sets A, B, C, we have:

(1) $A \sim A$
(2) If $A \sim B$, then $B \sim A$
(3) If $A \sim B$ and $B \sim C$, then $A \sim C$

You are asked to prove these properties in Exercise C.1.1.

Finite Sets

Example 3 suggests the way in which we count finite sets. For $n \in Z^+$, let us denote by C_n the set of positive integers less than or equal to n; that is,

$$C_n = \{1, 2, \ldots, n\}$$

Definition C.1.2 A set A is ***finite*** provided that either $A = \varnothing$ or there is some $n \in Z^+$ such that

$$A \sim C_n$$

in the latter case, A is said to have n elements.

Note: The number n in Definition C.1.2 can be shown to be unique; that is, if $A \sim C_n$ and $A \sim C_m$, then $m = n$ (see Exercise C.1.5).

Now if A and B both have n elements, it follows from properties (2) and (3) above that $A \sim B$. Conversely, if A is finite and has n elements, and if $B \sim A$, then B is also finite and has n elements. We state this simple result as a theorem:

Theorem C.1.1 *Two nonempty finite sets are cardinally equivalent if and only if they have the same number of elements.*

Infinite Sets

Definition C.1.3 A set A is ***infinite*** provided that it is not finite; that is, a set A is infinite provided that A is not empty and there is no positive integer n such that $A \sim C_n$.

It can be shown that Z^+ is infinite, according to Definition C.1.3 (see Exercise C.1.4). There are many other familiar examples of infinite sets—to name a few, Z, Q, and R are infinite sets that are discussed later in this chapter.

In Example 4 we found that the set E of *even* positive integers is cardinally equivalent to the set Z^+ of *all* positive integers; thus, we have a set cardinally equivalent to a proper subset of itself. The scientist Galileo was surprised to observe this phenomenon and regarded it as a paradox; to him, it violated the principle that "the whole is greater than any one of its proper parts". Example 5 illustrates a similar situation because it shows that the interval (0, 2) is cardinally equivalent to its subinterval (0, 1). Indeed, according to the following theorem, Galileo's "paradox" characterizes infinite sets.

GALILEO GALILEI

(b. February 15, 1564; d. January 8, 1642)

First a student of medicine, then a physicist and astronomer, Galileo was appointed professor of mathematics at Pisa and then at Padua. By using the deductive method in the study of motion and mechanics, he developed relations between distance, velocity and acceleration and derived the parabolic orbit of a projectile. His writings indicate some understanding of cardinality, a concept of some controversy more than two centuries later. His achievements seem to have created a balance between theory and experiment but his work was followed with such great interest that he was challenged by Church authorities. The joy of his accomplishments covering 40 years was overcome by the power of the Inquisition, which tried him, forced him to recant, and sentenced him to house imprisonment.

Theorem C.1.2 **Dedekind's Theorem** *A set is infinite if and only if it is cardinally equivalent to a proper subset of itself.*

The proof of Dedekind's Theorem is outlined in Exercise C.1.6. Because the condition given in the theorem is both necessary and sufficient, we know not only that every infinite set is equivalent to a proper subset of itself, but also that no finite set has this property. Some authors use this property to define *infinite*.

As an application of Dedekind's Theorem, we see that the open interval (0, 2) is infinite because, by Example 5, it is equivalent to its proper subset (0, 1).

Denumerable Sets

It is clear that if a set A is cardinally equivalent to an infinite set B, then A is also infinite. In particular, all sets cardinally equivalent to Z^+ are infinite, and we give a special designation to such sets.

Definition C.1.4 A set A is ***denumerably infinite*** (or ***denumerable***) provided that $A \sim Z^+$.

The result of Example 4 is that the set E of even positive integers is denumerably infinite.

Example 6: Show that the following sets are denumerable: (a) the set T of all integers greater than 10; (b) the set Z of all integers.

Solution:

(a) The mapping $f : T \rightarrow Z^+$ defined for $n \in T$ by $f(n) = n - 10$ is a bijection.

(b) The mapping $g : Z \rightarrow Z^+$ defined by

$$g(n) = \begin{cases} 2n + 1 & \text{if } n \in Z,\ n \geq 0 \\ -2n & \text{if } n \in Z,\ n < 0 \end{cases}$$

is a bijection.

(Verification that f and g have the required properties is left for you to work out in Exercise C.1.2.)

A denumerable set may be thought of as one whose elements can be listed: a first element, a second element, . . . , an nth element, and so on, in such a way that there is one element for each positive integer n, with no repetition. That is, A is denumerable if and only if

$$A = \{a_1, a_2, \ldots, a_n, \ldots\}$$

where $a_m \neq a_n$ if $m \neq n$. This interpretation is useful in the proof of the following theorem.

Theorem C.1.3 *If A and B are denumerable sets, then $A \times B$ is also denumerable.*

PROOF: Since A and B are denumerable, we can list their elements, $A = \{a_1, a_2, \ldots, a_n, \ldots\}$, $B = \{b_1, b_2, \ldots, b_n, \ldots\}$, in such a way that if $m \neq n$, then $a_m \neq a_n$ and $b_m \neq b_n$. The elements of $A \times B$ are the ordered pairs (a_i, b_j) where $i, j \in Z^+$. We arrange these ordered pairs in the following array:

$$\begin{array}{ccccc} (a_1,b_1) & (a_1,b_2) & (a_1,b_3) & (a_1,b_4) & \cdots \\ (a_2,b_1) & (a_2,b_2) & (a_2,b_3) & (a_2,b_4) & \cdots \\ (a_3,b_1) & (a_3,b_2) & (a_3,b_3) & (a_3,b_4) & \cdots \\ (a_4,b_1) & (a_4,b_2) & (a_4,b_3) & (a_4,b_4) & \cdots \\ & & \cdots & & \end{array}$$

It follows from the equality property for ordered pairs that no two of these ordered pairs are equal. It is sufficient then to devise a way of listing them. A simple procedure is

$$A \times B = \{c_1, c_2, c_3, \ldots, c_n, \ldots\}$$

where

$$c_1 = (a_1, b_1)$$
$$c_2 = (a_2, b_1),\ c_3 = (a_1, b_2)$$
$$c_4 = (a_3, b_1),\ c_5 = (a_2, b_2),\ c_6 = (a_1, b_3)$$
$$c_7 = (a_4, b_1),\ c_8 = (a_3, b_2),\ c_9 = (a_2, b_3),\ c_{10} = (a_1, b_4)$$
$$\cdots$$

In Figure C.1, we start at the upper left corner of the array, then follow along the upward slanting lines as indicated by the arrows. Each ordered pair is assigned a unique positive integer, and each positive integer determines a unique ordered pair. In Exercise C.1.7, you are asked to find an algebraic formula, in terms of i and j, for the positive integer n that corresponds to (a_i, b_j). ■

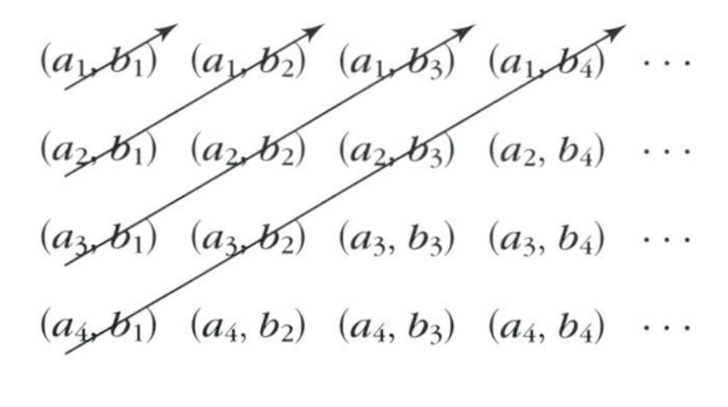

Figure C.1
Listing $A \times B$

Corollary C.1.1 *The sets $Z^+ \times Z^+$, $Z \times Z$, and $Z \times Z^+$ are denumerable.*

EXERCISES C.1

1. Show that cardinal equivalence is reflexive, symmetric, and transitive.

2. Verify that the mappings given in parts (a) and (b) of Example 6 are bijections.

3. Show that the following sets are denumerable.
   (a) The set of reciprocals of the positive integers
   (b) The set of all positive integral multiples of 5
   (c) The set of even integers greater than 9

4. This exercise shows that Z^+ is infinite.
   (a) Prove by induction: If $S = \{m_1, m_2, \ldots, m_n\}$ is any set of n positive integers, then S has a greatest element.
   (b) Use part (a) to prove that there does not exist a positive integer n such that Z^+ is equivalent to $\{1, 2, \ldots, n\}$.

5. Show that if $m, n \in Z^+$ and $m < n$, then C_m is not cardinally equivalent to C_n. Explain how this fact justifies the note following Definition C.1.2.

6. The following steps outline a proof of Dedekind's Theorem.
   (a) Show that every denumerable set is equivalent to a proper subset of itself.
   (b) Show that every infinite set has a denumerable subset.

(c) Show that every infinite set is equivalent to a proper subset of itself.
(d) Show that a finite set is not equivalent to any of its proper subsets.

7. Find a formula in terms of i and j for the positive integer n that corresponds to the ordered pair (a_i, b_j) in the proof of Theorem C.1.3.

C.2 *Uncountable Sets*

Comparing Cardinals

When A and B are cardinally equivalent sets, we say that "A and B have the same number of elements." Because of Theorem C.1.1, this manner of speaking is consistent with the behavior of finite sets. We also say that "A and B have the same cardinal number", or even that "the cardinal number of A is equal to the cardinal number of B." It is also convenient to write $|A| = |B|$, where the vertical bars are read "cardinal number of." Note that we have *not* defined the term *cardinal number* nor the symbol $|A|$ itself. The various statements above are just descriptive ways of saying that $A \sim B$.

If A and B are sets such that A is equivalent to a subset of B, then we say that "the cardinal number of A is less than or equal to the cardinal number of B," and we write $|A| \leq |B|$. It is easy to show that this manner of speaking is again consistent with inequalities involving cardinal numbers of finite sets.

Example 1: Let A be a finite set; then $A \sim C_n$ for some $n \in Z^+$. But C_n is a subset of Z^+, so we have $|A| \leq |Z^+|$. Therefore, the cardinal number of any finite set is less than or equal to the cardinal number of Z^+. Moreover, since every infinite set has a denumerable subset, it follows that $|Z^+|$ is less than or equal to the cardinal number of any infinite set.

An important result of set theory is known as Bernstein's Theorem (also as the Cantor–Bernstein Theorem and as the Schröder–Bernstein Theorem).

Theorem C.2.1 **Bernstein's Theorem** *If A and B are sets with the property that each is cardinally equivalent to a subset of the other, then A and B are cardinally equivalent.*

In terms of mappings, the theorem says that if there exists a one-to-one mapping from A into B, and also a one-to-one mapping from B into A, then there exists a one-to-one mapping from *all* of A *onto all* of B—that is, a bijection $f : A \to B$. This result is rather deep and we will not prove it here (if you are interested in a proof, see the elegant one by Birkhoff and MacLane*).

*Garrett Birkhoff and Saundeers MacLane, *A Survey of Modern Algebra,* 4th ed. (New York: Macmillan, 1977), 387–388.

Bernstein's Theorem can be expressed in terms of cardinal inequalities in a very natural and easily remembered form: If A and B are sets such that

$$|A| \leq |B| \qquad \text{and} \qquad |B| \leq |A|$$

then

$$|A| = |B|$$

The theorem offers a two-step method for establishing the cardinal equivalence of two sets A and B: prove (1) that $|A| \leq |B|$, then prove (2) that $|B| \leq |A|$. We apply this method in the following theorem.

Theorem C.2.2 *The set Q of rational numbers is denumerable.*

PROOF: Our goal is to prove that $|\mathsf{Q}| = |\mathsf{Z}^+|$. Since Z^+ is a subset of Q and is cardinally equivalent to itself, the inequality $|\mathsf{Z}^+| \leq |\mathsf{Q}|$ is immediate. To obtain the reverse inequality, we first define $f: \mathsf{Q} \to \mathsf{Z} \times \mathsf{Z}^+$ as follows: for $r \in \mathsf{Q}$, we let $f(r)$ be the ordered pair (p, q) where $p \in \mathsf{Z}$, $q \in \mathsf{Z}^+$, and p/q represents r in lowest terms (see Example 6 in Section P.1). It is clear that f is injective. Now by Corollary C.1.1, $\mathsf{Z} \times \mathsf{Z}^+$ is denumerable, so there exists a bijection $g: \mathsf{Z} \times \mathsf{Z}^+ \to \mathsf{Z}^+$. The composite mapping $h = g \circ f$ maps Q into Z^+ and is injective. Therefore, $|\mathsf{Q}| \leq |\mathsf{Z}^+|$, and, by Bernstein's Theorem, $|\mathsf{Q}| = |\mathsf{Z}^+|$. ■

Uncountable Sets

A set A that is either finite or denumerable is *countable*; otherwise A is *uncountable*. Most of the sets we have discussed so far are countable, and so you may be wondering whether it is possible to find uncountable sets. The following remarkable theorem, due to Cantor, gives us a start.

Theorem C.2.3 **Cantor's Theorem** *The interval $U = (0, 1)$ is an uncountable subset of* R.

PROOF: It is easy to show that U is not finite. The mapping $f: U \to \left(0, \frac{1}{2}\right)$ defined by $f(x) = \frac{x}{2}$ is bijection. Thus, U is equivalent to a proper subset of itself, and by Dedekind's Theorem, U is infinite.

Now we show that U is not denumerable either. Every real number $x \in (0, 1)$ can be expressed as an unending decimal* of the form

$$x = 0.d_1\, d_2\, d_3 \cdots d_n \cdots$$

where each d_n is one of the digits 0, 1, 2, . . . , 9. It can be shown that this decimal representation is unique, except for the "terminating" decimals, which can be expressed either with a final string of 9s or a final string of 0s; for

*Decimal expansion of real numbers is discussed in Section 2.8.

GEORG CANTOR

(b. March 3, 1845; d. January 6, 1918)

Born in Russia, Cantor studied philosophy, mathematics and physics in Germany to teach at the University of Halle from 1869 to 1905. He contributed to number theory and the foundations of analysis, using convergent sequences of rational numbers to define irrational numbers. Cantor's theory of "aggregates" (infinite sets) introduced different orders of infinity and led quickly to a crisis in the foundations of mathematics precipitated by the discovery of paradoxes. Some mathematicians took a stand with Cantor and his new ideas and others rejected them as illegitimate. Cantor suffered personally through the uproar he had caused but continued to produce coherent work on sets and numbers, including the ordering and the arithmetic of transfinite numbers.

example, $0.134999\ldots = 0.1350000\ldots$. To avoid ambiguity we will use only the form with an ending string of 0s.

Suppose the set U is denumerable. Then there exists a bijection $f: Z^+ \to U$. By letting $f(1) = x_1$, $f(2) = x_2, \ldots, f(n) = x_n, \ldots$, we can arrange the elements of U in a list, giving them their decimal representations as indicated above.

$$\begin{aligned} x_1 &= 0.d_{11}d_{12}d_{13}\cdots d_{1n}\cdots \\ x_2 &= 0.d_{21}d_{22}d_{23}\cdots d_{2n}\cdots \\ x_3 &= 0.d_{31}d_{32}d_{33}\cdots d_{3n}\cdots \\ &\ \ \vdots \\ x_n &= 0.d_{n1}d_{n2}d_{n3}\cdots d_{nn}\cdots \\ &\cdots \end{aligned}$$

Since $f: Z^+ \to U$ is surjective, every element of U must appear on the list as some x_n. That this does not happen is seen as follows: We construct a number $y \in U$ such that $y = 0.e_1\, e_2\, e_3 \ldots e_n \ldots$, where e_1 differs from d_{11} and is not 9, e_2 differs from d_{22} and is not 9, ..., and e_n differs from d_{nn} and is not 9. For instance, one specific rule for finding e_n is

$$e_n = \begin{cases} d_{nn} + 1 & \text{if } d_{nn} \neq 8, 9 \\ d_{nn} - 1 & \text{if } d_{nn} = 8 \text{ or } 9 \end{cases}$$

Now the decimal y thus constructed is in the proper form and represents a real number in U. It differs from every decimal in the list $\{x_1, x_2, \ldots, x_n, \ldots\}$ because the nth digit of y differs from the nth digit of x_n. Thus,

$f : Z^+ \to U$ cannot be surjective, contradicting the assumption that U is denumerable. ∎

By knowing that U is uncountable, we can obtain other uncountable sets by finding sets that are equivalent to U. In the exercises below you are asked to show that any open interval (a, b) in R, with $a < b$, is equivalent to U (Exercise C.2.1a). Also, R itself is equivalent to U (Exercise C.2.1c), as are various subsets of the plane.

Example 2: The *unit square* in the plane is the set

$$U \times U = \{(x, y) : 0 < x < 1, 0 < y < 1\}$$

We show that $U \times U \sim U$. We let x and y be elements of U expressed in decimal notation (with the convention described in the proof of Cantor's Theorem) as follows:

$$x = 0.x_1x_2x_3 \ldots x_n \ldots \qquad \text{and} \qquad y = 0.y_1y_2y_3 \ldots y_n \ldots$$

Then we associate with $(x, y) \in U \times U$ the real number

$$w = 0.x_1y_1x_2y_2x_3y_3 \ldots x_ny_n \ldots$$

The mapping $f : U \times U \to U$ defined by $f(x, y) = w$ is injective (not surjective, however), and so $U \times U$ is equivalent to a subset of U: $|U \times U| \leq |U|$. On the other hand, we can map U into $U \times U$ injectively by means of the function $g(x) = \left(x, \frac{1}{2}\right)$, so that U is equivalent to a subset of $U \times U : |U| \leq |U \times U|$. By Bernstein's Theorem, $U \times U \sim U$.

When $|A| \leq |B|$ but A and B are not equivalent, we say that "the cardinal number of A is less than the cardinal number of B", and write $|A| < |B|$. In the sense of this definition, the cardinal number of Z^+ is less than the cardinal number of U. We have seen in this section examples of two orders of infinity; the fact that there are even higher orders of infinity is brought out in Exercise C.2.8 below.

EXERCISES C.2

1. (a) Show that any open interval (a, b) in R (with $a < b$) is equivalent to the open interval $U = (0, 1)$ and is therefore uncountable.
   (b) Use part (a) to show that any two nonempty bounded open intervals in R are equivalent.
   (c) Show that R is equivalent to U. [*Hint:* Consider $f(x) = x/(|x| + 1)$ for $x \in R$.]
   (d) Show that U is equivalent to the *closed* interval $[0, 1]$ by using Bernstein's Theorem.
   (e) Show that U is equivalent to $[0, 1]$ by finding a bijection.

2. Let $U = (0, 1)$. Use Bernstein's Theorem to establish the following equivalences:
   (a) $Z^+ \sim Q \cap U$
   (b) $Q \cap U \sim Q$
   (c) $U \sim (0,1) \cup (1, 2)$

3. (a) Show that if A and B are sets such that $A \subseteq B$, then $|A| \leq |B|$.
   (b) Show that any subset of a countable set is countable.
   (c) Show that if A, B, and C are sets such that $|A| \leq |B|$ and $|B| \leq |C|$, then $|A| \leq |C|$. (*Note:* This is a statement about mappings!)

4. Prove or disprove each of the following statements:
   (a) The union of two countable sets is countable
   (b) The intersection of two countable sets is countable
   (c) The union of two uncountable sets is uncountable
   (d) The intersection of two uncountable sets is uncountable
   (e) The set of irrational numbers is uncountable

5. If $A_1, A_2, \ldots, A_n$ are sets, their *Cartesian product* $A_1 \times A_2 \times \ldots \times A_n$ is defined to be the set of all n-tuples

   $\{(a_1, a_2, \ldots, a_n) : a_1 \in A_1, a_2 \in A_2, \ldots, a_n \in A_n\}$

   Show by induction that a Cartesian product of n denumerable sets $A_1, A_2, \ldots A_n$ is also denumerable.

   [*Note:* Two n-tuples $(a_1, a_2, \ldots, a_n)$ and $(b_1, b_2, \ldots, b_n)$ are equal if and only if $a_i = b_i$ for each i, $1 \le i \le n$.]

6. Let $\{A_1, A_2, \ldots, A_n, \ldots\}$ be a denumerable class of sets, one set for each $n \in Z^+$. The union of such a class of sets is defined to be

   $$\bigcup_{n=1}^{\infty} A_n = \{x : x \in A_n \text{ for some } n \in Z^+\}$$

   Show that if each A_n is countable then the union is countable.

7. Let S be the set of all functions f that map Z^+ into Z^+. Show that S is uncountable.

8. The power class of a set A is the class $P(A)$ of all subsets of A. Show that $|A| < |P(A)|$ for every set A.

PROJECT: Denumerability of the Algebraic Numbers

Definition

An ***algebraic*** number is a real number that is a root of a polynomial equation

$$a_n x^n + a_{n-1}x^{n-1} + \cdots + a_0 = 0$$

where the a_i are integers. A real number that is not algebraic is *transcendental.*

Example: The square root of 2 is algebraic since it satisfies the equation $x^2 - 2 = 0$. The numbers π and e are known to be transcendental, although it is not easy to show that. (A proof that e is transcendental can be found in Simmons.*) The result of (d) below is that there are very many transcendental numbers indeed.

Use the fact that a polynomial equation of degree n has at most n roots, together with the results of Exercise 4 above, to establish the following.

(a) The set of polynomials of degree n (fixed) with integer coefficients is denumerable.
(b) The set of roots of polynomial equations of degree n (fixed) with integer coefficients is denumerable.
(c) The set of algebraic numbers is denumerable.
(d) The set of transcendental numbers is uncountable.

*George Simmons, *Calculus Gems* (New York: McGraw-Hill, 1992), 286–293.

REAL ANALYSIS

CHAPTER ONE

COMPLETENESS OF R

1.1 *Upper and Lower Bounds*

The system of real numbers has an important property that distinguishes it from other ordered fields, such as the rationals. This property, known as *completeness*, is based upon upper and lower bounds of subsets of R, and so we begin our discussion by recalling the terminology associated with boundedness specifically as it relates to sets of real numbers.

A set S of real numbers is *bounded below* provided that there exists a real number a such that the inequality $a \leq x$ holds for every $x \in S$. In this case, a is a *lower bound* of S, and S is *bounded below* by a. Similarly, S is *bounded above* provided there exists a real number b such that $x \leq b$ for every $x \in S$; then b is an *upper bound* of S, and S is *bounded above* by b. Finally, a set S of real numbers is *bounded* provided it is bounded above *and* below.

Example 1:

(i) The set of positive real numbers is bounded below by 0; it is not bounded above and is therefore not bounded.

(ii) The open interval* $(0, 1)$ is bounded below by 0 and above by 1; it is therefore bounded.

(iii) The set R of all real numbers is neither bounded above nor bounded below.

*Intervals are discussed in Section R.2.

Example 2: Is the set Z^+ of positive integers bounded above? Most students will argue immediately that the answer is no because, for any $n \in Z^+$, the integer $n + 1$ is greater than n. This argument shows that there is no largest positive integer, but that is not the question! The question is whether or not there is some real number that is larger than every positive integer. Could it be the case that the integers occur along a number line as in Figure 1.1 in which they are crowded into the part of the line to the left of some large real number B? We will return to this question in the next section.

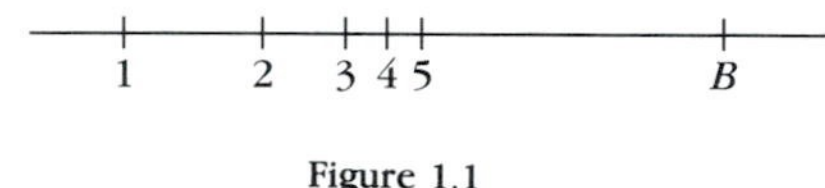

Figure 1.1
Is Z^+ bounded above?

If S is a subset of R, then the set $-S = \{y : y = -x \text{ for some } x \in S\}$ is the reflection of S through the origin, a sort of "mirror image" of S (see Figure 1.2).

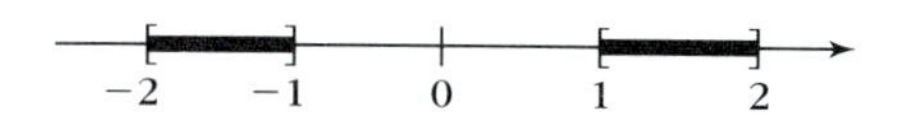

Figure 1.2
Reflection of a set through the origin

Theorem 1.1.1 **Reflection Property** *Let S be a nonempty subset of R, and let $-S = \{y : y = -x \text{ for some } x \in S\}$. Then a real number a is a lower bound of S if and only if $-a$ is an upper bound of $-S$. Equivalently, b is an upper bound of S if and only if $-b$ is a lower bound of $-S$.*

PROOF: Suppose that a is a lower bound of S. Then

$$a \leq x \qquad \text{for every } x \in S$$

so that

$$-a \geq -x \qquad \text{for every } x \in S$$

But this inequality is equivalent to

$$-a \geq y \qquad \text{for every } y \in -S$$

so that $-a$ is an upper bound of $-S$. This shows that if a is a lower bound of S, then $-a$ is an upper bound of $-S$. The converse can be established by reversing these steps. ■

Supremum and Infimum

Let $S \subseteq R$ be a set with an upper bound b. Then S has many other upper bounds because any real number greater than b is also an upper bound of S.

A more interesting matter is whether S has any upper bounds that are *less* than b.

Definition 1.1.1 Let $S \subseteq R$ and let $l \in R$. Then l is the ***least upper bound***, or ***supremum***, of S provided that the following two conditions hold:

(i) l is an upper bound of S
(ii) l is less than or equal to every upper bound of S

The least upper bound of S is denoted sup S.

It is easy to see that a set can have no more than one least upper bound, so that we can refer to "the" least upper bound of S when it exists.

To show that a number l is the least upper bound of S we must show that conditions (i) and (ii) in the definition hold. It is sometimes helpful to establish condition (ii) by showing that no real number less than l is an upper bound of S.

Example 3: If $S = (0, 1)$, we claim that $\sup S = 1$. Clearly 1 is an upper bound of S, so condition (i) holds. Moreover, no real number less than 1 is an upper bound of S. Certainly no negative number is an upper bound of S, and if $0 \leq b < 1$, then $(b + 1)/2$ is an element of S that is greater than b. It follows that any upper bound of S is greater than or equal to 1, so condition (ii) holds.

Example 3 brings out the fact that the least upper bound of a set need not belong to the set.

Parallel to the concept of least upper bound is that of greatest lower bound:

Definition 1.1.2 Let $S \subseteq R$ and let $g \in R$. Then g is the ***greatest lower bound***, or ***infimum***, of S provided that the following two conditions hold:

(i) g is a lower bound of S
(ii) g is greater than or equal to every lower bound of S

The greatest lower bound of S is denoted inf S.

A connection between supremum and infimum is established by the following theorem.

Theorem 1.1.2 *Let S be a nonempty subset of R and let $-S = \{y : y = -x \text{ for some } x \in S\}$. Then:*

(1) If sup S exists and equals l, then inf $(-S)$ exists and equals $-l$.
(2) If inf S exists and equals g, then sup $(-S)$ exists and equals $-g$.

PROOF: To prove part (1), let $l = \sup S$; we show that $-l = \inf(-S)$. Since l is an upper bound of S, we know by Theorem 1.1.1 that $-l$ is a lower bound of $-S$. It remains for us to show that no real number greater than $-l$ is a lower bound of $-S$. Suppose the contrary, that some real number $m > -l$ is a lower bound of $-S$. Then, by Theorem 1.1.1 again, we see that $-m < l$ would be an upper bound of S—contradicting the fact that l is the least upper bound of S. So no real number greater than $-l$ can be a lower bound of $-S$, and we conclude that $-l = \inf(-S)$. You are asked to prove part (2) in Exercise 1.1.5. ■

In Example 3 we saw that if $\sup S$ exists, it is not necessarily an element of S; however, if it is an element of S, it is the greatest element. Similarly, $\inf S$, if it belongs to S, is the least element.

Example 4: If S is the closed interval $[0, 1]$, then $\inf S = 0$, $\sup S = 1$, and both of these belong to S.

We will often need to use the least upper bound idea as characterized in the following theorem:

Theorem 1.1.3 *Let S be a nonempty subset of R. A real number l is the least upper bound of S if and only if the following two conditions hold:*

(1) l is an upper bound of S
(2) For an arbitrary $\varepsilon > 0$, there exists an $x \in S$ such that $x > l - \varepsilon$.

PROOF: Suppose $l = \sup S$. Then condition (1) obviously holds. If ε is positive, $l - \varepsilon$ is less than l and is therefore not an upper bound of S. But then there must exist an element of S greater than $l - \varepsilon$, and so condition (2) holds. To prove the converse, suppose that l satisfies conditions (1) and (2). Then l is an upper bound of S, and if any number $m < l$ were an upper bound of S, we could choose $\varepsilon = l - m > 0$ and by condition (2) there would exist an $x \in S$ with $x > l - (l - m) = m$, thereby contradicting the fact that m is an upper bound of S. ■

There is, of course, a theorem just like Theorem 1.1.3 for greatest lower bounds. You are asked to formulate and prove such a theorem in Exercise 1.1.6.

We conclude this section by presenting several useful theorems concerning suprema and infima.

Theorem 1.1.4 *If S and T are nonempty subsets of R having least upper bounds s and t, respectively, and if $S + T$ denotes the set $\{x + y : x \in S \text{ and } y \in T\}$, then*

$$\sup(S + T) = s + t$$

PROOF: It is clear that $s + t$ is an upper bound of $S + T$. To show that $s + t$ is the *least* upper bound, we use Theorem 1.1.3. Let $\varepsilon > 0$; then $\varepsilon/2$ is also a positive number. Since $s = \sup S$, part (2) of Theorem 1.1.3 tells us that we can find an $x \in S$ such that $x > s - \frac{\varepsilon}{2}$. Similarly, we can find a $y \in T$ such that $y > t - \frac{\varepsilon}{2}$. Then

$$x + y > \left(s - \frac{\varepsilon}{2}\right) + \left(t - \frac{\varepsilon}{2}\right) = (s + t) - \varepsilon \qquad \blacksquare$$

Theorem 1.1.5 *If S is a nonempty subset of R with $s = \sup S$, and if cS denotes the set $\{cx : x \in S\}$ where c is a fixed real number, then*

(1) If $c > 0$, then $\sup (cS) = cs$
(2) If $c < 0$, then $\inf (cS) = cs$

PROOF: For part (1), since $s = \sup S$, we have for every $x \in S$, $x \leq s$; because $c > 0$ it follows that $cx \leq cs$ for every $x \in S$, and therefore cs is an upper bound of cS. We now use part (2) of Theorem 1.1.3 to establish the fact that cs is the *least* upper bound of cS: Let $\varepsilon > 0$; then ε/c is also a positive number, and since s is by hypothesis the least upper bound of S, there must exist an element x of S such that

$$x > s - \frac{\varepsilon}{c}$$

Then

$$cx > cs - \varepsilon$$

and by part (2) of Theorem 1.1.3, we conclude the $\sup (cS) = cs$.

For part (2), if $c = -1$, then cS is just the set $-S$ in Theorem 1.1.2, so

$$\inf (cS) = \inf (-S) = -\sup S = cs$$

Thus, the equation in part (2) holds for $c = -1$. For any other negative number, we can write $c = (-1)|c|$, and then apply part (1). You should supply the details. $\blacksquare$

Theorems 1.1.3, 1.1.4, and 1.1.5 have counterparts for infima. You are asked to state and prove these counterparts in Exercises 1.1.6, 1.1.7, and 1.1.8.

EXERCISES 1.1

1. Which of the following sets are bounded above? Bounded below? Bounded?
   (a) $S = \{x : x \in R \text{ and } x^2 < 1\}$
   (b) $S = \left\{x : x = \frac{1}{n} \text{ for some } n \in Z^+\right\}$
   (c) $T = \{r : r \in Q \text{ and } r^2 < 2\}$
   (d) $S = \{y : y \in R \text{ and } y < 0\}$
   (e) $T = \{x : x \in R \text{ and } x > 4\}$

2. Show that if a set $S \subseteq R$ has a greatest element M, then M is the least upper bound of S.

3. Let S be the interval $(-5, 9]$. By verifying conditions (i) and (ii) of the appropriate definition, show that
   (a) $\sup(-5, 9] = 9$
   (b) $\inf(-5, 9] = -5$

4. Determine $\sup S$ and $\inf S$ (if they exist) for each of the following sets.
   (a) $S = \{x : x \in R \text{ and } x < 3\}$
   (b) $S = \{x : x \in R \text{ and } x \leq 0\}$
   (c) $S = \{x : |x| < 1\}$
   (d) $S = \{x : x^2 \geq 9\}$
   (e) $S = Z^+$

5. Prove part (2) of Theorem 1.1.2.

6. State and prove a theorem like Theorem 1.1.3 for greatest lower bounds.

7. State and prove a theorem like Theorem 1.1.4 for greatest lower bounds.

8. Suppose that in the hypothesis of Theorem 1.1.5, "$s = \sup S$" is replaced by "$s = \inf S$". How should the conclusion be changed?

9. Prove or disprove each of the following. (The sets cS and $S + T$ are as defined in Theorems 1.1.4 and 1.1.5.)
   (a) If S is bounded, then every subset of S is bounded.
   (b) If every subset of S is bounded, then S is bounded.
   (c) If c is a real number and S is bounded above, then cS is bounded above.
   (d) If c is a real number and S is bounded, then cS is bounded.
   (e) If c is a real number and cS is bounded, then S is bounded.
   (f) If S and T are nonempty and bounded above, then $S + T$ is bounded above.
   (g) If S and T are nonempty, and if T is not bounded above, then $S + T$ is not bounded above.

1.2 *Axiom of Completeness*

The properties of the real numbers discussed in Sections R.1 and R.2 (algebraic properties and properties of order) are more or less familiar from your study of elementary mathematics. We now look at an additional property of R that is not as likely to be familiar:

> **Axiom of Completeness:** If S is any nonempty set of real numbers that is bounded above, then S has a least upper bound in R.

Note that the conclusion of the Axiom of Completeness says that the least upper bound of S exists in R; it may or may not be an element of S.

Although the completeness axiom sounds simple, it is nevertheless a very powerful assumption and has far-reaching consequences, as we shall see.

Archimedean Law

Our first theorem has roots that can be traced to classical geometry.

Theorem 1.2.1 **Archimedean Law** *If a and b are real numbers, $a > 0$, then there exists a positive integer n such that $na > b$.*

Before giving a proof of the Archimedean Law let us consider a geometric interpretation. Imagine two line segments, one of length a (thought of as short), the other of length b (thought of as long). The Archimedean Law says that, by repeatedly marking off copies of the short segment end-to-end along the longer segment, we can eventually overreach it (see Figure 1.3).

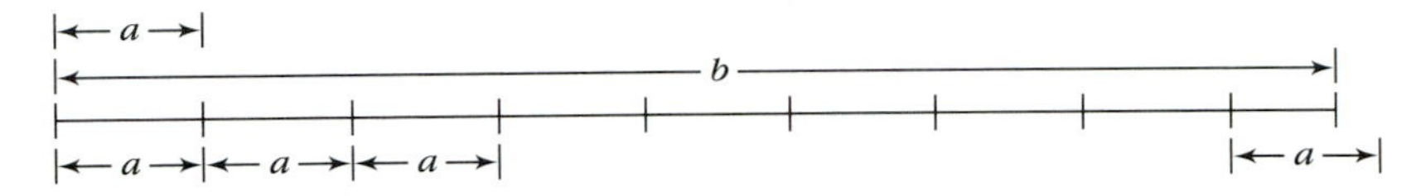

Figure 1.3
Archimedean Law

PROOF: We use the Axiom of Completeness and the method of proof by contradiction. Let $a > 0$. Suppose that there does not exist a positive integer n such that $na > b$. Then, for every $n \in Z^+$, we have $na \leq b$, so that b is an upper bound of the nonempty set $S = \{na : n \in Z^+\}$. Invoking the Axiom of Completeness, we let $l = \sup S$, and it follows that

$$na \leq l \qquad \text{for every } n \in Z^+$$

But $n + 1$ is a positive integer whenever n is, so that we also have, for $n \in Z^+$,

$$(n + 1)\,a \leq l \qquad \text{for every } n \in Z^+$$

From this it follows that

$$\begin{aligned} na + a &\leq l \qquad \text{for every } n \in Z^+ \\ na &\leq l - a \qquad \text{for every } n \in Z^+ \end{aligned}$$

and we see that $l - a$ is an upper bound of S—a contradiction since $l - a < l$. This contradiction establishes the theorem. ∎

Now we can resolve the question left unsettled in Example 2 in Section 1.1.

Corollary 1.2.1 *The set Z^+ is not bounded above in R.*

PROOF: We apply the Archimedean Law with $a = 1$. For any real number b there exists an $n \in Z^+$ such that $n \cdot 1 > b$. Thus, no real number can serve as an upper bound of Z^+, and so Z^+ is not bounded above. ∎

Just as the positive integers become "arbitrary large" in R (Corollary 1.2.1), their reciprocals become "arbitrarily small."

Corollary 1.2.2 *If ε is an arbitrary positive number, no matter how small, there exists a positive integer n such that $\frac{1}{n} < \varepsilon$.*

PROOF: The inequality $\frac{1}{n} < \varepsilon$ is equivalent to $n > \frac{1}{\varepsilon}$. The existence of an n satisfying the latter is guaranteed by Corollary 1.2.1. ■

The Square Root of 2

In Section L.2 it was shown that there does not exist a rational number whose square is 2 (see Theorem L.2.4). In other words, in the field Q of rationals there is no square root of 2. How then can we be sure that there is a square root of 2 in R? We now apply the completeness property to show that there is.

Theorem 1.2.2 *There exists a positive real number whose square is equal to 2.*

PROOF: Consider the set S of all positive real numbers whose squares are less than 2:

$$S = \{x : x \in R, x > 0 \text{ and } x^2 < 2\}.$$

S is nonempty and is bounded above (for example, $\frac{3}{2}$ is an upper bound). We can invoke the Axiom of Completeness and declare that S has a least upper bound, which we call s. We note that $s > 1$. We will see that s has the required property, namely, that $s^2 = 2$. Our plan is to show that neither of the alternatives $s^2 < 2$ or $s^2 > 2$ is possible.

Case (i): Suppose that $s^2 < 2$. We show how to find a positive number t such that $t > s$ and $t^2 < 2$, thus contradicting the fact that s is an upper bound of S. To find it, we consider numbers of the form $s + \frac{1}{n}$, where n is a positive integer. By squaring such a number, we have

$$\left(s + \frac{1}{n}\right)^2 = s^2 + 2s \cdot \frac{1}{n} + \frac{1}{n^2} \leq s^2 + 2\frac{s}{n} + \frac{1}{n} = s^2 + (2s + 1)\frac{1}{n}$$

We are assuming here that $s^2 < 2$, so the number $2 - s^2$ is positive. By Corollary 1.2.2, we can choose a particular n so large that

$$\frac{1}{n} < \frac{2 - s^2}{2s + 1}$$

Then, for such an n, we have

$$\left(s + \frac{1}{n}\right)^2 \leq s^2 + (2s + 1) \cdot \frac{1}{n} < s^2 + (2s + 1) \cdot \frac{2 - s^2}{2s + 1} = 2$$

Setting $t = s + \frac{1}{n}$, we have obtained a positive number $t > s$ such that $t^2 < 2$, establishing the desired contradiction.

Case (ii): Suppose $s^2 > 2$. We show how to find a positive number t such that $t < s$ and t is an upper bound of S, thus contradicting the fact that s is the *least* upper bound of S. To find t this time, we consider numbers of the form $s - \frac{1}{n}$, where n is a positive integer. By squaring we get

$$\left(s - \frac{1}{n}\right)^2 = s^2 - 2s \cdot \frac{1}{n} + \frac{1}{n^2} > s^2 - \frac{2s}{n}$$

Now we choose a particular n such that

$$s^2 - \frac{2s}{n} > 2$$

This choice is also possible by Corollary 1.2.2, because the desired inequality is equivalent to

$$\frac{1}{n} < \frac{s^2 - 2}{2s}$$

and $s^2 - 2$ is assumed to be positive in this case. Then, with $t = s - \frac{1}{n}$ for this choice of n, we have $t > 0$ and

$$t^2 = \left(s - \frac{1}{n}\right)^2 > s^2 - \frac{2s}{n} > 2 > x^2$$

for every $x \in S$. But from $t^2 > x^2$ it follows that $t > x$, so that t is an upper bound of S, and the desired contradiction has been obtained.

Since the alternatives $s^2 < 2$ and $s^2 > 2$ are impossible, we conclude that $s^2 = 2$. ■

Rational Density Theorem

In Section 2.6, we will see that every positive real number has a square root in R; later, in Section 3.6, we will prove that every positive real number has an nth root for every $n \in Z^+$. However, many of these roots, and many other real numbers as well, fail to be rational. The rational numbers therefore do not fill out the number line; that is, if we plotted only the rationals, there would be many "holes". Nevertheless the rationals are spread out so as to be "dense" on the line, in the sense that every interval, no matter how short, contains some rationals. We will make this argument informally at first, then give a proof.

In order to visualize the manner in which the rationals are distributed on the number line, we begin by marking off the multiples of the unit fractions: $\frac{1}{2}, \frac{1}{3}, \frac{1}{4}, \ldots$. First the multiples of $\frac{1}{2}$ (see Figure 1.4).

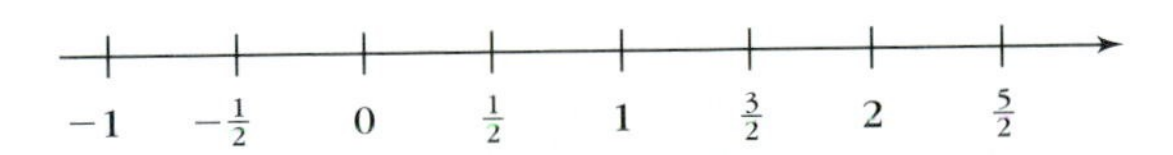

Figure 1.4
Multiples of $\frac{1}{2}$

Next, we mark off any multiples of $\frac{1}{3}$ not already marked off (see Figure 1.5).

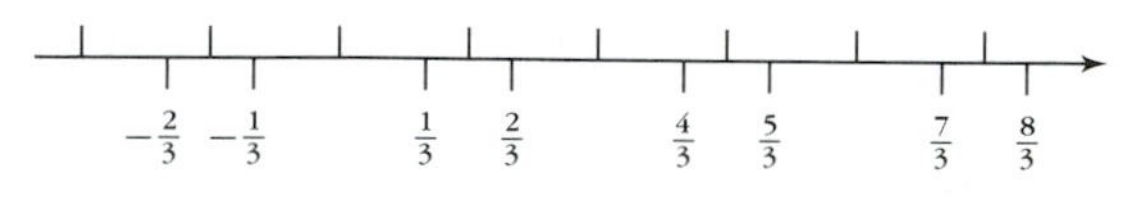

Figure 1.5
Multiples of $\frac{1}{2}$ and $\frac{1}{3}$

Then we mark any multiples of $\frac{1}{4}$ not already marked off (see Figure 1.6).

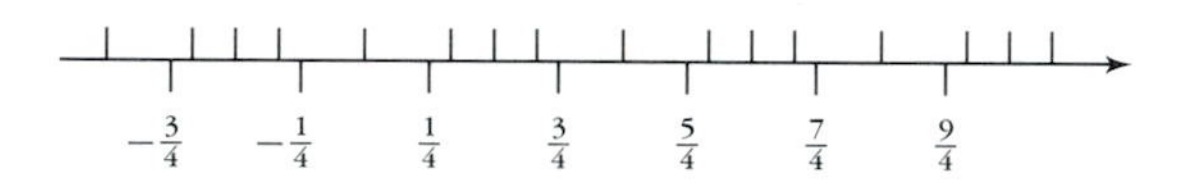

Figure 1.6
Multiples of $\frac{1}{2}$, $\frac{1}{3}$, and $\frac{1}{4}$

Then the multiples of $\frac{1}{5}$, and so on. As we continue the process, our markings divide the line into shorter and shorter segments. It is then intuitively clear that if (a, b) is any nonempty open interval, no matter how short, eventually some of these markings will fall within (a, b).

Theorem 1.2.3 **Rational Density** *If $a, b \in R$ and $a < b$, then there exists a rational number r such that $a < r < b$.*

PROOF: Let $a, b \in R$ with $a < b$. Then the length of the interval (a, b) is $b - a$. If q is a positive integer and if we divide the number line by marking off the points $0, \pm\frac{1}{q}, \pm\frac{2}{q}$, then the resulting segments are of length $\frac{1}{q}$. Certainly if we choose q large enough that $\frac{1}{q} < b - a$ (using Corollary 1.2.2 once more), then at least one of the points marked off should fall inside the interval (a, b).

Let us see how to determine such a point. First, observe that there is no harm in assuming that $a \geq 0$. Let S be the set of all positive integers n such that $\frac{n}{q} > a$. The Archimedean Law guarantees that S is nonempty. By the well-orderedness of Z^+ (see Section P.1), S has a least element, which we call p. Then

$$\frac{p}{q} > a \geq 0$$

but, by the choice of p,

$$\frac{p-1}{q} \leq a$$

from which it follows that

$$\frac{p}{q} \leq a + \frac{1}{q} < a + (b - a) = b$$

So we have obtained a rational number $r = \frac{p}{q}$ satisfying $a < r < b$. ∎

Corollary 1.2.3 **Approximation of Reals by Rationals** *Let $x \in R$. Then for any $\varepsilon > 0$, there exists an $r \in Q$ such that $|x - r| < \varepsilon$.*

You are asked to prove Corollary 1.2.3 in Exercise 1.2.3.

Like the rationals, the irrationals are also dense. You are asked to prove this in Exercise 1.2.2b.

Theorem 1.2.4 **Irrational Density** *If $a, b \in R$ and $a < b$, then there exists an irrational number x such that $a < x < b$.*

Greatest Integer Function

The following theorem relates to the definition of the greatest integer function (see Example 11 in Section R.3).

Theorem 1.2.5 *If $x \in R$, the set $S = \{n : n \in Z \text{ and } n \leq x\}$ has a greatest element.*

PROOF: It follows from Corollary 1.2.1 that S is not empty (work Exercise 1.2.6); also S is bounded above by x. By the Axiom of Completeness, sup S exists, and we set $m = \sup S$. It is sufficient to show that $m \in S$. Suppose $m \notin S$. Since $m - 1$ is not an upper bound of S, there exists an element n_1 of S such that $m - 1 < n_1 < m$. But then n_1 is not an upper bound of S either, and there exists an element n_2 of S such that $n_1 < n_2 < m$. Consequently,

$$n_1 < n_2 < m < n_1 + 1$$

and we find the integer n_2 between the consecutive integers n_1 and $n_1 + 1$, contradicting Corollary P.1.1. We conclude that $m \in S$, and so m is the greatest element of S. ■

For a given $x \in R$, the number m in the proof of Theorem 1.2.5 is clearly unique, and by its definition, we have

$$m \leq x < m + 1$$

By letting $d = x - m$, we have $0 \leq d < 1$, and so x is uniquely expressible in the form

$$x = m + d \tag{1}$$

where m is an integer and $0 \leq d < 1$. The integer m is $[x]$—the greatest integer less than or equal to x.

By adding 1 to both sides of Equation (1), we find that

$$x + 1 = (m + 1) + d \tag{2}$$

with $0 \leq d < 1$. Because of the uniqueness of the expression in Equation (1), Equation (2) tells us that $[x + 1] = m + 1$, so for all $x \in R$,

$$[x + 1] = [x] + 1 \tag{3}$$

Completeness in Terms of Infimum

The following theorem is equivalent to the Axiom of Completeness. You are asked to prove it in Exercise 1.2.5.

Theorem 1.2.6 *If S is any nonempty set of real numbers that is bounded below, then S has a greatest lower bound in R.*

★ *Complete Ordered Fields*

In Section R.2, upper and lower bounds are defined in an arbitrary ordered field. The concepts of supremum and infimum can also be introduced in that setting; for example:

Definition 1.2.1 Let F be an ordered field and let $S \subseteq F$. Then an element $l \in F$ is the ***least upper bound***, or ***supremum***, of S provided that

(i) l is an upper bound of S
(ii) l is less than or equal to every upper bound of S

It now makes sense to ask whether a given ordered field satisfies a condition like the Axiom of Completeness or like the Archimedean Law (Theorem 1.2.1).

Definition 1.2.2 An ordered field F is ***complete*** provided that every nonempty subset of F that is bounded above has a least upper bound in F.

Definition 1.2.3 An ordered field F is ***Archimedean*** provided that if $a, b \in F$ and a is positive, there exists an $n \in Z^+$ such that $na > b$ (where na is understood to mean $a + a + \cdots + a$, with n terms all equal to a).

It follows (exactly as in the proof of Theorem 1.2.1) that every complete ordered field is necessarily Archimedean. However, the converse fails; for example, the ordered field of rationals is Archimedean but not complete.

A project following Exercises 1.2 develops a non-Archimedean ordered field whose elements are familiar to anyone who knows high school algebra. Of course, this field is not complete.

When it comes to complete ordered fields, it can be shown that there is only one, in the sense that any such field is structurally identical to R. This fact, known as the Isomorphism Theorem, is explained and proved in the appendix. Because of the Isomorphism Theorem, there is little point in studying complete ordered fields in general; we may as well just study R.

EXERCISES 1.2

1. Let $S = \{x : x = \frac{1}{n} \text{ for some } n \in Z^+\}$. Determine sup S and inf S (if they exist) and prove your claims.

2. (a) Show that the sum of a rational number and an irrational number is irrational.
   (b) Use part (a) to prove Theorem 1.2.4.
   (c) Prove or disprove: The sum of any two irrational numbers is irrational.
   (d) Prove or disprove: The product of any two irrational numbers is irrational.

3. Prove Corollary 1.2.3.

4. **Approximation of Reals by Decimals** A ***decimal*** is a rational number of the form $p/10^n$, where $p \in Z$ and $n \in Z^+$.
   (a) Show that the decimals are dense; that is, that if $a, b \in R$ and $a < b$, then there exists a decimal d such that $a < d < b$.
   (b) Show that if $x \in R$ and $\varepsilon > 0$, then there exists a decimal d such that $|x - d| < \varepsilon$.

5. Prove Theorem 1.2.6.

6. Explain how it follows from Corollary 1.2.1 that the set S in the proof of Theorem 1.2.5 is not empty.

7. Prove that there exists a real number whose square is 3.

★8. Show that the ordered field Q of rational numbers is not complete. (*Hint:* The set $\{r : r \in Q, r \geq 0$ and $r^2 < 2\}$ is bounded above but does not have a least upper bound in Q.)

★9. If A and B are nonempty subsets of an ordered field, we define $A \leq B$ to mean that each element of A is less than or equal to every element of B. Show that if A and B are nonempty subsets of a complete ordered field F, with $A \leq B$, then there exists an element $c \in F$ such that $A \leq \{c\} \leq B$.

★10. **Dedekind Cut Property** Let F be an ordered field. A ***Dedekind cut*** in F is an ordered pair (A, B) of nonempty subsets of F such that $A \cup B = F$ and each element of A is less than or equal to every element of B. Show that if F is complete and (A, B) is a Dedekind cut in F, then either A has a greatest element or B has a least element.

★ PROJECT: A Non-Archimedean Ordered Field

A rational expression in x is a quotient of polynomials, such as

$$\frac{x}{x^2 + 1}, \frac{4 - x}{1 + 3x + x^2}, \frac{x^3 + 1}{x}, \text{ and so on.}$$

By using high school algebra, it is easy to show that the set $R(x)$ of all rational expressions in x forms a field under the operations of addition and multiplication. Now $R(x)$ can be ordered by means of the following definition of "positive." First, a polynomial is *positive* provided its leading coefficient is a positive real number. Then a rational expression $p(x)/q(x)$ is *positive* provided the product $p(x)q(x)$ is positive.

(1) Show that $R(x)$ is an ordered field with the positive elements described above. (See Section R.1.)

(2) The rational expressions formed by "quotients" $n/1$, where $n \in Z^+$, are the positive integers. Show that this subset of $R(x)$ is bounded above, and conclude that $R(x)$ is not Archimedean.

CHAPTER TWO

SEQUENCES AND SERIES

2.1 *Definition and Examples of Sequences*

A *sequence* is an ordered, infinite list of terms, having a first term, a second term, . . . , an nth term, and so on. The following definition formulates this idea mathematically.

Definition 2.1.1 A ***sequence*** is a function whose domain is the set Z^+ of positive integers.

Notation for sequences differs slightly from that for other functions in the following way: if s is a sequence, then the value of s at a positive integer n is denoted by s_n rather than by $s(n)$. We call s_n the *nth term* or the *general term* of the sequence s, and write

$$s = \{s_n\}_{n=1}^{\infty} \qquad \text{or, more briefly,} \qquad s = \{s_n\}$$

If we wish to display the first few terms, it is convenient to write

$$s_1, s_2, s_3, \ldots, s_n, \ldots$$

Example 1: The equation

$$s_n = \frac{1}{n} \qquad n \in Z^+$$

defines the sequence s of reciprocals of the positive integers,

$$1, \frac{1}{2}, \frac{1}{3}, \ldots, \frac{1}{n}, \ldots$$

$$s = \left\{\frac{1}{n}\right\}_{n=1}^{\infty}$$

The symbol n used in defining the terms of a sequence is called an *index* and can be replaced by another symbol; m or k is often used as an index instead of n. When a letter is used as the index of a sequence, its values are restricted to positive integers.

Example 2: The equation

$$E_m = 2m \qquad m \in Z^+$$

defines the sequence of even integers

$$2, 4, 6, \ldots, 2m, \ldots$$

$$E = \{2m\}_{m=1}^{\infty}$$

Example 3: The equation

$$t_k = \frac{k}{k+1} \qquad k \in Z^+$$

defines the sequence $\frac{1}{2}, \frac{2}{3}, \frac{3}{4}, \ldots, k/(k+1), \ldots$.

Example 4: If c is a fixed real number, the equation

$$s_n = c \qquad n \in Z^+$$

defines a *constant* sequence in which every term is c.

Example 5: For $n \in Z^+$, let S_n be the open interval $\left(-\frac{1}{n}, \frac{1}{n}\right)$. Then $\{S_n\}_{n=1}^{\infty}$ is a sequence whose terms are sets. (Some properties of this sequence appear in Exercise 2.1.10.)

Example 6: For $n \in Z^+$, let f_n be the function $f_n(x) = x^n$ $(0 \le x \le 1)$. Then $\{f_n\}_{n=1}^{\infty}$ is a sequence whose terms are real functions with domain $[0, 1]$. (See Exercise 2.1.11.)

Examples 5 and 6 show that the terms of a sequence need not be real numbers. We can have sequences of sets, functions, vectors, matrices, and so on. In this chapter, however, we focus on sequences of real numbers, so unless stated otherwise, the word *sequence* should be understood as "sequence of real numbers". Such a sequence is a special kind of real function, one whose domain is the set Z^+.

Monotone Sequences

Definition 2.1.2 Let s be a sequence. Then

(i) s is ***increasing**** provided that $s_{n+1} \geq s_n$ for $n \in Z^+$
(ii) s is ***decreasing*** provided that $s_{n+1} \leq s_n$ for $n \in Z^+$
(iii) s is ***strictly increasing*** provided that $s_{n+1} > s_n$ for $n \in Z^+$
(iv) s is ***strictly decreasing*** provided that $s_{n+1} < s_n$ for $n \in Z^+$

A sequence with one of these four properties is called ***monotone***.

Note: It can be shown by induction that $\{s_n\}_{n=1}^{\infty}$ is increasing if and only if $s_n \leq s_m$ whenever n and m are positive integers with $n < m$. Similar remarks apply to decreasing, strictly increasing, and strictly decreasing sequences.

Let us examine some of the examples above.

Example 7: $\left\{\dfrac{1}{n}\right\}_{n=1}^{\infty}$ is strictly decreasing because $1/(n + 1) < 1/n$ for $n \in Z^+$.

Example 8: $\{2m\}_{m=1}^{\infty}$ is strictly increasing because $2(m + 1) > 2m$ for $m \in Z^+$.

Example 9: $t = \{k/(k + 1)\}_{k=1}^{\infty}$ is strictly increasing. One way to verify this is to show that $t_{k+1} - t_k$ is greater than 0 for $k \in Z^+$:

$$\frac{k+1}{k+2} - \frac{k}{k+1} = \frac{1}{(k+2)(k+1)} > 0$$

A second way is to show that t_{k+1}/t_k is greater than 1 for $k \in Z^+$ (see Exercise 2.1.5).

Example 10: A constant sequence $\{c\}_{n=1}^{\infty}$ is both increasing and decreasing. It is easy to show that the only sequences with this property are the constant sequences.

*Some texts say *nondecreasing*.

Example 11: The sequence $\left\{\left[\frac{n}{2}\right]\right\}_{n=1}^{\infty}$ (where the square bracket notation indicates the greatest integer function) is increasing but not strictly increasing:

$$0, 1, 1, 2, 2, \ldots, \left[\frac{n}{2}\right], \ldots.$$

Note: Example 11 gives us a chance to emphasize how we "count" the number of terms of a sequence. The second and third terms, s_2 and s_3, both have the value 1, but they are different terms of the sequence. The first five terms are $s_1 = 0$, $s_2 = 1$, $s_3 = 1$, $s_4 = 2$, $s_5 = 2$.

Bounded Sequences

Any sequence of real numbers is a real function and so we may ask whether or not it is bounded in accordance with the definitions in Section R.3.

(i) A sequence s is bounded above if and only if there is a real number b such that $s_n \leq b$ for $n \in Z^+$.

(ii) A sequence s is bounded below if and only if there is a real number a such that $s_n \geq a$ for $n \in Z^+$.

(iii) A sequence is bounded if and only if it is bounded above and bounded below; it is unbounded if and only if it fails to be either bounded above or bounded below.

Note: It is easy to show that a sequence $\{s_n\}_{n=1}^{\infty}$ is bounded if and only if there is a positive number M such that $|s_n| \leq M$ for $n \in Z^+$.

Example 12: The sequence $\left\{\frac{1}{n}\right\}_{n=1}^{\infty}$ is bounded above since $\frac{1}{n} \leq 1$ for $n \in Z^+$. It is also bounded below since $\frac{1}{n} > 0$ for $n \in Z^+$. Thus, $\left\{\frac{1}{n}\right\}_{n=1}^{\infty}$ is bounded.

Example 13: The sequence $\{n\}_{n=1}^{\infty}$ is unbounded since it is not bounded above. (Why?)

Example 14: The sequence $\{(-1)^n n^2\}_{n=1}^{\infty}$ is neither bounded above nor bounded below.

Subsequences

Suppose we want to select certain terms of a sequence s

$$s_1, s_2, s_3, \ldots, s_n, \ldots$$

and arrange them in a list in such a way as to maintain their original order. The resulting *subsequence* will have the form

$$s_{n_1}, s_{n_2}, s_{n_3}, \ldots, s_{n_k}, \ldots$$

where n_1 is the subscript of the first selected term of s, n_2 is the subscript of the next selected term of s, and so on. Clearly the subscripts n_k are positive integers satisfying the inequalities

$$n_1 < n_2 < n_3 < \ldots < n_k < \ldots$$

and the kth term of our subsequence is defined by

$$t_k = s_{n_k} \qquad k \in Z^+$$

Definition 2.1.3 Let $s = \{s_n\}_{n=1}^{\infty}$ be a sequence and let $N = \{n_k\}_{k=1}^{\infty}$ be a strictly increasing sequence of positive integers. Then the sequence $t = \{t_k\}_{k=1}^{\infty}$, defined by $t_k = s_{n_k}$ for $k \in Z^+$, is a ***subsequence*** of s. In terms of composition of functions,

$$t = s \circ N$$

Example 15: Let $s = \left\{\frac{1}{n}\right\}_{n=1}^{\infty}$ and let $N = \{2k\}_{k=1}^{\infty}$. We then obtain a subsequence $t = \{s_{n_k}\}_{k=1}^{\infty}$ consisting of terms of s with even subscripts:

$$\frac{1}{2}, \frac{1}{4}, \frac{1}{6}, \ldots, \frac{1}{2k}, \ldots.$$

Example 16: If $s = \{s_n\}_{n=1}^{\infty}$ is a sequence and if m is a fixed positive integer, then the sequence

$$s_m, s_{m+1}, \ldots, s_{m+k-1}, \ldots$$

obtained by omitting the first $m - 1$ terms of s is a subsequence of s; here we may let $N = \{n_k\}_{k=1}^{\infty}$, where $n_k = m + k - 1$ for $k \in Z^+$. We denote this subsequence by

$$\{s_n\}_{n=m}^{\infty}$$

and call it a *tail* of the original sequence.

EXERCISES 2.1

1. Find the first five terms of each of the following sequences.

(a) $\left\{1 - \frac{1}{n}\right\}_{n=1}^{\infty}$

(b) $\{(-1)^n n\}_{n=1}^{\infty}$

(c) $\left\{\left(1 + \frac{1}{k}\right)^k\right\}_{k=1}^{\infty}$

(d) $\left\{\left[\frac{n+2}{3}\right]\right\}_{n=1}^{\infty}$ (the brackets denote the greatest integer function)

(e) $\{(m-1)(m-2)(m-3)\}_{m=1}^{\infty}$

(f) $\{s_n\}_{n=1}^{\infty}$ where $s_n = \begin{cases} n+1 & \text{for } n \text{ odd} \\ 0 & \text{for } n \text{ even} \end{cases}$

(g) $\{t_k\}_{k=1}^{\infty}$ where $t_k = |k - 3|$

2. Find an equation for the nth term of a sequence whose first few terms are as indicated. (Answers are not unique.)

(a) $6, 7, 8, \ldots$

(b) $\frac{1}{8}, -\frac{1}{16}, \frac{1}{32}, \ldots$

(c) $1, 0, -1, 0 \ldots$

(d) $1, 1, 2, 3, \ldots$

3. Determine whether the given sequence is monotone, and explain your conclusions.

(a) $\left\{1 - \frac{1}{n}\right\}_{n=1}^{\infty}$

(b) $\{(-1)^n n\}_{n=1}^{\infty}$

(c) $\{n^2 - n\}_{n=1}^{\infty}$

(d) $\left\{\dfrac{1}{1+k}\right\}_{k=1}^{\infty}$
(e) $\{(1-n)^n\}_{n=1}^{\infty}$
(f) $\left\{\left[\dfrac{n+2}{3}\right]\right\}_{n=1}^{\infty}$ (the brackets denote the greatest integer function)
(g) $\left\{\dfrac{1-n}{n^2}\right\}_{n=1}^{\infty}$
(h) $\left\{\dfrac{2^n 3^n}{5^{n+1}}\right\}_{n=1}^{\infty}$

4. Determine whether the given sequence is bounded above, bounded below, or bounded.

(a) $\{(-1)^n\}_{n=1}^{\infty}$
(b) $\{(-1)^n n\}_{n=1}^{\infty}$
(c) $\left\{\dfrac{1}{1+k}\right\}_{k=1}^{\infty}$
(d) $\{s_n\}_{n=1}^{\infty}$ where $s_n = \begin{cases} n+1 & \text{for } n \text{ odd} \\ 0 & \text{for } n \text{ even} \end{cases}$

5. Show that if $t = \{t_k\}_{k=1}^{\infty}$ is a sequence of *positive* numbers such that $t_{k+1}/t_k > 1$ for $k \in Z^+$, then t is strictly increasing. (Why is it necessary to require that t_k be positive for $k \in Z^+$?) Apply this result to $\{t_k\}_{k=1}^{\infty}$ in Example 9.

6. If $s = \{s_n\}_{n=1}^{\infty}$ and $t = \{t_n\}_{n=1}^{\infty}$ are sequences, then (as defined in Section R.3) the sum $s + t$ is the sequence whose nth term is $s_n + t_n$. That is

$$s + t = \{s_n + t_n\}_{n=1}^{\infty}$$

Similarly, the difference, product, and quotient are given by

$$s - t = \{s_n - t_n\}_{n=1}^{\infty}$$
$$st = \{s_n t_n\}_{n=1}^{\infty}$$
$$\frac{s}{t} = \left\{\frac{s_n}{t_n}\right\}_{n=1}^{\infty} \quad \text{(provided } t_n \neq 0 \text{ for } n \in Z^+\text{)}$$

Prove or disprove the following statements.

(a) If s and t are both increasing, then so is their sum.
(b) If s and t are both increasing, then so is their difference.
(c) If s and t are both increasing sequences of positive numbers, then so is their product.
(d) If s and t are both increasing sequences of positive numbers, then so is their quotient.
(e) If s and t are both bounded, then so is their difference.
(f) If s and t are both bounded sequences of positive numbers, then so is their quotient.

7. Show that if $\{s_n\}_{n=1}^{\infty}$ is increasing and if $\{\sigma_n\}_{n=1}^{\infty}$ is defined by

$$\sigma_n = \frac{s_1 + s_2 + \cdots + s_n}{n} \qquad n \in Z^+$$

then $\{\sigma_n\}_{n=1}^{\infty}$ is also increasing.

8. Given $\{s_n\}_{n=1}^{\infty}$ and $\{n_k\}_{k=1}^{\infty}$, find a formula for the kth term of the subsequence $t = \{s_{n_k}\}_{k=1}^{\infty}$, and compute t_1, t_2, and t_3.

(a) $s_n = \dfrac{1}{n}$; $n_k = 4k + 3$
(b) $s_n = (-1)^{n+1} n$; $n_k = 3k$
(c) $s_n = \dfrac{(-1)^n}{n^2}$; $n_k = 2k - 1$
(d) $s_n = \begin{cases} n^2 & \text{for } n \text{ even} \\ 0 & \text{for } n \text{ odd} \end{cases}$; $n_k = 2k$
(e) $s_n = \begin{cases} n^2 & \text{for } n \text{ even} \\ 0 & \text{for } n \text{ odd} \end{cases}$; $n_k = 2k + 1$

9. (a) Show that a subsequence of an increasing sequence is increasing.
(b) Show that a subsequence of a bounded sequence is bounded.
(c) Show that if t is a subsequence of s, and u is a subsequence of t, then u is a subsequence of s.

10. Let $\{S_n\}_{n=1}^{\infty}$ be the sequence of open intervals $S_n = \left(-\dfrac{1}{n}, \dfrac{1}{n}\right)$.

(a) Show that $\{S_n\}_{n=1}^{\infty}$ is decreasing in the sense that $S_{n+1} \subseteq S_n$ for $n \in Z^+$.
(b) Show that there is one and only one number that belongs to S_n for every $n \in Z^+$.

11. Let $\{f_n\}_{n=1}^{\infty}$ be the sequence of functions $f_n(x) = x^n$ $(0 \le x \le 1)$.

(a) Show that the graphs of these functions have two points in common.
(b) Show that $\{f_n\}_{n=1}^{\infty}$ is decreasing in the sense that $f_{n+1}(x) \le f_n(x)$ for every $n \in Z^+$ and every $x \in [0, 1]$.

2.2 *Null Sequences*

You will recall from calculus that the limit concept is fundamental. In this and the following section we develop the concept of limit for sequences. In Chapter 3 we extend this idea to other real functions.

We begin here with a special kind of sequence in which the limit is zero. Such a sequence will be called "null". A null sequence has the property that its terms eventually become and remain arbitrarily small in absolute value. Before making a formal definition let's consider an example.

Example 1: The sequence $1, \frac{1}{4}, \frac{1}{9}, \ldots, \frac{1}{n^2}, \ldots$ defined by

$$s = \left\{\frac{1}{n^2}\right\}_{n=1}^{\infty}$$

is null. Everyone would agree that the terms eventually become "small", and this is true no matter what standard of "smallness" is applied. John may say "small" means "less than 1/100" while Maria requires "less than 1/10,000". Then John will be satisfied that the terms of the sequence beginning with the eleventh term ($n \geq 11$) are small because

$$|s_n| = \frac{1}{n^2} \leq \frac{1}{11^2} < \frac{1}{100} \qquad \text{whenever } n \geq 11$$

And Maria will accept as small the terms beginning with the 101st because

$$|s_n| = \frac{1}{n^2} \leq \frac{1}{101^2} < \frac{1}{10{,}000} \qquad \text{whenever } n \geq 101.$$

Now, to get at the idea of "arbitrarily" small, let ε denote an arbitrary positive number. (Think of ε as being as small as desired.) Someone who considers "small" to mean "less than ε" will agree that s_n is small whenever

$$\frac{1}{n^2} < \varepsilon$$

The terms of the sequence are this small whenever $n > \sqrt{1/\varepsilon}$, and if N is a fixed positive integer greater than $\sqrt{1/\varepsilon}$,* then

$$|s_n| = \frac{1}{n^2} \leq \frac{1}{N^2} < \varepsilon \qquad \text{whenever } n \geq N$$

Now it is in this sense that we say that the terms of $\{s_n\}_{n=1}^{\infty}$ eventually become and remain arbitrarily small: If ε is any positive number, no matter how small, there exists a positive integer N, depending on ε, such that $|s_n| < \varepsilon$ for $n = N$ and for all larger values of n.†

*The fact that such an N exists is guaranteed by the Archimedean Law.

†Keep in mind that the variable n is understood to denote a positive integer.

Definition 2.2.1 A sequence $\{s_n\}_{n=1}^{\infty}$ is ***null*** provided that for every $\varepsilon > 0$ there is a positive integer N such that

$$|s_n| < \varepsilon \qquad \text{whenever } n \geq N$$

To prove that a sequence $\{s_n\}_{n=1}^{\infty}$ is null, we take ε to be an arbitrary positive number and show how to determine a positive integer N such that $|s_n| < \varepsilon$ whenever $n \geq N$.

Example 2: Let $s_n = (-1)^n/\sqrt{n}$. We show that $\{s_n\}_{n=1}^{\infty}$ is null. Let $\varepsilon > 0$ (i.e., let ε be an arbitrary positive number). We wish to find $N \in Z^+$ such that

$$\left|\frac{(-1)^n}{\sqrt{n}}\right| < \varepsilon \qquad \text{whenever } n \geq N$$

But $|(-1)^n/\sqrt{n}| = 1/\sqrt{n}$, so it is sufficient to have $\sqrt{n} > 1/\varepsilon$ or $n > 1/\varepsilon^2$. If we choose for N a positive integer greater than $1/\varepsilon^2$, and if $n \geq N$, then $n > 1/\varepsilon^2$ also, so that $|(-1)^n/\sqrt{n}| < \varepsilon$.

Example 3: Let $s_n = (3n^2 - 2)/(2n^3 + 4n)$. We show that $\{s_n\}_{n=1}^{\infty}$ is null. Notice first that

$$|s_n| = \frac{3n^2 - 2}{2n^3 + 4n} \leq \frac{3n^2}{2n^3 + 4n} \leq \frac{3n^2}{2n^3} = \frac{3}{2n}$$

Thus, for $\varepsilon > 0$, the inequality $|s_n| < \varepsilon$ holds whenever $\frac{3}{2n} < \varepsilon$. If we choose for N a positive integer greater than $\frac{3}{2\varepsilon}$, then $|s_n| < \varepsilon$ whenever $n \geq N$.

Example 4: Let $t_n = (n^2 + 1)/(2n^3 - n)$. We show that $\{t_n\}_{n=1}^{\infty}$ is null. First notice that, for $n \in Z^+$,

$$n^2 + 1 \leq n^2 + n^2 = 2n^2$$

and

$$2n^3 - n \geq 2n^3 - n^3 = n^3$$

so that

$$|t_n| = \frac{n^2 + 1}{2n^3 - n} \leq \frac{2n^2}{n^3} = \frac{2}{n}$$

Given $\varepsilon > 0$, we may let N be a positive integer greater than $2/\varepsilon$; then $|t_n| < \varepsilon$ whenever $n \geq N$.

Example 5: The sequence $\{(n^2 + 1)/(n^2 + 4n)]_{n=1}^{\infty}$ is not null. For all $n \in Z^+$, we have

$$s_n = \frac{n^2 + 1}{n^2 + 4n} \geq \frac{n^2}{n^2 + 4n} = \frac{1}{1 + (4/n)} \geq \frac{1}{5}$$

Consequently, for $\varepsilon = \frac{1}{5}$, there does not exist an $N \in Z^+$ with the property required in the definition of "null".

The theorems below establish several properties of null sequences that should come as no surprise to persons who have studied calculus: sums, differences, and products of null sequences are null, and any sequence that is "squeezed" between two null sequences must be null itself.

Theorem 2.2.1 *If $\{s_n\}_{n=1}^{\infty}$ and $\{t_n\}_{n=1}^{\infty}$ are null, then $\{s_n + t_n\}_{n=1}^{\infty}$ is also null.*

Remark: Before we give a formal proof, let us appeal to common sense. We are *given* that s_n and t_n can be made as small as we want, and required to make $s_n + t_n$ small, say, less than some positive number ε. The sum will surely be less than ε if each term is kept less than $\varepsilon/2$. It remains only to express these ideas in the form of a proof.

PROOF: Let $\varepsilon > 0$ be given. Since $\{s_n\}_{n=1}^{\infty}$ is null, there exists an $N_1 \in Z^+$ such that $|s_n| < \varepsilon/2$ whenever $n \geq N_1$. Since $\{t_n\}_{n=1}^{\infty}$ is null, there is an $N_2 \in Z^+$ such that $|t_n| < \varepsilon/2$ whenever $n \geq N_2$. Now let N be the maximum of the two positive integers N_1 and N_2. If $n \geq N$, then the inequalities $n \geq N_1$ and $n \geq N_2$ both hold, so that $|s_n| < \varepsilon/2$ and $|t_n| < \varepsilon/2$.

Finally, by the Triangle Inequality,*

$$|s_n + t_n| \leq |s_n| + |t_n| < \frac{\varepsilon}{2} + \frac{\varepsilon}{2} = \varepsilon$$

for $n \geq N$. ■

Theorem 2.2.2 *If $\{s_n\}_{n=1}^{\infty}$ is bounded and $\{t_n\}_{n=1}^{\infty}$ is null, then $\{s_n t_n\}_{n=1}^{\infty}$ is null.*

PROOF: Let $B > 0$ be a real number such that $|s_n| \leq B$ for all $n \in Z^+$. Let $\varepsilon > 0$; corresponding to the positive number ε/B there is an $N \in Z^+$ such that $|t_n| < \varepsilon/B$ whenever $n \geq N$. Then, for $n \geq N$, we have

$$|s_n t_n| = |s_n||t_n| < B\left(\frac{\varepsilon}{B}\right) = \varepsilon \qquad ■$$

Example 6: Since the sine function is bounded, the sequence $\{\sin n\}_{n=1}^{\infty}$ is bounded. Example 1 shows that $\{1/n^2\}_{n=1}^{\infty}$ is null, and so $\{(\sin n)/n^2\}_{n=1}^{\infty}$ is null.

The following corollaries are consequences of Theorems 2.2.1 and 2.2.2.

Corollary 2.2.1 *If $\{t_n\}_{n=1}^{\infty}$ is null and $\{c\}_{n=1}^{\infty}$ is a constant sequence, then $\{ct_n\}_{n=1}^{\infty}$ is null.*

*See part (viii) of Theorem R.2.1.

Corollary 2.2.2 *If $\{s_n\}_{n=1}^{\infty}$ and $\{t_n\}_{n=1}^{\infty}$ are null, then $\{s_n - t_n\}_{n=1}^{\infty}$ is null.*

Theorem 2.2.3 *Every null sequence is bounded.*

PROOF: Let $\{s_n\}_{n=1}^{\infty}$ be null. Corresponding to $\varepsilon = 1 > 0$, there exists a positive integer N such that $|s_n| < 1$ whenever $n \geq N$. Then let

$$B = \max\{|s_1|, |s_2|, \ldots, |s_{N-1}|, 1\}^*$$

Clearly, $|s_n| \leq B$ for $n \in Z^+$. ■

Corollary 2.2.3 *If $\{s_n\}_{n=1}^{\infty}$ and $\{t_n\}_{n=1}^{\infty}$ are null, then $\{s_nt_n\}_{n=1}^{\infty}$ is null.*

Theorem 2.2.4 **Squeeze Theorem** *If $\{r_n\}_{n=1}^{\infty}$ and $\{t_n\}_{n=1}^{\infty}$ are null, and if $r_n \leq s_n \leq t_n$ for $n \in Z^+$, then $\{s_n\}_{n=1}^{\infty}$ is null.*

PROOF: Let $\varepsilon > 0$. Since $\{t_n\}_{n=1}^{\infty}$ is null, there is an $N_1 \in Z^+$ such that $|t_n| < \varepsilon$ for $n \geq N_1$. Then, for $n \geq N_1$,

$$s_n \leq t_n \leq |t_n| < \varepsilon$$

Similarly, since $\{r_n\}_{n=1}^{\infty}$ is null, there is an N_2 in Z^+ such that $|r_n| < \varepsilon$ *for* $n \geq N_2$. Then, for $n \geq N_2$,

$$-s_n \leq -r_n \leq |r_n| < \varepsilon$$

For $n \geq N = \max\{N_1, N_2\}$, we have $-s_n < \varepsilon$ and $s_n < \varepsilon$; hence, $|s_n| < \varepsilon$. ■

The proof of the following useful lemma is left for you to work out in Exercise 2.2.7a.

Lemma 2.2.1 *A sequence $\{s_n\}_{n=1}^{\infty}$ is null if and only if $\{|s_n|\}_{n=1}^{\infty}$ is null.*

Theorem 2.2.5 **Geometric Sequence** *If $|r| < 1$, then the sequence $\{r^n\}_{n=1}^{\infty}$ is null.*

PROOF: If $0 < r < 1$, we proceed as follows: set $s = 1/r$; then $s > 1$, and so $s = 1 + p$ for some positive number p. It follows that $s^n \geq 1 + np$ for $n \in Z^+$ (work Exercise 2.2.6), and so

$$0 < r^n \leq \frac{1}{1 + np} < \frac{1}{np}$$

Now $\{r^n\}_{n=1}^{\infty}$ is null because it is squeezed between the null sequences $\{0\}_{n=1}^{\infty}$ and $\{1/(np)\}_{n=1}^{\infty}$.

*It can be shown by induction that any nonempty finite set of real numbers has a greatest element (denoted max) and a least element (denoted min).

If $-1 < r < 0$, then by what we have already shown, the sequence $\{|r|^n\}_{n=1}^{\infty} = \{|r^n|\}_{n=1}^{\infty}$ is null, and so by Lemma 2.2.1, $\{r^n\}_{n=1}^{\infty}$ is also null. Obviously $\{r^n\}_{n=1}^{\infty}$ is null when $r = 0$, and so the conclusion of the theorem holds for all r with $|r| < 1$. ∎

Theorem 2.2.6 *If $|r| < 1$, then the sequence $\{nr^n\}_{n=1}^{\infty}$ is null.*

PROOF: You are asked to prove this theorem in Exercise 2.2.8a. ∎

EXERCISES 2.2

1. Use Definition 2.2.1 to show that the sequence $\{s_n\}_{n=1}^{\infty}$ is null, given the value of s_n.
   (a) $\dfrac{1}{\sqrt{n}}$
   (b) $\dfrac{3n-2}{n^2}$
   (c) $\dfrac{n^2+2n+1}{4n^3-2n^2}$
   (d) $\dfrac{\sqrt{n}}{n+1}$
   (e) $\dfrac{\cos n\pi}{n}$
   (f) 0

2. Show that the sequence $\{s_n\}_{n=1}^{\infty}$ is not null, given the value of s_n.
   (a) 1
   (b) $\sqrt{n}$
   (c) $\dfrac{n}{2n+1}$
   (d) $(-1)^n$
   (e) $s_n = \begin{cases} 0 & \text{for } n \text{ even} \\ (-1)^k & \text{for } n = 2k-1,\ k \in Z^+ \end{cases}$

3. Prove Corollary 2.2.3.

4. Show that if p is a fixed positive integer and c is a fixed real number, then $\{c/n^p\}_{n=1}^{\infty}$ is null.

5. Show that the only constant sequence that is null is the sequence $\{0\}_{n=1}^{\infty}$.

6. Prove Bernoulli's inequality, that if $s = 1 + p$ for some $p > 0$, then $s^n \geq 1 + np$ for $n \in Z^+$.

7. Show that each of the following statements is equivalent to "$\{s_n\}_{n=1}^{\infty}$ is null".
   (a) $\{|s_n|\}_{n=1}^{\infty}$ is null
   (b) $\{-s_n\}_{n=1}^{\infty}$ is null
   (c) $\{s_n^2\}_{n=1}^{\infty}$ is null

8. (a) Prove Theorem 2.2.6. [*Hint:* Adapt the proof of Theorem 2.2.5 by showing that if
   $$s = 1 + p \qquad p > 0$$
   then, for $n > 1$,
   $$s^n \geq 1 + np + \frac{n(n-1)}{2}p^2.\Big]$$
   (b) Prove that if $|r| < 1$, then $\{n^2 r^n\}_{n=1}^{\infty}$ is null.

9. Prove or disprove: If $\{s_n t_n\}_{n=1}^{\infty}$ is null, then either $\{s_n\}_{n=1}^{\infty}$ or $\{t_n\}_{n=1}^{\infty}$ is null.

10. Show by means of an example that an infinite set of real numbers does not necessarily have a maximum or a minimum.

2.3 *Convergent Sequences*

A sequence whose terms eventually become and remain arbitrarily close to a fixed real number L is said to "converge to L" or to have L as a "limit". This idea can be expressed in terms of null sequences.

Definition 2.3.1 A sequence $\{s_n\}_{n=1}^{\infty}$ ***converges*** to a real number L provided that the sequence $\{s_n - L\}_{n=1}^{\infty}$ is null. We then write

$$s_n \to L$$

where the arrow is read "converges to". The number L is called the ***limit*** of $\{s_n\}_{n=1}^{\infty}$ and is denoted

$$L = \lim_{n\to\infty} s_n$$

A sequence is ***convergent*** provided there is a real number L to which it converges; otherwise it ***diverges***.

Example 1: The constant sequence $\{c\}_{n=1}^{\infty}$ converges to c because $\{c - c\}_{n=1}^{\infty} = \{0\}_{n=1}^{\infty}$ is a null sequence.

Example 2: Let $s_n = n/(n + 1)$. Clearly the terms of $\{s_n\}_{n=1}^{\infty}$ become close to the number 1. To prove that $\{s_n\}_{n=1}^{\infty}$ converges to 1, we must show $\{s_n - 1\}_{n=1}^{\infty}$ is null. Observe that

$$\left|s_n - 1\right| = \left|\frac{n}{n+1} - 1\right| = \frac{1}{n+1}$$

Let $\varepsilon > 0$ and choose N to be a positive integer greater than $\frac{1}{\varepsilon} - 1$. A little calculation shows that if $n \geq N$, then $1/(n + 1) < \varepsilon$. Therefore, $|s_n - 1| = 1/(n + 1) < \varepsilon$ for $n \geq N$ and so $\{s_n - 1\}_{n=1}^{\infty}$ is null and

$$\lim_{n\to\infty} s_n = 1$$

It follows immediately from Definition 2.3.1 that a sequence converges to zero if and only if it is null. The results of the examples of Section 2.2 can therefore be stated as

$$\lim_{n\to\infty} \frac{1}{n^2} = 0, \qquad \lim_{n\to\infty} \frac{(-1)^n}{\sqrt{n}} = 0$$

and so on.

The ε-N Formulation

The definition of convergence can be formulated without reference to null sequences:

> $s_n \to L$ if and only if corresponding to an arbitrary $\varepsilon > 0$ there is a positive integer N such that $|s_n - L| < \varepsilon$ whenever $n \geq N$.

Example 3: We show with the ε-N formulation that

$$\lim_{n\to\infty} \frac{3n^2 + n + 4}{6n^2} = \frac{1}{2}$$

Let $\varepsilon > 0$; we must find $N \in Z^+$ such that

$$\left|\frac{3n^2 + n + 4}{6n^2} - \frac{1}{2}\right| < \varepsilon \qquad \text{for } n \geq N$$

But elementary algebra shows that

$$\left|\frac{3n^2 + n + 4}{6n^2} - \frac{1}{2}\right| = \frac{n + 4}{6n^2} \leq \frac{n + 4n}{6n^2} = \frac{5}{6n}$$

Now we see that if we take N to be a positive integer greater than $5/(6\varepsilon)$, then, for $n \geq N$, we have

$$\left|\frac{3n^2 + n + 4}{6n^2} - \frac{1}{2}\right| < \varepsilon$$

Uniqueness

A sequence may not converge at all, but if it does, there is only one real number to which it can converge.

Theorem 2.3.1 *The limit of a convergent sequence is unique.*

PROOF: Suppose $\{s_n\}_{n=1}^{\infty}$ is a sequence such that $s_n \to L_1$ and $s_n \to L_2$. Then $\{s_n - L_1\}_{n=1}^{\infty}$ and $\{s_n - L_2\}_{n=1}^{\infty}$ are both null, and so is their difference:

$$\{(s_n - L_1) - (s_n - L_2)\}_{n=1}^{\infty} = \{L_2 - L_1\}_{n=1}^{\infty}$$

Now it is easy to see that the only constant sequence that is null is $\{0\}_{n=1}^{\infty}$. Therefore, $L_2 - L_1 = 0$ and $L_2 = L_1$. ∎

Because of Theorem 2.3.1, we are justified in speaking of *the* limit of a convergent sequence.

Subsequences of a Convergent Sequence

Theorem 2.3.2 *If a sequence converges to the number L, then the same is true of every subsequence.*

PROOF: Let $s = \{s_n\}_{n=1}^{\infty}$ be convergent and let t be a subsequence of s. Then $t_k = s_{n_k}$, where $\{n_k\}_{k=1}^{\infty}$ is a strictly increasing sequence of positive integers:

$$n_1 < n_2 < \cdots < n_k < \cdots$$

Suppose $\lim\limits_{n\to\infty} s_n = L$. We show that $\lim\limits_{k\to\infty} t_k = L$. Let $\varepsilon > 0$. There exists an $N \in Z^+$ such that $|s_n - L| < \varepsilon$ for $n \geq N$. Now if $k \geq N$, then $n_k \geq k \geq N$, and so $|s_{n_k} - L| < \varepsilon$ for $k \geq N$. That is, $|t_k - L| < \varepsilon$ for $k \geq N$. ∎

As an application of Theorem 2.3.2, we can say that if $\{s_n\}_{n=1}^{\infty}$ is convergent, then the subsequences $\{s_{2k-1}\}_{k=1}^{\infty}$ and $\{s_{2k}\}_{k=1}^{\infty}$ of odd- and even-subscripted terms must both converge to the limit of $\{s_n\}_{n=1}^{\infty}$. Consequently, if these subsequences converge to different limits (or if either of them diverges), then the sequence $\{s_n\}_{n=1}^{\infty}$ diverges.

Example 4: The sequence $\{(-1)^n\}_{n=1}^{\infty}$ is divergent because the subsequences $\{(-1)^{2k-1}\}_{k=1}^{\infty} = \{-1\}_{k=1}^{\infty}$ and $\{(-1)^{2k}\}_{k=1}^{\infty} = \{1\}_{k=1}^{\infty}$ converge to different limits.

Boundedness of a Convergent Sequence

Theorem 2.3.3 *Every convergent sequence is bounded.*

PROOF: Let $\{s_n\}_{n=1}^{\infty}$ be convergent and let $L = \lim_{n\to\infty} s_n$. Since $\{s_n - L\}_{n=1}^{\infty}$ is null, it is bounded (Theorem 2.2.3), and there is a positive number B such that

$$|s_n - L| \leq B \qquad \text{for } n \in Z^+$$

But

$$s_n = (s_n - L) + L$$

and by the Triangle Inequality, we have

$$|s_n| \leq |s_n - L| + |L| \leq B + |L| \qquad \text{for } n \in Z^+$$

Accordingly, $\{s_n\}_{n=1}^{\infty}$ is bounded. ∎

Example 5: $\{n\}_{n=1}^{\infty}$ is divergent because it is not bounded.

The converse of Theorem 2.3.3 fails. The sequence $\{(-1)^n\}_{n=1}^{\infty}$ is bounded, but we have seen that it diverges.

Significance of a Positive Limit

Theorem 2.3.4 *If a sequence $\{s_n\}_{n=1}^{\infty}$ converges to a positive (negative) limit, then its terms eventually become and remain positive (negative).*

PROOF: We prove the formulation of the theorem for positive. Given that $s_n \to L > 0$, we must show that there is an $N \in Z^+$ such that $s_n > 0$ for $n \geq N$. Corresponding to $\varepsilon = L$ there is an N such that $|s_n - L| < L$ whenever $n \geq N$. For $n \geq N$, we therefore have

$$-L < s_n - L < L$$

or

$$0 < s_n < 2L$$

The terms s_n are therefore positive for $n \geq N$. ∎

Corollary 2.3.1 *A sequence of nonpositive (nonnegative) numbers cannot converge to a positive (negative) limit.*

Absolute Value of a Convergent Sequence

Theorem 2.3.5 *If $\{s_n\}_{n=1}^{\infty}$ converges to L, then $\{|s_n|\}_{n=1}^{\infty}$ converges to $|L|$.*

PROOF: Let $\varepsilon > 0$. Then there is an $N \in Z^+$ such that $|s_n - L| < \varepsilon$ whenever $n \geq N$. By Theorem R.2.1, part (ix),

$$\big| |s_n| - |L| \big| \leq |s_n - L|$$

so $\big| |s_n| - |L| \big| < \varepsilon$ for $n \geq N$. ■

Square Root of a Convergent Sequence

Theorem 2.3.6 *If $s_n \geq 0$ for $n \in Z^+$ and if $s_n \to L$, then $\sqrt{s_n} \to \sqrt{L}$.*

PROOF: L is the limit of a nonnegative sequence, so $L \geq 0$ and $\sqrt{L}$ exists. Let $\varepsilon > 0$. In case $L = 0$, there is an $N \in Z^+$ such that $s_n < \varepsilon^2$ for $n \geq N$, and so $\sqrt{s_n} < \varepsilon$ for $n \geq N$. In case $L > 0$, we have

$$|\sqrt{s_n} - \sqrt{L}| = \frac{|s_n - L|}{\sqrt{s_n} + \sqrt{L}} < \frac{|s_n - L|}{\sqrt{L}}$$

If we choose $N \in Z^+$ such that $|s_n - L| < \varepsilon\sqrt{L}$ for $n \geq N$, then

$$|\sqrt{s_n} - \sqrt{L}| < \varepsilon \qquad \text{for } n \geq N \qquad ■$$

A Special Limit

We consider next another limit involving nth roots. Like Theorems 2.2.5 and 2.2.6, the derivation employs the Binomial Theorem in an interesting way.

Theorem 2.3.7 $\sqrt[n]{n} \to 1$

PROOF: Since $n \geq 1$, we must have $\sqrt[n]{n} \geq 1$ for $n \in Z^+$. Thus, for each n, there exists a real number $p_n \geq 0$ such that

$$\sqrt[n]{n} = 1 + p_n$$

By raising both sides to the nth power and using the Binomial Theorem, we obtain

$$n = (1 + p_n)^n = 1 + np_n + \frac{n(n-1)}{2}p_n^2 + \cdots + p_n^n$$

Since the terms on the right are nonnegative, we see that for $n > 1$,

$$n \geq \frac{n(n-1)}{2} p_n^2$$

It follows that

$$0 \leq p_n \leq \sqrt{\frac{2}{n-1}} \qquad \text{for } n > 1$$

By the Squeeze Theorem, $\{p_n\}_{n=1}^{\infty} = \{\sqrt[n]{n} - 1\}_{n=1}^{\infty}$ is null, and so $\sqrt[n]{n} \to 1$. ■

The next theorem generalizes the Squeeze Theorem for null sequences (Theorem 2.2.4) to convergent sequences.

Theorem 2.3.8 **Squeeze Theorem** *If the sequences $\{r_n\}_{n=1}^{\infty}$ and $\{t_n\}_{n=1}^{\infty}$ both converge to L and if $r_n \leq s_n \leq t_n$ for $n \in Z^+$, then $\{s_n\}_{n=1}^{\infty}$ converges to L.*

PROOF: You are asked to prove this theorem in Exercise 2.3.6. ■

Neighborhood Formulation of Convergence

Suppose that $s_n \to L$. Then, for each $\varepsilon > 0$, there is an $N \in Z^+$ such that $|s_n - L| < \varepsilon$ for $n \geq N$. All terms in the tail $s_N, s_{N+1}, s_{N+2}, \ldots$ must then lie in the ε-neighborhood of L, $N_\varepsilon(L)$.* Therefore, only a finite number of terms can lie outside $N_\varepsilon(L)$. Suppose, conversely, that for each $\varepsilon > 0$, all but a finite number of terms of $\{s_n\}_{n=1}^{\infty}$ lie inside $N_\varepsilon(L)$. Then there must be an $N \in Z^+$ such that $N_\varepsilon(L)$ contains all terms in the tail beginning with s_N. So for $n \geq N$, we must have $|s_n - L| < \varepsilon$.

Theorem 2.3.9 *A sequence $\{s_n\}_{n=1}^{\infty}$ converges to L if and only if every neighborhood of L contains s_n for all except a finite number of values of n.*

Theorem 2.3.9 means that if a sequence converges to L, then the terms "crowd around" L on the number line in such a way that if we should remove a small neighborhood of L, then only a finite number of terms would remain, and this is true no matter how small a neighborhood we remove (see Figure 2.1).

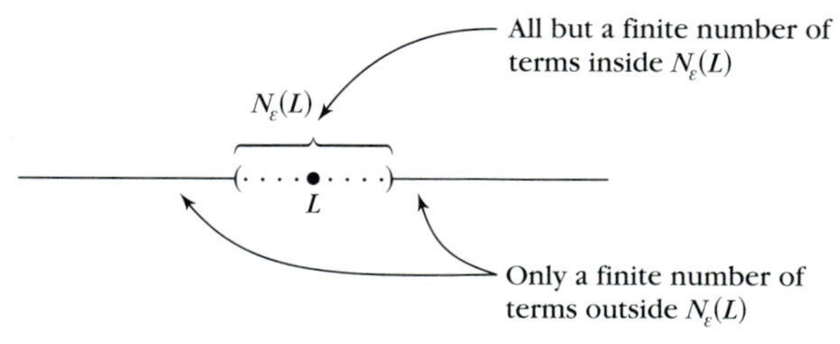

Figure 2.1

*The ε-neighborhood of L is the open interval $(L - \varepsilon, L + \varepsilon)$. (See the discussion of intervals and neighborhoods in Section R.2.)

Example 6: The sequence $\{n/(n+1)\}_{n=1}^{\infty}$ converges to 1. Consider $N_\varepsilon(1)$ when $\varepsilon = \frac{1}{10}$. Since

$$\left|\frac{n}{n+1} - 1\right| = \frac{1}{n+1}$$

we have

$$\frac{n}{n+1} \in N_{1/10}(1)$$

whenever $1/(n+1) < 1/10$; that is, whenever $n > 9$. The terms of the tail $s_{10}, s_{11}, s_{12}, \ldots$ all lie inside $N_{1/10}(1)$, and only the terms $s_1, s_2, \ldots, s_9$ lie outside this neighborhood.

Example 7: Here is an alternative proof of the uniqueness of the sequential limit using Theorem 2.3.9: Suppose L_1 and L_2 ($L_1 \neq L_2$) both are limits of a sequence $\{s_n\}_{n=1}^{\infty}$. We can then choose neighborhoods of L_1 and L_2 that are disjoint—for if $\varepsilon = |L_2 - L_1|/2$ then, $N_\varepsilon(L_1) \cap N_\varepsilon(L_2) = \varnothing$. It is clearly impossible for both of these neighborhoods to contain all but a finite number of terms of $\{s_n\}_{n=1}^{\infty}$.

The following corollary is an immediate consequence of Theorem 2.3.9.

Corollary 2.3.2 *$\{s_n\}_{n=1}^{\infty}$ converges to L if and only if, for every $m \in Z^+$, the tail $\{s_n\}_{n=m}^{\infty}$ converges to L.*

EXERCISES 2.3

1. Show that the sequence whose general term is given below converges to the indicated number L.

   (a) $\dfrac{n+3}{n+4}$; $L = 1$

   (b) $\dfrac{n}{n^2+1}$; $L = 0$

   (c) $\dfrac{4n^2-3}{3n^2+4}$; $L = \dfrac{4}{3}$

   (d) $\left(1 + \dfrac{1}{n}\right)^3$; $L = 1$

   (e) $(-1)^n \dfrac{\pi}{n}$; $L = 0$

   (f) $s_n = \begin{cases} 1 & \text{for } n \text{ odd} \\ \dfrac{n+1}{n} & \text{for } n \text{ even}; \end{cases}$ $L = 1$

2. Show that the sequence whose general term is given below is divergent.

   (a) $(-1)^n \dfrac{n}{n+1}$

   (b) $3n + 5$

   (c) $\cos \dfrac{n\pi}{2}$

   (d) $n - \dfrac{1}{n}$

   (e) $s_n = \begin{cases} \dfrac{n+3}{n+4} & \text{for } n \text{ odd} \\ \dfrac{n+3}{2n+1} & \text{for } n \text{ even} \end{cases}$

3. Show that if $\{s_n\}_{n=1}^{\infty}$ converges to a negative limit, then its terms eventually become and remain negative.

4. Show that if a sequence $\{s_n\}_{n=1}^{\infty}$ has the property that $s_{2k} \to L$ and $s_{2k-1} \to L$, then $s_n \to L$.

5. Show by an example that the converse of Theorem 2.3.5 is false.

6. Prove Theorem 2.3.8.

7. Prove Theorem 2.3.4 using the neighborhood formulation of convergence.
8. Criticize the following argument: Every neighborhood of 1 contains all terms of the sequence $\{(-1)^n\}_{n=1}^{\infty}$ except perhaps the term -1. Therefore, every neighborhood of 1 contains all except a finite number of terms of the sequence, so, by Theorem 2.3.9, the limit is 1.

2.4 *The Limit Theorems*

The objective of this section is to prove some theorems that greatly facilitate the evaluation of limits.

Theorem 2.4.1 *Let $\{s_n\}_{n=1}^{\infty}$ and $\{t_n\}_{n=1}^{\infty}$ be convergent sequences with limits L and M, respectively. Then $\{s_n + t_n\}_{n=1}^{\infty}$, $\{s_n - t\}_{n=1}^{\infty}$, and $\{s_nt_n\}_{n=1}^{\infty}$ are also convergent, and we have*

(i) $\lim_{n\to\infty}(s_n + t_n) = L + M = \lim_{n\to\infty} s_n + \lim_{n\to\infty} t_n$

(ii) $\lim_{n\to\infty} (s_n - t_n) = L - M = \lim_{n\to\infty} s_n - \lim_{n\to\infty} t_n$

(iii) $\lim_{n\to\infty} (s_n t_n) = LM = \lim_{n\to\infty} s_n \cdot \lim_{n\to\infty} t_n$

PROOF:

(i) From $\{(s_n + t_n) - (L + M)\}_{n=1}^{\infty} = \{(s_n - L) + (t_n - M)\}_{n=1}^{\infty}$ we see that $\{(s_n + t_n) - (L + M)\}_{n=1}^{\infty}$ is a sum of null sequences and is null itself by Theorem 2.2.1.

(ii) Since $\{(s_n - t_n) - (L - M)\}_{n=1}^{\infty}$ is a difference of null sequences, the result follows from Corollary 2.2.2.

(iii) We can write $\{s_nt_n - LM\}_{n=1}^{\infty} = \{(s_n - L)t_n + L(t_n - M)\}_{n=1}^{\infty}$. It follows by Theorems 2.2.1, 2.2.2, and 2.3.3 that $\{s_nt_n - LM\}_{n=1}^{\infty}$ is null. ■

According to Theorem 2.4.1, the limit of a sum of convergent sequences is the sum of their limits, and similar statements hold for the difference and product. With quotients we must avoid division by zero; we begin our consideration of quotients by examining the behavior of reciprocals.

Lemma 2.4.1 *If $t_n \neq 0$ for $n \in Z^+$ and $t_n \to M \neq 0$, then the sequence $\{1/t_n\}_{n=1}^{\infty}$ is bounded.*

PROOF: Since $t_n \to M$, we know by Theorem 2.3.5 that $|t_n| \to |M|$. Then, corresponding to the positive number $\varepsilon = |M|/2$, there is an $N \in Z^+$ such that

$$|M|/2 < |t_n| < 3\,|M|/2$$

for $n \geq N$. It follows that $|1/t_n| < 2/|M|$ for $n \geq N$ and so

$$\max\left\{\frac{1}{|t_1|}, \frac{1}{|t_2|}, \ldots, \frac{1}{|t_{N-1}|}, \frac{2}{|M|}\right\}$$

is an upper bound of $\{|1/t_n|\}_{n=1}^{\infty}$ for $n \in Z^+$. ■

Theorem 2.4.2 *Let $\{s_n\}_{n=1}^{\infty}$ and $\{t_n\}_{n=1}^{\infty}$ be convergent sequences with limits L and M, respectively. Assume further that $t_n \neq 0$ for $n \in Z^+$ and that $M \neq 0$. Then $\{s_n/t_n\}_{n=1}^{\infty}$ is convergent and $\lim_{n\to\infty}(s_n/t_n) = L/M$.*

PROOF: The calculation

$$\begin{aligned}\frac{s_n}{t_n} - \frac{L}{M} &= \frac{Ms_n - Lt_n}{t_n M} \\ &= \frac{M(s_n - L) + L(M - t_n)}{M} \cdot \frac{1}{t_n}\end{aligned}$$

expresses $\{(s_n/t_n) - (L/M)\}_{n=1}^{\infty}$ as the product of a null sequence and a bounded sequence; by Theorem 2.2.2, $\{(s_n/t_n) - (L/M)\}_{n=1}^{\infty}$ is null. ■

Example 1: To evaluate $\lim_{n\to\infty} s_n$ for

$$s_n = \frac{4n^2 + 3}{5n^2 + 2n + 1}$$

we first divide each term of the numerator and denominator by n^2, obtaining an algebraically equivalent form for s_n:

$$s_n = \frac{4 + \frac{3}{n^2}}{5 + \frac{2}{n} + \frac{1}{n^2}}$$

Theorem 2.4.2 tells us that the limit of s_n is the quotient of $\lim_{n\to\infty}\left(4 + \frac{3}{n^2}\right)$ by $\lim_{n\to\infty}\left(5 + \frac{2}{n} + \frac{1}{n^2}\right)$ *provided* that these limits both exist and that the second limit is nonzero. But it is easy to see by using Theorem 2.4.1 that $\lim_{n\to\infty}\left(4 + \frac{3}{n^2}\right) = 4$ and that $\lim_{n\to\infty}\left(5 + \frac{2}{n} + \frac{1}{n^2}\right) = 5$, so that Theorem 2.4.2 can be applied and therefore

$$\lim_{n\to\infty} s_n = \lim_{n\to\infty} \frac{4 + \frac{3}{n^2}}{5 + \frac{2}{n} + \frac{1}{n^2}} = \frac{\lim_{n\to\infty}\left(4 + \frac{3}{n^2}\right)}{\lim_{n\to\infty}\left(5 + \frac{2}{n} + \frac{1}{n^2}\right)} = \frac{4}{5}$$

Example 2: Let

$$s_n = \frac{5n + 8}{3n^2 + 1}$$

We divide each term by n^2, obtaining

$$\frac{5n + 8}{3n^2 + 1} = \frac{\frac{5}{n} + \frac{8}{n^2}}{3 + \frac{1}{n^2}}$$

In this form we see from Theorem 2.4.1 that the numerator converges to 0 and the denominator converges to 3. By Theorem 2.4.2, $s_n \to 0$.

Example 3: Let

$$s_n = \frac{n^2 + 1}{n}$$

By division we get

$$s_n = n + \frac{1}{n}$$

from which it follows that $\{s_n\}_{n=1}^{\infty}$ is unbounded. Hence, $\{s_n\}_{n=1}^{\infty}$ is divergent by Theorem 2.3.3.

Note: In applying Theorems 2.4.1 and 2.4.2 (as well as other limit theorems), it is important to keep in mind that the given sequences $\{s_n\}_{n=1}^{\infty}$ and $\{t_n\}_{n=1}^{\infty}$ are required to be *convergent*, and that in Theorem 2.4.2 the limit of the denominator must be nonzero.

In future evaluation of limits it will often be convenient to apply a theorem tentatively, contingent upon the convergence of $\{s_n\}_{n=1}^{\infty}$, $\{t_n\}_{n=1}^{\infty}$, and so on. Only after such convergence has been established is the application of the theorem justified. With this understanding we can write the argument in Example 2 briefly, as follows:

$$\lim_{n\to\infty} \frac{5n + 8}{3n^2 + 1} = \lim_{n\to\infty} \frac{\frac{5}{n} + \frac{8}{n^2}}{3 + \frac{1}{n^2}} = \frac{0}{3} = 0$$

However, consider again the divergent sequence $\{(n^2 + 1)/n\}_{n=1}^{\infty}$ in Example 3, which can be regarded as the product of $\{n^2 + 1\}_{n=1}^{\infty}$ and $\left\{\frac{1}{n}\right\}_{n=1}^{\infty}$. To argue that the limit of $(n^2 + 1)/n$ is 0 because of part (iii) of Theorem 2.4.1, and the fact that $\lim_{n\to\infty} \frac{1}{n} = 0$, is nonsense! The fallacy, of course, is that $\{n^2 + 1\}_{n=1}^{\infty}$ is divergent, so part (iii) of Theorem 2.4.1 does not apply.

Rational Functions of n

Let s_n be a rational function of n, that is, let

$$s_n = \frac{a_0 n^p + a_1 n^{p-1} + \cdots + a_p}{b_0 n^q + b_1 n^{q-1} + \cdots + b_q}$$

where p and q are nonnegative integers and the a's and b's are constants with a_0 and b_0 nonzero. Examples 1, 2, and 3 have involved rational functions. The method of Examples 1 and 2 can be used to show that

$$\text{If } p = q \quad \text{then} \quad \lim_{n\to\infty} s_n = a_0/b_0$$

$$\text{If } p < q \quad \text{then} \quad \lim_{n\to\infty} s_n = 0$$

In case $p > q$ (as in Example 3), the sequence $\{s_n\}_{n=1}^{\infty}$ is divergent, as we shall prove in the next section. Thus, the behavior of a rational function can be determined quickly by inspecting the degrees of the numerator and denominator.

Example 4: By inspection, we see that:

$$\frac{3n^2 - 5n}{5 + 8n^2} \to \frac{3}{8} \quad \text{and} \quad \frac{8n}{n^2 + 5} \to 0$$

Also, by inspection,

$$\left\{\frac{n^3 - 5}{n^2 + 1}\right\}_{n=1}^{\infty}$$

is divergent.

Example 5:

$$\lim_{n\to\infty}\left(\frac{n + 4}{2n + 5}\right)^2 = \left(\lim_{n\to\infty} \frac{n + 4}{2n + 5}\right)^2 \qquad \text{[Theorem 2.4.1, part (iii)]}$$

$$= \left(\frac{1}{2}\right)^2 = \frac{1}{4}$$

In the following examples, we apply Theorems 2.4.1 and 2.4.2 in combination with earlier results.

Example 6:

$$\lim_{n\to\infty}\sqrt{\frac{n + 4}{2n + 5}} = \sqrt{\lim_{n\to\infty}\left(\frac{n + 4}{2n + 5}\right)} \qquad \text{(Theorem 2.3.6)}$$

$$= \sqrt{\frac{1}{2}} = \frac{\sqrt{2}}{2}$$

Example 7:

$$\lim_{n\to\infty}\frac{1 + 2^n}{1 + 3^n} = \lim_{n\to\infty}\frac{\frac{1}{3^n} + \frac{2^n}{3^n}}{\frac{1}{3^n} + 1}$$

$$= \frac{\lim_{n\to\infty}\left(\frac{1}{3}\right)^n + \lim_{n\to\infty}\left(\frac{2}{3}\right)^n}{\lim_{n\to\infty}\left(\frac{1}{3}\right)^n + 1}$$

$$= \frac{0 + 0}{0 + 1} = 0 \qquad \text{(Theorem 2.2.5)}$$

Example 8:

$$\lim_{n\to\infty} \frac{1 - \sqrt{n}}{1 + \sqrt{n}} = \lim_{n\to\infty} \frac{(1/\sqrt{n}) - 1}{(1/\sqrt{n}) + 1} = \frac{0 - 1}{0 + 1} = -1$$

since $\{1/\sqrt{n}\}_{n=1}^{\infty}$ is null.

Theorem 2.4.3 *If a is a positive real number, then the sequence $\{\sqrt[n]{a}\}_{n=1}^{\infty}$ converges to 1.*

PROOF: There are three cases: $a > 1$, $a = 1$, and $a < 1$. The case $a = 1$ is trivial. Suppose next that $a > 1$. Then $\sqrt[n]{a} > 1$ for $n \in Z^+$ (why?), and so

$$\sqrt[n]{a} = 1 + p_n$$

where $p_n > 0$ for $n \in Z^+$. By raising both sides to the nth power and using the Binomial Theorem, we obtain

$$a = (1 + p_n)^n = 1 + np_n + \cdots + p_n^n$$

where all the terms on the right are positive. Consequently, for $n \in Z^+$, $np_n < a$, and so $p_n < a/n$. It follows that $\{p_n\}_{n=1}^{\infty}$ is null, and since $p_n = \sqrt[n]{a} - 1$, we have shown that $\sqrt[n]{a} \to 1$.

Finally, the case $0 < a < 1$ can be handled by letting $b = 1/a$, so that $b > 1$. Now $\sqrt[n]{b} \to 1$ by the first part, and so

$$\sqrt[n]{a} = \frac{1}{\sqrt[n]{b}} \to \frac{1}{1} = 1$$

by Theorem 2.4.2. ∎

Example 9: We use part (iii) of Theorem 2.4.1, Theorem 2.4.3, and Theorem 2.3.7 to get

$$\lim_{n\to\infty} \sqrt[n]{2n} = \lim_{n\to\infty} \sqrt[n]{2} \cdot \lim_{n\to\infty} \sqrt[n]{n} = 1$$

Splitting a Sequence

Let $\{x_n\}_{n=1}^{\infty}$ be a sequence. Let $\{n_k\}_{k=1}^{\infty}$ and $\{m_l\}_{l=1}^{\infty}$ be strictly increasing sequences of positive integers such that each $n \in Z^+$ is either n_k for some k, or m_l for some l, but not both. Then the sequence $\{x_n\}_{n=1}^{\infty}$ has both $\{x_{n_k}\}_{k=1}^{\infty}$ and $\{x_{m_l}\}_{l=1}^{\infty}$ as subsequences; moreover, each term of $\{x_n\}_{n=1}^{\infty}$ is a term in one or the other of the two subsequences but not both. We say that $\{x_n\}_{n=1}^{\infty}$ is *split* into the subsequences $\{x_{n_k}\}_{k=1}^{\infty}$ and $\{x_{m_l}\}_{l=1}^{\infty}$. If $\{x_n\}_{n=1}^{\infty}$ converges to a limit L, then both of these subsequences also converge to L by Theorem 2.3.2. The converse is also true.

Theorem 2.4.4 *If the sequence $\{x_n\}_{n=1}^{\infty}$ can be split into two subsequences $\{x_{n_k}\}_{k=1}^{\infty}$ and $\{x_{m_l}\}_{l=1}^{\infty}$ such that $\{x_{n_k}\}_{k=1}^{\infty}$ and $\{x_{m_l}\}_{l=1}^{\infty}$ both converge to L, then $\{x_n\}_{n=1}^{\infty}$ converges to L.*

PROOF: We define sequences $\{y_n\}_{n=1}^{\infty}$ and $\{z_n\}_{n=1}^{\infty}$ as follows:

$$y_n = \begin{cases} x_{n_k} & \text{if } n = n_k \text{ for some } k \\ L & \text{otherwise} \end{cases}$$
$$z_n = \begin{cases} x_{m_l} & \text{if } n = m_l \text{ for some } l \\ L & \text{otherwise} \end{cases}$$

Since $\{x_{n_k}\}_{k=1}^{\infty}$ converges to L, and the sequence $\{y_n\}_{n=1}^{\infty}$ is obtained from $\{x_{n_k}\}_{k=1}^{\infty}$ by inserting terms that are equal to L, it is clear that $y_n \to L$. Similarly, $z_n \to L$. Now for every $n \in Z^+$, we have

$$y_n + z_n = x_n + L$$

so that, by Theorem 2.4.1,

$$\lim_{n\to\infty} x_n = \lim_{n\to\infty}(y_n + z_n - L) = L \qquad \blacksquare$$

In Section 2.3 we observed that if a sequence $\{s_n\}_{n=1}^{\infty}$ converges to L, then so do the subsequences $\{s_{2k}\}_{k=1}^{\infty}$ and $\{s_{2k-1}\}_{k=1}^{\infty}$. The converse of this statement is an immediate consequence of Theorem 2.4.4. By combining these two results we obtain the following corollary.

Corollary 2.4.1 *If $\{s_n\}_{n=1}^{\infty}$ is a sequence and L a real number, then $s_n \to L$ if and only if $s_{2k} \to L$ and $s_{2k-1} \to L$.*

Preservation of Inequalities

If the terms of a convergent sequence all lie in a closed bounded interval, then their limit also lies in that interval.

Theorem 2.4.5 *Let L be the limit of a convergent sequence $\{s_n\}_{n=1}^{\infty}$ whose terms satisfy $a \le s_n \le b$ ($n \in Z^+$). Then $a \le L \le b$.*

PROOF: By Theorem 2.4.1, $\lim_{n\to\infty}(s_n - a) = L - a$. But $s_n - a \ge 0$, and by Corollary 2.3.1, we must have $L - a \ge 0$ or $a \le L$. Similarly, $L \le b$. $\blacksquare$

Because of Theorem 2.4.5 we can say that limits preserve inequalities of the type $\le$ and $\ge$. However, they do not preserve $<$ or $>$, as the example $\left\{1 + \frac{1}{n}\right\}_{n=1}^{\infty}$ shows:

$$1 + \frac{1}{n} > 1 \qquad n \in Z^+$$

but

$$\lim_{n\to\infty}\left(1 + \frac{1}{n}\right) = 1$$

EXERCISES 2.4

In Exercises 1–15, determine whether the given sequence converges; if it does, find the limit. Justify your conclusions.

1. $\dfrac{n^2 - 1}{n^2 + 1}$

2. $\dfrac{n^3 - 1}{1 + n^4}$

3. $\dfrac{3n^2 + 4n + 5}{1 - n + 4n^2}$

4. $\left(\dfrac{2n}{n + 1}\right)^3$

5. $\sqrt{\dfrac{n^2 + 4}{9n^2 + 3}}$

6. $\dfrac{m^3 + 3m^2 + m + 1}{5m^2}$

7. $\dfrac{\sqrt{n^4 + 3n^2}}{n^2 + 1}$

8. (a) $\sqrt{n^2 + 1} - n$ (b) $\sqrt{n^2 + n} - n$

9. $\dfrac{1 + 2^n}{2^{n+1}}$

10. $\dfrac{1 + 3^n}{1 + 2^n}$

11. $\dfrac{1 + 2^n}{3^n + 2^n}$

12. $\dfrac{1 + 2^n}{3^n - 2^n}$

13. $\dfrac{1 + k^2}{3^k - 2^k}$

14. $\sqrt[n]{n + 1}$

15. $\sqrt[n]{n^2 + 1}$

In Exercises 16–22, prove or disprove the given statement.

16. If s is convergent and t is divergent, then $s + t$ is divergent.

17. If $s + t$ is convergent and s is divergent, then t is divergent.

18. If s and t are divergent, then $s + t$ is divergent.

19. If $s + t$ and $s - t$ are convergent, then s and t are convergent.

20. If $s + t$ and $s - t$ are divergent, then s and t are divergent.

21. If s is convergent and t is divergent, then st is divergent.

22. If L is the limit of a convergent sequence $\{s_n\}_{n=1}^{\infty}$ and if $a < s_n < b$ $(n \in Z^+)$, then $a < L < b$.

23. (a) Criticize the following argument:

$$\lim_{n\to\infty} \frac{\sin n}{n} = \lim_{n\to\infty}(\sin n) \cdot \lim_{n\to\infty} \frac{1}{n} = \left(\lim_{n\to\infty} \sin n\right) \cdot 0 = 0$$

(b) Give a valid argument showing that

$$\lim_{n\to\infty} \frac{\sin n}{n} = 0.$$

24. Show that if $\{s_n\}_{n=1}^{\infty}$ is a sequence of positive real numbers such that $\{(-1)^n s_n\}_{n=1}^{\infty}$ is convergent, then $\{s_n\}_{n=1}^{\infty}$ is null.

25. Show that if $\{s_n\}_{n=1}^{\infty}$ is convergent, then the sequence $\{\sigma_n\}_{n=1}^{\infty}$ defined by

$$\sigma_n = \frac{s_1 + \cdots + s_n}{n}$$

is convergent and $\lim_{n\to\infty} \sigma_n = \lim_{n\to\infty} s_n$.

2.5 *Divergence to Infinity*

Unbounded sequences are necessarily divergent (Theorem 2.3.3). In this section we investigate two special types of unbounded sequences.

Positive Infinity

A sequence whose terms eventually become and remain arbitrarily large is said to diverge to positive infinity.

Definition 2.5.1 A sequence $\{s_n\}_{n=1}^{\infty}$ ***diverges to positive infinity*** provided that, corresponding to an arbitrary $B > 0$, there exists a positive integer N such that

$$s_n > B \qquad \text{whenever } n \geq N$$

When this is the case, we write $s_n \to \infty$.

We think of B in this definition as being as large as desired. The terms of $\{s_n\}_{n=1}^{\infty}$ then become larger than B ($s_N > B$) and remain larger than B for $n > N$.

Many familiar sequences diverge to positive infinity.

Example 1: Let $s_n = \sqrt{n}$. We show that $s_n \to \infty$. Let $B > 0$. We know that $\sqrt{n}$ must be greater than B whenever $n > B^2$. Thus, if we let N be a positive integer greater than B^2, then $s_n > B$ for $n \geq N$.

Example 2: Let $s_n = n^3 - 2n^2$. We show that $s_n \to \infty$. Let $B > 0$. Now $s_n = n^2(n - 2)$, and for $n \geq 3$ we have $s_n \geq n^2$. Let N_1 be a positive integer greater than $\sqrt{B}$, so that $n^2 > B$ for $n \geq N_1$. If $N = \max\{3, N_1\}$, then $s_n > B$ for $n \geq N$.

Example 3: The divergent sequence $\{[1 + (-1)^n]n\}_{n=1}^{\infty}$

$$0, 4, 0, 8, 0, 12, \ldots$$

does not diverge to positive infinity. Although the terms may be made arbitrarily large, they do not *remain* large.

The following properties can be proved directly from Definition 2.5.1 (work Exercises 2.5.2 and 2.5.3).

Theorem 2.5.1 *If $s_n \to \infty$ and $t_n \to \infty$, then $s_n + t_n \to \infty$ and $s_n t_n \to \infty$.*

Theorem 2.5.2 *If $s_n \to \infty$ and $\{t_n\}_{n=1}^{\infty}$ is bounded, then $s_n + t_n \to \infty$.*

Corollary 2.5.1 *If $s_n \to \infty$ and $\{t_n\}_{n=1}^{\infty}$ is convergent, then $s_n + t_n \to \infty$.*

Negative Infinity

Definition 2.5.2 A sequence $\{s_n\}_{n=1}^{\infty}$ ***diverges to negative infinity*** provided that, corresponding to an arbitrary $B < 0$, there exists a positive integer N such that

$$s_n < B \qquad \text{whenever } n \geq N$$

When this is the case, we write $s_n \to -\infty$.

Here we think of B as being large in absolute value but negative. The terms of $\{s_n\}_{n=1}^{\infty}$ become less than B ($s_N < B$) and remain less than B for $n > N$.

Example 4: If $s_n = 10n - n^2$, then $s_n \to -\infty$. Let $B < 0$. We are to show that $10n - n^2 < B$ for n sufficiently large. Now $10n - n^2 = -n(n - 10)$, and the inequality $s_n < B$ is equivalent to

$$-n(n - 10) < B \qquad \text{or} \qquad n(n - 10) > -B$$

If $n \geq 11$, then $n - 10 \geq 1$, and so $n(n - 10) \geq n$. Let N_1 be a positive integer greater than $-B$ and let $N = \max\{11, N_1\}$. If $n \geq N$, then $n - 10 \geq 1$, and $n(n - 10) \geq n \geq N_1 > -B$. It follows that $s_n < B$ for $n \geq N$.

Proofs of the following properties are left for you to work out in Exercises 2.5.4, 2.5.5, and 2.5.6.

Theorem 2.5.3 $s_n \to -\infty$ *if and only if* $-s_n \to \infty$.

Theorem 2.5.4 *If* $s_n \to -\infty$ *and* $t_n \to -\infty$, *then* $s_n + t_n \to -\infty$ *and* $s_n t_n \to \infty$.

Theorem 2.5.5 *If* $s_n \to -\infty$ *and* $\{t_n\}_{n=1}^{\infty}$ *is bounded, then* $s_n + t_n \to -\infty$.

Corollary 2.5.2 *If* $s_n \to -\infty$ *and* $\{t_n\}_{n=1}^{\infty}$ *is convergent, then* $s_n + t_n \to -\infty$.

Notation

The symbols ∞ and $-\infty$ do not stand for real numbers but are part of a notation that describes a special type of behavior. In contrast to the statement "$\frac{1}{n} \to 0$", which refers to the real number zero, the statement "$n^2 \to \infty$" does not refer to any real number at all.

There is a temptation to perform algebraic operations with the symbols ∞ and $-\infty$. For instance, we might interpret the equations

$$\infty + \infty = \infty \qquad \text{and} \qquad \infty \cdot \infty = \infty$$

as a very brief way of expressing the properties in Theorem 2.5.1. This interpretation is fine, but to offer these equations as a *proof* of the theorem is to beg the question completely.

We cannot give a sensible interpretation to the symbol $\infty - \infty$. The reason is that if $\{s_n\}_{n=1}^{\infty}$ and $\{t_n\}_{n=1}^{\infty}$ both diverge to ∞, we cannot draw any conclusion whatever concerning the behavior of $\{s_n - t_n\}_{n=1}^{\infty}$. For instance, the following examples:

$$\begin{array}{lll} s_n = n^2 & t_n = n & s_n - t_n \to \infty \\ s_n = n & t_n = n^2 & s_n - t_n \to -\infty \\ s_n = n + 2 & t_n = n & s_n - t_n \to 2 \end{array}$$

show that, under the hypothesis that $s_n \to \infty$ and $t_n \to \infty$, $\{s_n - t_n\}_{n=1}^{\infty}$ may behave in a variety of different ways.

Similar difficulties arise if we try to interpret symbols such as ∞/∞, $\infty/-\infty$, $0 \cdot \infty$, and so on.

Theorem 2.5.6 *Let $s_n \to \infty$ $(-\infty)$ and $t_n \to L$. If $L > 0$, then $s_n t_n \to \infty$ $(-\infty)$; if $L < 0$, then $s_n t_n \to -\infty$ (∞); if $L = 0$, then no conclusion can be drawn about the behavior of $\{s_n t_n\}_{n=1}^{\infty}$.*

PROOF: We consider the case in which $s_n \to \infty$ and $L > 0$. Because $L = \lim_{n\to\infty} t_n > 0$, there exists an $N_1 \in Z^+$ such that $t_n > L/2$ whenever $n \geq N_1$. Then, for $n \geq N_1$,

$$s_n t_n > s_n \cdot \frac{L}{2}$$

Let $B > 0$ be given. Since $s_n \to \infty$, there is an $N_2 \in Z^+$ such that $s_n > 2B/L$ whenever $n \geq N_2$. Finally, if $N = \max\{N_1, N_2\}$, we conclude that

$$s_n t_n > \frac{2B}{L} \cdot \frac{L}{2} = B \qquad n \geq N$$

The remaining cases are left for you to work out in Exercises 2.5.8 and 2.5.11. ∎

Polynomials in n

Theorem 2.5.6 enables us to determine the behavior of a polynomial function of n by examining its leading coefficient.

Theorem 2.5.7 *Let s_n be a polynomial in n of degree $p \geq 1$,*

$$s_n = a_0 n^p + a_1 n^{p-1} + a_2 n^{p-2} + \cdots + a_p \; (a_0 \neq 0)$$

Then $s_n \to \infty$ if $a_0 > 0$ and $s_n \to -\infty$ if $a_0 < 0$.

PROOF: By factoring n^p out of all terms in the expression for s_n, we have

$$s_n = n^p\left(a_0 + a_1\frac{1}{n} + a_2\frac{1}{n^2} + \cdots + a_p\frac{1}{n^p}\right)$$

Now $n^p \to \infty$, and the factor in parentheses converges to the nonzero number a_0. Our conclusion follows from Theorem 2.5.6. ∎

Example 5: By Theorem 2.5.7, $3n - n^2 \to -\infty$.

Rational Functions

We return to the discussion of rational functions of n begun in Section 2.4, taking up now the case in which the degree of the numerator is greater than that of the denominator:

$$s_n = \frac{a_0 n^p + a_1 n^{p-1} + \cdots + a_p}{b_0 n^q + b_1 n^{q-1} + \cdots + b_q} \qquad a_0, b_0, \neq 0; p > q$$

By using the long division process we express s_n in the form

$$s_n = t_n + r_n$$

where t_n is a polynomial in n of degree ≥ 1 and r_n is a rational function of n in which the degree of the numerator is less than the degree of the denominator. We already know that $r_n \to 0$ and, depending upon the sign of its leading coefficient, the polynomial t_n diverges either to ∞ or to $-\infty$. By Corollaries 2.5.1 and 2.5.2, we see that s_n behaves in exactly the same manner as t_n.

Example 6: Consider

$$s_n = \frac{6n^3 - 7n^2 + 10n - 1}{2n - 1}$$

By the division process we obtain

$$\frac{6n^3 - 7n^2 + 10n - 1}{2n - 1} = (3n^2 - 2n + 4) + \frac{3}{2n - 1}$$

so that $t_n = 3n^2 - 2n + 4$ and $r_n = 3/(2n - 1)$. Our discussion shows that s_n behaves exactly like t_n; that is, $s_n \to \infty$.

Reciprocals

If $s_n > 0$ and $s_n \to \infty$, then the sequence of reciprocals $\{1/s_n\}_{n=1}^{\infty}$ is null. For $\varepsilon > 0$, we can take N large enough that $s_n > 1/\varepsilon$ for $n \geq N$; thus, $1/s_n < \varepsilon$ for $n \geq N$. Conversely, if $\{1/s_n\}_{n=1}^{\infty}$ is null and $s_n > 0$ for $n \in Z^+$, we see that $s_n \to \infty$. Thus, for sequences of positive terms, $s_n \to \infty$ if and only if $\{1/s_n\}_{n=1}^{\infty}$ is null. By considering the absolute value sequence, and recalling that a sequence is null if and only if the sequence of absolute values is null (Lemma 2.2.1), we obtain the following useful result.

Theorem 2.5.8 *If $\{s_n\}_{n=1}^{\infty}$ is a sequence of nonzero terms, then $|s_n| \to \infty$ if and only if $\{1/s_n\}_{n=1}^{\infty}$ is null.*

Geometric Sequences

Example 7: Consider the geometric sequence $\{r^n\}_{n=1}^{\infty}$ where r is a fixed real number. We already know that if $|r| < 1$, then $\{r^n\}_{n=1}^{\infty}$ is null (Theorem 2.2.5). For $r = 1$ it is clear that $r^n \to 1$; while if $r = -1$, then $r^n = (-1)^n$ diverges (Example 4 in Section 2.3).

We are left with the cases $r > 1$ and $r < -1$, in both of which it is true that $|r| > 1$. Then $|r| = 1 + p$ for some $p > 0$, and $|r|^n =$

$(1 + p)^n \geq 1 + np$ for $n \in Z^+$ (see Exercise 2.2.6). From this it readily follows that $|r|^n \to \infty$, and $\{r^n\}_{n=1}^{\infty}$ diverges. When $r > 1$, $\{r^n\}_{n=1}^{\infty}$ diverges to ∞; when $r < -1$, $\{r^n\}_{n=1}^{\infty}$ alternates between positive and negative values and is unbounded.

Theorem 2.5.9 **Geometric Sequence** *For the sequence $\{r^n\}_{n=1}^{\infty}$, we have:*

(i) *If $|r| < 1$, then $r^n \to 0$*
(ii) *If $r = 1$, then $r^n \to 1$*
(iii) *If $r > 1$, then $\{r^n\}_{n=1}^{\infty}$ diverges to ∞*
(iv) *If $r \leq -1$, then $\{r^n\}_{n=1}^{\infty}$ diverges*

Subsequences

Divergence to infinity is similar in some respects to convergence. An example is the following theorem.

Theorem 2.5.10 *If a sequence diverges to ∞ (or $-\infty$), then the same is true of every subsequence.*

PROOF: The proof is left for you to work out in Exercise 2.5.12. ■

Divergence to ∞, or to $-\infty$, is often referred to as *proper* divergence.

EXERCISES 2.5

1. Establish the following statements directly from the appropriate definition.

(a) $n^3 \to \infty$
(b) $4n - n^3 \to -\infty$
(c) $\dfrac{n^3 + 4}{n} \to \infty$
(d) $\dfrac{4 - n^3}{n^2 + 1} \to -\infty$
(e) $2^n \to \infty$ (*Hint:* Show that $2^n > n$.)
(f) $\dfrac{3^n + 2^n}{2^n} \to \infty$ $\left(\textit{Hint: } \left(\dfrac{3}{2}\right)^n \geq 1 + \dfrac{n}{2}\right)$

2. Prove Theorem 2.5.1.

3. Prove Theorem 2.5.2 and Corollary 2.5.1.

4. Prove Theorem 2.5.3.

5. Prove Theorem 2.5.4.

6. Prove Theorem 2.5.5 and Corollary 2.5.2.

7. Give examples of sequences $\{s_n\}_{n=1}^{\infty}$ and $\{t_n\}_{n=1}^{\infty}$ for which $s_n \to \infty$, $t_n \to \infty$, and:

(a) $s_n/t_n \to \infty$
(b) $s_n/t_n \to 0$
(c) $s_n/t_n \to 2$

8. Verify the assertion made in Theorem 2.5.6, that if $s_n \to \infty$ and $t_n \to 0$, no conclusion can be drawn about the behavior of $\{s_n t_n\}_{n=1}^{\infty}$, by giving examples of such sequences for which

(a) $s_n t_n \to \infty$
(b) $s_n t_n \to 0$
(c) $s_n t_n \to 2$

9. Prove or disprove the following statements.

(a) If $s_n \to \infty$ and c is constant, then $cs_n \to \infty$.
(b) If $s_n \geq 0$ for $n \in Z^+$ and $s_n \to \infty$, then $\sqrt{s_n} \to \infty$.
(c) If $s_n \geq 0$ for $n \in Z^+$ and $\sqrt{s_n} \to \infty$, then $s_n \to \infty$.

(d) If $s_n \to \infty$ and $t_n \to -\infty$, then $s_n t_n \to -\infty$.
(e) If $s_n \to \infty$ and $t_n \to \infty$, then $s_n - t_n \to 0$.
(f) If $s_n \to -\infty$ and $t_n \to 0$, then $s_n t_n \to 0$.
(g) If $s_n \to 0$, then $1/s_n \to \infty$.
(h) If $s_n > 0$ and $t_n > 0$ for $n \in Z^+$ and if $s_n t_n \to \infty$, then either $s_n \to \infty$ or $t_n \to \infty$.

10. Determine whether the sequence converges or diverges. If the sequence converges, find the limit. If it diverges to ∞ or $-\infty$, specify. Justify your conclusions.

(a) $7n^2 - 4n$
(b) $7n^2 - 4n^3$
(c) $2n^2 + \dfrac{3n + 4}{n^2 + 1}$
(d) $1 + n - n^2 + \dfrac{1}{2n + 1}$
(e) $\dfrac{(-1)^n(n^3 + 2)}{n + 1}$
(f) $\dfrac{(-1)^{n+1}(3n - 4)}{6 + n^2}$
(g) $\dfrac{n(n + 1)}{(n + 2)(n + 3)}$

11. Prove Theorem 2.5.6 for the case:

(a) $s_n \to \infty$ and $L < 0$
(b) $s_n \to -\infty$ and $L > 0$

12. Prove Theorem 2.5.10.

2.6 *Monotone Sequences*

Monotone sequences have nice properties. This section is devoted to several theorems that bear this out. Our first theorem is an extremely important one and follows directly from the completeness property of *R*.

Theorem 2.6.1

(i) *If $\{s_n\}_{n=1}^\infty$ is increasing and bounded above, then $\{s_n\}_{n=1}^\infty$ is convergent, and $\lim_{n\to\infty} s_n = \sup\{s_n : n \in Z^+\}$.*

(ii) *If $\{s_n\}_{n=1}^\infty$ is decreasing and bounded below, then $\{s_n\}_{n=1}^\infty$ is convergent, and $\lim_{n\to\infty} s_n = \inf\{s_n : n \in Z^+\}$.*

(iii) *A bounded monotone sequence is convergent.*

PROOF:

(i) Let $\{s_n\}_{n=1}^\infty$ be increasing and bounded above. We show that the limit of $\{s_n\}_{n=1}^\infty$ exists. Consider the range S of $\{s_n\}_{n=1}^\infty$; that is, the set

$$S = \{s_n : n \in Z^+\}$$

Clearly S is not empty; it is bounded above by hypothesis. Accordingly, S has a least upper bound, which we denote by L:

$$L = \sup S = \sup\{s_n : n \in Z^+\}$$

Now we show that $s_n \to L$. Let $\varepsilon > 0$ be given. Because L is the least upper bound of S, $L - \varepsilon$ is not an upper bound of S and there must exist an $N \in Z^+$ such that $s_N > L - \varepsilon$. Because $\{s_n\}_{n=1}^\infty$ is increasing, we see that

$$L - \varepsilon < s_N \le s_{N+1} \le s_{N+2} \le \cdots$$

But all terms of $\{s_n\}_{n=1}^{\infty}$ are less than or equal to L, so we now have

$$L - \varepsilon < s_N \le s_{N+1} \le s_{N+2} \le \cdots \le L$$

showing that $|s_n - L| < \varepsilon$ for $n \ge N$. Thus, $s_n \to L$.

(ii) You are asked to prove this part in Exercise 2.6.1.

(iii) This statement follows immediately from (i) and (ii). ■

The Number e

Example 1: Let $s_n = \left(1 + \frac{1}{n}\right)^n$. We apply part (i) of Theorem 2.6.1 to prove that $\{s_n\}_{n=1}^{\infty}$ converges. By the Binomial Theorem, s_n can be expressed as a sum of $n + 1$ positive terms, as follows:

$$\begin{aligned} s_n &= \left(1 + \frac{1}{n}\right)^n \\ &= 1 + n \cdot \frac{1}{n} + \frac{n(n-1)}{2!} \cdot \frac{1}{n^2} + \frac{n(n-1)(n-2)}{3!} \cdot \frac{1}{n^3} + \cdots + \frac{n(n-1)\cdots 1}{n!} \cdot \frac{1}{n^n} \\ &= 1 + n \cdot \frac{1}{n} + \frac{n(n-1)}{n \cdot n} \cdot \frac{1}{2!} + \frac{n(n-1)(n-2)}{n \cdot n \cdot n} \cdot \frac{1}{3!} + \cdots + \frac{n(n-1)\cdots 1}{n \cdot n \cdot \cdots \cdot n} \cdot \frac{1}{n!} \\ s_n &= 1 + 1 + \left(1 - \frac{1}{n}\right)\frac{1}{2!} + \left(1 - \frac{1}{n}\right)\left(1 - \frac{2}{n}\right)\frac{1}{3!} + \cdots + \left(1 - \frac{1}{n}\right)\left(1 - \frac{2}{n}\right)\cdots\left(1 - \frac{n-1}{n}\right)\frac{1}{n!} \qquad (1) \end{aligned}$$

Now consider s_{n+1}; we obtain a formula for s_{n+1} by replacing n by $n + 1$ in Equation (1). The result is a sum of $n + 2$ positive terms, in which each of the first $n + 1$ terms is greater than or equal to the corresponding term of s_n. For example, the fourth term of s_{n+1}:

$$\left(1 - \frac{1}{n+1}\right)\left(1 - \frac{2}{n+1}\right)\frac{1}{3!}$$

is greater than the fourth term of s_n:

$$\left(1 - \frac{1}{n}\right)\left(1 - \frac{2}{n}\right)\frac{1}{3!}$$

It follows that $s_{n+1} \ge s_n$ for every n, so that $\{s_n\}_{n=1}^{\infty}$ is increasing.

To see that $\{s_n\}_{n=1}^{\infty}$ is bounded above, we continue to investigate the expression (1) for s_n, first replacing all factors of the form $\left(1 - \frac{k}{n}\right)$ by 1 (which is larger), then replacing $k!$ by 2^{k-1} (which is smaller) for $k = 2, 3, \ldots, n$:

$$\begin{aligned} s_n &\le 1 + 1 + \frac{1}{2!} + \frac{1}{3!} + \cdots + \frac{1}{n!} \\ &\le 1 + \left(1 + \frac{1}{2} + \frac{1}{2^2} + \cdots + \frac{1}{2^{n-1}}\right) = 1 + \frac{1 - (1/2^n)}{1 - \frac{1}{2}} \end{aligned}$$

We have used Equation (6), Section P.2, to simplify the sum in parenthesis. It now follows that

$$\begin{aligned} s_n &\le 1 + 2\left(1 - \frac{1}{2^n}\right) \\ &= 3 - \frac{1}{2^{n-1}} < 3 \end{aligned}$$

Thus, $\{s_n\}_{n=1}^{\infty}$ is bounded above, and by part (i) of Theorem 2.6.1, it is convergent; we denote its limit by e.

Definition 2.6.1

$$e = \lim_{n \to \infty} \left(1 + \frac{1}{n}\right)^n$$

Note: This limit defines the number e introduced by Euler and used in calculus as the base of natural logarithms.

Leonhard Euler

(b. April 15, 1707; d. September 18, 1783)

From the University of Basel to Saint Petersburg to Berlin and back to Saint Petersburg, Euler was the key figure in eighteenth century mathematics. He wrote more than 700 books and papers. In 1733 an illness deprived him of the use of one eye, and in 1766 a cataract took the sight from the other. Though blind, he performed complex calculations in his head and continued producing manuscripts until his death. Through his work, both trigonometry and calculus were released from their ties to geometry; he established much of the acceptable notation in use today, including the symbols for π and e (the latter he named after himself). His *Introductio in Analysin Infinitorum*, a two-volume work of great scope, proved to be of considerable importance to those who followed him.

Since 3 is an upper bound of $\{s_n : n \in Z^+\}$ and e is its least upper bound, we see that $e \leq 3$; of course, this is only a rough approximation. It can be shown that e is an irrational number; to five decimals, $e = 2.71828$.

Example 2:

$$\lim_{n \to \infty} \left(1 + \frac{1}{n}\right)^{2n} = \left[\lim_{n \to \infty} \left(1 + \frac{1}{n}\right)^n\right]^2 = e^2$$

by Theorem 2.4.1, part (iii).

Example 3:

$$\lim_{n \to \infty} \left(1 + \frac{1}{2n}\right)^{2n} = e$$

by Theorem 2.3.2 because $\left\{\left(1 + \frac{1}{2k}\right)^{2k}\right\}_{k=1}^{\infty}$ is a subsequence of $\left\{\left(1 + \frac{1}{n}\right)^n\right\}_{n=1}^{\infty}$.

A Square Root Algorithm

An interesting method for obtaining square roots is known in elementary school mathematics as the "divide and average" algorithm. It is important in our theoretical development because it establishes the existence of square roots of positive real numbers.

Let b be any positive real number. Define a sequence $\{x_n\}_{n=1}^{\infty}$ recursively (by induction) as follows:

(i) $x_1 = 1$

(ii) $x_{n+1} = \dfrac{1}{2}\left(x_n + \dfrac{b}{x_n}\right) \qquad n \in Z^+$

If $b = 2$, for instance, we get $x_1 = 1$, $x_2 = \frac{3}{2} = 1.5$, $x_3 = \frac{17}{12} = 1.416\ldots$, and so on.

We will show that, for each positive real number b, the sequence $\{x_n\}_{n=1}^{\infty}$ defined by (i) and (ii) converges and that its limit L satisfies the equation $L^2 = b$. Thus, L is a square root of b.

Clearly the terms of $\{x_n\}_{n=1}^{\infty}$ are all positive, so that $\{x_n\}_{n=1}^{\infty}$ is bounded below by zero. To prove convergence it is sufficient to show that $\{x_n\}_{n=1}^{\infty}$ is decreasing. First, observe that any real number y of the form $y = \frac{1}{2}\left(x + \frac{b}{x}\right)$ satisfies the inequality $y^2 \geq b$:

$$\begin{aligned} y^2 - b &= \frac{1}{4}\left(x + \frac{b}{x}\right)^2 - b \\ &= \frac{1}{4}\left(x^2 + 2b + \frac{b^2}{x^2}\right) - b \\ &= \frac{1}{4}\left(x^2 - 2b + \frac{b^2}{x^2}\right) \\ &= \frac{1}{4}\left(x - \frac{b}{x}\right)^2 \\ &\geq 0 \end{aligned}$$

Now since each term of the sequence $\{x_n\}_{n=1}^{\infty}$ after the first is of the form $\frac{1}{2}\left(x + \frac{b}{x}\right)$ for some x, we know that $x_n^2 \geq b$ for $n > 1$. Therefore, for $n > 1$,

$$x_{n+1} = \frac{x_n^2 + b}{2x_n} \leq \frac{x_n^2 + x_n^2}{2x_n} = x_n$$

It follows that the tail $\{x_n\}_{n=2}^{\infty}$ is a decreasing sequence and therefore converges. By Corollary 2.3.2, $\{x_n\}_{n=1}^{\infty}$ converges.

Let $L = \lim_{n\to\infty} x_n$. Clearly, $L \geq 0$. Now by taking the limit on both sides of the equation

$$2x_{n+1}\, x_n = x_n^2 + b$$

[derived from (ii) above], we obtain $2L^2 = L^2 + b$ or $L^2 = b$.

Theorem 2.6.2 *Every positive real number has a positive square root in* R.

The question of the existence of nth roots ($n \in Z^+$) is considered in Section 3.6.

Behavior of Monotone Sequences

Theorem 2.6.3

(i) An increasing sequence either converges or diverges to positive infinity.
(ii) A decreasing sequence either converges or diverges to negative infinity.
(iii) A monotone sequence is either convergent or properly divergent.
(iv) Every subsequence of a monotone sequence has the same limiting behavior as the sequence itself.

PROOF:

(i) Let $\{s_n\}_{n=1}^{\infty}$ be increasing. Either $\{s_n\}_{n=1}^{\infty}$ is bounded above or it is not. If $\{s_n\}_{n=1}^{\infty}$ is bounded above, then by part (i) of Theorem 2.6.1 it converges. If $\{s_n\}_{n=1}^{\infty}$ is not bounded above, then for any $B > 0$ there exists a positive integer N such that $s_N > B$; and since $\{s_n\}_{n=1}^{\infty}$ is increasing, we see that

$$B < s_N \leq s_{N+1} \leq \cdots$$

and $s_n > B$ for all $n \geq N$. Thus $s_n \to \infty$.

(ii) You are asked to prove part (ii) in Exercise 2.6.2.

(iii) This statement follows immediately from parts (i) and (ii).

(iv) This part follows from the previous parts together with Theorems 2.3.2 and 2.5.10. ∎

Supremum as a Limit

We conclude this section with some theorems expressing the supremum of a set of real numbers as a sequential limit; similar theorems hold for the infimum.

Theorem 2.6.4 *If S is a nonempty set of real numbers that is bounded above, and if $L = \sup S$, then there exists a sequence $\{s_n\}_{n=1}^{\infty}$ of points of S such that $s_n \to L$.*

PROOF: For any $\varepsilon > 0$, it follows by the definition of L that $L - \varepsilon$ is not an upper bound of S; there must therefore be an element of S that is greater than $L - \varepsilon$. Now apply this observation to the numbers $L - \frac{1}{n}$ for each $n \in Z^+$, choosing $s_n \in S$ such that $L - \frac{1}{n} < s_n \leq L$. By the Squeeze Theorem (Theorem 2.3.8), $s_n \to L$. ∎

If the supremum of a set does not belong to the set, then it is possible to find a *strictly increasing* sequence converging to the supremum.

Theorem 2.6.5 *If S is a nonempty set of real numbers that is bounded above, and if $L = \sup S \notin S$, then there exists a strictly increasing sequence $\{s_n\}_{n=1}^{\infty}$ of points of S such that $s_n \to L$.*

PROOF: You are asked to prove this theorem in Exercise 2.6.8. ∎

EXERCISES 2.6

1. Prove part (ii) of Theorem 2.6.1.
2. Prove part (ii) of Theorem 2.6.3.
3. Verify the following limits by referring to appropriate theorems.
   (a) $\lim_{n\to\infty}\left(1+\frac{1}{n}\right)^{n+1} = e$
   (b) $\lim_{n\to\infty}\left(1+\frac{1}{n+1}\right)^{n} = e$
4. Show that $\lim_{n\to\infty}\left(1+\frac{2}{n}\right)^n = e^2$. [*Hint:* Verify that
$$\left(1+\frac{1}{n+1}\right)^2 \le 1+\frac{2}{n} \le \left(1+\frac{1}{n}\right)^2$$
for $n \in Z^+$.
Take the nth power and then use the Squeeze Theorem.]
5. Show that $\lim\left(1-\frac{1}{n}\right)^n = 1/e$. [*Hint:* Let $s_n = \left(1-\frac{1}{n}\right)^n$ and $t_n = 1/s_n$ for $n > 1$; then
$$t_{n+1} = \left(\frac{n+1}{n}\right)^{n+1} = \left(1+\frac{1}{n}\right)^{n+1}$$
so $t_n \to e$.]
6. Show that
$$\lim_{n\to\infty}\left(1+\frac{1}{2n}\right)^n = \sqrt{e}$$
7. State and prove a theorem similar to Theorem 2.6.4 for the infimum of a set of real numbers.
8. (a) What sequence $\{s_n\}_{n=1}^{\infty}$ does the proof of Theorem 2.6.4 provide if S is the set $(0, 1) \cup \{2\}$?
   (b) Prove Theorem 2.6.5.
9. Let $\{x_n\}_{n=1}^{\infty}$ be defined recursively by (i) $x_1 = 1$ and (ii) $x_{n+1} = \sqrt{1+x_n}$ $(n \in Z^+)$.
   (a) Use part (i) of Theorem 2.6.1 to show that $\{x_n\}_{n=1}^{\infty}$ is convergent.
   (b) Find $\lim_{n\to\infty} x_n$.
10. Let $\{x_n\}_{n=1}^{\infty}$ be defined recursively by (i) $x_1 = 2$ and (ii) $x_{n+1} = \sqrt{2+x_n}$. Show that $\{x_n\}_{n=1}^{\infty}$ converges, and find its limit.

PROJECT: Cantor's Theorem on Nested Intervals

For $n \in Z^+$, let I_n be a closed, bounded interval $[a_n, b_n]$ $(a_n \le b_n)$ in R and denote by $|I_n|$ the *length* of I_n, $b_n - a_n$.

(1) Prove the following theorem.

Theorem Cantor's Theorem *If $\{I_n\}_{n=1}^{\infty}$ is a sequence of nonempty closed bounded intervals such that*

(i) $I_{n+1} \subseteq I_n$ *for* $n \in Z^+$

and

(ii) $|I_n| \to 0$

then there exists a unique real number that belongs to I_n for every $n \in Z^+$.

(2) Show that the conclusion of Cantor's Theorem may fail if the intervals I_n are not closed.

(3) Show that the conclusion of Cantor's Theorem may fail if the sequence $\{|I_n|\}_{n=1}^{\infty}$ does not converge to zero.

(4) Show that a sequence of closed but unbounded intervals $\{I_n\}_{n=1}^{\infty}$ satisfying condition (i) may have an empty intersection.

2.7 *The Cauchy Criterion*

We now investigate some theoretical results of considerable interest. Our first step is to show that from an arbitrary sequence of real numbers we can extract a monotone subsequence. It then follows easily that every bounded sequence has a convergent subsequence; this is the famous Bolzano-Weierstrass Theorem, formulated for sequences. Finally we obtain a condition, known as the Cauchy Criterion, that is necessary and sufficient for a sequence of real numbers to converge.

Lemma 2.7.1 *Every sequence of real numbers has a monotone subsequence.*

PROOF: Let s be an arbitrary sequence of real numbers,

$$s: \quad s_1, s_2, s_3, \ldots$$

Now consider the sequences

$$\begin{array}{rl} s^{(1)}: & s_1, s_2, s_3, \ldots \\ s^{(2)}: & s_2, s_3, s_4, \ldots \\ s^{(3)}: & s_3, s_4, s_5, \ldots \\ \vdots & \\ s^{(n)}: & s_n, s_{n+1}, s_{n+2}, \ldots \\ \vdots & \end{array}$$

For each $n \in Z^+$, $s^{(n)}$ is a subsequence of s containing all terms in the tail that starts with s_n. Now we use a proof by cases: Either (i) every one of these subsequences has a greatest term, or (ii) at least one of these subsequences fails to have a greatest term.

Case (i): Assume that every $s^{(n)}$ has a greatest term. Now we must recognize that a greatest term of a sequence may occur more than once; we say that s_m is *the* greatest term of a sequence when m is the *first* (least) subscript for which s_m is greater than or equal to all terms of the sequence.

Let n_1 be the subscript of the greatest term of $s^{(1)}$. Then let n_2 be the subscript of the greatest term of $s^{(n_1+1)}$, noting that $n_2 > 1$ and $s_{n_2} \leq s_{n_1}$. Next let n_3 be the subscript of the greatest term of $s^{(n_2+1)}$, noting that $n_3 > n_2$ and $s_{n_3} \leq s_{n_2}$. Having chosen n_k with $n_k > n_{k-1}$ and $s_{n_k} \leq s_{n_{k-1}}$, we let n_{k+1} be the subscript of the greatest term of $s^{(n_k+1)}$. Then $n_{k+1} > n_k$ and $s_{n_{k+1}} \leq s_{n_k}$. By induction we obtain a sequence $\{n_k\}_{k=1}^{\infty}$ of positive integers such that $n_1 < n_2 < n_3 < \cdots$ and $s_{n_1} \geq s_{n_2} \geq s_{n_3} \geq \cdots$. Therefore, the sequence $t = \{t_k\}_{k=1}^{\infty}$ defined by $t_k = s_{n_k}$ is a decreasing subsequence of s.

Case (ii): Suppose that, for a certain $N \in Z^+$, the subsequence $s^{(N)}$

$$s_N, s_{N+1}, s_{N+2}, \ldots$$

fails to have a greatest term. It follows that for any term of $s^{(N)}$ there is a term of greater value having a larger subscript (work Exercise 2.7.3).

We select a subsequence t of s in the following way: We let $n_1 = N$ and define $t_1 = s_N$. Next let n_2 be the least subscript greater than n_1 for which $s_{n_2} > s_{n_1}$. Proceed by induction. Having chosen $n_k > n_{k-1}$ in such a way that $s_{n_k} > s_{n_{k-1}}$, let n_{k+1} be the least subscript greater than n_k for which $s_{n_{k+1}} > s_{n_k}$. The sequence of positive integers $\{n_k\}_{k=1}^{\infty}$ obtained in this way clearly satisfies $n_1 < n_2 < n_3 < \cdots$ and $s_{n_1} < s_{n_2} < s_{n_3} < \cdots$. The sequence $t = \{t_k\}_{k=1}^{\infty}$ defined by $t_k = s_{n_k}$, is a strictly increasing subsequence of s. ■

Theorem 2.7.1 **Bolzano-Weierstrass Theorem for Sequences** *Every bounded sequence of real numbers has a convergent subsequence.*

PROOF: Let s be a bounded sequence. By Lemma 2.7.1, there exists a monotone subsequence t of s. Being a subsequence of s, t is also bounded. Then, by part (iii) of Theorem 2.6.1, t is convergent. ■

Cauchy Sequences

Roughly speaking, a Cauchy sequence is one whose terms eventually become arbitrarily close to one another.

Definition 2.7.1 A sequence $\{s_n\}_{n=1}^{\infty}$ is ***Cauchy*** provided that for every $\varepsilon > 0$ there exists a positive integer N such that if m and n are subscripts satisfying $m \geq N$ and $n \geq N$, then $|s_m - s_n| < \varepsilon$; briefly,

$$|s_m - s_n| < \varepsilon \qquad \text{whenever } m, n \geq N$$

AUGUSTIN-LOUIS CAUCHY

(b. August 21, 1789; d. May 23, 1857)

In 1814 Cauchy published a work on definite integrals that became the basis of complex analysis. It brought favorable interest from his contemporaries, and he followed with 788 other papers and several books expounding rigor and covering infinite series, differential equations, theory of numbers, real and complex functions, determinants, and probability, as well as topics in mathematical physics. His life spans years of great political upheaval in France, and Cauchy involved himself in the affairs of Church and State. Except for 18 years of intermittent political exile (1830–1848), he did his work from a teaching position at L'École Polytechnique. Cauchy's name is attached to tests, inequalities, and theorems throughout calculus and analysis.

Example 1: Let $s_n = (n + 2)/n$. We prove that $\{s_n\}_{n=1}^{\infty}$ is a Cauchy sequence. Let $\varepsilon > 0$. We must determine $N \in Z^+$ such that for $m, n \geq N$ we will have $|s_m - s_n| < \varepsilon$. Let m be the larger of the two subscripts; then

$$|s_m - s_n| = \left|\frac{2}{m} - \frac{2}{n}\right| = \frac{2}{n} - \frac{2}{m} < \frac{2}{n}$$

and if N is a positive integer greater than $2/\varepsilon$, then for $m, n \geq N$, we have $|s_m - s_n| < \varepsilon$.

It is easy to see that every convergent sequence is Cauchy, and we will prove soon that every Cauchy sequence is convergent. Thus, the property described in Definition 2.7.1 is another way of formulating the idea of convergence; it is of importance theoretically because it gives us a way of dealing with convergence without specific knowledge of the limit.

Lemma 2.7.2 *Every Cauchy sequence is bounded.*

PROOF: Let $\{s_n\}_{n=1}^{\infty}$ be Cauchy. Invoking the definition with $\varepsilon = 1$, there must exist an $N \in Z^+$ such that $|s_m - s_n| < 1$ when $m, n \geq N$. In particular, letting $m = N$ and $n \geq N$, we get $|s_N - s_n| < 1$. But $s_n = (s_n - s_N) + s_N$, so we have for $n \geq N$,

$$\begin{aligned} |s_n| &\leq |s_n - s_N| + |s_N| = |s_N - s_n| + |s_N| \\ &< 1 + |s_N| \end{aligned}$$

Thus, the number $B = \max\{|s_1|, |s_2|, \ldots, |s_{N-1}|, 1 + |s_N|\}$ is an upper bound of $\{s_n\}_{n=1}^{\infty}$. ∎

Lemma 2.7.3 *If $\{s_n\}_{n=1}^{\infty}$ is Cauchy and has a subsequence $\{t_k\}_{k=1}^{\infty}$ that converges to some real number L, then $\{s_n\}_{n=1}^{\infty}$ itself converges to L.*

PROOF: Suppose that $\{s_n\}_{n=1}^{\infty}$ is Cauchy and that $\{t_k\}_{k=1}^{\infty} = \{s_{n_k}\}_{k=1}^{\infty}$ converges to L; we want to show that $s_n \to L$. Let $\varepsilon > 0$; we must determine $N \in Z^+$ such that $|s_n - L| < \varepsilon$ whenever $n \geq N$. First observe that for any n_k, we can use the Triangle Inequality to write

$$|s_n - L| \leq |s_n - s_{n_k}| + |s_{n_k} - L|$$

Since $\{s_n\}_{n=1}^{\infty}$ is Cauchy, there is an $N_1 \in Z^+$ such that $|s_m - s_n| < \frac{\varepsilon}{2}$ for $m, n \geq N_1$; since $\{s_{n_k}\}_{n=1}^{\infty}$ converges to L, there is a $K \in Z^+$ such that $|s_{n_k} - L| < \frac{\varepsilon}{2}$ for $k \geq K$. We choose n_k to be a *fixed* subscript such that $k \geq K$ and $n_k \geq N_1$. The positive integer N_1 is a suitable choice for N, because for $n \geq N_1$, we have $|s_n - s_{n_k}| < \frac{\varepsilon}{2}$ and $|s_{n_k} - L| < \frac{\varepsilon}{2}$ so that $|s_n - L| < \varepsilon$. ∎

Theorem 2.7.2 **Cauchy Criterion** *A sequence of real numbers is convergent if and only if it is a Cauchy sequence.*

PROOF: Let $\{s_n\}_{n=1}^{\infty}$ be a Cauchy sequence. By Lemma 2.7.2, $\{s_n\}_{n=1}^{\infty}$ is bounded. By Theorem 2.7.1, $\{s_n\}_{n=1}^{\infty}$ has a convergent subsequence, say, $\{s_{n_k}\}_{k=1}^{\infty}$. If $s_{n_k} \to L$, then by Lemma 2.7.3, $\{s_n\}_{n=1}^{\infty}$ itself converges to L. ■

The proof that every convergent sequence is Cauchy is left for you to work out in Exercise 2.7.6.

EXERCISES 2.7

1. Determine whether the sequence whose general term is given below satisfies case (i) or case (ii) in the proof of Lemma 2.7.1. Then find the first five terms of the monotone subsequence constructed in the proof.

   (a) $(-1)^{n+1}$ (b) $\sin \frac{n\pi}{2}$

   (c) $(-1)^n \frac{n}{n+1}$ (d) $(-1)^n\left(1 - \frac{1}{n^2}\right)$

   (e) $n - n^2$ (f) $\left(1 + \frac{1}{n}\right)^n$

2. Prove or disprove the following statements.
   (a) Every sequence has an increasing subsequence.
   (b) Every sequence has a bounded subsequence.
   (c) Every unbounded sequence has a subsequence that diverges to ∞ or to $-\infty$.
   (d) If a sequence has a greatest term, then every subsequence of that sequence has a greatest term.

3. Let s be a sequence that fails to have a greatest term. Show that for any term s_n there exists a term s_m with $m > n$ and $s_m > s_n$.

4. Show that if s_n is the greatest term of a sequence s, and s_{n_k} is the greatest term of a subsequence of s, then $s_{n_k} \leq s_n$. (Can you also conclude that $n_k \geq n$?)

5. Justify the following alternative formulations of the Cauchy property.
   (a) A sequence $\{s_n\}_{n=1}^{\infty}$ is Cauchy if and only if, for every $\varepsilon > 0$, there exists an $N \in Z^+$ such that for all pairs of positive integers m, n, with $m \geq n \geq N$, it follows that $|s_m - s_n| < \varepsilon$.
   (b) A sequence $\{s_n\}_{n=1}^{\infty}$ is Cauchy if and only if, for every $\varepsilon > 0$, there exists an $N \in Z^+$ such that for $k \in Z^+$ and $n \geq N$, $|s_{n+k} - s_n| < \varepsilon$.

6. Show that every convergent sequence is Cauchy. (*Hint:* $|s_m - s_n| = |s_m - L + L - s_n|$, where L is the limit.)

7. Show that the sum of two Cauchy sequences is also Cauchy.

⋆ PROJECT: Construction of R from Q Using Cauchy Sequences

Note: This rather ambitious project is based on the idea of Cauchy sequences. In addition, it involves the construction of a number system, similar in certain respects to the constructions in Section L.5. Before undertaking this project it would be a good idea to study that material, as well as Section 2.7 and the topic of equivalence relations in Section L.3.

In constructing the rationals from the system Z of integers (Section L.5), an equivalence relation was defined on the set of ordered pairs of integers, and the cells in the partition corresponding to that equivalence relation became the elements of the new system. Then operations (addition and multiplication) for these new elements were defined and it was shown that the desired algebraic and order properties hold. Now in the construction of the reals outlined below, we begin with the set of Cauchy sequences of rationals and define an equivalence relation on that set; the cells in the resulting partition will become the elements of the "new" system R (the reals), and operations will be defined for these cells in such a way as to obtain a field that is not only ordered but complete.

Archimedean Property for Q

In this project we start with the assumption that the system Q of rational numbers is an ordered field. Later we will need to use the fact that Q is Archimedean—that if r and s are rational numbers with $r > 0$, then there exists a positive integer n such that $nr > s$.

Of course, the Archimedean property for Q is an immediate consequence of Theorem 1.2.1. But the proof of the latter was based upon the Axiom of Completeness, and since our major objective here is to establish the completeness of R, we cannot use any facts derived from it. Therefore we must find a direct proof that Q is Archimedean. Fortunately it is not at all hard to do so.

(1) Show that the ordered field Q is Archimedean. (It is sufficient to show that for every rational number s there exists an $n \in Z^+$ such that $n > s$.)

Fundamental Sequences

We consider the set of all sequences in Q that have the Cauchy property; such a sequence will be called a *fundamental* sequence.

Definition

A ***fundamental*** sequence is a sequence $\{r_k\}_{k=1}^{\infty}$ of rational numbers ($r_k \in Q$ for $k \in Z^+$) such that for every rational $\varepsilon > 0$ there exists a $K \in Z^+$ such that $|r_k - r_l| < \varepsilon$ for $k, l \geq K$.

Notice that this definition refers only to facts and relations among rational numbers.

(2) Show that every fundamental sequence is bounded in Q; that is, if $\{r_k\}_{k=1}^{\infty}$ is a fundamental sequence, then there exists a rational number r such that $|r_k| \leq r$ for $k \in Z^+$. (*Hint:* The proof of this fact can be modeled after the proof of Lemma 2.7.2.)

Operations on Fundamental Sequences

Definition

The ***sum***, ***difference***, and ***product*** of two fundamental sequences $\{r_k\}_{k=1}^{\infty}$ and $\{s_k\}_{k=1}^{\infty}$ are defined as follows:

$$\{r_k\}_{k=1}^{\infty} + \{s_k\}_{n=1}^{\infty} = \{r_k + s_k\}_{k=1}^{\infty}$$
$$\{r_k\}_{n=1}^{\infty} - \{s_k\}_{n=1}^{\infty} = \{r_k - s_k\}_{k=1}^{\infty}$$
$$\{r_k\}_{n=1}^{\infty} \cdot \{s_k\}_{n=1}^{\infty} = \{r_k s_k\}_{k=1}^{\infty}$$

(3) Show that the sum, the difference, and the product of any two fundamental sequences are also fundamental sequences.

Definition

A fundamental sequence is ***null*** provided that for every rational $\varepsilon > 0$, there exists a $K \in Z^+$ such that $|r_k| < \varepsilon$ for $k \geq K$.

(4) Show that the sum, the difference, and the product of two null fundamental sequences are also null.

It is an important fact that each non-null fundamental sequence is "bounded away" from the zero sequence. We will use this property (stated more precisely in the following exercise) when we define *inverse* and, later *positive*.

(5) Show that if a fundamental sequence is not null, then either there exists a rational $\varepsilon > 0$ and $K \in Z^+$ such that $r_k \geq \varepsilon$ for $k \geq K$, or there exists a rational $\varepsilon > 0$ and $K \in Z^+$ such that $r_k \leq -\varepsilon$ for $k \geq K$.

It follows immediately from Exercise 5 that a non-null fundamental sequence has at most a finite number of zero terms. We now make the following definition:

Definition If a fundamental sequence $\{s_k\}_{k=1}^{\infty}$ is not null, then its ***inverse*** is the sequence $\{t_k\}_{k=1}^{\infty}$ where

$$t_k = \begin{cases} s_k^{-1} & \text{if } s_k \neq 0 \\ 1 & \text{if } s_k = 0 \end{cases}$$

In light of the comment preceding the definition it is clear that the product $\{s_k t_k\}_{k=1}^{\infty}$ consists of terms that are equal to 1 from some term on.

(6) Show that the inverse of a non-null fundamental sequence is also a non-null fundamental sequence.

Division by a non-null sequence is now possible.

Definition If $\{r_k\}_{n=1}^{\infty}$ and $\{s_k\}_{k=1}^{\infty}$ are fundamental sequences, and if $\{s_k\}_{k=1}^{\infty}$ is not null, we define

$$\{r_k\}_{k=1}^{\infty} \div \{s_k\}_{k=1}^{\infty} = \{r_k t_k\}_{k=1}^{\infty}$$

where $\{t_k\}_{k=1}^{\infty}$ is the inverse of $\{s_k\}_{k=1}^{\infty}$ as defined following Exercise 5.

Equivalence of Fundamental Sequences

Definition Let $\{r_k\}_{k=1}^{\infty}$ and $\{s_k\}_{k=1}^{\infty}$ be fundamental sequences. Then $\{r_k\}_{k=1}^{\infty}$ is ***equivalent*** to $\{s_k\}_{k=1}^{\infty}$, written $\{r_k\}_{k=1}^{\infty} \sim \{s_k\}_{k=1}^{\infty}$, provided that the difference $\{r_k - s_k\}_{k=1}^{\infty}$ is null.

(7) Show that the relation $\sim$ is an equivalence relation in the set of all fundamental sequences.

The Set R and Its Operations

Definition A ***real number*** is a cell in the partition of the set of fundamental sequences corresponding to $\sim$. The set of all these cells is denoted by R.

Now that we have the *set* R of real numbers, we need to define addition and multiplication:

Definition Let $x, y \in R$. Suppose that x is the cell containing the fundamental sequence $\{r_k\}_{k=1}^{\infty}$, while y is the cell containing the fundamental sequence $\{s_k\}_{k=1}^{\infty}$. Then the ***sum*** $x + y$ is the cell containing the fundamental sequence $\{r_k + s_k\}_{k=1}^{\infty}$, and the ***product*** xy is the cell containing the fundamental sequence $\{r_k s_k\}_{k=1}^{\infty}$. That is, if $x = \overline{\{r_k\}_{k=1}^{\infty}}$ and $y = \overline{\{s_k\}_{k=1}^{\infty}}$, then

$$x + y = \overline{\{r_k + s_k\}_{k=1}^{\infty}}$$
$$xy = \overline{\{r_k s_k\}_{k=1}^{\infty}}$$

Of course, it is necessary to show that the sum and product of cells are well defined.

(8) Show that if x is the cell containing the fundamental sequences $\{r_k\}_{k=1}^{\infty}$ and $\{r_k'\}_{k=1}^{\infty}$, and if y is the cell containing the fundamental sequences $\{s_k\}_{k=1}^{\infty}$ and $\{s_k'\}_{k=1}^{\infty}$, then $\{r_k + s_k\}_{k=1}^{\infty} \sim \{r_k' + s_k'\}_{k=1}^{\infty}$, so that addition of real numbers is well defined. Show similarly that multiplication is well defined.

R Is a Field

We can now show that R is a field with the operations defined above.

(9) Verify the commutative, associative, and distributive laws for the operations on R. Show that the cell $\overline{\{0\}_{k=1}^{\infty}}$ containing the zero sequence is in fact the class of all null fundamental sequences, and that it serves as the zero of R. Show also that the cell $\overline{\{1\}_{k=1}^{\infty}}$ is the unity of R. Show that if $x = \overline{\{r_k\}_{k=1}^{\infty}}$, then the cell $\overline{\{-r_k\}_{k=1}^{\infty}}$ is the opposite of x, and that if x is not zero, then the cell containing the inverse of $\{r_k\}_{k=1}^{\infty}$ is the inverse of x.

Order

Definition A fundamental sequence $\{r_k\}_{k=1}^{\infty}$ is ***positive*** provided that for some $\varepsilon > 0$ ($\varepsilon \in Q$) there exists a $K \in Z^+$ such that $r_k \geq \varepsilon$ for $k \geq K$.

It follows from Exercise 5 that every non-null fundamental sequence either is positive itself or has an opposite that is positive. Moreover, the sum and the product of two positive fundamental sequences are also positive.

(10) Verify the remarks in the preceding paragraph. Show also that if $\{r_k\}_{k=1}^{\infty}$ and $\{r'_k\}_{k=1}^{\infty}$ are equivalent and if one is positive, then so is the other.

The set of all fundamental sequences can be ordered by defining $\{r_k\}_{k=1}^{\infty}$ to be greater than $\{s_k\}_{k=1}^{\infty}$ provided that $\{r_k - s_k\}_{k=1}^{\infty}$ is positive; that is, provided that there is some $\varepsilon > 0$ ($\varepsilon \in Q$) and some $K \in Z^+$ such that $r_k - s_k > \varepsilon$ for $k \geq K$.

Positivity and order can now be introduced into R.

Definition A real number $x = \overline{\{r_k\}_{k=1}^{\infty}}$ is ***positive*** provided the fundamental sequence $\{r_k\}_{k=1}^{\infty}$ is positive.

The definition of "positive" for real numbers makes sense because of Exercise 10, and it is clear that R satisfies the postulates for order (Postulates 9 and 10 in Section R.1). Thus, R is an ordered field in which $x > y$ means that $\{r_k\}_{k=1}^{\infty} > \{s_k\}_{k=1}^{\infty}$, where $x = \overline{\{r_k\}_{k=1}^{\infty}}$ and $y = \overline{\{s_k\}_{k=1}^{\infty}}$.

"Rational Real" Numbers

If r is a fixed element of Q, we will call the real number $\overline{\{r\}_{k=1}^{\infty}}$ a "rational real number". Of course, $\overline{\{r\}_{k=1}^{\infty}}$ is not an element of Q; it has originated from the element $r \in Q$ and is just the cell containing the constant sequence $\{r\}_{k=1}^{\infty}$. Nevertheless it is convenient to think of the rational numbers as being "embedded" in R by identifying $\overline{\{r\}_{k=1}^{\infty}}$ with r, and so we will write r instead of $\overline{\{r\}_{k=1}^{\infty}}$. This abbreviation will do no harm because the equation

$$r + s = t$$

holds for $r, s, t \in Q$ if and only if

$$\overline{\{r\}_{k=1}^{\infty}} + \overline{\{s\}_{k=1}^{\infty}} = \overline{\{t\}_{k=1}^{\infty}}$$

holds in R.

Similarly,

$$rs = t$$

holds in Q if and only if

$$\overline{\{r\}_{k=1}^{\infty}}\ \overline{\{s\}_{k=1}^{\infty}} = \overline{\{t\}_{k=1}^{\infty}}$$

holds in R. These equivalences say that the rational real numbers behave algebraically just like their counterparts in Q.* Moreover it is easy to see that the inequality $r < s$ in Q holds if and only if $\overline{\{r\}_{k=1}^{\infty}} < \overline{\{s\}_{k=1}^{\infty}}$ in R, and so order is also preserved by this embedding. Therefore we will no longer distinguish the rational real numbers from their counterparts in Q.

Completeness of R

Under the order relation we have established, R is Archimedean. For if $B \in R$; it follows from Exercise 2 that there exists a rational number r such that $r \geq B$. But by Exercise 1, there is a positive integer n such that $n > r$. Therefore, for every $B \in R$ there is a positive integer n such that $n > B$.

Finally we prove that R is complete—that is, that every nonempty subset of R that has an upper bound in R has a least upper bound there. The proof outlined below is adapted from Chapter 11 of van de Waerden's *Algebra*.†

Let S be a nonempty set of real numbers that is bounded above, say, by the real number B; then $x \leq B$ for $x \in S$. By the Archimedean property, there is a positive integer n such that $n > B$. Now let x be a fixed element of S; we can choose an integer m such that $m < x$, giving us $m < x < n$. For each $k \in Z^+$, consider the set M_k of all multiples of 2^{-k} that lie between m and n inclusive:

$$M_k = \{q \cdot 2^{-k} : q \in Z, m \leq q \cdot 2^{-k} \leq n\}$$

The set M_k is finite and we can choose $y_k \in Q$ to be the least element of M_k that is an upper bound of S. Observe that $y_k - 2^{-k}$ is not an upper bound of S.

(11) (a) Show that if $l > k$, $l \in Z^+$, then

$$y_k - 2^{-k} < y_l \leq y_k$$

so that $|y_k - y_l| < 2^{-k}$.

(b) Show that $\{y_k\}_{k=1}^{\infty}$ is a fundamental sequence.

(c) Show that the real number $y = \overline{\{y_k\}_{k=1}^{\infty}}$ is an upper bound of S.

(d) Show that y is the least upper bound of S.

★ 2.8 *Infinite Series*

In this section we see how the limit concept can be used to give meaning to a sum of a countably infinite number of terms.

Definition 2.8.1 An ***infinite series*** is an expression of the form

$$\sum_{n=1}^{\infty} a_n = a_1 + a_2 + \cdots + a_n + \cdots \tag{1}$$

where the $a_n (n \in Z^+)$ are real numbers, known as the ***terms*** of the series. The sequence of ***partial sums*** of the series (1) is the sequence $\{S_n\}_{n=1}^{\infty}$ defined recursively by

$$S_1 = a_1$$
$$S_{k+1} = S_k + a_{k+1} \qquad k \in Z^+$$

*If you have studied abstract algebra, you will recognize this embedding as an isomorphism.
†B. L. van de Waerden, *Algebra*, vol. I (New York: Springer-Verlag, 1991), pp. 240–241.

We can write the general term of the sequence $\{S_n\}_{n=1}^{\infty}$ as

$$S_n = a_1 + a_2 + \cdots + a_n = \sum_{k=1}^{n} a_k$$

The series (1) is said to ***converge*** if and only if its sequence $\{S_n\}_{n=1}^{\infty}$ of partial sums converges; in case of convergence, the limit of $\{S_n\}_{n=1}^{\infty}$ is the ***sum*** of the series (1). A series that does not converge is said to ***diverge***. If $\{S_n\}_{n=1}^{\infty}$ diverges to $\infty(-\infty)$ we say that the ***sum*** of the series (1) is $\infty(-\infty)$.

Remark: Before going further note that there are two sequences involved with an infinite series: the sequence $\{a_n\}_{n=1}^{\infty}$ consists of the terms being added; the sequence $\{S_n\}_{n=1}^{\infty}$ consists of sums obtained by adding blocks of consecutive terms from the beginning. It is essential to distinguish these two sequences carefully.

Geometric Series

Example 1: Consider the series

$$\sum_{n=1}^{\infty} \frac{1}{2^n} = \frac{1}{2} + \frac{1}{4} + \cdots + \frac{1}{2^n} + \cdots$$

The terms of this series are the real numbers $a_n = 1/2^n (n \in Z^+)$. Its partial sums are given by

$$\begin{aligned} S_n &= \frac{1}{2} + \frac{1}{4} + \cdots + \frac{1}{2^n} \\ &= \frac{1}{2}\left(1 + \frac{1}{2} + \cdots + \frac{1}{2^{n-1}}\right) \\ &= \frac{1}{2} \cdot \frac{1 - \frac{1}{2^n}}{1 - \frac{1}{2}} = 1 - \frac{1}{2^n} \qquad \text{by Equation (6), Sec. P.2} \end{aligned}$$

Clearly the sequence $\{S_n\}_{n=1}^{\infty}$ converges to 1, and so the given series converges and its sum is 1.

The following example generalizes Example 1.

Example 2: Let a and r be real numbers such that $a \neq 0$ and $|r| < 1$. The *geometric series*

$$\sum_{n=1}^{\infty} ar^{n-1} = a + ar + \cdots + ar^{n-1} + \cdots$$

has terms $ar^{n-1}(n \in Z^+)$ and partial sums

$$S_n = a(1 + r + \cdots + r^{n-1}) = a\left(\frac{1 - r^n}{1 - r}\right) \qquad \text{by Equation (6), Sec. P.2}$$

$$= \frac{a}{1 - r} - \frac{ar^n}{1 - r}$$

Because $|r| < 1$, the sequence $\{r^n\}_{n=1}^{\infty}$ is null (Theorem 2.2.5), and it follows that $S_n \to a/(1 - r)$. Thus, the geometric series converges when $|r| < 1$ to the sum $a/(1 - r)$.

In order for a series to converge, its terms must form a null sequence.

Theorem 2.8.1 *If the series* $\sum_{n=1}^{\infty} a_n$ *converges, then* $a_n \to 0$

PROOF: Let $\{S_n\}_{n=1}^{\infty}$ be the sequence of partial sums of $\sum_{n=1}^{\infty} a_n$ and let $S = \lim_{n\to\infty} S_n$ denote its sum. This limit exists by our hypothesis. Then for $n \in \mathbb{Z}^+$, $a_{n+1} = S_{n+1} - S_n$, and it follows that

$$\lim_{n\to\infty} a_{n+1} = \lim_{n\to\infty} (S_{n+1} - S_n) = \lim_{n\to\infty} S_{n+1} - \lim_{n\to\infty} S_n = S - S = 0.$$

But $\{a_{n+1}\}_{n=1}^{\infty}$ is a tail of $\{a_n\}_{n=1}^{\infty}$, so we conclude that $\{a_n\}_{n=1}^{\infty}$ is also null. ∎

Consider the series $\sum_{n=1}^{\infty} ar^{n-1}$ where $a \neq 0$ but $|r|$ is not less than 1. Then the sequence $\{ar^{n-1}\}_{n=1}^{\infty}$ is not null, and by Theorem 2.8.1 the series cannot converge. By combining this observation with the result of Example 2, we have the following theorem.

Theorem 2.8.2 **Geometric Series** *The geometric series* $\sum_{n=1}^{\infty} ar^{n-1} (a \neq 0)$ *converges if and only if* $|r| < 1$; *in the case of convergence, its sum is* $a/(1 - r)$.

Comment on Notation: The symbol n used in defining a series $\sum_{n=1}^{\infty} a_n$ is called an *index*; clearly it can be replaced by another symbol, such as m or k, without changing the meaning. Thus,

$$\sum_{n=1}^{\infty} a_n = \sum_{m=1}^{\infty} a_m = \sum_{k=1}^{\infty} a_k$$

and so on. If the context makes it clear what the index is, and if the summation extends from 1 to ∞, then it is convenient to write simply $\sum$ instead of $\sum_{n=1}^{\infty}$. Sometimes we wish to begin a series with $n = 0$ or some other integer; in such cases that will be explicitly indicated on the sigma sign.

The next theorem follows from the corresponding theorems about sequences; the proof is left for you to work out in Exercise 2.8.23.

Theorem 2.8.3 *Let $\sum a_n$ and $\sum b_n$ be convergent series with sums A and B, respectively. Then $\sum (a_n + b_n)$ converges and its sum is $A + B$. If c is any constant, then $\sum ca_n$ converges and its sum is cA.*

Series with Nonnegative Terms

In the remainder of this section we consider only series whose terms are all nonnegative. Let $\sum a_n$ be such a series and let $\{S_n\}_{n=1}^{\infty}$ be its sequence of partial sums. Then since $S_{n+1} = S_n + a_{n+1}$ and $a_{n+1} \geq 0$, we see that $\{S_n\}_{n=1}^{\infty}$ is increasing. By part (i) of Theorem 2.6.3, we conclude that $\{S_n\}_{n=1}^{\infty}$ either converges or diverges to ∞. Thus, we have the following theorem.

Theorem 2.8.4 *A series of nonnegative terms either converges or diverges to ∞. Such a series converges if its sequence of partial sums is bounded above; it diverges otherwise.*

The following examples illustrate the two possible outcomes in Theorem 2.8.4.

Example 3: The *harmonic series*

$$\sum \frac{1}{n} = 1 + \frac{1}{2} + \frac{1}{3} + \cdots + \frac{1}{n} + \cdots$$

diverges to ∞ in spite of the fact that its terms approach zero. To see why, consider the subsequence of the sequence of partial sums corresponding to powers of 2:

$$\begin{aligned}
S_1 &= 1 \\
S_2 &= 1 + \tfrac{1}{2} \\
S_4 &= 1 + \tfrac{1}{2} + \left(\tfrac{1}{3} + \tfrac{1}{4}\right) \geq 1 + \tfrac{1}{2} + \tfrac{2}{4} = 1 + \tfrac{1}{2} + \tfrac{1}{2} \\
S_8 &= 1 + \tfrac{1}{2} + \tfrac{1}{3} + \tfrac{1}{4} + \tfrac{1}{5} + \tfrac{1}{6} + \tfrac{1}{7} + \tfrac{1}{8} \geq 1 + \tfrac{1}{2} + \left(\tfrac{1}{4} + \tfrac{1}{4}\right) + \left(\tfrac{1}{8} + \tfrac{1}{8} + \tfrac{1}{8} + \tfrac{1}{8}\right) \\
&= 1 + \tfrac{1}{2} + \tfrac{2}{4} + \tfrac{4}{8} \\
&= 1 + \tfrac{3}{2}
\end{aligned}$$

By induction it can be shown that $S_{2^k} \geq 1 + \frac{k}{2}$, and so $\{S_{2^k}\}_{k=1}^{\infty}$ is unbounded and therefore diverges to ∞. But $\{S_{2^k}\}_{k=1}^{\infty}$ is a subsequence of $\{S_n\}_{n=1}^{\infty}$ and so by part (iv) of Theorem 2.6.3 $\{S_n\}_{n=1}^{\infty}$ diverges to ∞. Therefore $\sum 1/n$ diverges.

Note: Example 3 shows that the converse of Theorem 2.8.1 fails.

Example 4: The series

$$\sum \frac{1}{n^2} = 1 + \frac{1}{4} + \frac{1}{9} + \cdots + \frac{1}{n^2} + \cdots$$

converges. Here we consider the subsequence $\{S_{2^k-1}\}_{k=1}^{\infty}$ of the sequence of partial sums $\{S_n\}_{n=1}^{\infty}$:

$$S_1 = 1$$
$$S_3 = 1 + \left(\frac{1}{2^2} + \frac{1}{3^2}\right) \leq 1 + \frac{1}{2^2} + \frac{1}{2^2} = 1 + \frac{2}{2^2} = 1 + \frac{1}{2}$$
$$S_7 = 1 + \left(\frac{1}{2^2} + \frac{1}{3^2}\right) + \left(\frac{1}{4^2} + \frac{1}{5^2} + \frac{1}{6^2} + \frac{1}{7^2}\right) \leq 1 + \frac{2}{2^2} + \frac{4}{4^2} = 1 + \frac{1}{2} + \frac{1}{4}$$

By induction it can be shown that

$$S_{2^k-1} \leq 1 + \frac{1}{2} + \frac{1}{4} + \cdots + \frac{1}{2^{k-1}}$$

and for $k \in Z^+$ we have [by Equation (6), Section P.2],

$$S_{2^k-1} \leq \frac{1 - \frac{1}{2^k}}{1 - \frac{1}{2}} = 2 - \frac{1}{2^{k-1}} < 2$$

Thus, $\{S_{2^k-1}\}_{k=1}^{\infty}$ is bounded above and converges. It follows from part (iv) Theorem 2.6.3 that $\{S_n\}_{n=1}^{\infty}$ also converges to the same limit. We see that $\sum 1/n^2$ converges to a sum that is less than 2.

Our method in Example 4 shows convergence, but unlike Examples 1 and 2, it does not provide the exact sum of the series. It is an interesting fact that the sum of the series $\sum 1/n^2$ is $\pi^2/6$, but we will not prove that here.*

Comparison Tests

Knowing that the series $\sum 1/n^2$ converges, we should expect the series $\sum 1/n^3$ to converge also because the terms of the latter series are less than those of $\sum 1/n^2$. This idea suggests the concept of "domination".

Definition 2.8.2 If $\sum a_n$ and $\sum b_n$ are series such that

$$0 \leq a_n \leq b_n$$

for $n \in Z^+$, then we say that $\sum b_n$ ***dominates*** $\sum a_n$ or that $\sum a_n$ is ***dominated*** by $\sum b_n$.

Theorem 2.8.5 **Comparison Test** *Let $\sum a_n$ and $\sum b_n$ be series of nonnegative terms such that $\sum b_n$ dominates $\sum a_n$. Then if $\sum b_n$ converges, so does $\sum a_n$; equivalently, if $\sum a_n$ diverges, then so does $\sum b_n$.*

*See George Simmons, "Euler's Discovery of the Formula $\sum_1^{\infty} \frac{1}{n^2} = \frac{\pi^2}{6}$" and "A Rigorous Proof of Euler's Formula $\sum_1^{\infty} \frac{1}{n^2} = \frac{\pi^2}{6}$," in *Calculus Gems* (New York: McGraw-Hill, 1992), pp. 267–271.

PROOF: Let $B_n = b_1 + b_2 + \cdots + b_n$ and $A_n = a_1 + a_2 + \cdots + a_n$ so that $\{B_n\}_{n=1}^{\infty}$ and $\{A_n\}_{n=1}^{\infty}$ are the sequences of partial sums of $\sum b_n$ and $\sum a_n$, respectively. These sequences are both increasing. Since $a_n \leq b_n$ for $n \in Z^+$, it follows that $A_n \leq B_n$ for $n \in Z^+$. Given that $\{B_n\}_{n=1}^{\infty}$ converges, we can write $\lim_{n\to\infty} B_n = B = \sup\{B_n : n \in Z^+\}$ by Theorem 2.6.1. Since $A_n \leq B_n$, we see that $\{A_n\}_{n=1}^{\infty}$ is bounded above by B. By Theorem 2.6.1, $\{A_n\}_{n=1}^{\infty}$ is also convergent; its limit A satisfies the condition

$$A = \sup\{A_n : n \in Z^+\} \leq B \qquad \blacksquare$$

Example 5: The series $\sum n/(n^3 + 4)$ is dominated by $\sum 1/n^2$ because

$$\frac{n}{n^3 + 4} \leq \frac{n}{n^3} = \frac{1}{n^2} \qquad n \in Z^+$$

Since $\sum 1/n^2$ is convergent, we conclude that $\sum n/(n^3 + 4)$ is also convergent.

Example 6: The series $\sum 1/2\sqrt{n}$ dominates the series $\sum 1/2n$ because

$$\frac{1}{2\sqrt{n}} \geq \frac{1}{2n} \qquad n \in Z^+$$

Since $\sum 1/2n$ is divergent (why?) we conclude that $\sum 1/2\sqrt{n}$ is divergent.

It is clear that the convergence or divergence of an infinite series is not affected by changing the first term—or for that matter by changing any *finite* number of terms. In fact, a series $\sum_{n=1}^{\infty} a_n$ converges if and only if, for every $m \in Z^+$, the tail $\sum_{n=m}^{\infty} a_n$ converges.

Example 7: Beginning with the fourth term, the series $\sum 1/n!$ is dominated by $\sum 1/2^n$, a convergent geometric series. Therefore, $\sum 1/n!$ converges.

Example 8: For $p > 0$, the series $\sum 1/n^p$ is called the *p-series* . Until now we have not formally discussed the meaning of n^p except when p is a positive integer. In case $p = \frac{1}{2}$, you know of course that $n^{1/2}$ means $\sqrt{n}$, that $n^{1/3}$ means $\sqrt[3]{n}$, and so on; however a general definition of n^p involves difficulties when p is irrational. Section 4.10 gives such a definition; it can then be proved (Exercise 4.10.5) that $n^{p_1} \leq n^{p_2}$ whenever $n \in Z^+$ and $0 \leq p_1 \leq p_2$. Thus, for $0 < p < 1$, the p-series $\sum 1/n^p$ dominates the harmonic series $\sum 1/n$ (Example 3) and therefore diverges by the Comparison Test. Similarly, we see that if $p \geq 2$, then $\sum 1/n^p$ is dominated by $\sum 1/n^2$ (Example 4) and therefore converges. There remains the question of the convergence or divergence of $\sum 1/n^p$ when $1 < p < 2$; this matter will be settled in Section 4.13 by showing that $\sum 1/n^p$ converges for all $p > 1$.

If the terms of a series $\sum a_n$ ($a_n > 0$) are no larger than M times the terms of another series $\sum b_n$, and if $\sum b_n$ converges, then so does $\sum a_n$. This simple observation leads to a very efficient test for convergence that uses a limit.

Theorem 2.8.6 **Limit Comparison Test** *Let $\sum a_n$ and $\sum b_n$ be series of positive terms such that $\lim_{n\to\infty} (a_n/b_n)$ exists. Then:*

(i) The convergence of $\sum b_n$ is a sufficient condition for the convergence of $\sum a_n$.
(ii) If in addition, $\lim_{n\to\infty} (a_n/b_n) \neq 0$, then the convergence of $\sum a_n$ is sufficient for the convergence of $\sum b_n$.

PROOF:

(i) Since the sequence $\{a_n/b_n\}_{n=1}^{\infty}$ is convergent, it is bounded, so there is a number M such that $0 < a_n/b_n \leq M$ for $n \in Z^+$. Thus, $0 < a_n \leq Mb_n$, and $\sum a_n$ is dominated by $\sum Mb_n$, which converges because $\sum b_n$ converges (Theorem 2.8.3).
(ii) The condition $\lim_{n\to\infty} (a_n/b_n) \neq 0$ requires that $\lim_{n\to\infty} (b_n/a_n)$ exists, and so we can interchange the roles of a_n and b_n in part (i) ∎

Example 9: $\sum 5n/(3n^3 + 4)$ converges by a limit comparison with $\sum 1/n^2$:

$$\lim_{n\to\infty}\left(\frac{5n}{3n^3 + 4} \div \frac{1}{n^2}\right) = \lim_{n\to\infty}\left(\frac{5n^3}{3n^3 + 4}\right) = \frac{5}{3}$$

Since $\sum 1/n^2$ is known to converge, the series $\sum 5n/(3n^3 + 4)$ must converge also.

Example 10: $\sum (n^2 + 2n + 8)/(3n^3 + 5n + 10)$ diverges by a limit comparison with $\sum 1/n$. Since

$$\lim_{n\to\infty}\left(\frac{n^2 + 2n + 8}{3n^3 + 5n + 10} \div \frac{1}{n}\right) = \frac{1}{3}$$

the two series either both converge or both diverge, and $\sum 1/n$ diverges.

Example 11: The series $\sum (\sqrt[n]{n}/n^p)$, where p is a fixed positive real number, converges if and only if $\sum 1/n^p$ converges because

$$\lim_{n\to\infty}\left(\frac{\sqrt[n]{n}}{n^p} \div \frac{1}{n^p}\right) = \lim_{n\to\infty} \sqrt[n]{n} = 1$$

Using the result stated in Example 8, we conclude that $\sum (\sqrt[n]{n}/n^p)$ converges if and only if $p > 1$.

Ratio Test

Theorem 2.8.7 **The Ratio Test** *Let $a_n > 0$ for $n \in Z^+$, and assume that $\lim_{n\to\infty} (a_{n+1}/a_n) = \rho$ exists. Then:*

(i) $\sum a_n$ *converges if* $\rho < 1$
(ii) $\sum a_n$ *diverges if* $\rho > 1$

PROOF:

(i) Since ρ is the limit of a positive sequence, $\rho \geq 0$; and we are assuming in this case that $\rho < 1$. Let ε be a fixed positive number less than $1 - \rho$. Then there is an $N \in Z^+$ such that $a_{n+1}/a_n < \rho + \varepsilon$ whenever $n \geq N$. Choose $r = \rho + \varepsilon$, and note that $0 < r < 1$, so that

$$a_{n+1} < ra_n \qquad n \geq N$$

By letting n take successively the values $N, N + 1, N + 2, \ldots$, we obtain

$$\begin{aligned} &a_{N+1} < ra_N \\ &a_{N+2} < ra_{N+1} < r^2 a_N \\ &a_{N+3} < ra_{N+2} < r^3 a_N \\ &\vdots \\ &a_{N+k} < ra_{N+k-1} < r^k a_N \\ &\vdots \end{aligned}$$

But this means that the series

$$a_{N+1} + a_{N+2} + \cdots + a_{N+k} + \cdots$$

is dominated by the convergent geometric series

$$ra_N + r^2 a_N + \cdots + r^k a_N + \cdots$$

and thus converges. Since a tail of $\sum a_n$ converges, we conclude that $\sum a_n$ itself converges.

(ii) In this case $\rho > 1$, and so we can choose $N \in Z^+$ such that for $n \geq N$ we have $a_{n+1}/a_n > 1$. It then follows that $a_N < a_{N+1} < a_{N+2} < \cdots$; clearly $\{a_n\}_{n=1}^{\infty}$ is not null, so by Theorem 2.8.1, the series $\sum a_n$ diverges. ■

Now consider the p-series $\sum 1/n^p (p > 0)$. For all values of p, we obtain

$$\rho = \lim_{n\to\infty} \left[\frac{1}{(n + 1)^p} \div \frac{1}{n^p}\right] = 1$$

When $p = 1$, the series diverges, but when $p = 2$, the series converges. Thus, we can draw no conclusion when $\rho = 1$.

Note: The number ρ in the Ratio Test can be regarded as a measure of the rate of growth of the terms of $\sum a_n$; for large n, a_{n+1} is approximately ρ times a_n. The theorem says that the series converges if the rate of growth is small ($\rho < 1$), it diverges if the rate is large ($\rho > 1$).

If $\rho = \infty$ (that is, if $\{(a_{n+1}/a_n)\}_{n=1}^{\infty}$ diverges to ∞), it is clear that $\sum a_n$ diverges.

Example 12: Consider $\sum n^2 2^n/3^{n+1}$. Here ρ is given by

$$\rho = \lim_{n\to\infty}\left(\frac{(n+1)^2}{3^{n+2}}2^{n+1} \div \frac{n^2 2^n}{3^{n+1}}\right) = \lim_{n\to\infty}\frac{2}{3}\left(\frac{n+1}{n}\right)^2 = \frac{2}{3}$$

Since $\rho < 1$, the series converges.

Example 13: Consider $\sum 2^n/n!$. The number ρ is given by

$$\rho = \lim_{n\to\infty}\left(\frac{2^{n+1}}{(n+1)!} \div \frac{2^n}{n!}\right) = \lim_{n\to\infty}\frac{2}{n+1} = 0$$

and so the series converges.

Example 14: Consider $\sum(2n)!/(n!)^2$. The number ρ is given by

$$\begin{aligned}\rho &= \lim_{n\to\infty}\left\{\frac{[2(n+1)]!}{[(n+1)!]^2} \div \frac{(2n)!}{(n!)^2}\right\}\\ &= \lim_{n\to\infty}\frac{2(n+1)(2n+1)}{(n+1)(n+1)} = 4\end{aligned}$$

Since $\rho > 1$, the series diverges.

Example 15: Consider $\sum [2 + (-1)^n]/(n^2 + 1)$. For odd n, we have

$$\frac{a_{n+1}}{a_n} = \frac{3}{(n+1)^2+1}\cdot\frac{n^2+1}{1}$$

and so a subsequence of $\{a_{n+1}/a_n\}_{n=1}^{\infty}$ converges to 3. But the subsequence corresponding to even n converges to $\frac{1}{3}$. Therefore, the sequence $\{a_{n+1}/a_n\}_{n=1}^{\infty}$ does not converge, and so the Ratio Test cannot be applied. However, the series does converge by comparison with $\sum (3/n^2)$.

Note: The Limit Comparison Test and the Ratio Test both involve the calculation of the limit of a ratio, and students sometimes confuse them. Keep in mind that the Limit Comparison Test compares two series, one of whose convergence or divergence is known; the comparison is carried out by means of the quotient of the general terms of the two series. The Ratio Test, on the other hand, is not a comparison test at all. It involves only the terms of the given series and is based on the rate at which these terms are growing, as determined by the limit of the ratio of consecutive terms.

★★ *Systems of Numeration: Base Ten*

Infinite series shed some light on our base-ten positional system of numeration (the "decimal" system). You know, for instance, that 543 means $5 \times 10^2 + 4 \times 10^1 + 3 \times 10^0$. It is a fact (work Exercise 2.8.26) that every positive integer M can be expressed, uniquely, in terms of powers of 10 as follows:

$$M = m_p 10^p + m_{p-1} 10^{p-1} + \cdots + m_1 10 + m_0$$

where the $m_i (i = 0, 1, \ldots, p)$ are *digits* (nonnegative integers less than 10). In decimal notation we write

$$M = m_p m_{p-1} \cdots m_1 m_0.$$

Real numbers less than 1 can be expressed as sums of powers $\frac{1}{10}$ (negative powers of 10). For example,

$$\frac{1}{8} = \frac{1}{10} + 2 \cdot \frac{1}{10^2} + 5 \cdot \frac{1}{10^3}$$

or, in decimal notation,

$$\frac{1}{8} = 0.125$$

However, some real numbers, such as $\frac{1}{3}$, cannot be expressed as a *finite* sum of powers of $\frac{1}{10}$; to obtain $\frac{1}{3}$ we must use an infinite series:

$$\frac{1}{3} = \frac{3}{10} + \frac{3}{10^2} + \frac{3}{10^3} + \cdots + \frac{3}{10^n} + \cdots = 0.333\ldots$$

In general, if $0 < r < 1$, there exist digits d_n $(n \in Z^+)$ with $0 \le d_n \le 9$ such that

$$r = \frac{d_1}{10} + \frac{d_2}{10^2} + \cdots + \frac{d_n}{10^n} + \cdots = \sum_{n=1}^{\infty} \frac{d_n}{10^n}$$

In decimal notation,

$$r = 0.d_1 d_2 \ldots d_n \ldots$$

That such an infinite series converges is easily established by a comparison test (work Exercise 2.8.25). It requires a bit more effort to prove that every real number between 0 and 1 can be obtained as such a series (try Exercise 2.8.27).

Terminating decimals are those $\left(\text{such as } \frac{1}{8}\right)$ that stop after a finite number of digits; in such cases we can consider the digits to be 0 from some point

★★ These portions of Section 2.8 are optional; the corresponding exercises are also marked with a double star.

on. A real number given by a terminating decimal can also be represented by a decimal that ends in an infinite string of 9s; for example,

$$0.125000\ldots = 0.124999\ldots$$

This curiosity can readily be verified by calculating the sum of the geometric series on the right.

The decimal expansion of a real number r between 0 and 1 is unique unless r is represented by a terminating decimal, in which case r can also be represented by a decimal that ends in an infinite string of 9s (Exercise 2.8.28).

The remarks above can be adapted to apply to an arbitrary positive number x by breaking it into a sum such as

$$x = m + r$$

where $m = [x]$ and $0 \le r < 1$.

Theorem 2.8.8 *Every positive real number x can be expressed in the form*

$$\begin{aligned} x &= m_p m_{p-1} \ldots m_1 m_0 \cdot d_1 d_2 \ldots d_n \ldots \\ &= m_p 10^p + m_{p-1} 10^{p-1} + \cdots + m_1 10 + m_0 + \sum_{n=1}^{\infty} \frac{d_n}{10^n} \end{aligned}$$

where the $m_i (i = 0, 1, \ldots, p)$ and $d_n (n \in \mathbb{Z}^+)$ are digits. This expression is unique unless the d_n are all zero from some point on.

★★ *Systems of Numeration: Other Bases*

Let B be any integer greater than 1. Then B can be used as the base for a system of numeration for the real numbers. In this case there are B "digits" (representing the integers from 0 to $B - 1$ inclusive), and every real number x can be expressed as a sum of powers of B:

$$x = m_p B^p + m_{p-1} B^{p-1} + \cdots + m_1 B + m_0 + \sum_{n=1}^{\infty} \frac{d_n}{B^n}$$

This expression for x "to the base B" is unique except for the case in which the infinite series terminates. The details of the theory are completely analogous to the special case in which B is ten.

Base two numeration is called *binary*; base three, *ternary*.

Example 16: In the ternary system, "digits" 0, 1, and 2 are used; the real number 3 is written 10; one-third is written 0.1; two-ninths is 0.02; thirty-five and five-ninths is 1022.12.

EXERCISES 2.8

In Exercises 1–21 test the series for convergence. If possible, find the sum.

1. $\sum \sqrt{\dfrac{n+1}{n}}$

2. $\sum \dfrac{n^2+1}{100n^2+1}$

3. $\sum \dfrac{5^{n+4}}{8^n}$

4. $\sum \dfrac{a^{2n}}{(a^2+1)^n}$ (a a constant)

5. $\sum \dfrac{4^n}{3^{n+5}}$

6. $\sum \dfrac{1}{2n}$

7. $\sum \dfrac{1}{2n-1}$

8. $\sum \dfrac{1}{An+B}$ (A, B positive constants)

9. $\sum \dfrac{n^2+1}{n^3}$

10. $\sum \dfrac{n+1}{n^4}$

11. $\sum \dfrac{1}{\sqrt{n^4+4}}$

12. $\sum (\sqrt{n+1}-\sqrt{n})$

13. $\sum \dfrac{n!}{10^n}$

14. (a) $\sum \dfrac{n!}{n^n}$ (b) $\sum \dfrac{n^n}{n!}$

15. $\sum \dfrac{1}{n\sqrt[n]{n}}$

16. $\sum \left(1+\dfrac{1}{n}\right)^n$

17. $\sum \dfrac{1}{\left(1+\frac{1}{n}\right)^n}$

18. $\sum \dfrac{1}{2^n}\left(1+\dfrac{1}{n}\right)^n$

19. $\sum \dfrac{2^n(n!)^2}{(2n)!}$

20. $\sum \dfrac{1}{\sqrt[n]{n^2+1}}$

21. $\sum \dfrac{1\cdot 3\cdot \cdots \cdot (2n-1)}{2\cdot 4\cdot \cdots \cdot (2n)}$

22. The partial sums of the series $\sum 1/n(n+1)$ are easily obtained by means of a device similar to the method of partial fractions in calculus. Observe that

$$\frac{1}{k(k+1)} = \frac{1}{k} - \frac{1}{k+1}$$

$$\begin{aligned} S_n &= \sum_{k=1}^{n} \frac{1}{k(k+1)} = \sum_{k=1}^{n}\left(\frac{1}{k}-\frac{1}{k+1}\right) \\ &= \left(1-\frac{1}{2}\right)+\left(\frac{1}{2}-\frac{1}{3}\right)+\cdots+\left(\frac{1}{n}-\frac{1}{n+1}\right) \\ &= 1-\frac{1}{n+1} \end{aligned}$$

Since $\lim_{n\to\infty} S_n = 1$, we see that this series converges and that the sum is 1.

Use the method suggested above to find an explicit expression for the partial sums of each series below and determine the sum.

(a) $\sum_{n=2}^{\infty} \dfrac{1}{n(n-1)}$ (b) $\sum_{n=3}^{\infty} \dfrac{1}{n^2-3n+2}$

(c) $\sum_{n=2}^{\infty} \dfrac{1}{n^2-1}$

23. Prove Theorem 2.8.3.

24. Prove or disprove the following statements.
   (a) If $\sum a_n$ and $\sum b_n$ diverge, then so does $\sum(a_n+b_n)$.
   (b) If $\sum(a_n+b_n)$ converges, then so do $\sum a_n$ and $\sum b_n$.
   (c) If $\sum(a_n+b_n)$ and $\sum(a_n-b_n)$ both converge, then so do $\sum a_n$ and $\sum b_n$.
   (d) If $a_n>0$ for $n\in Z^+$ and if $\sum a_n$ converges, then $\sum a_n^2$ converges.
   (e) If $a_n>0$ for $n\in Z^+$ and if $\sum a_n^2$ converges, then $\sum a_n$ converges.
   (f) If $a_n>0$, $b_n>0$ for $n\in Z^+$, if $\sum b_n$ diverges, and if $\lim_{n\to\infty}(a_n/b_n)$ exists, then $\sum a_n$ diverges.

25. Prove the convergence of a series of the form $\sum_{n=1}^{\infty} d_n/10^n$, given that each $d_n (n \in Z^+)$ is an integer with $0 \le d_n < 10$. Show that the sum S satisfies $0 \le S \le 1$.

★★ 26. (a) Show that every positive integer M can be expressed in the form

$$M = m_p 10^p + m_{p-1}10^{p-1} + \cdots + m_1 10 + m_0$$

where p is a positive integer and the m_i $(i = 0, 1, \ldots, p)$ are integers between 0 and 9. (*Hint:* Start with the fact that the sequence $\{10^n\}_{n=1}^{\infty}$ is not bounded above.)

(b) Show that the expression for M in part (a) is unique.

★★ 27. Show that if $0 < r < 1$, then r is the sum of a series $\sum_{n=1}^{\infty} d_n/10^n$ where the d_n are digits $(n \in Z^+)$. [*Hint:* Define $d_1 = [10r]$, note that d_1 is a digit, and define $r_1 = r - (d_1/10)$. Show that $r = (d_1/10) + r_1$, with $0 \le 10r_1 < 1$. Next, define $d_2 = [10^2 r_1]$ and $r_2 = r_1 - (d_2/10^2)$. Show that $r = (d_1/10) + (d_2/10^2) + r_2$, with $0 \le 10^2 r_2 < 1$. Continue by induction, showing how to obtain d_{k+1} from d_k; then show that the resulting series $\sum_{n=1}^{\infty} d_n/10^n$ converges to r.]

★★ 28. Show that if $0 < r < 1$ and if r is expressible in more than one way as a decimal; that is, if

$$r = \sum_{n=1}^{\infty} \frac{d_n}{10^n} = \sum_{n=1}^{\infty} \frac{e_n}{10^n}$$

then there is an $N \in Z^+$ such that $d_n = e_n$ for $n < N$, d_N and e_N differ by 1, and if d_N is the larger of d_N and e_N, then $d_N = 0$ for $n > N$ and $e_n = 9$ for $n > N$.

★★ 29. Express the following real numbers in the base 3 (ternary) system:

(a) one hundred (b) one one-hundredth
(c) one-fourth

30. Show that the harmonic series $\sum \frac{1}{n}$ diverges by proving that its sequence of partial sums is not Cauchy.

★★ 31. The *Cantor set C* was first described by Georg Cantor in 1883. One way of defining this set uses the base three numeration system: C is the set of all real numbers in the closed unit interval that have ternary expansions in which the digit 1 does not occur. Thus (in ternary notation), we have

$0.2020 \in C$
$0.1001 \notin C$
$0.1 \in C$ because $0.1 = 0.0222\ldots$

Show that C is uncountable. (*Note:* Another way of defining the Cantor set is given in a project following Section 3.10.)

★ 32. Let a sequence $\{x_n\}_{n=1}^{\infty}$ be defined recursively as follows:

(i) $x_{n+1} = 1/(1 + x_n) (n \in Z^+)$
(ii) $x_1 = 1$

(a) Show that $x_{n+1} = F_n/F_{n+1}$ for $n \in Z^+$, where $\{F_n\}_{n=1}^{\infty}$ is the Fibonacci sequence defined in the project following Section P.2.

(b) Use the formula obtained in part (d) of that project,

$$F_n^2 - F_{n-1}F_{n+1} = (-1)^n \qquad n > 1$$

to prove that

$$|x_{n+1} - x_n| = \frac{1}{F_{n+1}F_n} \le \frac{1}{n^2}$$

Hence, by the Triangle Inequality,

$$|x_{n+k} - x_n| \le \frac{1}{(n+k-1)^2} + \frac{1}{(n+k-2)^2} + \cdots + \frac{1}{n^2} \qquad n > 1$$

(c) Show that, for $n, k > 1$,

$$|x_{n+k} - x_n| \le S_{n+k-1} - S_{n-1}$$

where $\{S_n\}_{n=1}^{\infty}$ is the sequence of partial sums of the series $\sum 1/n^2$. It follows from Exercise 2.7.5b that $\{x_n\}_{n=1}^{\infty}$ is Cauchy. (Why?)

(d) Find $\lim_{n\to\infty} x_n$.

33. Let a sequence $\{x_n\}_{n=1}^{\infty}$ be defined recursively as follows:

(i) $x_1 = 0$ and $x_2 = 1$
(ii) $x_{n+2} = \frac{1}{2}(x_n + x_{n+1})$ for $n \in Z^+$

(a) Show that $\{x_n\}_{n=1}^{\infty}$ is Cauchy. (Plot several terms of the sequence to see what is happening.)

(b) Find $\lim_{n\to\infty} x_n$. (*Hint:* Show that the odd subscripted terms satisfy

$$x_{2k+1} = \frac{1}{2^{2k-1}} + x_{2k-1}$$

and find $\lim_{k\to\infty} x_{2k+1}$.)

★ 2.9 Series with Positive and Negative Terms

We turn our attention in this section to series whose terms may be of mixed sign. In the first instance we consider the special case in which the signs alternate according to one of the patterns $+ - + - + - \cdots$, or $- + - + - + \cdots$.

Definition 2.9.1 An ***alternating series*** is one whose terms are alternately positive and negative, that is, a series of one of the forms

$$\sum (-1)^{n+1} c_n = c_1 - c_2 + c_3 - c_4 + \cdots$$

or

$$\sum (-1)^n c_n = -c_1 + c_2 - c_3 + c_4 - \cdots$$

in which $c_n > 0$ for $n \in Z^+$.

Clearly it is sufficient to study alternating series of the first type. Concerning such series we have the following theorem.

Theorem 2.9.1 **Alternating Series Test** *Let $\{c_n\}_{n=1}^{\infty}$ be a decreasing, null sequence of positive numbers. Then the alternating series $\sum (-1)^{n+1} c_n$ is convergent.*

PROOF: Let $\{S_n\}_{n=1}^{\infty}$ be the sequence of partial sums of the series $\sum (-1)^{n+1} c_n$. We consider the subsequences $\{S_{2k-1}\}_{k=1}^{\infty}$ and $\{S_{2k}\}_{k=1}^{\infty}$:

$$\begin{array}{ll} S_1 = c_1 & S_2 = c_1 - c_2 \\ S_3 = c_1 - c_2 + c_3 & S_4 = c_1 - c_2 + c_3 - c_4 \\ \vdots & \vdots \\ S_{2k-1} = c_1 - c_2 + \cdots + c_{2k-1} & S_{2k} = c_1 - c_2 + \cdots - c_{2k} \\ \vdots & \vdots \end{array}$$

The subsequence $\{S_{2k-1}\}_{k=1}^{\infty}$ is decreasing because

$$S_{2k+1} = S_{2k-1} - (c_{2k} - c_{2k+1})$$

and $c_{2k} - c_{2k+1} \geq 0$. It is bounded below by zero as may be seen by grouping terms:

$$S_{2k-1} = (c_1 - c_2) + (c_3 - c_4) + \cdots + (c_{2k-3} - c_{2k-2}) + c_{2k-1}$$

We conclude that $\{S_{2k-1}\}_{k=1}^{\infty}$ is convergent, and we set $L = \lim_{k \to \infty} S_{2k-1}$. On the other hand, the subsequence $\{S_{2k}\}_{k=1}^{\infty}$ is increasing and bounded above (why?) and therefore also convergent; we set $M = \lim_{k \to \infty} S_{2k}$.

Now for $k \in Z^+$, $c_{2k} = S_{2k-1} - S_{2k}$, and because $\{c_n\}_{n=1}^{\infty}$ is null, we have

$$0 = \lim_{k\to\infty} c_{2k} = \lim_{k\to\infty} (S_{2k-1} - S_{2k}) = L - M$$

Thus, $L = M$ and the conclusion of the theorem follows from Corollary 2.4.1. ■

Example 1 The *alternating harmonic series*

$$\sum (-1)^{n+1} \frac{1}{n}$$

satisfies the conditions of the Alternating Series Test: $\left\{\frac{1}{n}\right\}_{n=1}^{\infty}$ is a decreasing, null sequence of positive numbers. Therefore, the series converges. (Its sum is the natural logarithm of 2, approximately 0.6931. See Section 4.16, Example 5.)

Example 2: Consider $\sum (-1)^{n+1} n/(n^2 + 1)$. The sequence $\{c_n\}_{n=1}^{\infty} = \{n/(n^2 + 1)\}_{n=1}^{\infty}$ is clearly null. It is also decreasing, as may be seen by considering c_{n+1}/c_n for $n \in Z^+$. We have

$$\frac{c_{n+1}}{c_n} = \frac{n + 1}{(n + 1)^2 + 1} \cdot \frac{n^2 + 1}{n} = \frac{n^3 + n^2 + n + 1}{n^3 + 2n^2 + 2n} < 1$$

The given series therefore converges.

Let $\sum (-1)^{n+1} c_n$ be a series satisfying the hypothesis of the Alternating Series Test, and let its sum be S. It follows from the proof of Theorem 2.9.1 that if n is odd, then

$$S_n \geq S \geq S_{n+1} = S_n - c_{n+1}$$

If n is even, then

$$S_n \leq S \leq S_{n+1} = S_n + c_{n+1}$$

In either case the partial sum S_n differs from the exact sum S by no more than $c_{n+1} : |S_n - S| \leq c_{n+1}$.

Estimate of Error: Let $\sum (-1)^{n+1} c_n$ be an alternating series satisfying the hypothesis of Theorem 2.9.1. If the sum of the series is estimated by calculating a partial sum, then the error of this estimate is no greater than the absolute value of the first term not included in that partial sum.

Example 3: Let us add the first five terms of the alternating harmonic series (Example 1):

$$S_5 = 1 - \frac{1}{2} + \frac{1}{3} - \frac{1}{4} + \frac{1}{5}$$
$$= \frac{47}{60}$$

The estimate of error tells us that the exact sum of the series differs from $\frac{47}{60}$ by no more than $\frac{1}{6}$.

Definition 2.9.2 A series $\sum a_n$ is ***absolutely convergent*** provided that the series $\sum |a_n|$ is convergent.

Example 4: $\sum (-1)^{n+1}/n^2$ is absolutely convergent.

Theorem 2.9.2 *If $\sum |a_n|$ is convergent, then $\sum a_n$ is convergent.*

PROOF: Let $A_n = |a_1| + |a_2| + \cdots + |a_n|$ denote the partial sums of $\sum |a_n|$ and let $S_n = a_1 + a_2 + \cdots + a_n$ denote the partial sums of $\sum a_n$. The convergence of $\sum |a_n|$ means that $\{A_n\}_{n=1}^{\infty}$ is convergent, hence Cauchy. We show that $\{S_n\}_{n=1}^{\infty}$ is also Cauchy. Let $\varepsilon > 0$ be given; then there is a positive integer N such that $|A_m - A_n| < \varepsilon$ whenever $m, n \geq N$. If m is the larger of m, n, then

$$\begin{aligned} |A_m - A_n| &= (|a_1| + \cdots + |a_m|) - (|a_1| + \cdots + |a_n|) \\ &= |a_{n+1}| + \cdots + |a_m| < \varepsilon \end{aligned}$$

But then for $m \geq n \geq N$,

$$\begin{aligned} |S_m - S_n| &= |(a_1 + \cdots + a_m) - (a_1 + \cdots + a_n)| \\ &= |a_{n+1} + \cdots + a_m| \\ &\leq |a_{n+1}| + \cdots + |a_m| < \varepsilon \end{aligned}$$

It follows that $\{S_n\}_{n=1}^{\infty}$ is Cauchy and thus convergent; therefore, $\sum a_n$ converges. ∎

Theorem 2.9.2 says that a series that converges absolutely is convergent in the ordinary sense.

Definition 2.9.3 A series $\sum a_n$ that converges, but not absolutely, is ***conditionally convergent***.

The series in Examples 1 and 2 are conditionally convergent.

Positive and Negative Parts of a Series

Consider an arbitrary series $\sum a_n$. Let

$$p_n = \begin{cases} a_n & \text{if } a_n \geq 0 \\ 0 & \text{if } a_n < 0 \end{cases}$$

$$q_n = \begin{cases} 0 & \text{if } a_n \geq 0 \\ -a_n & \text{if } a_n < 0 \end{cases}$$

Thus, $\sum p_n$ and $\sum q_n$ are series of nonnegative terms. $\sum p_n$ is called the *positive part* of $\sum a_n$; $\sum q_n$ is called the *negative part*.

Example 5: For the alternating harmonic series, the positive and negative parts begin as follows:

$$\sum p_n = 1 + 0 + \frac{1}{3} + 0 + \frac{1}{5} + 0 + \frac{1}{7} + \cdots$$
$$\sum q_n = 0 + \frac{1}{2} + 0 + \frac{1}{4} + 0 + \frac{1}{6} + 0 + \cdots$$

Notice the presence of the zeros.

We continue to examine the positive and negative parts of $\sum a_n$, noting that, for $n \in Z^+$,

$$a_n = p_n - q_n$$
$$|a_n| = p_n + q_n$$

Thus, the partial sums of $\sum a_n$, $\sum |a_n|$, $\sum p_n$, and $\sum q_n$ are given by

$$S_n = a_1 + \cdots + a_n$$
$$A_n = |a_1| + \cdots + |a_n|$$
$$P_n = p_1 + \cdots + p_n$$
$$Q_n = q_1 + \cdots + q_n$$

It follows that

$$0 \leq P_n \leq A_n$$
$$0 \leq Q_n \leq A_n$$
$$S_n = P_n - Q_n$$
$$A_n = P_n + Q_n$$

If $\sum a_n$ converges absolutely, then both series $\sum p_n$ and $\sum q_n$ are dominated by $\sum |a_n|$ and consequently they converge. Conversely, if both $\sum p_n$ and $\sum q_n$ converge, then by the last equation above, $\sum |a_n|$ converges. Thus, we have the following theorem:

Theorem 2.9.3 *A series is absolutely convergent if and only if its positive and negative parts both converge.*

If $\sum a_n$ converges, then the convergence of either of $\sum p_n$ or $\sum q_n$ implies the convergence of the other. So if $\sum a_n$ converges conditionally, its positive and negative parts both diverge. Thus:

Theorem 2.9.4: *The positive and negative parts of a conditionally convergent series both diverge to* ∞.

Rearrangements of Series

A series $\sum b_n$ is a *rearrangement* of $\sum a_n$ provided there is a one-to-one correspondence between the terms of the two series in which corresponding terms have equal values. More formally:

Definition 2.9.4 The series $\sum b_n$ is a ***rearrangement*** of $\sum a_n$ provided there exists a bijection $\phi : Z^+ \to Z^+$ such that $b_{\phi(n)} = a_n$.

For instance,

$$\frac{1}{2} + 1 + \frac{1}{4} + \frac{1}{3} + \frac{1}{6} + \frac{1}{5} + \cdots$$

is a rearrangement of the harmonic series; the one-to-one correspondence simply interchanges the $(2n - 1)$st and $2n$th terms.

Theorem 2.9.4 has remarkable consequences in connection with rearrangements. If a series converges conditionally to a finite sum, then there exists a rearrangement of that series that diverges to ∞. In the following informal discussion we see how such a rearrangement can be effected. First observe that there is no harm in assuming that the series has no zero terms.

Stage 1: Take just enough nonzero terms from the beginning of the positive part of the series that the accumulated sum is greater than 1, then subtract just enough nonzero terms from the beginning of the negative part of the series that the accumulated sum is now less than 1. These steps can be taken because the positive and negative parts both diverge to ∞.

Stage 2: Now add just enough nonzero terms from the beginning of the remainder of the positive part of the series that the accumulated sum is greater than 2; then subtract just enough nonzero terms from the beginning of the remainder of the negative part of the series that the accumulated sum is less than 2.

At stage N, we have taken terms whose accumulated sum is less than N but differs from N by no more than the last term of the negative part that was included. Since the terms of the negative part of the series form a null sequence, they eventually become and remain less than 1, so for N sufficiently large, the accumulated sum at the Nth stage is greater than $N - 1$. It follows that the rearranged series diverges to ∞.

Clearly this argument can be modified to obtain a rearrangement of the given series that diverges to $-\infty$. And if S is an arbitrarily chosen real number, it can be shown that a rearrangement exists that converges to S.

Theorem 2.9.5 *If $\sum a_n$ is a conditionally convergent series, then it is possible to find rearrangements of $\sum a_n$ that*

- *(i)* *Diverge to ∞*
- *(ii)* *Diverge to $-\infty$*
- *(iii)* *Converge to an arbitrarily chosen sum S*

The situation for absolutely convergent series is quite different.

Lemma 2.9.1 *Let $\sum b_n$ be a rearrangement of a series $\sum a_n$ of nonnegative terms. If $\sum a_n$ converges to a finite sum S, then $\sum b_n$ also converges to S; if $\sum a_n$ diverges to ∞, then so does $\sum b_n$.*

PROOF: We consider the case in which $\sum a_n$ converges to a finite sum. We let $\{S_n\}_{n=1}^{\infty}$ and $\{T_n\}_{n=1}^{\infty}$ be the sequences of the partial sums of $\sum a_n$ and $\sum b_n$, respectively. Both these sequences are increasing, and $S_n \to S$, where $S = \sup\{S_n : n \in Z^+\}$. Now for each $m \in Z^+$, the terms in the partial sum T_m are all contained in some partial sum $S_{m'}$. Thus, $T_m \leq S_{m'} \leq S$, showing that the sequence $\{T_n\}_{n=1}^{\infty}$ of partial sums of $\sum b_n$ is bounded above by S; now $\{T_n\}_{n=1}^{\infty}$ is increasing, and it follows that $T = \lim_{n\to\infty} T_n \leq S$. But since $\sum a_n$ is also a rearrangement of $\sum b_n$, the same argument shows that $S \leq T$, and so $S = T$. The proof of the unbounded case is left for you to develop in Exercise 2.9.17. ■

Theorem 2.9.6 *Every rearrangement of an absolutely convergent series also converges absolutely and to the same sum.*

PROOF: Let $\sum b_n$ be a rearrangement of the absolutely convergent series $\sum a_n$. The positive part of $\sum b_n$ is a rearrangement of the positive part of $\sum a_n$, and by Lemma 2.9.1 converges to the same sum. Similarly, the negative parts of $\sum a_n$ and $\sum b_n$ converge to the same sum. It follows that $\sum b_n$ converges absolutely to the same sum as $\sum a_n$. ■

Ratio Test for Absolute Convergence

The Ratio Test (Theorem 2.8.7) applies to series of positive terms. However, it can be used to prove *absolute* convergence (hence convergence) of a series of terms of mixed signs.

Example 6: The series $\sum \left(-\frac{2}{3}\right)^n$ converges absolutely because ρ for the absolute series is given by

$$\rho = \lim_{n\to\infty} \frac{\left(\frac{2}{3}\right)^{n+1}}{\left(\frac{2}{3}\right)^n} = \frac{2}{3} < 1$$

Series Whose Terms Are Functions

The Ratio Test is often used to examine series in which the terms are functions of some variable, say, x. The question in such cases is: For what values of x (if any) does the series converge?

Example 7: For the series $\sum (\sin^n x)/n$, we can calculate the test ratio for the absolute series:

$$\left|\frac{\sin^{n+1} x}{n+1}\right| \div \left|\frac{\sin^n x}{n}\right| = \left|\frac{n}{n+1} \sin x\right|$$

Its limit (a function of x) is

$$\rho(x) = \lim_{n\to\infty}\left|\frac{n}{n+1} \sin x\right| = \left|\sin x\right|$$

The original series converges absolutely whenever $|\sin x| < 1$. But $|\sin x|$ is never greater than 1, so there remain the cases in which $|\sin x| = 1$. For $x = \frac{\pi}{2} + 2n\pi$ $(n \in Z)$, $\sin x = 1$ and the original series is $\sum \frac{1}{n}$ (divergent); for $x = \frac{3\pi}{2} + 2n\pi$ $(n \in Z)$, $\sin x = -1$ and the original series is $\sum(-1)^n/n$ (conditionally convergent). A summary is found in the following table:

For	The Series $\sum (\sin^n x)/n$
$x = \frac{\pi}{2} + 2n\pi$	Diverges
$x = \frac{3\pi}{2} + 2n\pi$	Converges conditionally
All other $x \in R$	Converges absolutely

Power Series

A *power series* is a series of the form

$$\sum_{n=0}^{\infty} a_n(x-c)^n$$

where the a_n are real constants depending on n, and c is a fixed real number. Written out, such a series has the form

$$a_0 + a_1(x-c) + a_2(x-c)^2 + \cdots + a_n(x-c)^n + \cdots$$

The summation extends from zero to infinity in order to allow for the constant term a_0.

Power series have many interesting properties, some of which we will encounter in Chapter 5. For the present, we will simply try to determine the set of x for which such a series converges. Often the Ratio Test is helpful in this regard.

Example 8: For the series $\sum_{n=0}^{\infty} \frac{x^n}{n!}$, in which $c = 0$ and $a_n = \frac{1}{n!}$, we apply the Ratio Test to the absolute series:

$$\rho(x) = \lim_{n\to\infty}\left[\frac{|x|^{n+1}}{(n+1)!} \div \frac{|x|^n}{n!}\right] = \lim_{n\to\infty}\frac{|x|}{n+1} = 0$$

Thus, $\rho(x) = 0$ for all x, and the original series converges for all $x \in R$.

Example 9: The series $\sum_{n=1}^{\infty}(2x - 3)^n/n$ converges absolutely whenever

$$\rho(x) = \lim_{n\to\infty}\left(\frac{|2x - 3|^{n+1}}{n + 1} \div \frac{|2x - 3|^n}{n}\right) < 1$$

Now $\rho(x) = |2x - 3|$ and $|2x - 3| < 1$ for $1 < x < 2$, and so we have absolute convergence in the interval $(1, 2)$. At $x = 1$, there is conditional convergence, and at $x = 2$, divergence. For all other values of x, $\rho(x) > 1$, and we have divergence because the terms of the given series do not converge to 0 (see Exercise 2.9.18).

EXERCISES 2.9

1. Give a proof of Theorem 2.9.2 that uses positive and negative parts.

In Exercises 2–12 determine whether the series is conditionally convergent, absolutely convergent, or divergent. Justify your conclusion.

2. $\sum \frac{(-1)^{n+1}n^2}{n^3 + 1}$

3. $\sum \frac{(-1)^{n+1}}{\sqrt{2n + 1}}$

4. $\sum \frac{(-1)^{n+1}}{\sqrt[n]{n}}$

5. $\sum \frac{(-1)^n n!}{(2n)!}$

6. $\sum(-1)^{n+1}(\sqrt{n + 1} - \sqrt{n})$

7. $\sum \frac{(-3)^n}{n!}$

8. $\sum\left(\frac{1}{\sqrt{n}} + \frac{(-1)^{n+1}}{n}\right)$

9. $\sum \frac{(-1)^{n+1}(n!)^2}{(2n)!}$

10. $\sum\left(\frac{-n}{n + 1}\right)^n$

11. $\sum \frac{\frac{1}{2} + (-1)^n}{n}$

12. $\sum(-1)^{n+1}\frac{1 \cdot 3 \cdot \cdots \cdot (2n - 1)}{2 \cdot 4 \cdot \cdots \cdot 2n}$

13. Prove or disprove the following statements.

(a) If $\sum \sqrt{a_n^2 + b_n^2}$ converges, then $\sum a_n$ and $\sum b_n$ converge absolutely.

(b) If the series $\sum a_n$ and $\sum b_n$ converge absolutely, then $\sum \sqrt{a_n^2 + b_n^2}$ converges.

(c) The sum of two absolutely convergent series is absolutely convergent.

(d) The sum of two conditionally convergent series is conditionally convergent.

(e) If the positive and negative parts of a series both diverge, then the series is conditionally convergent.

14. Describe a procedure for rearranging a conditionally convergent series so that the rearrangement diverges to $-\infty$.

15. Describe a procedure for rearranging a conditionally convergent series so that the rearrangement converges to an arbitrarily chosen sum S.

16. When parentheses are introduced into a series, the resulting series is called a *grouping* of the original. Thus,

$$\left(1 - \frac{1}{2}\right) + \left(\frac{1}{3} - \frac{1}{4}\right) + \left(\frac{1}{5} - \frac{1}{6}\right) + \left(\frac{1}{7} - \frac{1}{8}\right) + \cdots$$

is a grouping of the alternating harmonic series.

(a) Prove or disprove: If a series converges, then any grouping of it also converges, and to the same sum.

(b) Prove or disprove: If a series diverges, then any grouping of it also diverges.

17. Show that if $\sum b_n$ is a rearrangement of a series $\sum a_n$ of nonnegative terms, and if $\sum a_n$ diverges to ∞, then so does $\sum b_n$.

18. Verify the remark in Example 9 that, when $\rho(x) > 1$, the terms of the series do not converge to zero.

19. Find the set of all real numbers x for which the given series of functions converges.

(a) $\sum_{n=1}^{\infty} \frac{\cos^n x}{n^2}$ (b) $\sum_{n=1}^{\infty} \sin^n x$

(c) $\sum_{n=1}^{\infty} \cos^n x$ (d) $\sum_{n=1}^{\infty} \frac{(x+1)^n}{\sqrt{n}}$

(e) $\sum_{n=1}^{\infty} n^2 x^n$ (f) $\sum_{n=1}^{\infty} \frac{(3x-4)^n}{n}$

(g) $\sum_{n=0}^{\infty} n! x^n$

20. Let $c > 0$. Test for convergence:

$$\sum_{n=1}^{\infty} (-1)^{n+1}(1 - \sqrt[n]{c})$$

21. Prove or disprove the following statements:

(a) If $\sum a_n$ converges, then $\sum (a_n + a_{n+1})$ converges.

(b) If $\sum (a_n + a_{n+1})$ converges, then $\sum a_n$ converges.

(c) If $\sum (|a_n| + |a_{n+1}|)$ converges, then $\sum |a_n|$ converges.

CHAPTER THREE

LIMITS AND CONTINUITY

3.1 *Cluster Points, Isolated Points*

The statement

$$\lim_{x\to a} f(x) = L$$

is familiar from elementary calculus. Here a and L are real numbers and f is a real function. The idea conveyed by the limit statement is that the function f can be made as close as desired to L by keeping x close enough to but different from a. Although a itself is perhaps not in the domain of f, there must exist elements in the domain that are arbitrarily close to but different from a.

Definition 3.1.1 Let $S \subseteq R$, $a \in R$. Then a is a ***cluster point*** of S provided that every deleted neighborhood of a contains one or more elements of S.

It makes sense to talk about $\lim_{x \to a} f(x)$ only when a is a cluster point of the domain of f. Before taking up the definition of limit itself, let us investigate the concept of cluster point.

Example 1: Let U be the open interval $(0, 1)$. Then 0 and 1 are both cluster points of U even though they do not belong to U. All elements of U itself are also cluster points of U.

Example 2: Let $S = (0, 1) \cup \{2\}$. Then 2 is not a cluster point of S because the deleted neighborhood $N_1^*(2)$ contains no elements of S.

Example 3: Let $S = \left\{\frac{1}{n} : n \in Z^+\right\}$. Then 0 is a cluster point of S because of the Archimedean property. In fact, it can be shown that 0 is the only cluster point of S (work Exercise 3.1.3).

Definition 3.1.2 Let $S \subseteq R$. The ***derived set*** of S is the set S' of all cluster points of S.

Example 4: Let $U = (0, 1)$. Then $U' \supseteq [0, 1]$ by Example 1. Moreover, no real number outside $[0, 1]$ is a cluster point of U, so $U' \subseteq [0, 1]$. It follows that $U' = [0, 1]$.

Example 5: Let $S = (0, 1) \cup \{2\}$; then $S' = [0, 1]$. Exercise 3.1.1 asks you to verify that $S' \supseteq [0, 1]$ and that $S' \subseteq [0, 1]$.

Example 6: Let $S = \left\{\frac{1}{n} : n \in Z^+\right\}$; then $S' = \{0\}$ by Example 3.

The following theorem shows that the cluster point property can be formulated in various ways.

Theorem 3.1.1 *Let $S \subseteq R$, $a \in R$. Then $a \in S'$ if and only if any one of the following conditions holds:*

- *(i) For every $\delta > 0$ there exists an $x \in S$ such that $0 < |x - a| < \delta$*
- *(ii) There is a sequence $\{x_n\}_{n=1}^{\infty}$ in $S \dashv \{a\}$ such that $x_n \to a$ (that is, there is a sequence $\{x_n\}_{n=1}^{\infty}$ in S such that $x_n \to a$ with $x_n \neq a$ for each $n \in Z^+$)*
- *(iii) There is a sequence $\{y_k\}_{k=1}^{\infty}$ of distinct points of S (that is, $y_k = y_l$ only if $k = l$) such that $y_k \to a$*
- *(iv) Every neighborhood of a contains infinitely many elements of S*

PROOF: Condition (i) is simply the definition of $a \in S'$. It is sufficient to establish a chain of conditional statements: (i) $\Rightarrow$ (ii) $\Rightarrow$ (iii) $\Rightarrow$ (iv) $\Rightarrow$ (i).

To show that (i) $\Rightarrow$ (ii), let $\delta_n = \frac{1}{n}$ and choose for x_n an element of S in the deleted δ_n neighborhood of a; clearly $x_n \to a$ and $x_n \neq a$ for $n \in Z^+$. To prove that (ii) $\Rightarrow$ (iii), let $\{x_n\}_{n=1}^{\infty}$ be a sequence in $S \dashv \{a\}$ with $x_n \to a$, and by induction select a subsequence of $\{x_n\}_{n=1}^{\infty}$ with the required properties (work Exercise 3.1.4a). That (iii) $\Rightarrow$ (iv) and (iv) $\Rightarrow$ (i) is clear. ■

Corollary 3.1.1 *A finite set has no cluster points.*

Definition 3.1.3 Let $S \subseteq R$ and $a \in R$. Then a is an ***isolated point*** of S provided that $a \in S$ and there is a deleted neighborhood of a that contains no elements of S.

A point of S is an isolated point of S if and only if it is not a cluster point of S.

Example 7: Let $S = (0, 1) \cup \{2\}$; then $\{2\}$ is an isolated point of S.

Example 8: Let $S = \left\{\frac{1}{n} : n \in Z^+\right\}$; then all elements of S are isolated points of S.

If we adjoin to a set S all of its cluster points, we obtain a set that is known as the "closure" of S.

Definition 3.1.4 Let $S \subseteq R$. The ***closure*** of S is the set $\bar{S} = S \cup S'$.

Example 9: The closure of the open interval $(0, 1)$ is the closed interval $[0, 1]$. The closure of $(0, 1) \cup \{2\}$ is $[0, 1] \cup \{2\}$.

Example 10: The closure of the set Z of integers is just Z itself: $\bar{Z} = Z$.

Example 11: The closure of the set $S = \left\{\frac{1}{n} : n \in Z^+\right\}$ is $\bar{S} = \left\{\frac{1}{n} : n \in Z^+\right\} \cup \{0\}$.

EXERCISES 3.1

1. Verify the assertions in Example 5 that if $S = (0, 1) \cup \{2\}$, then $S' \supseteq [0, 1]$ and $S' \subseteq [0, 1]$.
2. For each set S below determine the derived set S' and the closure $\bar{S}$.
   (a) $S = [0, 1]$
   (b) $S = \{1, 2, 3\}$
   (c) $S = Z^+$
   (d) $S = \{0\} \cup [1, 2]$
   (e) $S = Q$
3. Verify the assertions in Examples 3 and 8 that 0 is the only cluster point of $S = \left\{\frac{1}{n} : n \in Z^+\right\}$, so that S consists only of isolated points.
4. (a) Refer to Theorem 3.1.1 and finish the proof that (ii) $\Rightarrow$ (iii).
   (b) Prove Corollary 3.1.1.

5. Prove or disprove the following statements about sets of real numbers.
   (a) If $A \subseteq B$, then $A' \subseteq B'$.
   (b) If $A' = B'$, then $A = B$.
   (c) $A' \cup B' \subseteq (A \cup B)'$
   (d) $A' \cup B' \supseteq (A \cup B)'$
   (e) $A' \cap B' \subseteq (A \cap B)'$
   (f) $A' \cap B' \supseteq (A \cap B)'$

6. For each of the sets S in Exercise 2, determine the set of all isolated points of S.

7. Prove:

   Bolzano-Weierstrass Theorem for sets *If S is a bounded, infinite subset of R, then there exists a real number x such that x is a cluster point of S.*

3.2 *Definition of Limit*

There are several ways of formulating the definition of the limit statement

$$\lim_{x \to a} f(x) = L$$

We have chosen a definition that is based on the sequential limit. Since we already know a great deal about sequences, many of our fundamental theorems will come as easy by-products of earlier results.

In calculus you may have used the "epsilon-delta" definition of limit; we will discuss that approach later. For real functions defined over intervals, the epsilon-delta definition is equivalent to the one we will now adopt.

In the sequential approach, we investigate the limiting behavior of $f(x)$ as x approaches a by taking "test" sequences $\{x_n\}_{n=1}^{\infty}$ meeting the following conditions:

(i) x_n belongs to the domain D of f $(n \in Z^+)$
(ii) x_n is different from a $(n \in Z^+)$
(iii) $x_n \to a$

(Sequences meeting these three conditions exist whenever $a \in \overline{D}$.) Now if there is a real number L with the property that $f(x_n) \to L$ whenever $\{x_n\}_{n=1}^{\infty}$ is any sequence that meets these three conditions, then that number L is the limit of the function f as x approaches a.

Before stating the definition formally, let us show how easily this procedure works with a familiar example.

Example 1: We wish to verify that $\lim_{x \to 3} x^2 = 9$. [Here $f(x) = x^2$, $a = 3$, and $L = 9$.] We start with an arbitrary sequence $\{x_n\}_{n=1}^{\infty}$ of real numbers converging to 3 but different from 3; that is, $x_n \to 3$ and $x_n \neq 3$. Then $f(x_n) = x_n^2$ and by Theorem 2.4.1,

$$\lim_{n \to \infty} f(x_n) = \lim_{n \to \infty} x_n \cdot \lim_{n \to \infty} x_n = 3 \cdot 3 = 9.$$

Therefore $\lim_{x \to 3} f(x) = 9$.

Definition 3.2.1 Let f be a real function with domain D and let $a \in \overline{D}$ and $L \in R$. Then

$$f(x) \to L \qquad \text{as } x \to a$$

means that, for every sequence $\{x_n\}_{n=1}^{\infty}$ of points of $D \dashv \{a\}$ that converges to a, the sequence $\{f(x_n)\}_{n=1}^{\infty}$ of images converges to L. The arrows are read "approaches." If it exists, the number L is called the ***limit*** of f at a, denoted $\lim_{x \to a} f(x)$.

Example 2: Let $f(x) = mx + b$ $(x \in R)$, where m and b are constants. If a is an arbitrary real number and $\{x_n\}_{n=1}^{\infty}$ is any sequence of real numbers converging to a but different from a, we have $mx_n + b \to ma + b$ by Theorem 2.4.1. Thus, $\lim_{x \to a}(mx + b) = ma + b$.

Example 3: Consider the greatest integer function

$$f(x) = [x] \qquad x \in R$$

If m is any integer, then $\lim_{x \to m} f(x)$ does not exist because the sequences $\{x_n\}_{n=1}^{\infty}$ and $\{y_n\}_{n=1}^{\infty}$, defined by

$$x_n = m + \frac{1}{n} \qquad \text{and} \qquad y_n = m - \frac{1}{n}$$

both converge to m, while $f(x_n) \to m$ and $f(y_n) \to m - 1$.

Example 3 illustrates the fact that the value of $\lim_{n \to \infty} f(x_n)$ must be independent of the sequence $\{x_n\}_{n=1}^{\infty}$ that is chosen to approach a.

Example 4: Let

$$f(x) = \frac{3x^2 - 12}{x - 2} \qquad x \neq 2$$

Although 2 is not in the domain $R \dashv \{2\}$, it is a cluster point of the domain. For any sequence $\{x_n\}_{n=1}^{\infty}$ of real numbers converging to 2, but different from 2, we have

$$f(x_n) = \frac{3\left(x_n^2 - 4\right)}{x_n - 2} = 3(x_n + 2) \to 12$$

So $\lim_{x \to 2} f(x) = 12$.

Example 5: Let $f(x) = \frac{1}{x}(x \neq 0)$. The number 0 is not in the domain of f, but it is a cluster point of the domain, and so it is legitimate to inquire whether $\lim_{x \to 0} f(x)$ exists. By taking $x_n = \frac{1}{n}$, we have $x_n \to 0$ with $x_n \neq 0$, but $f(x_n) = n$ so that $\{f(x_n)\}_{n=1}^{\infty}$ diverges. Thus, $\lim_{x \to 0} \frac{1}{x}$ does not exist.

Following is an example in which the function is defined only on one side of the limiting value of x.

Example 6: Let $f(x) = \sqrt{x}\,(x \geq 0)$ and consider $\lim_{x \to 0} f(x)$. Even though the domain D of f contains only nonnegative numbers, 0 is a cluster point of D, and sequences $\{x_n\}_{n=1}^{\infty}$ with $x_n > 0$ and $x_n \to 0$ do exist. For every such sequence, $\lim_{n \to \infty} f(x_n) = \lim_{n \to \infty} \sqrt{x_n} = 0$ (by Theorem 2.3.6). Therefore, $\lim_{x \to 0} \sqrt{x} = 0$.

It is important to realize that the limit approached by $f(x)$ as x approaches a is determined by values of x in the domain that are *near* a but *different from* a.

Example 7: Let $f(x) = (x^3 + x)/x$ for $x \neq 0$, and let $f(0) = 2$. Then $\lim_{x \to 0} f(x)$ exists and equals 1. The value of f *at* $x = 0$ has no bearing on the limit.

Even though most functions we deal with are defined over intervals or unions of intervals in R, the limit concept applies more generally as the following two examples show.

Example 8: Let $f(r) = r^2$ for $r \in Q$. By the Rational Density Theorem, every real number a is a cluster point of Q, and so it is meaningful to consider $\lim_{r \to a} f(r)$. We let $\{r_n\}_{n=1}^{\infty}$ be a sequence of rationals such that $r_n \to a$ with $r_n \neq a$. Then the sequence $\{f(r_n)\}_{n=1}^{\infty} = \{r_n^2\}_{n=1}^{\infty}$ converges to a^2, so $\lim_{r \to a} f(r) = a^2$.

Example 9: Every rational number r can be expressed in lowest terms as a quotient p/q where $p \in Z$, $q \in Z^+$ (see Example 6 in Section P.1). The positive integer q is unique for a given r, and so the equation $f(r) = \frac{1}{q}(r \in Q)$ defines a function with domain Q. It is an interesting fact that $\lim_{r \to a} f(r) = 0$ for every *real* number a. Following is a proof.

Given an arbitrary sequence $\{r_n\}_{n=1}^{\infty}$ satisfying the conditions $r_n \in Q$, $r_n \to a$, and $r_n \neq a$, we must show that $f(r_n) \to 0$. Specifically, we must prove that for $\varepsilon > 0$ there is an $N \in Z^+$ such that $|f(r_n)| < \varepsilon$ for $n \geq N$. Suppose M is a positive integer such that $\frac{1}{M} \leq \varepsilon$. Since the sequence $\{r_n\}_{n=1}^{\infty}$ is convergent, it is bounded, and so its terms are contained in some bounded interval I. Now there are only a finite number of rational numbers in I with denominators less than or equal to M; thus, there is an $N \in Z^+$ such that for $n \geq N$, the denominator of r_n is greater than M so that $f(r_n) < \frac{1}{M} \leq \varepsilon$.

Uniqueness of Limits

Theorem 3.2.1 *If a is a cluster point of the domain of f, and if $\lim_{x \to a} f(x)$ exists, then it is unique.*

PROOF: This theorem is an immediate consequence of the uniqueness of sequential limits. ■

Basic Properties of Limits

In the following theorems we assume that the domain of f contains a deleted neighborhood of a, say $N_\delta^*(a)$, where δ is positive. This assumption requires that f be defined on open intervals on both sides of a: $(a - \delta, a)$ and $(a, a + \delta)$; f may or may not be defined at a itself. In any event, a is a cluster point of the domain of f.

Theorem 3.2.2 *Let f be a real function whose domain contains a deleted neighborhood of a. If $\lim_{x \to a} f(x)$ exists, then there is a deleted neighborhood of a on which f is bounded.*

PROOF: First, we note that the conclusion of the theorem can be expressed with quantifiers as follows:

$$(\exists \delta > 0)(\exists M > 0)(|f(x)| < M \text{ for all } x \in N_\delta^*(a))$$

Now suppose that $\lim_{x \to a} f(x)$ exists but that the conclusion of the theorem fails. Then no matter what positive δ is chosen, $|f(x)|$ assumes arbitrarily large values in $N_\delta^*(a)$. Let δ_n have the values $\frac{1}{n}$ for $n \in Z^+$, and choose $x_n \in N_{\delta_n}^*(a)$ with $|f(x_n)| \geq n$. Then the terms of the sequence $\{x_n\}_{n=1}^\infty$ belong to the domain of f with $x_n \neq a$ and $x_n \to a$. But the image sequence $\{f(x_n)\}_{n=1}^\infty$ is unbounded, so it does not converge, contradicting the existence of $\lim_{x \to a} f(x)$. ■

Theorem 3.2.3 *Let f and g be real functions whose domains both contain some deleted neighborhood of a, and suppose that $f(x) = g(x)$ for all x in that deleted neighborhood. If $\lim_{x \to a} f(x)$ exists, then so does $\lim_{x \to a} g(x)$, and these two limits are equal.*

Remark: We are not assuming that $f(x)$ and $g(x)$ are equal everywhere; they may be defined differently at a (or undefined there), and they may behave quite differently outside the given deleted neighborhood of a.

PROOF: Let δ be a positive number such that $f(x)$ and $g(x)$ are defined and equal for $x \in N_\delta^*(x)$; and let $\lim_{x \to a} f(x) = L$. We claim that $\lim_{x \to a} g(x) = L$. Let $\{x_n\}_{n=1}^\infty$ be a sequence of points in the domain of g such that $x_n \to a$ with $x_n \neq a$ $(n \in Z^+)$. There is an $N \in Z^+$ such that $x_n \in N_\delta^*(a)$ for $n \geq N$; hence, $f(x_n)$ is defined and equal to $g(x_n)$ for $n \geq N$. Since the sequence

$$f(x_N), f(x_{N+1}), f(x_{N+2}), \ldots$$

converges to L, so does the sequence $\{g(x_n)\}_{n=1}^\infty$. ■

Example 10: The functions $f(x) = 1 - x^2$ and $g(x) = |x^2 - 1|$ have equal values in $N_1^*(0)$. According to Theorem 3.2.3, then

$$\lim_{x \to 0} g(x) = \lim_{x \to 0} f(x) = 1$$

Example 11: Theorem 3.2.3 permits us to say that

$$\lim_{x \to 4} \frac{x - 4}{\sqrt{x} - 2} = \lim_{x \to 4} (\sqrt{x} + 2)$$

The function on the right approaches 4 as x approaches 4, and therefore so does $(x - 4)/(\sqrt{x} - 2)$.

Theorem 3.2.4 *Let f be a real function whose domain contains a deleted neighborhood of a and suppose that f has a negative (positive) limit as x approaches a. Then there exists a deleted neighborhood of a in which the values of f are all negative (positive).*

PROOF: We prove the theorem for the case in which $\lim_{x \to a} f(x) < 0$. If the conclusion of the theorem were not true, then for every $n \in Z^+$, f would take some nonnegative values in the deleted neighborhood $N_{\delta_n}^*(a)$, where $\delta_n = \frac{1}{n}$. We choose $x_n \in N_{\delta_n}^*(a)$ such that $f(x_n) \geq 0$. Clearly $x_n \to a$ with $x_n \neq a$, and so $\{f(x_n)\}_{n=1}^{\infty}$ converges to $\lim_{x \to a} f(x)$, which by hypothesis is less than 0. However, $f(x_n) \geq 0$ for $n \in Z^+$; and by Corollary 2.3.1, $\{f(x_n)\}_{n=1}^{\infty}$ cannot converge to a negative limit. This contradiction establishes the theorem. ■

Corollary 3.2.1 *Let f be a function whose domain contains a deleted neighborhood of a, and suppose that f has a nonzero limit at a. Then there exists a deleted neighborhood of a in which the values of f are nonzero.*

The Epsilon-Delta Formulation of Limit

In elementary calculus the statement $\lim_{x \to a} f(x) = L$ is often formulated as in the following theorem.

Theorem 3.2.5 *Let f be a real function whose domain contains a deleted neighborhood of the real number a, and let L be a real number. Then*

(i) $\lim_{x \to a} f(x) = L$

if and only if

(ii) For every $\varepsilon > 0$ there exists a $\delta > 0$ such that $|f(x) - L| < \varepsilon$ whenever $0 < |x - a| < \delta$

Note: In logical notation with quantifiers, condition (ii) can be expressed as follows:

$$(\forall \varepsilon > 0)(\exists \delta > 0)(0 < |x - a| < \delta \Rightarrow |f(x) - L| < \varepsilon)$$

PROOF: We prove that condition (i) implies (ii) (you are asked to prove the converse in Exercise 3.2.7). We assume that $\lim_{x \to a} f(x) = L$, and assume further that condition (ii) fails to hold. The negation of (ii) is

$$(\exists \varepsilon > 0)(\forall \delta > 0)(\exists x)(0 < |x - a| < \delta \text{ and } |f(x) - L| \geq \varepsilon)$$

So if (ii) fails, there exists a positive number ε such that, for every positive δ, an x exists with

$$0 < |x - a| < \delta \qquad \text{and} \qquad |f(x) - L| \geq \varepsilon$$

We let δ have the values of $\delta_n = \frac{1}{n}$ for $n \in Z^+$ and choose x_n such that $0 < |x_n - a| < \frac{1}{n}$ and $|f(x_n) - L| \geq \varepsilon$. Now clearly, $x_n \to a$ with $x_n \neq a$, and so we should have $f(x_n) \to L$. But the condition $|f(x_n) - L| \geq \varepsilon$ prevents $\{f(x_n)\}_{n=1}^{\infty}$ from converging to L and contradicts $\lim_{x \to a} f(x) = L$. ■

In terms of neighborhoods, Theorem 3.2.5 says that $\lim_{x \to a} f(x) = L$ if and only if for each neighborhood $N_\varepsilon(L)$ of L there is a deleted neighborhood of a, $N_\delta^*(a)$, such that $f(x) \in N_\varepsilon(L)$ whenever $x \in N_\delta^*(a)$.

In Figure 3.1 you can compare various equivalent formulations of the limit statement (sequential, ε-δ, and neighborhood).

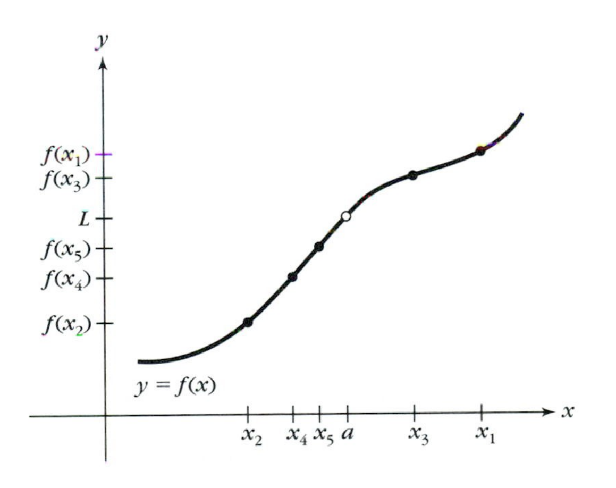

(a) Definition of $\lim_{x \to a} f(x) = L$

For an arbitrary test sequence $\{x_n\}_{n=1}^{\infty}$ with $x_n \in D$, $x_n \to a$, $x_n \neq a$ we must have $f(x_n) \to L$

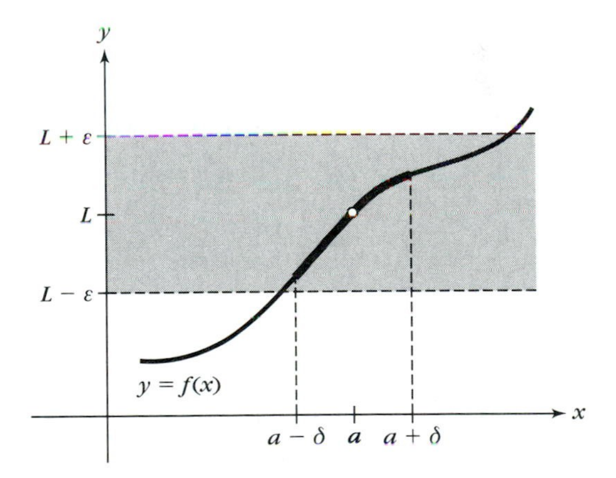

(b) ε - δ formulation of $\lim_{x \to a} f(x) = L$

For every $\varepsilon > 0$, there exists $\delta > 0$, such that $|f(x) - L| < \varepsilon$ whenever $0 < |x - a| < \delta$

Figure 3.1(a)(b)

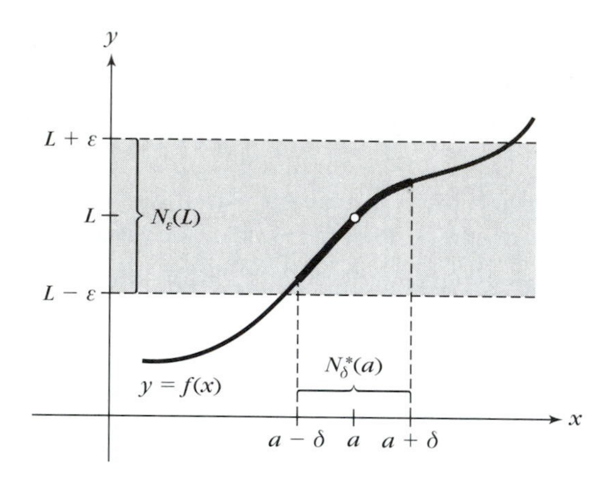

(c) Neighborhood formulation of $\lim_{x\to a} f(x) = L$

For every neighborhood $N_\varepsilon(L)$ there is a deleted neighborhood $N_\delta^*(a)$ such that $f(N_\delta^*(a)) \subseteq N_\varepsilon(L)$

Figure 3.1(c)

Example 12: If $f(x) = 3x + 5$, we have $\lim_{x\to 2} f(x) = 11$. To illustrate Theorem 3.2.5, suppose $\varepsilon > 0$ has been given. We wish to find a $\delta > 0$ as described in the theorem. The inequality $|f(x) - 11| < \varepsilon$ is equivalent to $|3x - 6| < \varepsilon$; clearly the latter holds whenever $|x - 2| < \frac{\varepsilon}{3}$. Thus, $\delta = \frac{\varepsilon}{3}$ will serve the purpose, since $|f(x) - 11| < \varepsilon$ when $0 < |x - 2| < \frac{\varepsilon}{3}$. (The inequality $|f(x) - 11| < \varepsilon$ also holds when $x = 2$, but that is irrelevant.)

Example 13: Let $f(x) = x^3 + 2x$. We use Theorem 3.2.5 to show that $\lim_{x\to 1} f(x) = 3$. Given $\varepsilon > 0$, we must find $\delta > 0$ such that $|f(x) - 3| < \varepsilon$ whenever $0 < |x - 1| < \delta$. The inequality $|f(x) - 3| < \varepsilon$ can be simplified as follows:

$$|x^3 + 2x - 3| < \varepsilon$$
$$|x - 1|\,|x^2 + x + 3| < \varepsilon$$

If we first stipulate that $|x - 1| < 1$, thereby restricting x to the interval $(0, 2)$, we see that the factor $|x^2 + x + 3|$ is less than 9 and that $|x^3 + 2x - 3| < 9|x - 1|$. Now if we further require that $|x - 1| < \frac{\varepsilon}{9}$, we get $|x^3 + 2x - 3| < 9 \cdot \frac{\varepsilon}{9} = \varepsilon$. A suitable choice of δ is $\min\{1, \frac{\varepsilon}{9}\}$; if $0 < |x - 1| < \delta$, we have both $|x - 1| < \frac{\varepsilon}{9}$ and $|x^2 + x + 3| < 9$, and so $|x^3 + 2x - 3| < \varepsilon$.

The following modification of Theorem 3.2.5 permits us to relax the assumption that f is defined throughout a deleted neighborhood of a. We do need to assume that a is a cluster point of the domain.

Theorem 3.2.6 *Let f be a real function with domain D and let a be a cluster point of D. Then*

(i) $\lim_{x \to a} f(x) = L$

if and only if

(ii) *For every* $\varepsilon > 0$ *there exists a* $\delta > 0$ *such that* $|f(x) - L| < \varepsilon$ *whenever* $x \in D$ *and* $0 < |x - a| < \delta$

PROOF: You are asked to prove this theorem in Exercise 3.2.8. ∎

The Limit Theorem

Theorem 3.2.7 *Let f and g be real functions whose domains both contain a deleted neighborhood of a,* $N^*_\delta(a)$, *with* $\delta > 0$. *Assume that both f and g have limits at a. Then*

(i) $\lim_{x \to a} [f(x) + g(x)] = \lim_{x \to a} f(x) + \lim_{x \to a} g(x)$

(ii) $\lim_{x \to a} [f(x) - g(x)] = \lim_{x \to a} f(x) - \lim_{x \to a} g(x)$

(iii) $\lim_{x \to a} [f(x)g(x)] = \lim_{x \to a} f(x) \lim_{x \to a} g(x)$

If, in addition, $\lim_{x \to a} g(x) \neq 0$, *then*

(iv) $$\lim_{x \to a} \frac{f(x)}{g(x)} = \frac{\lim_{x \to a} f(x)}{\lim_{x \to a} g(x)}$$

PROOF: We prove only the first and fourth parts.

(i) Let $L = \lim_{x \to a} f(x)$ and $M = \lim_{x \to a} g(x)$. The deleted neighborhood $N^*_\delta(a)$ in which both functions are defined is contained in the domain D_1 of $f + g$, so that a is a cluster point of D_1. If $\{x_n\}_{n=1}^\infty$ is an arbitrary sequence of points in $D_1 \dashv \{a\}$ with $x_n \to a$, we know $f(x_n) \to L$ and $g(x_n) \to M$. Thus, $(f + g)(x_n) = f(x_n) + g(x_n) \to L + M$ by Theorem 2.4.1; so $\lim_{x \to a}[f(x) + g(x)] = L + M$.

(iv) By Corollary 3.2.1, there exists a deleted neighborhood of a in which $g(x)$ is nonzero. The intersection of this deleted neighborhood of a with $N^*_\delta(a)$ is a deleted neighborhood of a and is contained in the domain D_2 of f/g; so a is a cluster point of D_2. If $\{x_n\}_{n=1}^\infty$ is a sequence in $D_2 \dashv \{a\}$ with $x_n \to a$, we have $(f/g)(x_n) = f(x_n)/g(x_n) \to L/M$ by Theorem 2.4.2; so

$$\lim_{x \to a} \frac{f(x)}{g(x)} = \frac{L}{M}$$ ∎

Example 14: We have

$$\lim_{x \to 2} \frac{x^4 - 16}{x^3 - 4x} = \lim_{x \to 2} \frac{x^2 + 4}{x} \qquad \text{Theorem 3.2.3}$$

$$= \frac{8}{2} = 4 \qquad \text{Theorem 3.2.7}$$

The following theorem is useful when a product of two functions is being considered and one factor has a limit of zero. The product approaches zero—unless the second factor becomes large.

Theorem 3.2.8 *Let f and g be real functions and let a be a real number such that the domains of f and g both contain the deleted neighborhood of a, $N_\delta^*(a)$, for some $\delta > 0$. If g is bounded on $N_\delta^*(a)$ and if $\lim_{x \to a} f(x) = 0$, then $\lim_{x \to a} f(x)g(x) = 0$.*

PROOF: You are asked to prove this theorem in Exercise 3.2.4.a. ∎

Example 15: Let the *signum function* sgn be defined by

$$\operatorname{sgn} x = \begin{cases} 1 & \text{if } x > 0 \\ -1 & \text{if } x < 0 \end{cases}$$

Clearly sgn x is bounded in every deleted neighborhood of 0. By Theorem 3.2.8, then,

$$\lim_{x \to 0} (x^2 + 3x)\operatorname{sgn} x = 0$$

EXERCISES 3.2

1. Use the sequential definition of limit to verify the following statements.
   (a) $\lim_{x \to -3} (x^2 + 2x) = 3$
   (b) $\lim_{x \to 4} \sqrt{x} = 2$
   (c) $\lim_{x \to 0} \frac{|x|}{x}$ does not exist
   (d) $\lim_{x \to 2} \frac{x^3 - 8}{x^2 - 4} = 3$

2. Let

$$f(x) = \begin{cases} 1 & \text{if } x \in Q \\ 0 & \text{if } x \in R - Q \end{cases}$$

   Show that $\lim_{x \to a} f(x)$ does not exist at any point $a \in R$.

3. Let

$$f(x) = \begin{cases} x & \text{if } x \in Q \\ -x & \text{if } x \in R - Q \end{cases}$$

   Show that $\lim_{x \to 0} f(x) = 0$ but that $\lim_{x \to a} f(x)$ does not exist for $a \neq 0$.

4. (a) Prove Theorem 3.2.8.
   (b) Evaluate $\lim_{x \to 0} \left(x \sin \frac{1}{x}\right)$.

5. Prove the second formulation of Theorem 3.2.4: If the domain of f contains a deleted neighborhood of a and if f has a positive limit as x approaches a, then there exists a deleted neighborhood of a in which the values of f are all positive.

6. Given the function f and the number a, find $L = \lim_{x \to a} f(x)$ if it exists, and if it does, find $\delta > 0$ for the given ε, as described in Theorem 3.2.5. Sketch the graph of f and show $N_\delta^*(a)$ and $N_\varepsilon(L)$ on your sketch.
   (a) $f(x) = \frac{1}{2}x + 4;\quad a = 2;\quad \varepsilon = \frac{1}{10}$
   (b) $f(x) = x^2;\quad a = 0;\quad \varepsilon = \frac{1}{4}$
   (c) $f(x) = x^2;\quad a = 1;\quad \varepsilon = \frac{1}{4}$

(d) $f(x) = x^2 + 2x$; $a = 1$; $\varepsilon = 1$
(e) $f(x) = [x]$; $a = \frac{4}{3}$; $\varepsilon = \frac{1}{2}$
(f) $f(x) = mx + b$; $a \in R$; ε an arbitrary positive number
(g) $f(x) = x^3$; $a = 2$; $\varepsilon = \frac{1}{10}$
(h) $f(x) = x^3 + 3x$; $a = -1$; $\varepsilon = \frac{1}{10}$

7. Prove the second part of Theorem 3.2.5, that if condition (ii) holds, then so does condition (i).

8. Prove Theorem 3.2.6.

3.3 *Variations of the Limit Concept*

One-Sided Limits

Let a and L be real numbers and let f be a real function with domain D. Assume further that every interval of the form $(a, a + h)$, where $h > 0$, contains points of D. Then it is possible to form sequences $\{x_n\}_{n=1}^{\infty}$ with the property that $x_n \in D$ with $x_n > a$ and $x_n \to a$. If, for every such sequence, it is true that $f(x_n) \to L$, then L is the *right-hand limit* of f at a and we write

$$f(x) \to L \qquad \text{as } x \to a^+$$

or simply,

$$L = \lim_{x \to a^+} f(x)$$

Example 1: Let f be the greatest integer function, $f(x) = [x]$. It is easy to verify that, for $m \in Z$,

$$\lim_{x \to m^+} [x] = m$$

Example 2: For the function $f(x) = \sqrt{x - 1}$, the right-hand limit of f at 1 is 0. Since f is defined only when $x \geq 1$, the concepts of right-hand limit at 1 and limit at 1 in this situation are identical. Thus,

$$\lim_{x \to 1} \sqrt{x - 1} = \lim_{x \to 1^+} \sqrt{x - 1} = 0$$

The left-hand limit of f at a is defined in an analogous way: Every interval of the form $(a - h, a)$, where $h > 0$, is assumed to contain points in the domain D of f. We then consider sequences $\{x_n\}_{n=1}^{\infty}$ such that $x_n \in D$ with $x_n < a$ and $x_n \to a$. If for every such sequence we have $f(x_n) \to L$, then L is the *left-hand limit* of f at a and we write

$$f(x) \to L \qquad \text{as } x \to a^- \qquad \text{or} \qquad L = \lim_{x \to a^-} f(x)$$

Uniqueness of the two one-sided limits follows immediately from the uniqueness of sequential limits.

Example 3: Consider again the greatest integer function. If $m \in Z$, then

$$\lim_{x \to m^-} [x] = m - 1$$

Example 4: $\lim_{x \to 1^-} \sqrt{x - 1}$ is meaningless because the domain contains no real numbers less than 1.

The following theorem provides a useful test for the existence of the ordinary limit at a when the domain of the function contains a deleted neighborhood of a.

Theorem 3.3.1 *Let $a, L \in R$ and let f be a real function whose domain contains a deleted neighborhood of a. Then*

(i) $\lim_{x \to a} f(x) = L$

if and only if

(ii) $\lim_{x \to a^+} f(x) = L$ *and* $\lim_{x \to a^-} f(x) = L$

PROOF: We prove that condition (ii) is sufficient for condition (i), leaving the converse for you to prove in Exercise 3.3.2a. Let $\{x_n\}_{n=1}^{\infty}$ be a sequence of points of $D \dashv \{a\}$ such that $x_n \to a$. We show that $f(x_n) \to L$. There are several cases to consider.

(1) If only a finite number of terms of $\{x_n\}_{n=1}^{\infty}$ are greater than a, then all the remaining terms are less than a, and since $\lim_{x \to a^-} f(x) = L$, we have $f(x_n) \to L$.
(2) If only a finite number of terms of $\{x_n\}_{n=1}^{\infty}$ are less than a, then all the remaining terms are greater than a, and since $\lim_{x \to a^+} f(x) = L$, we have $f(x_n) \to L$.
(3) If infinitely many terms of $\{x_n\}_{n=1}^{\infty}$ are greater than a, and infinitely many are less than a, then $\{x_n\}_{n=1}^{\infty}$ can be split into two subsequences: $\{x_{n_k}\}_{k=1}^{\infty}$, consisting of the terms of $\{x_n\}_{n=1}^{\infty}$ that are greater than a, and $\{x_{m_l}\}_{l=1}^{\infty}$, with those terms that are less than a. Because of condition (ii), we have $f(x_{n_k}) \to L$ and $f(x_{m_l}) \to L$, and by Theorem 2.4.4, $f(x_n) \to L$. ∎

Example 5: Consider $\lim_{x \to 2} |x^2 - 4|$. Since $|x^2 - 4| = x^2 - 4$ for $x \in (2, 3)$, we have $\lim_{x \to 2^+} |x^2 - 4| = \lim_{x \to 2^+} (x^2 - 4) = 0$. For $x \in (1, 2)$, we have $|x^2 - 4] = 4 - x^2$, and so $\lim_{x \to 2^-} |x^2 - 4| = \lim_{x \to 2^-} (4 - x^2) = 0$. By Theorem 3.3.1, $\lim_{x \to 2} |x^2 - 4| = 0$.

Example 6: In Examples 1 and 3 we have seen that if m is an integer, then $\lim_{x \to m^+} [x]$ and $\lim_{x \to m^-} [x]$ are different. According to Theorem 3.3.1, therefore, $\lim_{x \to m} [x]$ does not exist.

The hypothesis of Theorem 3.3.1 can be weakened slightly by assuming that the domain of f contains points on both sides of a in every deleted neighborood of a.

One-sided limits behave in many respects like ordinary limits. The following two theorems are analogs of Theorems 3.2.5 and 3.2.4.

Theorem 3.3.2 *Let f be a real function and let $a, L \in R$. Then:*

(i) If f is defined in an interval $(a, a + h)$ for some $h > 0$, then $\lim_{x \to a^+} f(x) = L$ if and only if for every $\varepsilon > 0$ there exists a $\delta > 0$ such that $|f(x) - L| < \varepsilon$ whenever $a < x < a + \delta$.

(ii) If f is defined in an interval $(a - h, a)$ for some $h > 0$, then $\lim_{x \to a^-} f(x) = L$ if and only if for every $\varepsilon > 0$ there exists a $\delta > 0$ such that $|f(x) - L| < \varepsilon$ whenever $a - \delta < x < a$.

Figure 3.2 enables us to compare the sequential definition of right-hand limit with the ε-δ formulation in part (i) of Theorem 3.3.2.

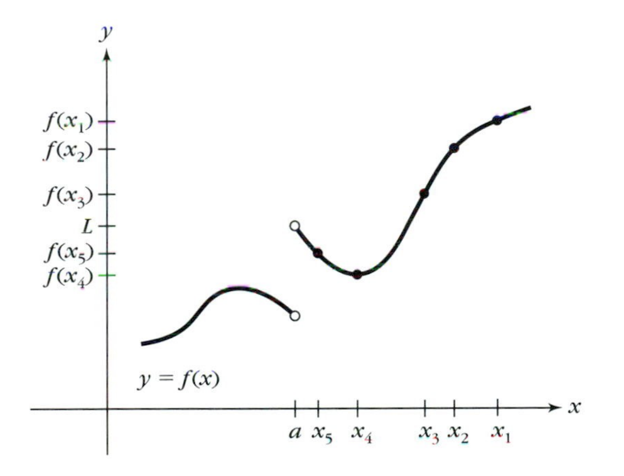

(a) Sequential definition of $\lim_{x \to a^+} f(x) = L$

For an arbitrary test sequence $\{x_n\}_{n=1}^{\infty}$ with $x_n \in D$, $x_n \to a$, $x_n > a$, we must have $f(x_n) \to L$

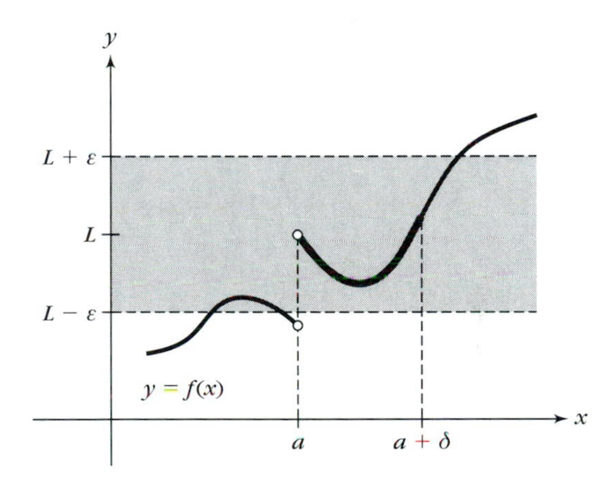

(b) ε - δ formulation of $\lim_{x \to a^+} f(x) = L$

For every $\varepsilon > 0$, there exists $\delta > 0$, such that $|f(x) - L| < \varepsilon$ whenever $a < x < a + \delta$

Figure 3.2

Theorem 3.3.3 *Let $a \in R$, and let f be a real function whose domain contains an interval $(a, a + h)$ for some $h > 0$. If $\lim_{x \to a^+} f(x)$ is positive (negative), then there is a $\delta > 0$ such that $f(x)$ is positive (negative) for $a < x < a + \delta$.*

Note: It follows from Theorem 3.3.3 that if $f(x) \geq 0$ (≤ 0) for $x \in (a, a + h)$ and if $\lim_{x \to a^+} f(x)$ exists, then $\lim_{x \to a^+} f(x) \geq 0$ (≤ 0). Similar remarks hold for limits from the left.

Infinite Behavior

Let f be a real function with domain D, and let a be a cluster point of D. Consider an arbitrary sequence $\{x_n\}_{n=1}^{\infty}$ with the properties $x_n \in D$, $x_n \neq a$ and $x_n \to a$. If for every such sequence we have $f(x_n) \to \infty$, then we say that $f(x)$ *approaches positive infinity as x approaches a*; that is,

$$f(x) \to \infty \qquad \text{as } x \to a$$

When this is the case, f does not have a limit at a—that is, $\lim_{x \to a} f(x)$ does not exist as a real number. It is nevertheless traditional to write $\lim_{x \to a} f(x) = \infty$.

Example 7: Let $f(x) = 1/x^2$ for $x \neq 0$, and let $x \to 0$. If $\{x_n\}_{n=1}^{\infty}$ is any sequence of nonzero real numbers such that $x_n \to 0$, then we have $f(x_n) = 1/x_n^2 \to \infty$. Hence, $\lim_{x \to 0} f(x) = \infty$.

Figure 3.3b suggests how to express the limit statement $f(x) \to \infty$ as $x \to a$ in terms of inequalities.

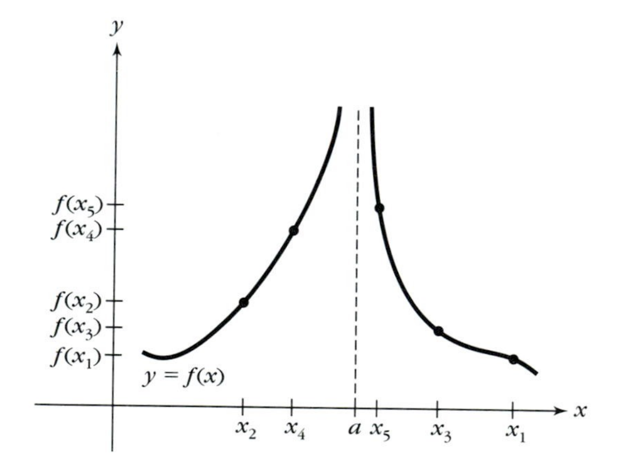

(a) Sequential definition of $\lim_{x \to a} f(x) = \infty$

For an arbitrary test sequence $\{x_n\}_{n=1}^{\infty}$ with $x_n \in D$, $x_n \to a$, $x_n \neq a$, we must have $f(x_n) \to \infty$

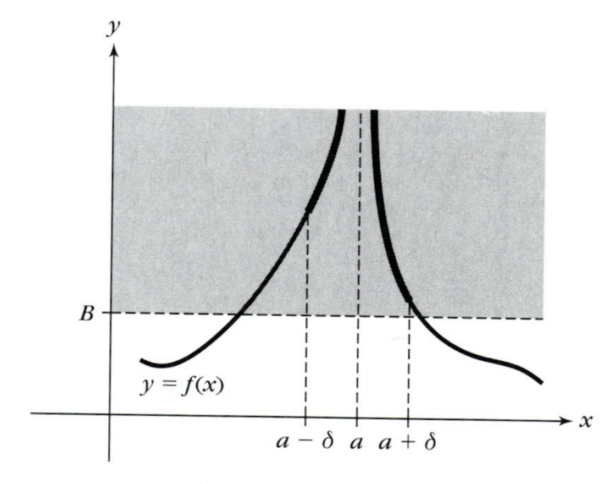

(b) B - δ formulation of $\lim_{x \to a} f(x) = \infty$

For every $B > 0$, there exists $\delta > 0$, such that $f(x) > B$ whenever $0 < |x - a| < \delta$

Figure 3.3

The statement

$$f(x) \to -\infty \quad \text{as } x \to a \qquad \text{or} \qquad \lim_{x \to a} f(x) = -\infty$$

is defined in a similar manner by replacing the condition $f(x_n) \to \infty$ by $f(x_n) \to -\infty$.

It is frequently the case with elementary functions that $f(x)$ approaches infinity on one side of a and negative infinity on the other side. You should be able to formulate definitions for the following statements:

$$\begin{array}{ll} f(x) \to \infty & \text{as } x \to a^+ \\ f(x) \to \infty & \text{as } x \to a^- \\ f(x) \to -\infty & \text{as } x \to a^+ \\ f(x) \to -\infty & \text{as } x \to a^- \end{array}$$

Example 8: Consider $f(x) = 1/(x^2 - 3x + 2)$ as $x \to 2^+$ and as $x \to 2^-$. If $\{x_n\}_{n=1}^{\infty}$ is an arbitrary sequence converging to 2 ($x_n \neq 2$), the sequence $\{x_n^2 - 3x_n + 2\}_{n=1}^{\infty}$ converges to zero. Therefore, $|f(x_n)| \to \infty$. Since $f(x) = 1/[(x - 2)(x - 1)]$, we see that $f(x)$ is positive for $x > 2$; thus, if $x_n \to 2$ with $x_n > 2$, we must have $f(x_n) \to \infty$, so that $f(x) \to \infty$ as $x \to 2^+$. For $1 < x < 2$, $f(x)$ is negative, so if $x_n \to 2$ with $x_n < 2$, then $f(x_n) \to -\infty$, so that $f(x) \to -\infty$ as $x \to 2^-$.

Behavior as x Becomes Infinite

Let f be a real function whose domain D contains points in every interval of the form (B, ∞) where $B > 0$. Suppose, for every sequence $\{x_n\}_{n=1}^{\infty}$ of points of D such that $x_n \to \infty$, we have $f(x_n) \to L$; then we say that $f(x)$ approaches L as x approaches ∞; this may be written

$$f(x) \to L \quad \text{as } x \to \infty \qquad \text{or} \qquad \lim_{x \to \infty} f(x) = L$$

This limit statement can also be formulated in terms of inequalities, as suggested in Figure 3.4.

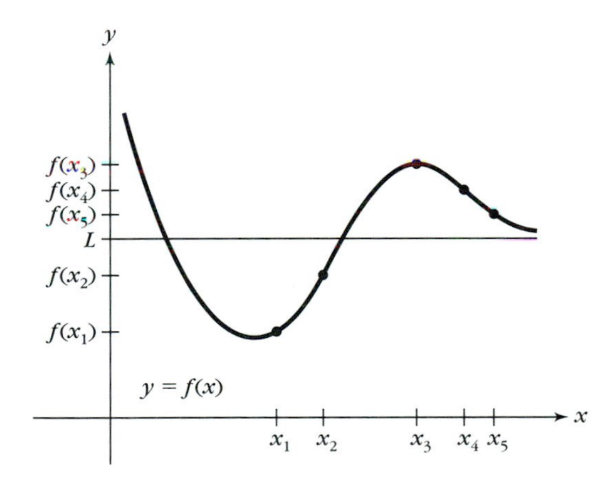

(a) Sequential definition of $\lim_{x \to \infty} f(x) = L$

For an arbitrary test sequence $\{x_n\}_{n=1}^{\infty}$ with $x_n \in D$, $x_n \to \infty$, we must have $f(x_n) \to L$

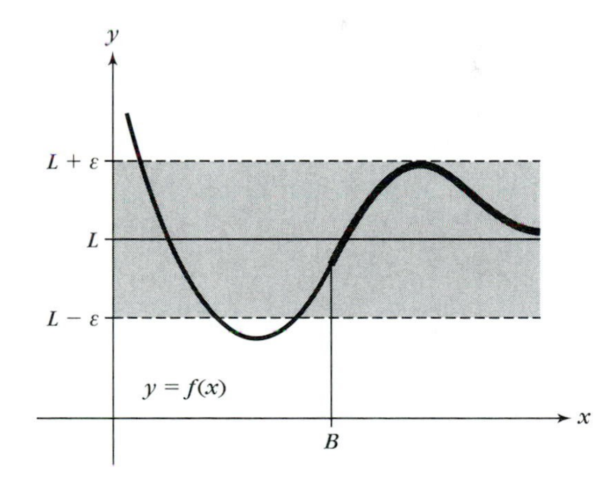

(b) ε - B formulation of $\lim_{x \to \infty} f(x) = L$

For every $\varepsilon > 0$, there exists $B > 0$, such that $|f(x) - L| < \varepsilon$ whenever $x > B$

Figure 3.4

If, for every sequence $\{x_n\}_{n=1}^{\infty}$ of points of D such that $x_n \to \infty$, we have $f(x_n) \to \infty$, then we say that $f(x)$ approaches ∞ as x approaches ∞; this may be written

$$f(x) \to \infty \quad \text{as } x \to \infty \qquad \text{or} \qquad \lim_{x\to\infty} f(x) = \infty$$

Example 9: $\displaystyle\lim_{x\to\infty} \frac{x}{x+1} = 1$

Example 10: $\displaystyle\lim_{x\to\infty}(x^2+1) = \infty$

Formulations of other limit statements, such as

$$\lim_{x\to\infty} f(x) = -\infty$$
$$\lim_{x\to-\infty} f(x) = L$$
$$\lim_{x\to-\infty} f(x) = \infty$$
$$\lim_{x\to-\infty} f(x) = -\infty$$

are left for you to work out in Exercise 3.3.6.

EXERCISES 3.3

1. Show that the functions below have the limiting behavior described. (The brackets indicate the greatest integer function.)

(a) $\displaystyle\lim_{x\to 3^-}[x] = 2$

(b) $\displaystyle\lim_{x\to(1/2)^+}[2x] = 1;\ \lim_{x\to(1/2)^-}[2x] = 0$

(c) $\displaystyle\lim_{x\to 2^+}\sqrt{x^2-4} = 0;\ \lim_{x\to 2}\sqrt{x^2-4} = 0$

(d) $\displaystyle\lim_{x\to 0^+}\frac{|x|}{x} = 1;\ \lim_{x\to 0^-}\frac{|x|}{x} = -1$

(e) $\displaystyle\lim_{x\to 0}\left[x+\tfrac{1}{2}\right] = 0$

(f) $\displaystyle\lim_{x\to 2^+}\frac{1}{x^2-4} = \infty;\ \lim_{x\to 2^-}\frac{1}{x^2-4} = -\infty$

(g) $\displaystyle\lim_{x\to -2^+}\frac{x}{x^2+x-2} = \infty$

(h) $\displaystyle\lim_{x\to 2}\frac{1}{x^2-4x+4} = \infty$

(i) $\displaystyle\lim_{x\to\infty}\frac{x}{x^2+1} = 0$

(j) $\displaystyle\lim_{x\to\infty}\sqrt{\frac{1+x^2}{x^2}} = 1$

2. (a) Finish the proof of Theorem 3.3.1 by showing that condition (i) is sufficient for condition (ii).

(b) Prove part (i) of Theorem 3.3.2.

(c) Use Theorem 3.3.2 to prove Theorem 3.3.3.

3. Prove that if the domains of f and g both contain an interval of the form $(a, a+h)$, where $h > 0$, and if $f(x) \to L$ and $g(x) \to M$ as $x \to a^+$, then the sum $f(x) + g(x) \to L + M$ as $x \to a^+$.

4. Let f and g be real functions whose domains both contain some interval of the form (B, ∞), where $B > 0$. Show by examples that nothing can be concluded about $f(x) - g(x)$ as $x \to \infty$ if it is known only that $f(x) \to \infty$ and $g(x) \to \infty$ as $x \to \infty$.

5. Let f and g be real functions whose domains both contain some interval of the form (B, ∞), where $B > 0$. What can be concluded about the limiting behavior of $f(x)g(x)$ as $x \to \infty$ if $f(x) \to L$ as $x \to \infty$ and $g(x) \to \infty$ as $x \to \infty$? (Consider the cases $L > 0$, $L < 0$, and $L = 0$.)

6. Formulate definitions of the following statements about limiting behavior. In each case give an example illustrating the behavior.

(a) $\displaystyle\lim_{x\to\infty} f(x) = -\infty$

(b) $\displaystyle\lim_{x\to-\infty} f(x) = L$

(c) $\displaystyle\lim_{x\to-\infty} f(x) = \infty$

(d) $\displaystyle\lim_{x\to-\infty} f(x) = -\infty$

7. Let $f(x)$ be nonzero for all x in some deleted neighborhood of a. Show that $f(x) \to 0$ as $x \to a$ if and only if $1/|f(x)| \to \infty$ as $x \to a$.

3.4 *Continuity of Real Functions*

A function that preserves limits is said to be *continuous.*

Definition 3.4.1 Let f be a real function with domain D, and let $a \in R$. Then f is ***continuous at a*** provided that $a \in D$ and that for every sequence $\{x_n\}_{n=1}^{\infty}$ of points of D such that $x_n \to a$, we have $f(x_n) \to f(a)$.

Example 1: The function $f(x) = x^2$ is continuous at 3. For 3 is in the domain of f and if $x_n \to 3$, then we have $x_n^2 \to 9$, so that $f(x_n) \to f(3)$.

The function in Example 1 is continuous at every real number.

Definition 3.4.2 A real function that is defined and continuous at every point of a set S is said to be ***continuous on S***. If the function is continuous at every real number, it is said to be ***continuous everywhere***.

Example 2: The absolute value function $f(x) = |x|$ is continuous everywhere because for every real number a, if $x_n \to a$, then $|x_n| \to |a|$ (by Theorem 2.3.5).

Example 3: Every straight line function $f(x) = mx + b$ (m, b constants) is continuous everywhere, as Example 2 in Section 3.2 shows. In particular, the constant function $f(x) = b$ and the identity function $f(x) = x$ are continuous everywhere.

Theorem 3.4.1 *Let f and g be real functions, and let $a \in R$. If f and g are continuous at a, so are the functions $f + g$, $f - g$, fg; and if also $g(a) \neq 0$, then f/g is continuous at a.*

PROOF: The proofs for sums, differences, and products are straightforward. In the case of the quotient, observe that any x for which $g(x) = 0$ cannot be in the domain D of f/g. Moreover, we are assuming here that $g(a) \neq 0$. Thus, if $x_n \to a$, $x_n \in D$, then $f(x_n)/g(x_n) \to f(a)/g(a)$ by Theorem 2.4.2. ∎

By induction, Theorem 3.4.1 can be generalized to sums and products of any finite number of terms.

Corollary 3.4.1 *If $f_1, f_2, \ldots, f_n$ are real functions, each of which is continuous at a real number a, then the sum $f_1 + f_2 + \cdots + f_n$ and the product $f_1 f_2 \cdots f_n$ are also continuous at a.*

Continuity of Polynomials and Rational Functions

If we combine continuous functions by means of addition and multiplication, we obtain other continuous functions. Starting with the identity function x, which is continuous everywhere, we see by Corollary 3.4.1 that x^n is continuous everywhere for every $n \in Z^+$.

Since a constant function is also continuous everywhere, it follows that all functions of the form $c_k x^k$ (c_k constant, $k \in Z^+$) are continuous. A polynomial function is given by a finite sum of such terms:

$$P(x) = c_0 + c_1 x + \cdots + c_n x^n$$

It follows by Corollary 3.4.1 that a polynomial function is continuous everywhere.

Theorem 3.4.2 *Every polynomial function is continuous everywhere.*

Theorem 3.4.3 *Every rational function*

$$f(x) = P(x)/Q(x)$$

where $P(x)$ and $Q(x)$ are polynomial functions, is continuous at all real numbers a except those for which $Q(a) = 0$.

Example 4: The rational function

$$f(x) = \frac{x^3 + 5x + 7}{x^2 - 4x + 3} = \frac{x^3 + 5x + 7}{(x - 1)(x - 3)}$$

is continuous at all real numbers except 1 and 3.

Like the polynomial and rational functions, most of the functions of elementary calculus are either continuous everywhere or continuous except at certain readily determined points.

Example 5: We accept for the present that the trigonometric functions sine and cosine are continuous everywhere. Their quotient

$$\tan x = \frac{\sin x}{\cos x}$$

is continuous except at the points where $\cos x = 0$; that is, except at the points $x = n\pi/2$, where n is an odd integer.

Continuity of Composite Functions

Theorem 3.4.4 *Let f and g be real functions such that the range of f is contained in the domain of g. If f is continuous at a and g is continuous at $f(a)$, then the composition $g \circ f$ is continuous at a.*

PROOF: Let $\{x_n\}_{n=1}^{\infty}$ be a sequence of points in the domain of f such that $x_n \to a$. Then $\{f(x_n)\}_{n=1}^{\infty}$ is a sequence in the domain of g for which $f(x_n) \to f(a)$. Now g is continuous at $f(a)$, so $g(f(x_n)) \to g(f(a))$, proving the continuity of $g \circ f$ at a. ■

Example 6: The function $h(x) = |x^2 - 4x|$ is the composition of continuous functions and is continuous everywhere.

Example 7: The function $f(x) = \sin\frac{1}{x}$ is continuous on the set of nonzero real numbers.

Isolated Points in the Domain

Recall that an isolated point of D is a point $a \in D$ with the property that some deleted neighborhood of a contains no points of D.

If a is an isolated point of the domain D of f, then f is continuous at a in a trivial way because any sequence $\{x_n\}_{n=1}^{\infty}$ of points of D converging to a must eventually end in an infinite string of a's, so that the image sequence $\{f(x_n)\}_{n=1}^{\infty}$ must end in an infinite string of terms equal to $f(a)$.

Testing for Continuity with Limits

If $a \in D$ and is not an isolated point of D, then it is meaningful to consider $\lim_{x\to a} f(x)$. Suppose f is continuous at a; if $\{x_n\}_{n=1}^{\infty}$ is a sequence such that $x_n \to a$ with $x_n \neq a$ and $x_n \in D$, then we must have $f(x_n) \to f(a)$—hence, $\lim_{x\to a} f(x) = \lim_{n\to\infty} f(x_n) = f(a)$. Conversely, if $\lim_{x\to a} f(x)$ exists and equals $f(a)$, then $f(x_n) \to f(a)$ whenever $x_n \to a$ with $x_n \in D$; so f is continuous at a. This argument establishes the following theorem:

Theorem 3.4.5 *Let f be a real function with domain D, and let a be a point of D that is not an isolated point of D. Then f is continuous at a if and only if*

$$\lim_{x\to a} f(x) = f(a)$$

Example 8: Let $f(x) = \sqrt{x}$ for $x \geq 0$ and let $a \geq 0$. It follows from Theorem 2.3.6 that $\lim_{x\to a} \sqrt{x} = \sqrt{a}$; that is, $\lim_{x\to a} f(x) = f(a)$. Thus, f is continuous for all nonnegative real numbers.

Limits by Substitution

If a is not an isolated point of the domain of f, and if f is continuous at a, then Theorem 3.4.5 says that the limit of $f(x)$ as x approaches a can be obtained by substitution.

Example 9: The rational function

$$f(x) = \frac{3x + 4}{x^3 - 1}$$

is continuous at every real number except 1. If we wish to find its limit, say at 2, we can simply substitute:

$$\lim_{x \to 2} f(x) = f(2) = \frac{3 \cdot 2 + 4}{2^3 - 1} = \frac{10}{7}$$

The Epsilon-Delta Formulation of Continuity

Theorem 3.4.6 *Let f be a real function with domain D and let $a \in D$. Then f is continuous at a if and only if for every $\varepsilon > 0$ there exists a $\delta > 0$ with the property $|f(x) - f(a)| < \varepsilon$ for all $x \in D$ with $|x - a| < \delta$.*

PROOF: If a is not an isolated point of D, this result is an immediate consequence of Theorems 3.4.5 and 3.2.6. If a is an isolated point of D, the conclusion follows trivially because for sufficiently small $\delta > 0$, the only $x \in D$ with $|x - a| < \delta$ is $x = a$ and $|f(x) - f(a)| = 0$. ∎

Corollary 3.4.2 *Let f be a real function with domain D and let $a \in D$. If f is continuous at a and $f(a) < 0$ (> 0), then there exists a $\delta > 0$ such that $f(x) < 0$ (> 0) for all $x \in D$ with $|x - a| < \delta$.*

PROOF: You are asked to prove this corollary in Exercise 3.4.6. ∎

Theorem 3.4.6 can be paraphrased in terms of neighborhoods.

Corollary 3.4.3 *Let f be a real function with domain D and let $a \in D$. Then f is continuous at a if and only if for every neighborhood $N_\varepsilon(f(a))$ there is a neighborhood $N_\delta(a)$ such that*

$$f(N_\delta(a) \cap D) \subseteq N_\varepsilon(f(a))$$

One-Sided Continuity

There are occasions when we wish to restrict our attention to the behavior of a function on one side of a point in its domain.

Definition 3.4.3 Let f be a real function with domain D and let $a \in R$. Then f is ***continuous from the right (left)*** at a provided that $a \in D$ and for every sequence $\{x_n\}_{n=1}^{\infty}$ such that $x_n \in D$ and $x_n \geq a$ ($x_n \leq a$), we have $f(x_n) \to f(a)$.

Example 10: The greatest integer function $f(x) = [x]$ is continuous from the right, but not from the left, at each integer m.

The one-sided counterparts for Theorems 3.4.5 and 3.4.6 are stated below. You are asked to prove them in the exercises.

Theorem 3.4.7 *Let f be a real function with domain D and let a be a point of D with the property that every interval of the form $(a, a + h)((a - h, a))$ with $h > 0$ contains points of D. Then f is continuous from the right (left) at a if and only if $\lim_{x \to a^+} f(x) = f(a)$ $(\lim_{x \to a^-} f(a) = f(a))$.*

Theorem 3.4.8 *Let f be a real function with domain D and let $a \in D$. Then f is continuous from the right (left) at a if and only if for every $\varepsilon > 0$ there exists a $\delta > 0$ such that $|f(x) - f(a)| < \varepsilon$ for all $x \in D$ with $a \le x < a + \delta$ $(a - \delta < x \le a)$.*

Corollary 3.4.4 *Let f be a real function with domain D and let $a \in D$. If f is continuous from the right at a and $f(a) < 0$ (> 0), then there exists a $\delta > 0$ such that $f(x) < 0$ (> 0) for all $x \in D$ with $a \le x < a + \delta$.*

A statement analogous to Corollary 3.4.4 can be proved for continuity from the left.

EXERCISES 3.4

1. Explain why the function is continuous on the given set.
   (a) $f(x) = x^2 \cos x, \quad x \in R$
   (b) $f(x) = x + \sin x, \quad x \in R$
   (c) $f(x) = |x - x^2|, \quad x \in R$
   (d) $f(x) = 1/(x^2 + 1), \quad x \in R$
   (e) $f(x) = \sin(1/x), \quad x \in R - \{0\}$
2. Show that the greatest integer function $f(x) = [x]$ is continuous at every real number that is not an integer.
3. Evaluate the following limits by using Theorem 3.4.5.
   (a) $\lim_{x \to 2} 3x/(x^2 + 5)$
   (b) $\lim_{x \to \pi} (\sin x + \cos 2x)$
   (c) $\lim_{x \to 3} [x + \frac{1}{2}]$
   (d) $\lim_{x \to \pi/4} \tan(\pi - x)$
4. Show that the function
$$f(x) = \begin{cases} x \sin \frac{1}{x} & x \ne 0 \\ 0 & x = 0 \end{cases}$$
is continuous everywhere.
5. (a) Show that if a real function f is continuous everywhere and if $f(x) = 0$ when x is rational, then $f(x) = 0$ for all real x.
   (b) Show that if f and g are continuous everywhere and if $f(x) = g(x)$ when x is rational, then $f(x) = g(x)$ for all real x.
6. Prove Corollary 3.4.2.
7. Prove Corollary 3.4.1.
8. Prove Theorem 3.4.7.
9. Prove Theorem 3.4.8.

10. Let f be a real function with the following properties:

(i) $f(x + y) = f(x) + f(y)$ for all $x, y \in R$

(ii) f is continuous at 0

(a) Show that f is continuous everywhere. [*Hint:* Show that $f(0) = 0$ and that

$$f(x - y) = f(x) - f(y).]$$

(b) Show that there exists an $m \in R$ such that for all rational numbers r, $f(r) = mr$.

(c) Show that $f(x) = mx$ for all $x \in R$. (Use Exercise 3.4.5.)

11. Let f be a real function with the following properties:

(i) $f(x + y) = f(x)f(y)$ for all $x, y \in R$

(ii) f is continuous at 0

Show that f is continuous everywhere. [*Hint:* Show that, unless $f = 0$ for all x, $f(0) = 1$ and $f(x - y) = f(x)/f(y)$.]

3.5 *Types of Discontinuity*

In examining a function for points of discontinuity, we will consider only points in the closure of its domain.

Definition 3.5.1 Let f be a real function with domain D and let $a \in \overline{D}$. Then f is ***discontinuous at a*** provided f is not continuous at a. The number a is a ***point of discontinuity*** of f.

In light of Theorem 3.4.5, there are several ways a function can be discontinuous at a. For one thing, f may simply be undefined at a; in other words, $a \notin D$. It may happen that $\lim_{x \to a} f(x)$ fails to exist. Or it may happen that $\lim_{x \to a} f(x)$ exists but is not equal to $f(a)$. The following examples illustrate various possibilities.

Removable Discontinuity

The mildest sort of discontinuity arises when the limit at a exists but the value of the function at a is either undefined or different from the limit. We can then "remove" the discontinuity at a by defining (or redefining) the value of the function at a to be the limit there.

Example 1: The function

$$f(x) = \frac{x^2 - 9}{x - 3} \qquad x \neq 3$$

is discontinuous at 3 because it is not defined at 3. However, the limit at 3 does exist; that is, $\lim_{x \to 3} f(x) = 6$. We extend f to the function F:

$$F(x) = \begin{cases} \dfrac{x^2 - 9}{x - 3} & \text{if } x \neq 3 \\ 6 & \text{if } x = 3 \end{cases}$$

which agrees with the original for $x \neq 3$ and is easily seen to be continuous at 3. In fact, the new function is really just $F(x) = x + 3$ [see parts (a) and (b) of Figure 3.5].

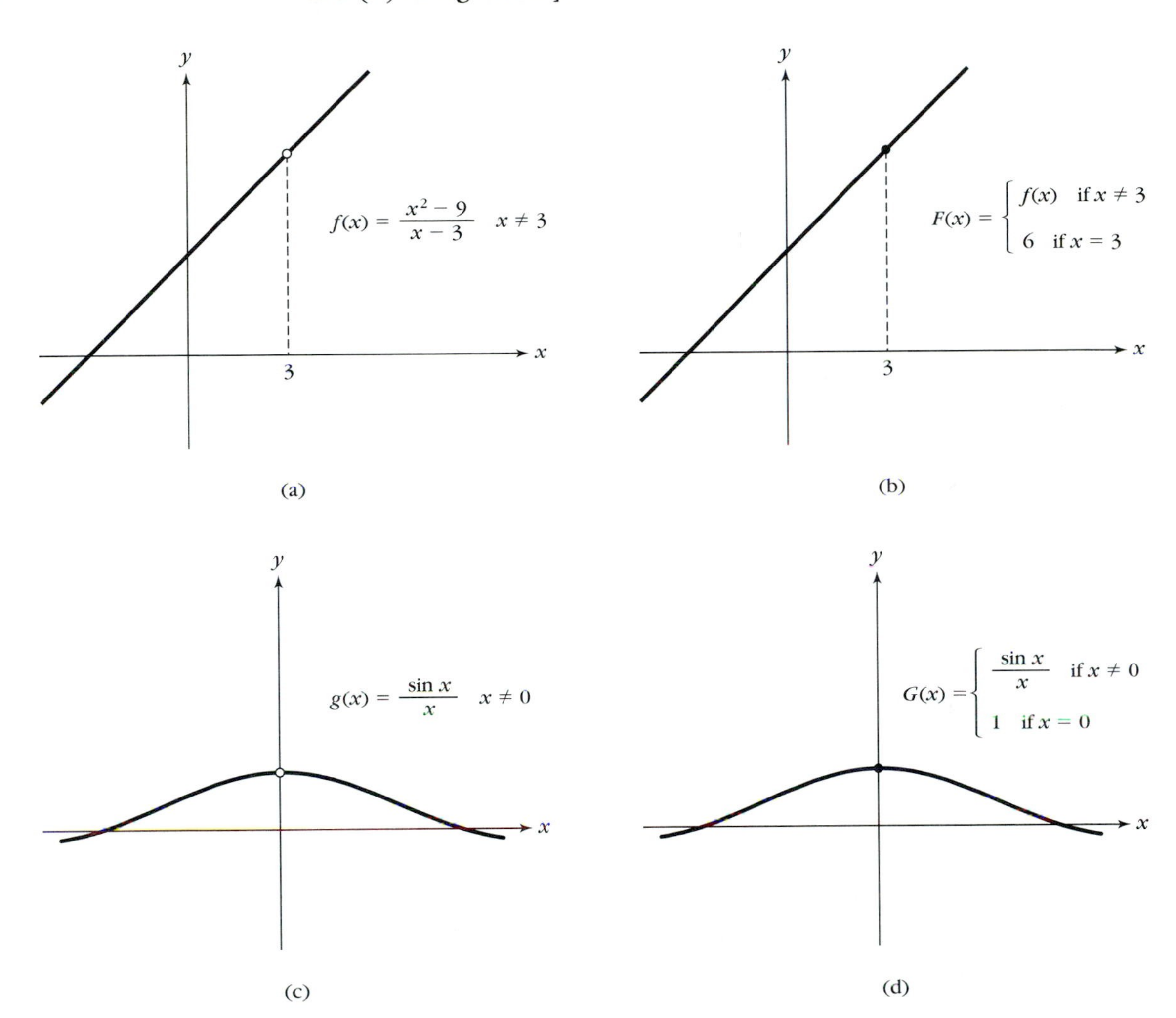

Figure 3.5
Removable discontinuities

You may object that the discontinuity in Example 1 is a result of deliberately expressing the function in a foolish manner—why not just start with the expression $x + 3$ instead of $(x^2 - 9)/(x - 3)$? So here is a less contrived example.

Example 2: The function

$$g(x) = \frac{\sin x}{x} \qquad x \neq 0$$

is undefined at 0, so it is discontinuous there; and no algebraic simplification will circumvent division by zero. However, it can be shown* that $\lim_{x \to 0}(\sin x/x) = 1$, and so we can remove the discontinuity at 0 by defining a new function:

$$G(x) = \begin{cases} \dfrac{\sin x}{x} & \text{if } x \neq 0 \\ 1 & \text{if } x = 0 \end{cases}$$

[see parts (c) and (d) of Figure 3.5].

Infinite Discontinuity

If $f(x)$ approaches infinity or negative infinity as x approaches a, then $\lim_{x \to a} f(x)$ does not exist and we say that f has an *infinite discontinuity* at a.

Example 3: The function

$$f(x) = \frac{1}{\sqrt{x}} \qquad x > 0$$

has an infinite discontinuity at 0 since $\lim_{x \to 0} f(x) = \infty$ (see Figure 3.6a).

Example 4: Consider

$$f(x) = \begin{cases} \dfrac{1}{(x-1)^2} & \text{if } x \neq 1 \\ 0 & \text{if } x = 1 \end{cases}$$

Since $\lim_{x \to 1} f(x) = \infty$, f has an infinite discontinuity at 1 (see Figure 3.6b).

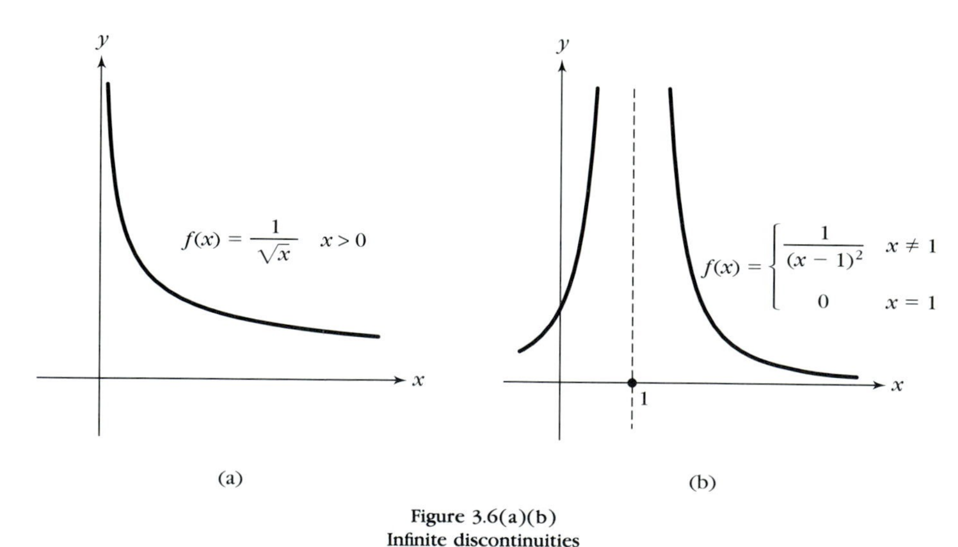

Figure 3.6(a)(b)
Infinite discontinuities

*You may wish to consult your calculus book for an informal geometric argument for this limit fact. We will prove it in Section 4.12 (Theorem 4.12.4).

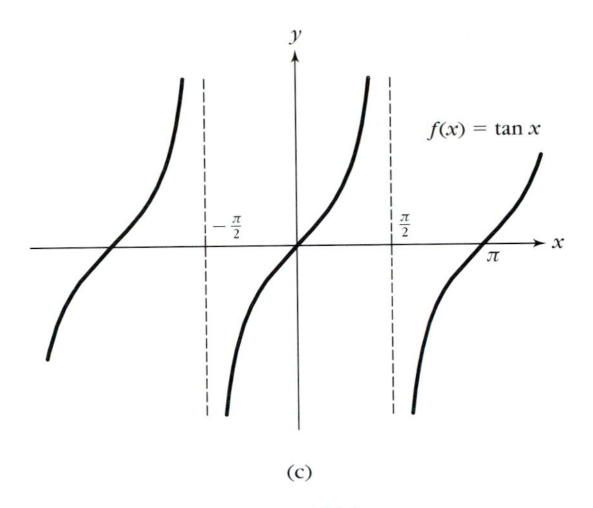

Figure 3.6(c)
Infinite discontinuities

Example 5: The function $f(x) = \tan x$ has infinite discontinuities at the numbers $(2n + 1)\pi/2$ for $n \in Z$ (see Figure 3.6c).

Jump Discontinuity

When the right- and left-hand limits at a exist but do not agree, the ordinary limit of the function does not exist at a (Theorem 3.3.1). This type of discontinuity is called a *jump discontinuity* because the graph exhibits a jump at a.

Example 6: The greatest integer function $f(x) = [x]$ has a jump discontinuity at each integer (see Figure 3.7).

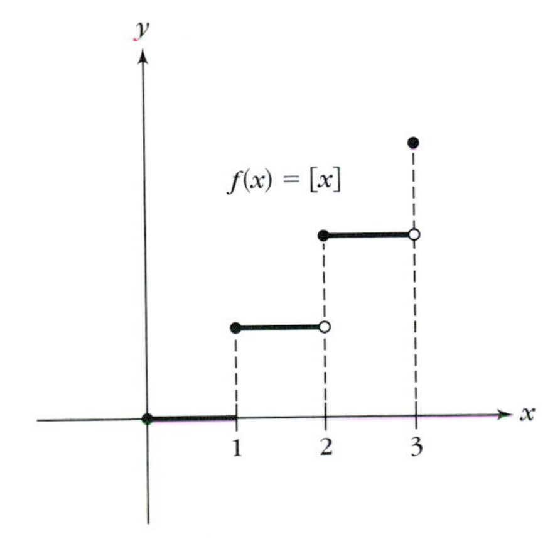

Figure 3.7
Jump discontinuities

Oscillation

There are many other modes of discontinuous behavior; for convenience we shall place them in one classification, which we call *oscillation* even though the behavior may vary considerably from one example to the next.

Example 7: Let $f(x) = \sin \frac{1}{x}$ $(x > 0)$. As x decreases from $\frac{1}{2\pi}$ to 0, the value of $\frac{1}{x}$ increases through positive values, without bound. Each time $\frac{1}{x}$ passes through an interval of length 2π, $f(x)$ assumes each value between -1 and 1 at least once. Thus, as x decreases from $\frac{1}{2\pi}$ to 0, the y-coordinate on the graph of $u = f(x)$ oscillates between -1 and 1 an infinite number of times. The sine waves on the graph of f are compressed into shorter and shorter horizontal intervals, and there are infinitely many of them between $\frac{1}{2\pi}$ and 0 (see Figure 3.8).

We can see that $f(x)$ has no limit as $x \to 0$ by considering two sequences. Let

$$x_n = \frac{1}{2n\pi} \qquad \text{and} \qquad y_n = \frac{1}{\left(2n + \frac{1}{2}\right)\pi}$$

Then $x_n \to 0$ and $y_n \to 0$, but since $f(x_n) = 0$ and $f(y_n) = 1$ for $n \in Z^+$, the sequences $\{f(x_n)\}_{n=1}^{\infty}$ and $\{f(y_n)\}_{n=1}^{\infty}$ do not converge to the same limit. Accordingly, $\lim_{x \to 0} f(x)$ does not exist, and f is discontinuous at zero. However, f is continuous everywhere else.

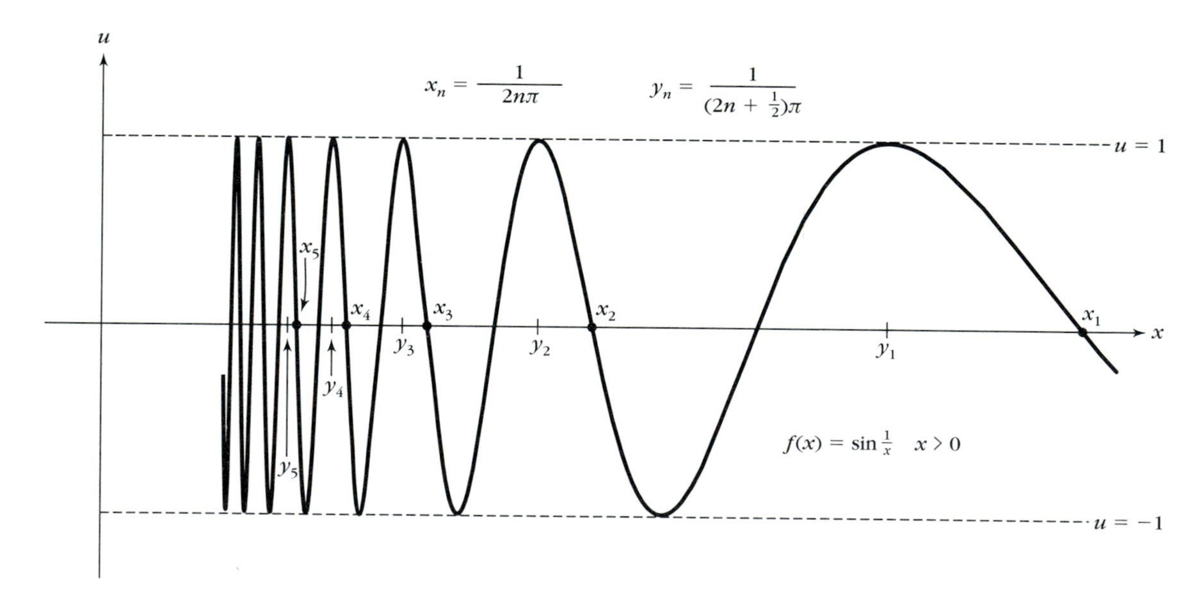

Figure 3.8
Oscillation

Example 8: Recall the function f defined for rationals by $f(r) = \frac{1}{q}$, where $r = \frac{p}{q}$ ($p \in Z$, $q \in Z^+$) is the representation of the rational number r in lowest terms. In Example 9 in Section 3.2 we showed that for every $a \in R$, $\lim_{x \to a} f(x) = 0$. Now the function

$$g(x) = \begin{cases} f(x) & \text{if } x \text{ is rational} \\ 0 & \text{if } x \text{ is irrational} \end{cases}$$

is defined for all real numbers, and it is easy to see that $\lim_{x \to a} g(x) = 0$ for every $a \in R$. It follows that g is discontinuous at every rational number r, because $\lim_{x \to r} g(x) = 0 \neq g(r)$; however, g is continuous at every irrational number.

EXERCISES 3.5

1. Find all discontinuities of the following functions. For each discontinuity, determine which type it represents and explain.
   (a) $f(x) = \dfrac{x}{x+1}$
   (b) $f(x) = \dfrac{x}{x^2 + x}$
   (c) $f(x) = \dfrac{x^2 - 3x}{x^2 - x - 6}$
   (d) $f(x) = [2x]$, where the brackets mean the greatest integer function.
   (e) $f(x) = \dfrac{|x|}{x}$
   (f) $f(x) = \dfrac{1}{\sqrt{x^2 - 1}}$
   (g) $f(x) = \sqrt{x^2 - 1}$
   (h) $f(x) = \dfrac{\tan x}{x}$
   (i) $f(x) = \dfrac{\sin^2 x}{x}$
   (j) $f(x) = \dfrac{\sin x}{\sqrt{x}}$
   (k) $f(x) = \dfrac{\sin x}{x^2}$
   (l) $f(x) = \cos \dfrac{1}{x}$
2. Suppose the domains of each of the real functions f and g contain a deleted neighborhood of a. Prove or disprove each of the following statements.
   (a) If f and g are discontinuous at a, then $f + g$ is discontinuous at a.
   (b) If f and g have jump discontinuities at a, then $f + g$ has a jump discontinuity at a.
   (c) If f and g have infinite discontinuities at a, then $f + g$ has an infinite discontinuity at a.
   (d) If f is discontinuous at a, then $|f|$ is discontinuous at a.
3. Is the following statement true or false? The function

$$f(x) = \begin{cases} 1 & \text{if } x \text{ is rational} \\ 0 & \text{if } x \text{ is irrational} \end{cases}$$

has infinitely many jump discontinuities.

3.6 *Continuity on a Closed Bounded Interval*

The philosopher and mathematician Descartes (see page 14) is said to have described a continuous function as one whose graph can be drawn without lifting the pencil from the paper. Descartes was anticipating in an intuitive way some of the properties of continuity brought out by the theorems of this section.

These theorems concern a real function f that is continuous on a closed interval $[a, b]$, where $a < b$. The domain of f may be just $[a, b]$ or it may be a larger set of real numbers containing $[a, b]$. In either case, the continuity of f at the endpoints will be understood to mean that $\lim_{x \to a^+} f(x) = f(a)$ and $\lim_{x \to b^-} f(x) = f(b)$.

Intermediate Values

Our first result can be paraphrased very simply in terms familiar to any calculus student: If the graph of a continuous function $y = f(x)$ is on one side of the x-axis when $x = a$, and on the other side when $x = b$, then the graph must cross the x-axis at least once between a and b.

Theorem 3.6.1 **Bolzano's Theorem** *Let f be continuous on the closed interval $[a, b]$ with $a < b$. If $f(a)$ and $f(b)$ are of opposite signs, then there exists a real number $\xi \in (a, b)$ such that $f(\xi) = 0$.*

PROOF: Suppose $f(a) < 0$ and $f(b) > 0$. Now consider the set

$$S = \{x : a \leq x \leq b \quad \text{and} \quad f(x) \leq 0\}$$

Clearly S is nonempty ($a \in S$) and bounded above (by b); so $\xi = \sup S$ exists. We claim that ξ satisfies the requirements of the theorem.

Clearly $\xi \in [a, b]$. Next, by the definition of ξ as $\sup S$, there exists (Theorem 2.6.4) a sequence of points $\{s_n\}_{n=1}^{\infty}$ with $s_n \in S$ $(n \in Z^+)$ and $s_n \to \xi$. Now

BERNHARD BOLZANO

(b. October 5, 1781; d. December 18, 1848)

Bolzano, a Bohemian priest and contemporary of Cauchy, was one of the advocates of greater rigor in mathematics. He dealt with continuity as a limit concept and realized that dy/dx was a symbol for a function, not to be interpreted as a ratio. Unlike Descartes, he recognized the need to prove the theorem here named after him; however, the real number system was not sufficiently developed at that time for him to give a proof that is rigorous by today's standards. Like Galileo, Bolzano anticipated set theory, observing the one-to-one correspondence of an infinite set and a proper subset in his treatise *Paradoxes of the Infinite*. The political and clerical pressures of the time cost him his position as professor of religion at the University of Prague although he continued his duties as a priest.

$f(s_n) \leq 0$ ($n \in Z^+$), and since f is continuous at ξ, we have $f(\xi) = \lim_{n\to\infty} f(s_n) \leq 0$.

It is clear now that $\xi < b$ because $f(b) > 0$. We can therefore obtain a sequence of points $\{t_n\}_{n=1}^{\infty}$ with $t_n \in (\xi, b](n \in Z^+)$ and $t_n \to \xi$. By the definition of ξ, $f(x) > 0$ for $x \in (\xi, b]$; hence, $f(t_n) > 0$ ($n \in Z^+$) and $f(\xi) = \lim_{n\to\infty} f(t_n) \geq 0$. It follows that $f(\xi) = 0$ and $a < \xi < b$. ■

Example 1: Consider the continuous function $f(x) = \cos x$ on the interval $[0, 3\pi]$. We see that $f(0) = 1$ and $f(3\pi) = -1$ are of opposite sign. There are three numbers in $[0, 3\pi]$ at which f assumes the value zero: $\pi/2$, $3\pi/2$, and $5\pi/2$.

Example 2: The polynomial function $f(x) = x^3 + 5x - 7$ has the values $f(1) = -1$, $f(2) = 11$. By Bolzano's Theorem, there exists a root of the equation $f(x) = 0$ somewhere between 1 and 2. By checking $f(1.5)$, we find $f(1.5) > 0$, indicating that a root lies between 1 and 1.5.

Bolzano's Theorem is easily generalized as follows:

Theorem 3.6.2 **Intermediate Value Theorem** *Let f be continuous on the closed interval $[a, b]$, with $a < b$. If μ is any real number between $f(a)$ and $f(b)$, then there exists a ξ between a and b such that $f(\xi) = \mu$.*

PROOF: We define $g(x) = f(x) - \mu$, note that g is continuous on $[a, b]$, and apply Bolzano's Theorem (complete the proof in Exercise 3.6.1). ■

Example 3: The function $f(x) = \sin x$ assumes the values -1 and 1, respectively, at $-\pi/2$ and $\pi/2$. By the Intermediate Value Theorem, for any real number μ between -1 and 1 there must exist a ξ between $-\pi/2$ and $\pi/2$ such that $\sin \xi = \mu$. It can be shown that the number ξ is unique for a given μ; ξ is called the *principal arcsine* of μ.

Example 4: The range of the greatest integer function $f(x) = [x]$ consists only of integers. We have $f(2) = 2$, $f(3) = 3$, but there is no x for which $f(x) = 2.5$; Theorem 3.6.2 does not apply because f is not continuous on $[2, 3]$.

The Intermediate Value Theorem is a powerful tool. Note the ease with which we can now prove the existence of nth roots of nonnegative real numbers:

Theorem 3.6.3 *Every nonnegative real number has a unique nonnegative real nth root ($n \in Z^+$).*

PROOF: The theorem clearly holds for the real number 0. Let $a \in R$ with $a > 0$. Now consider the function $f(x) = x^n$. Since $f(x) \to \infty$ as $x \to \infty$, there is a real number b such that $f(x) > a$ for $x \geq b$. Now f is continuous on the interval $[0, b]$ and we have $f(0) = 0 < a < f(b)$. We apply Theorem 3.6.2 with $\mu = a$, giving us a $\xi \in (0, b)$ such that $f(\xi) = a$. Thus, ξ is a positive number with the property that $\xi^n = a$. (You are asked to prove the uniqueness of ξ in Exercise 3.6.7.) ∎

Boundedness and Extreme Values

Our next goal is a theorem that guarantees the existence of maxima and minima for a function continuous on a closed bounded interval. As a first step we show that such a function is bounded.

Theorem 3.6.4 *If f is continuous on a closed bounded interval $[a, b]$ with $a < b$, then f is bounded on $[a, b]$.*

PROOF: Suppose that f is continuous but not bounded above on $[a, b]$. Then for every $n \in Z^+$ there exists an $x_n \in [a, b]$ such that $f(x_n) > n$. The sequence $\{x_n\}_{n=1}^{\infty}$ is bounded since its terms lie in $[a, b]$. By the Bolzano-Weierstrass Theorem (Theorem 2.7.1), $\{x_n\}_{n=1}^{\infty}$ has a convergent subsequence $\{x_{n_k}\}_{k=1}^{\infty}$. Let $c = \lim_{k\to\infty} x_{n_k}$. Then $c \in [a, b]$. (Why?) Since $x_{n_k} \to c$, we know by the continuity of f at c that $f(x_{n_k}) \to f(c)$. But this is not possible because $f(x_n) \to \infty$. A similar argument can be devised to contradict the assumption that f is not bounded below (work Exercise 3.6.8). ∎

Definition 3.6.1 Let f be a real function defined on a set D. Then:

(i) If there exists a point $\alpha \in D$ such that $f(\alpha) \leq f(x)$ for all $x \in D$, then f is said to assume an ***absolute minimum*** on D at α, and $f(\alpha)$ is the ***absolute minimum value*** of f on D.

(ii) If there exists a point $\beta \in D$ such that $f(x) \leq f(\beta)$ for all $x \in D$, then f is said to assume an ***absolute maximum*** on D at β, and $f(\beta)$ is the ***absolute maximum value*** of f on D.

(iii) The word ***extremum*** is used to mean maximum or minimum.

Example 5: Consider $f(x) = 1/(1 + x^2)$ on $[-1, 2]$. We see that f assumes an absolute maximum on $[-1, 2]$ at 0 because for every $x \in [-1, 2]$ (in fact, for every $x \in R$), we have $1 + x^2 \geq 1$, and so $f(x) = 1/(1 + x^2) \leq 1 = f(0)$. The absolute maximum value of f on $[-1, 2]$ is $f(0) = 1$. It can be shown that f assumes an absolute minimum on $[-1, 2]$ at 2 and that the absolute minimum value of f is $f(2) = \frac{1}{5}$ (work Exercise 3.6.9).

Example 6: The general quadratic function is defined for $x \in R$ by

$$f(x) = Ax^2 + Bx + C$$

where A, B, C are real constants and $A \neq 0$. This function assumes an absolute minimum if $A > 0$, and an absolute maximum if $A < 0$. The extremum occurs at the point $-B/2A$, as can be seen by completing the square:

$$\begin{aligned} f(x) &= A\left(x^2 + \frac{B}{A}x + \frac{B^2}{4A^2}\right) + C - \frac{B^2}{4A} \\ &= A\left(x + \frac{B}{2A}\right)^2 + C - \frac{B^2}{4A} \end{aligned}$$

The extreme value is $C - (B^2/4A)$.

Theorem 3.6.5 **Extreme Value Theorem** *If f is continuous on a closed bounded interval $[a, b]$, with $a < b$, then f assumes an absolute maximum and an absolute minimum there.*

PROOF: We show here that f assumes an absolute maximum on $[a, b]$ (you are asked to provide the proof for absolute minimum in Exercise 3.6.11). By Theorem 3.6.4, f is bounded above on $[a, b]$, so we can let $M = \sup\{f(x) : a \leq x \leq b\}$. It is sufficient to prove that M is a value of f. By Theorem 2.6.4, there is a sequence $\{y_n\}_{n=1}^{\infty}$ in the range of f such that $y_n \to M$. For $n \in Z^+$, let x_n be a point in $[a, b]$ such that $f(x_n) = y_n$. The sequence $\{x_n\}_{n=1}^{\infty}$ is bounded and by the Bolzano-Weierstrass Theorem (Theorem 2.7.1) has a convergent subsequence $\{x_{n_k}\}_{k=1}^{\infty}$, say $x_{n_k} \to \beta$. Then $\beta \in [a, b]$ (why?) and, by the continuity of f, $f(x_{n_k}) \to f(\beta)$. But on the other hand, $f(x_{n_k}) = y_{n_k} \to M$ (why?). Because limits are unique, we conclude that $f(\beta) = M$ and that f assumes its absolute maximum on $[a, b]$ at β. ■

The Extreme Value Theorem is an *existence* theorem—it does not tell us how to find extrema, only that they exist. Methods of finding extrema are discussed in Chapter 4.

Preservation of Closed Bounded Intervals

Let f be continuous on $[a, b]$. By the Extreme Value Theorem, f assumes a minimum and a maximum there; say that $m = f(\alpha)$ is the absolute minimum value and that $M = f(\beta)$ is the absolute maximum value of f on $[a, b]$ with α, $\beta \in [a, b]$. Thus, the image of $[a, b]$ under f is contained in the closed interval $[m, M]$. But by the Intermediate Value Theorem, if μ is any point between m and M, then there is a ξ between α and β such that $f(\xi) = \mu$. Consequently, the image of $[a, b]$ is in fact all of the interval $[m, M]$. A continuous function therefore maps a closed bounded interval onto a closed bounded interval (see Figure 3.9).

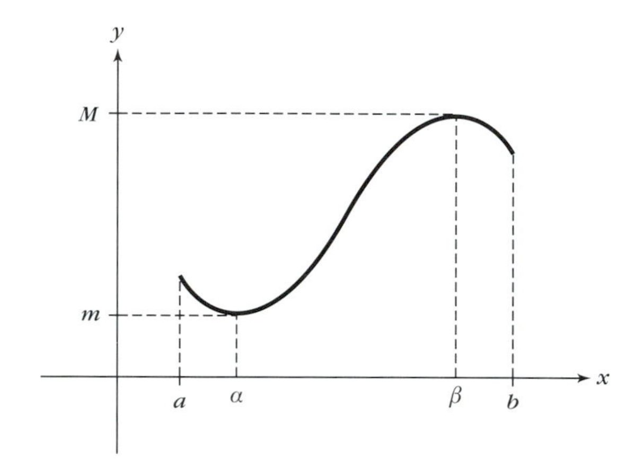

Figure 3.9
Continuity preserves closed bounded intervals

Theorem 3.6.6 *If f is continuous on a closed bounded interval $I = [a, b]$ with $a < b$, then $J = f(I)$ is also a closed bounded interval.*

Example 7: It follows from Example 5 that $f(x) = 1/(1 + x^2)$ maps the interval $[-1, 2]$ onto the interval $\left[\frac{1}{5}, 1\right]$.

EXERCISES 3.6

1. Complete the proof of Theorem 3.6.2.

2. Below, you are given a function f continuous on an interval $[a, b]$ and a real number μ. Show that μ is between $f(a)$ and $f(b)$, and find all numbers ξ, $a < \xi < b$, such that $f(\xi) = \mu$.

(a) $f(x) = x^2 + 3x$; $[1, 4]$; $\mu = 10$
(b) $f(x) = x/(1 + x^2)$; $[0, 1]$; $\mu = \frac{1}{3}$
(c) $f(x) = |x^2 - 4|$; $[-3, 2]$; $\mu = 4$

3. Let $f(x) = \frac{1}{x}$ for $0 < |x| \le 1$. Verify that, although $f(1)$ and $f(-1)$ are of opposite sign, there is no ξ between -1 and 1 with $f(\xi) = 0$. Explain why this does not contradict Theorem 3.6.1.

4. Let $f(x) = x/(x^2 - 4)$ for $x \neq \pm 2$. Show that $\mu = \frac{1}{2}$ is between $f(-1)$ and $f(3)$ but that there is no ξ between -1 and 3 with $f(\xi) = \frac{1}{2}$. Explain why this does not contradict Theorem 3.6.2.

5. Show that the equation $x^3 + 17x + 20 = 0$ has at least one real root between -2 and -1.

6. Show that if $P(x)$ is any polynomial function of odd degree, then the equation $P(x) = 0$ has at least one real root.

7. Prove the uniqueness of the nonnegative nth root of a nonnegative real number.

8. Complete the proof of Theorem 3.6.2 by showing that a function continuous on a closed bounded interval is bounded below.

9. Let $f(x) = 1/(1 + x^2)$. Show that f assumes an absolute minimum on $[-1, 2]$ at 2.

10. Let $f(x) = x/(x^2 + 1)$. Show that the absolute extrema of f on $[-2, 2]$ are assumed at -1 and 1.

11. Complete the proof of Theorem 3.6.5 by showing that a function continuous on a closed bounded interval assumes an absolute minimum there.

12. (a) Show by an example that a function continuous on a closed but unbounded interval may fail to assume an absolute maximum there.

(b) Where would the proof of Theorem 3.6.5 break down if the closed interval on which f is continuous were not bounded?

13. (a) Show by an example that a function continuous on a bounded but not closed interval may fail to assume an absolute maximum there.

(b) Where would the proof of Theorem 3.6.5 break down if the interval were bounded but not closed?

14. Show by an example that a function can assume absolute maxima (minima) on a set D at several different points. Show, however, that the absolute maximum (minimum) value is unique.

15. Prove or disprove each of the following statements, in which f and g are real functions defined on a set $D \subseteq R$.

(a) If f assumes an absolute maximum on D at β and if $g(x) = 2f(x)$ for $x \in D$, then g assumes an absolute maximum on D at β.

(b) If f assumes an absolute maximum on D at β and if $g(x) = [f(x)]^2$ for $x \in D$, then g assumes an absolute maximum on D at β.

(c) If f assumes an absolute maximum on D at β and if $g(x) = [f(x)]^3$ for $x \in D$, then g assumes an absolute maximum on D at β.

(d) If f assumes an absolute maximum on D at β and if $g(x) = |f(x)|$ for $x \in D$, then g assumes an absolute maximum on D at β.

16. (a) Show that if f is continuous on $[a, b]$ with $a < b$, then the range of f is either a single real number or uncountable.

(b) Show that if f is continuous on $[a, b]$ with $a < b$ and if f assumes only rational values, then f is constant.

17. Suppose f is continuous on $[0, 1]$ and that the range of f is contained in $[0, 1]$. Prove that there exists a number $\xi \in [0, 1]$ such that $f(\xi) = \xi$. (Such a point ξ is called a *fixed point* of f.) [*Hint:* Consider $g(x) = f(x) - x$.]

★ 3.7 *Roots and Rational Exponents*

The simplest of the elementary functions are the polynomials and rational functions, whose definitions are based only on the processes available in any field. On the other hand, square roots, cube roots, and so on, do not always exist within a given field (see Section R.3). It is an important consequence of Theorem 3.6.3 that nth roots do exist for the nonnegative elements of R; then, by using nth roots, we can introduce fractional exponents in the customary way. This section includes a brief summary of their properties.

Roots

Let $a \geq 0$ and $n \in Z^+$. Theorem 3.6.3 guarantees the existence of a unique nonnegative nth root of a, which is denoted $\sqrt[n]{a}$ (see Section R.3). In fact, if n is odd, every real number, positive or negative, has a unique nth root; if $a \neq 0$, then a and its nth root $\sqrt[n]{a}$ have the same sign (work Exercise 3.7.6). But if n is even, then each positive number a has two nth roots, $\sqrt[n]{a}$ and $-\sqrt[n]{a}$, while negative numbers do not have nth roots at all.

As a consequence of these facts, algebraic expressions such as $\sqrt[3]{x + 1}$, $\sqrt{(x^2 + 1)^3}$ are meaningful for all real numbers and can be used to define real functions such as:

$$f(x) = \sqrt[3]{x + 1} \qquad x \in R$$
$$g(x) = \sqrt{(x^2 + 1)^3} \qquad x \in R$$

Also,

$$h(x) = \sqrt{x + 1} \qquad x \geq -1$$

is a well-defined function on the given interval because $x + 1 \geq 0$ there.

Rational Exponents

If a is a positive real number, a^n can be defined recursively for $n \in Z^+$:

(i) $a^1 = a$
(ii) $a^{k+1} = a \cdot a^k \qquad k \in Z^+$

By induction it is easy to show (work Exercise 3.7.2) that the following exponential laws hold:

(1) $$a^n > 0 \qquad n \in Z^+$$

(2) $$a^{m+n} = a^m a^n \qquad m, n \in Z^+$$

(3) $$a^{m-n} = \frac{a^m}{a^n} \qquad m, n \in Z^+, m > n$$

(4) $$(a^m)^n = a^{mn} \qquad m, n \in Z^+$$

If we agree to define a^0 to be 1:

(5) $$a^0 = 1$$

then property (3) holds even when $m = n$; and if we further agree to define

(6) $$a^{-n} = \frac{1}{a^n}$$

then a^n has meaning for all integers n, and it can be shown that properties (1) through (6) hold for all $m, n \in Z$ (you are asked to show this in Exercise 3.7.3).

Theorem 3.6.3 permits us to use any rational number as an exponent; if r is a rational number, with $r = p/q$ ($p \in Z$, $q \in Z^+$) in lowest terms, then we define

(7) $$a^r = \sqrt[q]{a^p}$$

Equivalently, a^r is the unique positive number x satisfying

$$x^q = a^p$$

Properties (1) through (6) can now be shown to hold even when the exponents are allowed to be rational numbers (you are asked to show this in Exercise 3.7.4). The positive number a is called the *base* of the system of rational exponents.

Exponential Function for Rationals

As we shall find in Section 4.10, the simplest base to use for exponents in calculus is Euler's number $e = 2.718\ldots$ (see Section 2.6); this base gives rise to the "natural" exponential system. For the present, e^r has been defined, as indicated above, for rational numbers; the function

$$f(r) = e^r \quad r \in Q$$

satisfies properties (1) through (7) in which a is replaced by e and the exponents can be any rational numbers. The laws are restated below for future reference:

(1) $$e^r > 0 \quad r \in Q$$

(2) $$e^{r+s} = e^r e^s \quad r, s \in Q$$

(3) $$e^{r-s} = \frac{e^r}{e^s} \quad r, s \in Q$$

(4) $$(e^r)^s = e^{rs} \quad r, s \in Q$$

(5) $$e^0 = 1$$

(6) $$e^{-r} = \frac{1}{e^r} \quad r \in Q$$

(7) $$e^r = \sqrt[q]{e^p} \quad r = \frac{p}{q}, p \in Z, q \in Z^+$$

The function $f(r) = e^r$ ($r \in Q$) will be called the *natural* exponential function for rational numbers. In Section 4.10 we will see how this function can be extended to all real numbers in such a way that the extended function is continuous everywhere.

EXERCISES 3.7

1. For what real values of x are the following expressions defined?

(a) $\sqrt{x^2 - 4}$

(b) $\sqrt[3]{x^2 - 4}$

(c) $\dfrac{1}{\sqrt[4]{x^2 - 16}}$

(d) $\sqrt{x^2 - 4} + \sqrt{4 - x^2}$

(e) $\sqrt{1 + \frac{1}{x}}$

2. Show that if $a > 0$, then the laws of exponents (1) through (4) hold for positive integers m,n. [*Hint for* (2): Let m be a fixed but arbitrary positive integer and then use induction on n.]

3. Show that if a^{-n} is defined to be $1/a^n$ for $n \in Z^+$, then the laws of exponents (1) through (4) hold for all integers m, n. [*Hint for* (2): If m,n have the same sign, or if one is zero, the result is immediate. If they are of opposite sign, consider various subcases; one such subcase would be $m > 0$, $n < 0$, $m > |n|$.]

4. (a) Show that if $a > 0$ and if $r = p/q$ ($p \in Z$, $q \in Z^+$), then $a^r = a^{p/q}$ is the unique solution of the equation $x^q = a^p$—even if the fraction representing r is not in lowest terms.

(b) Show that if $a > 0$, then the laws of exponents (1) through (6) hold for all rational m,n.

5. (a) Show that if $a > 1$ and if r and s are rational, with $r < s$, then $a^r < a^s$.

(b) What if $a < 1$ in part (a)?

6. Let n be an odd integer. Show that every real number has a unique nth root and that if $a \neq 0$, then a and its nth root have the same sign.

3.8 *Monotone Functions*

Recall from Chapter 2 that monotone sequences are especially well behaved: for example, they are either convergent or properly divergent. This section introduces the concept of monotonicity for real functions defined on intervals. Monotone functions, like monotone sequences, have nice properties.

Monotone Functions

Definition 3.8.1 Let f be a real function and let I be a nonempty interval contained in the domain of f. Then f is ***increasing (decreasing)*** on I provided that for $x_1, x_2 \in I$,

$$f(x_1) \leq f(x_2) \quad (f(x_1) \geq f(x_2)) \qquad \text{whenever } x_1 < x_2$$

If, for $x_1, x_2 \in I$,

$$f(x_1) < f(x_2) \quad (f(x_1) > f(x_2)) \qquad \text{whenever } x_1 < x_2$$

then f is ***strictly increasing (strictly decreasing)*** on I. A function that is increasing or decreasing on an interval is said to be ***monotone*** there.

Example 1: The function $f(x) = x^2$ is strictly increasing on the interval $[0, \infty)$ because $x_1^2 < x_2^2$ whenever $0 \leq x_1 < x_2$. It is strictly decreasing on $(-\infty, 0]$. It is not monotone on any interval that contains both positive and negative numbers.

Chapter 4 discusses calculus-based methods of finding the intervals on which a function is monotone.

Continuity of the Inverse

The following theorem shows that a function that is continuous and strictly monotone on an interval has an inverse with similar properties.

Theorem 3.8.1 *If f is a real function whose domain is an interval I, and if f is strictly monotone on I, then f is invertible. Moreover, if f is also continuous on I, then f^{-1} is continuous on the interval $J = f(I)$ and strictly monotone in the same sense as f.*

PROOF: The invertibility of a strictly monotone function on an interval is an easy exercise (work Exercise 3.8.1a).

Now suppose that f is not only strictly monotone but also continuous on an interval I. The domain J of f^{-1} is the range $f(I)$ of f, and the fact that $f(I)$ is an interval is a consequence of the Intermediate Value Theorem (Theorem 3.6.2).

It is easy to verify that f^{-1} is strictly monotone in the same sense as f (work Exercise 3.8.1b). Suppose now that f and f^{-1} are both strictly increasing.

To establish the continuity of f^{-1} let y_0 be a point in the domain J of f^{-1}, and let x_0 be the real number in I such that $f(x_0) = y_0$; that is, let $x_0 = f^{-1}(y_0)$. Assume for the moment that x_0 is not an endpoint of I. Now let $\varepsilon > 0$ be given, with ε small enough that $x_0 - \varepsilon$ and $x_0 + \varepsilon$ are both elements of I. The interval $(f(x_0 - \varepsilon), f(x_0 + \varepsilon))$ is contained in J and contains the neighborhood $N_\delta(f(x_0))$, where $\delta = \min\{f(x_0 + \varepsilon) - f(x_0), f(x_0) - f(x_0 - \varepsilon)\}$ (see Figure 3.10). For this choice of δ, we have $|f^{-1}(y) - f^{-1}(y_0)| < \varepsilon$ whenever $|y - y_0| < \delta$. In case x_0 is an endpoint of I, then y_0 is an endpoint of J, and the argument can be made with half-neighborhoods.

The case in which f and f^{-1} are both strictly decreasing can be proved similarly (work Exercise 3.8.6). ■

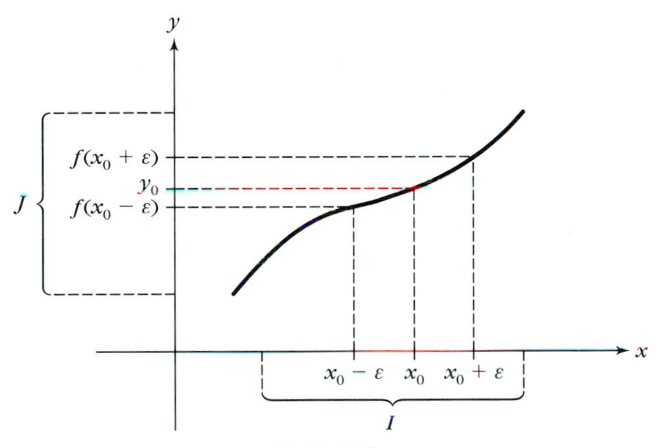

Figure 3.10
Continuity of the inverse of a monotone function

Example 2: For each $n \in Z^+$, the function $f(x) = x^n$ is continuous and strictly increasing on $[0, \infty)$ and maps $[0, \infty)$ onto $[0, \infty)$. Its inverse, the nth-root function, is therefore also continuous and strictly increasing on $[0, \infty)$ (see Figure 3.11b).

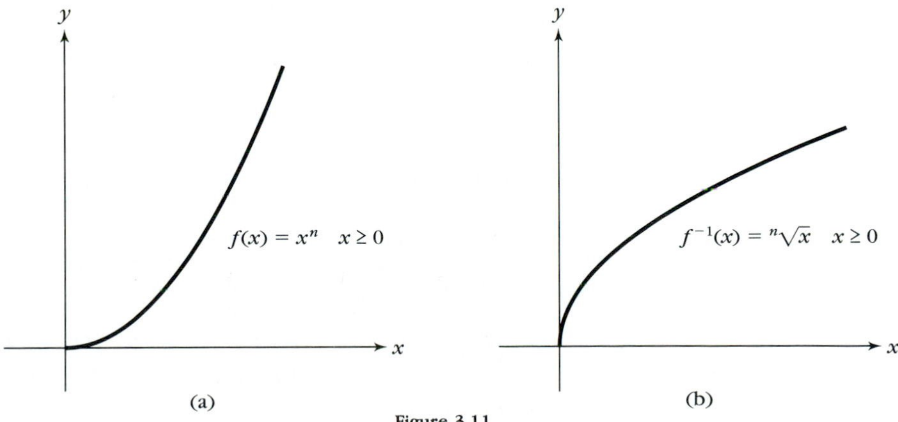

Figure 3.11
The nth-root function

Corollary 3.8.1 *The nth-root function is increasing and continuous on $[0, \infty)$.*

Example 3: In Chapter 4 we will see that the tangent function maps the interval $I = (-\pi/2, \pi/2)$ onto R and that it is strictly increasing and continuous on I. Its inverse, the *arctangent* function, is therefore strictly increasing and continuous on R (see Figure 3.12b).

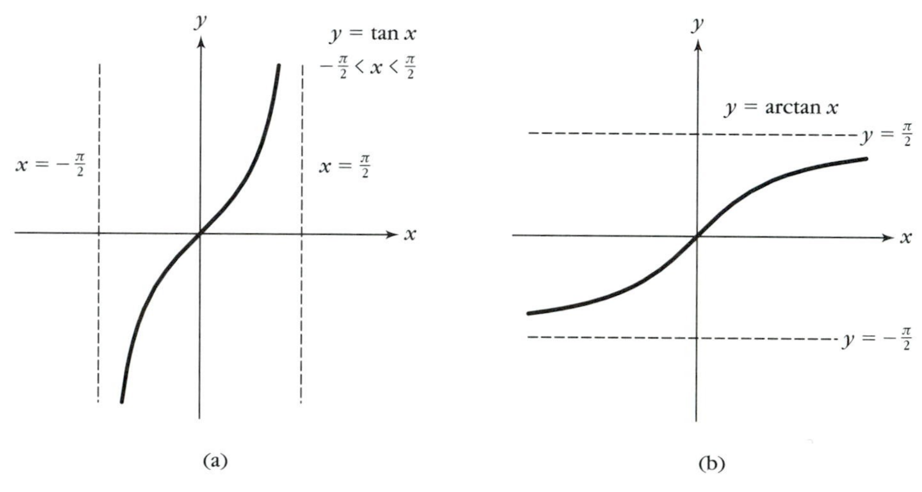

Figure 3.12
The arctangent function

Now let f be a function whose domain is an interval I. If f is strictly monotone on I, then we know that f is invertible. Does the converse hold? No, it is possible to find an example of an invertible function that is not monotone (work Exercise 3.8.3 to see). But if we stipulate that f be continuous as well as invertible on I, then we find that f must be strictly monotone on I (work Exercise 3.8.4a). The importance in this discussion of working with intervals is brought out in Exercise 3.8.4b.

Limits of Monotone Functions

The theorems that follow give valuable information about the limiting behavior of a function that is monotone on an interval. For simplicity we assume the function to be increasing, but parallel results hold for decreasing functions.

Theorem 3.8.2 *If a real function f is increasing on an open interval I and if $c \in I$, then the one-sided limits of f at c both exist and*

$$\lim_{x \to c^-} f(x) = \sup\{f(x) : x \in I,\ x < c\} \le f(c) \le \inf\{f(x) : x \in I,\ x > c\} = \lim_{x \to c^+} f(x)$$

PROOF: Consider the sets

$$A = \{f(x) : x \in I, x < c\}$$
$$B = \{f(x) : x \in I, x > c\}$$

Clearly A is nonempty and bounded above by $f(c)$, so that sup A exists. We denote sup A by L and prove that $L = \lim_{x \to c^-} f(x)$. Let $\{x_n\}_{n=1}^{\infty}$ be an arbitrary sequence in I with $x_n \to c$, and $x_n < c$. Let $\varepsilon > 0$. By part (ii) of Theorem 1.8.5, we can choose an $x \in I$, $x < c$, such that $f(x) > L - \varepsilon$. Since $x_n \to c$ and $x < c$, there exists $N \in Z^+$ such that $x_n > x$ for $n \geq N$. By the increasing property of f, we then have $f(x_n) \geq f(x) > L - \varepsilon$ for $n \geq N$; hence, $L - \varepsilon < f(x_n) \leq L$ for $n \geq N$. Thus, $f(x_n) \to L$, and so $\lim_{x \to c^-} f(x) = L = \sup A$. Clearly, $L \leq f(c)$.

A parallel argument shows that $R = \inf B$ exists and that $R = \lim_{x \to c^+} f(x) \geq f(c)$ (work Exercise 3.8.5). Finally, we have

$$\lim_{x \to c^-} f(x) = L \leq f(c) \leq R = \lim_{x \to c^+} f(x) \qquad \blacksquare$$

Discontinuities of Monotone Functions

Let us pause now to investigate an interesting consequence of Theorem 3.8.2 that tells us about the nature and the number of the discontinuities of a monotone function. If f is increasing on an open interval I and if $c \in I$, then either f is continuous at c [in case $\lim_{x \to c^-} f(x) = \lim_{x \to c^+} f(x)$] or f must have a jump discontinuity at c [in case $\lim_{x \to c^-} f(x) < \lim_{x \to c^+} f(x)$]. Therefore, each point c of discontinuity of f determines a nonempty open interval (L, R) on the y-axis, where L and R are the limits of f from the left and right, respectively (see Figure 3.13).

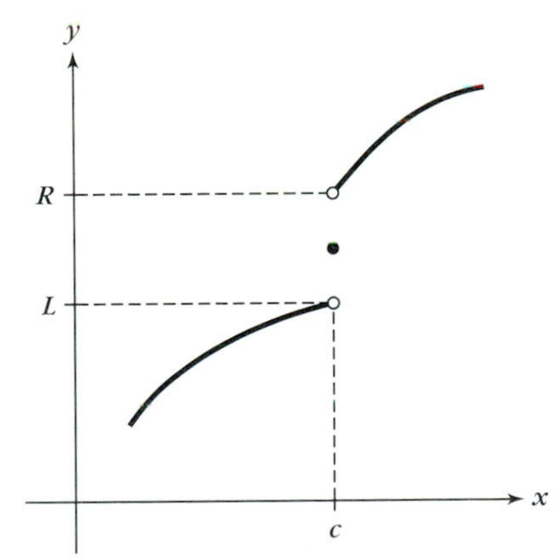

Figure 3.13
Jump discontinuity of a monotone function

It is not hard to show that different discontinuities of f determine disjoint open intervals (work Exercise 3.8.7). Thus, the set S of discontinuities of f is cardinally equivalent to a certain collection T of pairwise disjoint open intervals in $\mathbb{R}$. Now by the Rational Density Theorem, we can select a rational number in each of these open intervals; since the intervals are pairwise disjoint, the rational numbers selected are all different. It follows that the collection T of intervals is equivalent to a subset of $\mathbb{Q}$, and therefore so is S. This establishes the following corollary of Theorem 3.8.2:

Corollary 3.8.2 *A real function that is monotone on an interval can have only jump discontinuities there, and the set of all such discontinuities is countable.*

Endpoint Behavior

We now consider the limiting behavior of a monotone function as x approaches an endpoint of the interval.

Theorem 3.8.3 *Let f be increasing on (a, b) with $a < b$.*

(i) If f is bounded above on (a, b), then

$$\lim_{x \to b^-} f(x) = \sup \{f(x) : a < x < b\}$$

otherwise

$$\lim_{x \to b^-} f(x) = \infty$$

(ii) If f is bounded below on (a, b), then

$$\lim_{x \to a^+} f(x) = \inf \{f(x) : a < x < b\}$$

otherwise

$$\lim_{x \to a^+} f(x) = -\infty$$

PROOF: You are asked to prove this theorem in Exercise 3.8.9. ∎

Behavior at Infinity

The next theorem shows that for an increasing function on (a, ∞), the supremum, the limit as $x \to \infty$, and a certain sequential limit all turn out to be the same.

Theorem 3.8.4 *If a real function f is increasing on (a, ∞), then f either has a finite limit or approaches infinity as $x \to \infty$. Specifically:*

(i) If f is bounded above on (a, ∞), then $\sup \{f(x) : x \in (a, \infty)\}$, $\lim_{x \to \infty} f(x)$, and $\lim_{n \to \infty} f(n)$ all exist and are equal.

(ii) *If f is not bounded above on (a, ∞), then $\lim_{x\to\infty} f(x) = \lim_{n\to\infty} f(n) = \infty$.*

PROOF:

(i) Assume that f is bounded above on (a, ∞); then the set

$$\{f(x) : x \in (a, \infty)\}$$

is bounded above. By the Axiom of Completeness we can define $L = \sup \{f(x) : x \in (a, \infty)\}$. Let $\{x_n\}_{n=1}^{\infty}$ be any sequence in (a, ∞) such that $x_n \to \infty$. If $\varepsilon > 0$ is given, then there exists an $x \in (a, \infty)$ such that $f(x) > L - \varepsilon$; since $x_n \to \infty$, there is an $N \in Z^+$ such that $x_n > x$ for $n \geq N$. Then, for $n \geq N$,

$$L \geq f(x_n) \geq f(x) > L - \varepsilon$$

so that $L = \lim_{n\to\infty} f(x_n)$. Since $\{x_n\}_{n=1}^{\infty}$ can be any sequence with $x_n \to \infty$, we have proved that $L = \lim_{x\to\infty} f(x)$. Moreover, if we choose $x_n = n$, we have $L = \lim_{n\to\infty} f(n)$.

(ii) You are asked to prove part (ii) in Exercise 3.8.10. ■

Corollary 3.8.3 *If a real function f is defined and increasing on an interval (a, ∞), then $\lim_{x\to\infty} f(x)$ exists if and only if $\lim_{n\to\infty} f(n)$ exists; when these limits exist, they are equal.*

EXERCISES 3.8

1. (a) Show that if the domain of f is an interval I and if f is strictly monotone on I, then f is invertible.

(b) Verify the claim in the proof of Theorem 3.8.1 that f^{-1} is strictly monotone in the same sense as f.

2. Use Definition 3.8.1 to show that the following functions are strictly monotone on the given interval I. Find the domain of the inverse and an expression for $f^{-1}(x)$.

(a) $f(x) = x^2 + x$; $I = [0, \infty)$

(b) $f(x) = mx + b$ where m,b are constants with $m \neq 0$; $I = R$

(c) $f(x) = \frac{1}{x}$; $I = (0, \infty)$

(d) $f(x) = \dfrac{x}{1 + x^2}$; $I = [-1, 1]$

(e) $f(x) = \dfrac{x}{1 + x^2}$; $I = [1, \infty)$

3. Find an invertible real function f whose domain is an interval I such that f is not monotone on I.

4. (a) Let f be a real function whose domain is an interval I. Show that if f is continuous and invertible, then f is strictly monotone on I, so that f^{-1} is continuous.

(b) Consider the real function whose domain is $[-1, 0) \cup [1, 2]$:

$$f(x) = \begin{cases} x + 1 & \text{if } -1 \leq x < 0 \\ x & \text{if } 1 \leq x \leq 2 \end{cases}$$

Show that f is continuous and invertible but that f^{-1} has a point of discontinuity. (Thus, in part (a) it is essential that the domain of f be an interval.)

5. Let f be increasing on an open interval I and let $c \in I$. Define

$$B = \{f(x) : x \in I \quad \text{and} \quad x > c\}$$

Explain why $R = \inf B$ exists, and show that $R = \lim_{x \to c^+} f(x)$ and $R \geq f(c)$.

6. Outline the steps of a proof of Theorem 3.8.1 for the case in which f is decreasing.

7. Let f be increasing on an interval I, let c_1 and c_2 $(c_1 < c_2)$ be points of discontinuity of f in I, and let R_1 and L_2 be defined as follows:

$$R_1 = \inf \{f(x) : x \in I \quad \text{and} \quad x > c_1\}$$
$$L_2 = \sup \{f(x) : x \in I \quad \text{and} \quad x < c_2\}$$

Show that $R_1 \leq L_2$ and explain how this justifies the assertion in the text that "different discontinuities determine disjoint open intervals".

8. Give an example of a real function monotone on a bounded interval I that has an infinite number of points of discontinuity.

9. Prove Theorem 3.8.3.

10. Prove part (ii) of Theorem 3.8.4.

3.9 *Uniform Continuity*

There is a special kind of continuity known as *uniform* continuity that has very important consequences in analysis. It is, as the name suggests, a stronger kind of continuity. We will see in Chapter 4 that the key to establishing the existence of the definite integral of a continuous function defined on an interval $[a, b]$ lies in its uniform continuity.

In order to draw the distinction carefully we begin with two examples, the first of which exhibits ordinary continuity, the second of which exhibits uniform continuity. Then we will formulate a definition.

Ordinary Continuity

Let f be a real function with domain D. To say that f is continuous at a particular point $a \in D$ means, in terms of ε and δ, that for each $\varepsilon > 0$ there exists a $\delta > 0$ such that $|f(x) - f(a)| < \varepsilon$ whenever x is an element of D satisfying $|x - a| < \delta$. Let us express this idea symbolically with quantifiers:

$$(\forall \varepsilon > 0)(\exists \delta > 0)(\forall x \in D)(|x - a| < \delta \Rightarrow |f(x) - f(a)| < \varepsilon)$$

Keep in mind that we have been talking about a *particular a*. If f is continuous at *every* $a \in D$, then we have

$$(\forall a \in D)(\forall \varepsilon > 0)(\exists \delta > 0)(\forall x \in D)(|x - a| < \delta \Rightarrow |f(x) - f(a)| < \varepsilon)$$

Here the δ whose existence is asserted may depend on a as well as on ε, as the following example shows.

Example 1: The function $f(x) = x^2$ is continuous everywhere. Suppose ε is assigned the value 1; we consider the problem of choosing δ for two particular values of a, namely, $a = 2$ and $a = 10$. For $a = 2$, a suitable choice is $\delta = \frac{1}{5}$. To see this, note first that

$$|f(x) - f(2)| = |x^2 - 4| = |x + 2|\,|x - 2|$$

so that if $|x - 2| < \frac{1}{5}$, then $|x + 2| < 4\frac{1}{5} < 5$. Therefore, if $|x - 2| < \frac{1}{5}$, we have

$$
\begin{aligned}
|f(x) - f(2)| &= |x + 2|\,|x - 2| \\
&< 5 \cdot \frac{1}{5} = 1
\end{aligned}
$$

The choice $\delta = \frac{1}{5}$ therefore "works" for $\varepsilon = 1$ when $a = 2$. But it does not work for $\varepsilon = 1$ when $a = 10$ because $x = 10.1$ satisfies $|x - 10| < \frac{1}{5}$ while $|f(x) - f(10)| = 2.01 > 1$. A smaller δ is required when $a = 10$. It is easy to see that $\delta = \frac{1}{21}$ works for $\varepsilon = 1$ when $a = 10$. Of course, the smaller δ also works for $\varepsilon = 1$ when $a = 2$, but we will show in Example 5 that there is no positive δ that works for $\varepsilon = 1$ simultaneously for *all* choices of the real number a.

Uniform Continuity

Example 2: The function $f(x) = 2x + 3$ is also continuous everywhere. Let $\varepsilon > 0$ and a be specified and consider the problem of finding a $\delta > 0$ such that $|f(x) - f(a)| < \varepsilon$ whenever $|x - a| < \delta$. Since

$$|f(x) - f(a)| = |(2x + 3) - (2a + 3)| = 2|x - a|$$

it is clear that no matter what a is, we will have $|f(x) - f(a)| < \varepsilon$ whenever $|x - a| < \frac{\varepsilon}{2}$. That is, for a given ε the choice $\delta = \frac{\varepsilon}{2}$ works simultaneously for all a, so that δ depends on ε but not on a.

In Example 2 it was possible to choose δ on the basis of ε alone, in such a way that δ works for the given ε for all a in the domain of the function; that is:

$$(\forall \varepsilon > 0)(\exists \delta > 0)(\forall a, x \in D)(|x - a| < \delta \Rightarrow |f(x) - f(a)| < \varepsilon)$$

The situation described in Example 2 illustrates the concept of uniform continuity. Uniform continuity can be defined on any subset of the domain of the function.

Definition 3.9.1 Let f be a real function and let S be a subset of its domain. Then f is ***uniformly continuous*** on S provided that for every $\varepsilon > 0$ there exists a $\delta > 0$ such that for all a and x belonging to S,

$$|f(x) - f(a)| < \varepsilon \qquad \text{whenever } |x - a| < \delta$$

Again, in quantifier notation: f is uniformly continuous on S if and only if

$$(\forall \varepsilon > 0)(\exists \delta > 0)(\forall a, x \in S)(|x - a| < \delta \Rightarrow |f(x) - f(a)| < \varepsilon) \qquad \textbf{(1)}$$

Note: It follows immediately from the definition that if f is uniformly continuous on S, then it is also uniformly continuous on every subset of S.

Example 3: The straight line function $f(x) = mx + b$ is uniformly continuous on R. In case $m = 0$, f is constant and $|f(x) - f(a)| = 0$ for all $a,x \in R$. Otherwise, let $\varepsilon > 0$ and observe that if $\delta = \varepsilon/|m|$, then $|f(x) - f(a)| < \varepsilon$ holds for all $a,x \in R$ satisfying $|x - a| < \delta$.

Example 4: Let us prove that $f(x) = x^2$ is uniformly continuous on $[0, 10]$. Let $\varepsilon > 0$ be given and observe that for any points $a,x \in [0, 10]$, we have

$$|f(x) - f(a)| = |x + a|\,|x - a| \leq 20\,|x - a|$$

It follows at once that $\delta = \frac{\varepsilon}{20}$ satisfies the definition: For $a,x \in [0, 10]$, $|f(x) - f(a)| < \varepsilon$ holds whenever $|x - a| < \frac{\varepsilon}{20}$.

How can we show that a function f fails to be uniformly continuous on a set S?

Example 5: Consider again the function $f(x) = x^2$. We claim that f is not uniformly continuous on R. To explain why uniform continuity does not hold here we show that there is an $\varepsilon > 0$ for which the definition is not satisfied. Specifically we show that the definition does not hold for $\varepsilon = 1$. We suppose the definition does hold for $\varepsilon = 1$ and obtain a contradiction. A positive δ must exist such that for all $a,x \in R$ with $|x - a| < \delta$, we have $|f(x) - f(a)| < 1$. Now let a be an arbitrary positive number and let $x = a + \delta/2$; clearly, $|x - a| = \delta/2 < \delta$, so we should have $|f(x) - f(a)| < 1$. But

$$|f(x) - f(a)| = \left|\left(a + \frac{\delta}{2}\right)^2 - a^2\right| = a\delta + \frac{\delta^2}{4}$$

and so we have

$$a\delta + \frac{\delta^2}{4} < 1$$

$$a\delta < 1 - \frac{\delta^2}{4} < 1$$

and δ must be a positive real number with the property that $\delta < \frac{1}{a}$ for *every* $a > 0$. But no positive real number exists with this property. Therefore, f is not uniformly continuous on R.

Let us try to understand why the function $f(x) = x^2$ is uniformly continuous on $[0, 10]$ but not on R. We saw in Example 1 that the δ corresponding to a particular $\varepsilon(\varepsilon = 1)$ could be chosen as $\frac{1}{5}$ when $a = 2$; but when $a = 10$, a smaller δ was required for the same ε. From Example 5 we can see that if we take larger and larger positive values of a, the corresponding δ's must decrease because of the inequality $\delta < \frac{1}{a}$. We see that as $a \to \infty$, $\delta \to 0$, and there is no positive δ that works for arbitrarily large values of a.

A geometric approach is helpful. The ratio of ε to δ can be thought of as (approximately) the absolute value of the slope m of the graph of f: $\varepsilon/\delta \approx m$. For $f(x) = x^2$, we know from elementary calculus that the slope m increases without bound as we move along the graph to the right. Now since ε is fixed in this process, the value of δ must approach zero (see Figure 3.14a).

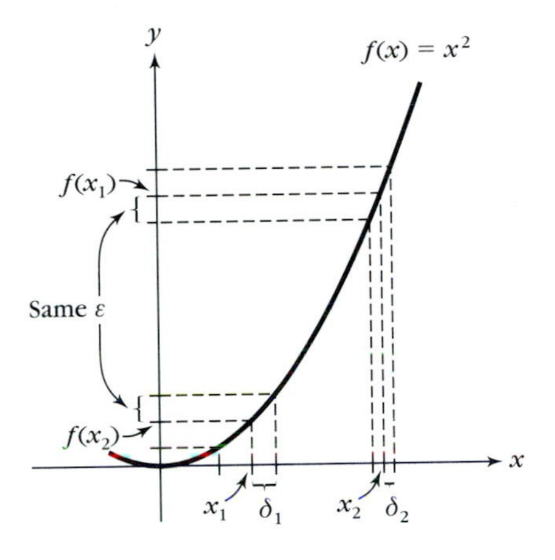

(a) Smaller δ required for x_2

(b) Same δ works for both x_1 and x_2

Figure 3.14
Relation of δ and ε in two examples

In contrast to this behavior, the graph of the function $f(x) = mx + b$ has a *constant* slope: $\varepsilon/\delta = |m|$. For a fixed ε, the choice $\delta = \varepsilon/|m|$ satisfies the requirement of uniform continuity (see Figure 3.14b).

In the next two examples we look at the function $f(x) = \frac{1}{x}$; first on the interval $[1, \infty)$, then on the interval $(0, 1]$.

Example 6: Consider $f(x) = \frac{1}{x}$ on the interval $[1, \infty)$. Let $\varepsilon > 0$ and choose $\delta = \varepsilon$. The inequality

$$|f(x) - f(a)| < \varepsilon$$

then holds for $a, x \in [1, \infty)$ whenever $|x - a| < \delta = \varepsilon$; that is:

$$|x - a| < \varepsilon \Rightarrow |f(x) - f(a)| = \left|\frac{a - x}{ax}\right| \leq \frac{|x - a|}{1} < \varepsilon$$

Thus, f is uniformly continuous on $[1, \infty)$.

Example 7: Now consider $f(x) = \frac{1}{x}$ on the interval $(0, 1]$. We show that the condition for uniform continuity fails in this situation. Let $\varepsilon = 1$, and suppose a $\delta > 0$ exists satisfying statement (1). Let $x_n = 1/n$ and let $a_n = 1/(n + 1)$ for $n \in Z^+$. The sequence

$$\{x_n - a_n\}_{n=1}^{\infty} = \left\{\frac{1}{n(n + 1)}\right\}_{n=1}^{\infty}$$

is null, and so for n sufficiently large, say, $n \geq N$, we have $|x_n - a_n| < \delta$; however, $|f(x_n) - f(a_n)| = 1$ for all $n \in Z^+$, contradicting (1).

Again, the slope idea helps us understand Examples 6 and 7: On $[1, \infty)$, the graph of $f(x) = \frac{1}{x}$ has bounded slope, but on $(0, 1]$ its slope is unbounded. We shall return to these considerations in Chapter 4 after we have developed the derivative concept.

Our main result in this section is that on a closed bounded interval ordinary continuity is sufficient for uniform continuity.

Theorem 3.9.1 *If a real function is continuous on a closed bounded interval, then it is uniformly continuous there.*

PROOF: Since this proof will be carried out by contradiction, let us write out in symbols the negation of statement (1): f is not uniformly continuous on S if and only if

$$(\exists \varepsilon > 0)(\forall \delta > 0)(\exists a, x \in S)(|x - a| < \delta \quad \text{and} \quad |f(x) - f(a)| \geq \varepsilon)$$

In words, f fails to be uniformly continuous on S if and only if there is some particular $\varepsilon > 0$ such that no matter what $\delta > 0$ is chosen, points $a, x \in S$ can be found for which $|x - a| < \delta$ but $|f(x) - f(a)| \geq \varepsilon$. Suppose that f is continuous on a closed bounded interval $[c, d]$ with $c < d$, but not uniformly continuous there. Then there exists an $\varepsilon > 0$ such that, for every $\delta > 0$, points x and a can be found in $[c, d]$ with $|x - a| < \delta$ and $|f(x) - f(a)| \geq \varepsilon$. Applying this statement with the values $\delta_n = \frac{1}{n}$ $(n \in Z^+)$, we can obtain sequences $\{x_n\}_{n=1}^{\infty}$ and $\{a_n\}_{n=1}^{\infty}$ in $[c, d]$ such that $|x_n - a_n| < \frac{1}{n}$ and $|f(x_n) - f(a_n)| \geq \varepsilon$. Consider $\{x_n\}_{n=1}^{\infty}$; by the Bolzano-Weierstrass Theorem for sequences (Theorem 2.7.1), $\{x_n\}_{n=1}^{\infty}$ has a convergent subsequence, say, $\{x_{n_k}\}_{k=1}^{\infty}$, converging to some point $L \in [c, d]$. It follows that $\{a_{n_k}\}_{k=1}^{\infty}$ also converges to L (work Exercise 3.9.8). From the continuity of f we see that $f(x_{n_k}) \to f(L)$ and $f(a_{n_k}) \to f(L)$; consequently, for all sufficiently large subscripts n_k, we must have $|f(x_{n_k}) - f(a_{n_k})| < \varepsilon$, contradicting the way we chose the sequences $\{x_n\}_{n=1}^{\infty}$ and $\{a_n\}_{n=1}^{\infty}$. ■

EXERCISES 3.9

1. Consider the function $f(x) = x^3$. Working with the definition of uniform continuity, determine whether f is uniformly continuous on the following intervals I.
   (a) $I = [0, 1]$
   (b) $I = [0, B]$ $\quad B > 0$
   (c) $I = [0, \infty)$

2. Consider the function $f(x) = 1/\sqrt{x}$. Working with the definition of uniform continuity, determine whether f is uniformly continuous on the following intervals I.
   (a) $I = [B, \infty)$ $\quad B > 0$
   (b) $I = (0, 1]$

3. Show that $f(x) = |x|$ is uniformly continuous on R.

4. (a) Show that if f is uniformly continuous on S and if $\{x_n\}_{n=1}^{\infty}$ is a Cauchy sequence in S, then $\{f(x_n)\}_{n=1}^{\infty}$ is also a Cauchy sequence.
   (b) Show that if f is uniformly continuous on its domain D and if x_0 is a cluster point of D, then $\lim_{x \to x_0} f(x)$ exists.
   (c) Show that if the domain of f is an open interval (c, d) and if f is uniformly continuous there, then there exists a function F that extends f continuously to $[c, d]$; that is, $F(x) = f(x)$ for $x \in (c, d)$ and F is continuous on $[c, d]$.
5. A real function f is said to satisfy a *Lipschitz condition* on S provided that there exists a positive number M such that $|f(x) - f(a)| \leq M\,|x - a|$ for $a,x \in S$. Show that if f satisfies a Lipschitz condition on S, then f is uniformly continuous there.
6. Prove or disprove the following statements.
   (a) If f and g are uniformly continuous on S, then so is $f + g$.
   (b) If f and g are uniformly continuous on S, then so is fg.
   (c) If f and g are uniformly continuous on S, and $g(x) \neq 0$ for $x \in S$, then f/g is uniformly continuous on S.
   (d) If f is uniformly continuous on S and uniformly continuous on T, then it is uniformly continuous on $S \cup T$.
7. Show that if f is continuous on R and periodic (see Definition R.3.3), then f is uniformly continuous on R.
8. Verify the claim made in the proof of Theorem 3.9.1, that $\{a_{n_k}\}_{k=1}^{\infty}$ converges to the same limit as $\{x_{n_k}\}_{k=1}^{\infty}$.

⋆ 3.10 Some Topological Concepts

The topological concepts of "open" and "closed", as applied to intervals, can be extended quite generally. In this section we discuss open and closed sets of real numbers. The definitions here are based upon the concepts of "neighborhood" and "cluster point".

Before proceding, however, we need to discuss unions and intersections of collections of sets.

Collections of Sets; Unions and Intersections

A set whose elements are sets is often referred to as a *collection* of sets.

Example 1: If $a \in R$, then $\{N_\delta(a) : \delta > 0\}$ is a collection of sets whose members are the neighborhoods of a.

Definition 3.10.1 If C is a collection of sets, then the ***union*** of C consists of all elements that belong to at least one of the members of C:

$$\begin{aligned} \cup C &= \cup\{S : S \in C\} \\ &= \{x : x \in S \text{ for some } S \in C\} \end{aligned}$$

The ***intersection*** of C consists of those elements belonging to every member of C:

$$\begin{aligned} \cap C &= \cap\{S : S \in C\} \\ &= \{x : x \in S \text{ for every } S \in C\} \end{aligned}$$

Example 2: For the collection in Example 1, we have

$$\cup\{N_\delta(a) : \delta > 0\} = R$$
$$\cap\{N_\delta(a) : \delta > 0\} = \{a\}$$

Example 3: If $I = \{I_n : n \in Z^+\}$, where I_n is the open interval $\left(0, \frac{1}{n}\right)$, then $\cup I = \cup\{I_n : n \in Z^+\} = (0, 1)$ and $\cap I = \varnothing$.

The *complement* of a set S of real numbers is the set S^c of real numbers that are not elements of $S : S^c = R \dashv S$. If C is a collection of sets of real numbers and if D is the collection consisting of the complements of the sets in S, then it follows from De Morgan's Laws (see Section L.1) that

$$(\cup C)^c = \cap D \qquad \text{and} \qquad (\cap C)^c = \cup D$$

Open Sets in R

Definition 3.10.2 A set G of real numbers is ***open in R*** (briefly, ***open***) provided that for each $x \in G$ there exists a $\delta > 0$ such that $N_\delta(x) \subseteq G$.

Example 4: An open interval (a, b) with $a < b$ is open in R. If $x \in (a, b)$, we can choose $\delta = \min\{x - a, b - x\}$, and it follows that $N_\delta(x) \subseteq (a, b)$.

Example 5: For $a, b \in R$, the intervals (a, ∞) and $(-\infty, b)$ are clearly open, as is $R = (-\infty, \infty)$ itself.

Example 6: The empty set $\varnothing$ is open; since $\varnothing$ has no elements whatever, there are no elements of $\varnothing$ that fail to have a neighborhood contained in $\varnothing$.

Example 7: It is an immediate consequence of the Irrational Density Theorem (Theorem 1.2.4) that the set of rational numbers is not open in R.

Theorem 3.10.1

(i) The union of any collection of open sets is open.
(ii) The intersection of a finite collection of open sets is open.

PROOF:

(i) Let G be a collection of open sets. A real number x belongs to the union $\cup G$ if and only if x belongs to at least one set in the collection G. Thus, if $x \in \cup G$, then $x \in G$ for some $G \in G$, and because G is open there is a $\delta > 0$ for which $N_\delta(x) \subseteq G$; this $N_\delta(x)$ is contained in $\cup G$.

(ii) Let G be a finite collection of open sets, say, $G = \{G_1, G_2, \ldots, G_n\}$. To say that x belongs to the intersection means that x belongs to each $G_i (i = 1, 2, \ldots, n)$. Since all of the sets G_i are open, there exists for each $i = 1, 2, \ldots, n$ a positive number δ_i such that $N_{\delta_i}(x) \subseteq G_i$. If δ is the minimum of $\delta_1, \delta_2, \ldots, \delta_n$, then $\delta > 0$ and $N_\delta(x)$ is contained in every $N_{\delta_i}(x)$; thus $N_\delta(x) \subseteq G_1 \cap G_2 \cap \cdots \cap G_n$. ■

Remark: Notice that the proof of part (ii) depends on the fact that the collection of open sets is finite, because an infinite set of positive numbers may fail to have a positive minimum. In Exercise 3.10.3 you are asked to show that the intersection of an infinite number of open sets can fail to be open.

Theorem 3.10.2 *A set of real numbers is open in R if and only if it is the union of a collection of open intervals.*

PROOF: Let G be open. Then for each $x \in G$, there is a $\delta_x > 0$ such that $N_{\delta_x} \subseteq G$. It follows that G is the union of all these open intervals N_{δ_x}. The converse follows immediately from Theorem 3.10.1 and the fact that an open interval is an open set. ■

A collection of intervals is *pairwise disjoint* provided that any two have empty intersection; that is, for I, J in the collection,

$$I \neq J \Rightarrow I \cap J = \varnothing$$

By using this concept it is possible to strengthen Theorem 3.10.2 considerably, as follows:

Theorem 3.10.3 *A set of real numbers is open in R if and only if it is the union of a countable collection of pairwise disjoint open intervals.*

PROOF: You are asked to prove Theorem 3.10.3 in Exercise 3.10.5. ■

Closed Sets in R

Definition 3.10.3 A set F of real numbers is ***closed in R*** (briefly, ***closed***) provided that every cluster point of F belongs to F.

Example 8: We know that a finite set of real numbers has no cluster points; thus, any finite set is closed.

Example 9: The set of rational numbers is not closed because its cluster points include all irrationals.

Observe from the examples above the concepts of "open" and "closed" do not create a dichotomy for subsets of R. That is, a given subset of R may be *neither* open *nor* closed (e.g., the rationals); and there exist subsets of R that are *both* open *and* closed (e.g., the empty set and R itself) (work Exercise 3.10.4). Nevertheless these two concepts are related in an important way through complements:

Theorem 3.10.4 *A set of real numbers is closed if and only if its complement is open.*

PROOF: Suppose F is closed and consider its complement $F^c = R - F$. If $x \in F^c$, then x is not an element of F, and hence it is not a cluster point of F; so there is a $\delta > 0$ such that $N_\delta(x)$ contains no points of F. Then $N_\delta(x) \subseteq F^c$. We see that F^c contains a neighborhood of each of its points and is therefore open.

Conversely, suppose F^c is open. If a real number x is not in F, it must be in the open set F^c, and some neighborhood of x must be contained in F^c. Then x cannot be a cluster point of F. It follows that all cluster points of F must belong to F, and F is therefore closed. ■

Example 10: By Theorem 3.10.4, we see that a closed interval $[a, b]$ with $a < b$ is closed in R because its complement is the union of open sets $(-\infty, a) \cup (b, \infty)$ and is therefore open.

Theorem 3.10.5

(i) The intersection of any collection of closed sets is closed.
(ii) The union of a finite collection of closed sets is closed.

PROOF:

(i) We use the fact that the complement of a union is the intersection of the complements. Suppose F is a collection of any number of closed sets; we want to show that $\cap F$ is closed. Consider its complement, $(\cap F)^c$, which is the same as $\cup G$, where G is the collection of all complements of the sets in F. Now the elements of G are open sets (Theorem 3.10.4) and so $\cup G$ is open (Theorem 3.10.1); we conclude that $\cap F = (\cup G)^c$ is closed (Theorem 3.10.4).
(ii) You are asked to prove part (ii) in Exercise 3.10.6. ■

Theorem 3.10.6 *A subset F of R is closed if and only if every convergent sequence of points in F converges to a point in F.*

PROOF: Let F be closed in R and let $x = \lim_{n\to\infty} x_n$ where $x_n \in F$ for $n \in Z^+$.

Suppose x is not an element of F; then it belongs to the open set F^c, and for some $\delta > 0$ we have $N_\delta(x) \subseteq F^c$. But then none of the x_n are within δ of x, contradicting the condition $x = \lim_{n\to\infty} x_n$. So $x \in F$. The proof of the converse is left for you to work out. ■

EXERCISES 3.10

1. Using the definitions, determine whether the following sets are open, closed, or neither.

(a) $\{x : x \in R \text{ and } x > 0\}$
(b) Z^+
(c) $\left\{\frac{1}{n} : n \in Z^+\right\}$
(d) $R \dashv Z$

2. Determine whether the following sets are open, closed, or neither.

(a) $\bigcup_{n=1}^{\infty}\left(\frac{1}{n}, 3 - \frac{1}{n}\right)$
(b) $\bigcap_{n=1}^{\infty}\left(\frac{1}{n}, 3 - \frac{1}{n}\right)$
(c) $\bigcap_{n=1}^{\infty}\left[0, \frac{1}{n}\right]$
(d) $\bigcup_{n=1}^{\infty}\left[0, \frac{1}{n}\right]$
(e) $\bigcap_{n=1}^{\infty}\left(0, \frac{1}{n}\right)$
(f) $\bigcup_{n=1}^{\infty}(n, n + 1)$
(g) $\bigcup_{n=1}^{\infty}(n, \infty)$
(h) $\bigcap_{n=1}^{\infty}[n, \infty)$

3. (a) Give an example showing that the intersection of an infinite collection of open sets is not necessarily open.
(b) Give an example showing that the union of an infinite collection of closed sets is not necessarily closed.

4. (a) Show that $\varnothing$ and R are open and closed subsets of R.
(b) Show that if S is a nonempty subset of R that is both open and closed, then $S = R$. (Hence, $\varnothing$ and R are the only subsets of R that are both open and closed.)

5. Prove Theorem 3.10.3.

6. Prove part (ii) of Theorem 3.10.5.

PROJECT: Interior, Closure, Boundary; Perfect Sets

Definition A real number x is an ***interior point*** of a set $S \subseteq R$ provided that there exists a $\delta > 0$ such that the δ neighborhood of x is contained in S. The ***interior*** of S is the set of all interior points of S:

$$S^0 = \{x : x \text{ is an interior point of } S\}$$

(1) Prove the following statements for subsets of R.

(a) For any set S, the interior S^0 of S is open.
(b) If $T \subseteq S$ and T is open, then $T \subseteq S^0$.
(c) The interior of S is the union of all open subsets of S.

(2) Show that if $S \subseteq R$, then its derived set S' is closed. (Recall that S' is the set consisting of all cluster points of S.)

The closure $\overline{S}$ of a set $S \subseteq R$ was defined in Section 3.1 as $S \cup S'$, where S' is the set of all cluster points of S.

(3) Prove the following statements for subsets of R.

(a) For any set S, $\overline{S}$ is closed.
(b) If $T \supseteq S$ and T is closed, then $T \supseteq \overline{S}$.
(c) The closure of S is the intersection of all closed sets that contain S.

(4) Show that for any set $S \subseteq R$, the closure of the complement of S is equal to the complement of the interior of S:

$$\overline{S^c} = (S^0)^c$$

Definition

A real number x is a ***boundary point*** of a set $S \subseteq R$ provided that every neighborhood of x contains at least one point in S and at least one point in S^c.

(5) Determine the boundaries of the following subsets of R.

(a) (a, b) $(a < b)$
(b) $[a, b]$ $(a < b)$
(c) Q
(d) R
(e) Z
(f) $\varnothing$

(6) Show that the boundary of a set is closed.

Perfect Sets

Definition

A set $S \subseteq R$ is ***perfect*** provided it is equal to its derived set $S = S'$.

(7) Show that a closed interval $[a, b]$ with $a < b$ is perfect.

The Cantor set has already been defined in Exercise 2.8.31. An alternative definition is given here. Starting with the closed unit interval $F_0 = [0, 1]$, we construct a sequence of sets $\{F_n\}_{n=1}^{\infty}$ recursively as follows. F_1 is obtained from F_0 by removing the open middle third of F_0. That is,

$$\begin{aligned} F_1 &= F_0 \dashv \left(\frac{1}{3}, \frac{2}{3}\right) \\ &= \left[0, \frac{1}{3}\right] \cup \left[\frac{2}{3}, 1\right] \end{aligned}$$

Next, F_2 is obtained from F_1 by removing the open middle third of each of the two closed intervals whose union is F_1:

$$\begin{aligned} F_2 &= F_1 \dashv \left[\left(\frac{1}{9}, \frac{2}{9}\right) \cup \left(\frac{7}{9}, \frac{8}{9}\right)\right] \\ &= \left[0, \frac{1}{9}\right] \cup \left[\frac{2}{9}, \frac{1}{3}\right] \cup \left[\frac{2}{3}, \frac{7}{9}\right] \cup \left[\frac{8}{9}, 1\right] \end{aligned}$$

Continue in this manner, obtaining F_{n+1} by removing the open middle third of each of the 2^n closed intervals whose union is F_n. Each $F_n (n \in Z^+)$ is a closed set. (Why?) Observe that the sequence $\{F_n\}_{n=1}^{\infty}$ is decreasing; that is, $F_1 \supseteq F_2 \supseteq \cdots \supseteq F_{n+1} \supseteq \cdots$. Now define the ***Cantor set*** to be $C = \bigcap_{n=1}^{\infty} F_n$. Thus, C is what remains of $[0, 1]$ after the removal of all the middle thirds.

(8) Show that C is closed.

(9) Show that the (infinite) sum of the lengths of the intervals removed is 1. (In a geometric sense, C is small.)

(10) Show that C is nondenumerable. (In a cardinal sense, C is large.)

(11) Show that C is perfect.

⋆ 3.11 *Compact Sets in R*

Another topological concept is "compactness". Of several possible formulations of this idea, let us choose one based on sequences:

Definition 3.11.1 A set S of real numbers is ***compact*** provided that every sequence in S has a subsequence that converges to an element of S.

Our first theorem gives a necessary and sufficient condition for a subset of R to be compact.

Theorem 3.11.1 *A set of real numbers is compact if and only if it is closed and bounded.*

PROOF: Let S be closed and bounded and let $\{x_n\}_{n=1}^{\infty}$ be any sequence in S. Then since S is bounded, so is the sequence $\{x_n\}_{n=1}^{\infty}$, and by the Bolzano-Weierstrass Theorem for Sequences (Theorem 2.7.1), $\{x_n\}_{n=1}^{\infty}$ has a subsequence converging to some point x. By Theorem 3.10.6, $x \in S$. Therefore, S is compact.

To prove the converse, we assume that S is compact and show that it is both bounded and closed. Suppose S is not bounded. It is then possible to obtain a sequence $\{x_n\}_{n=1}^{\infty}$ with $x_n \in S$ and $|x_n| > n$ $(n \in Z^+)$. This sequence has no convergent subsequence, contradicting the hypothesis; so S is bounded. Suppose S is not closed. Then a cluster point x of S exists such that $x \notin S$. It follows that there is a sequence $\{x_n\}_{n=1}^{\infty}$ with $x_n \in S$ and $x_n \to x$. (Why?) By our hypothesis, some subsequence $\{x_{n_k}\}_{k=1}^{\infty}$ converges to an element of S; but $\{x_{n_k}\}_{k=1}^{\infty}$ converges to x, which is *not* in S. This contradiction shows that S is closed. ∎

Example 1: Clearly, every closed bounded interval in R is compact; so is every finite set of real numbers.

Example 2: The set Z^+ of positive integers is not compact because it is not bounded.

Example 3: The interval $[0, 1)$ is not compact because it is not closed.

Example 4: The set $\left\{0, 1, \frac{1}{2}, \ldots, \frac{1}{n}, \ldots\right\}$ is compact (see Example 3 in Section 3.1).

Another condition equivalent to compactness involves the concept of an "open cover" of a set; that is, a collection of open sets whose union contains the given set.

Definition 3.11.2 Let S be a set of real numbers. An ***open cover*** of S is a collection C of open sets such that $S \subseteq \cup C$, and such a collection C is said to ***cover*** S.

Example 5: Let $S = [0, 1]$. Then the collection $C = \left\{\left(-\frac{1}{2}, \frac{1}{2}\right), \left(\frac{1}{3}, \frac{2}{3}\right), \left(\frac{1}{2}, \frac{3}{2}\right)\right\}$ is an open cover of S.

Example 6: Let $S = [0, 1]$, and let δ be a fixed positive number. Then the collection $C = \{N_\delta(x) : x \in S\}$ is an open cover of S.

It should be clear from Example 6 that it is not difficult to cover a given set by means of open sets, especially if we do not object to having a large number of sets in the cover. A question that frequently arises in analysis is the following: If a set S is covered by an infinite collection C of open sets, is it possible to select a *finite* number of sets belonging to the collection C in such a way that the selected sets themselves cover S?

Definition 3.11.3 A set S is said to have the ***Heine-Borel property*** provided that if C is an arbitrary open cover of S, then there exists a finite subcollection C_0 of C such that C_0 is an open cover of S.

Briefly, S has the Heine-Borel property provided that every open cover of S contains a finite subcover.

Example 7: The set $S = \left\{0, 1, \frac{1}{2}, \ldots, \frac{1}{n}, \ldots\right\}$ in Example 4 has the Heine-Borel property. Let C be any open cover of S. We must show that there is a *finite* subcollection of C that covers S. Since $0 \in S$ and C covers S, there is an open set $G_0 \in C$ such that $0 \in G_0$. Because G_0 is open, it contains a neighborhood of 0, say, $N_\delta(0)$ for some $\delta > 0$. If we let N be a positive integer $\geq \frac{1}{\delta}$, we see that for $n > N$, the numbers $\frac{1}{n}$ all lie in $N_\delta(0)$ and hence in G_0. For each of the numbers $\frac{1}{k}$ $(k = 1, 2, \ldots, N)$, we know that there is a $G_k \in C$ with $\frac{1}{k} \in G_k$. It follows that S is covered by $C_0 = \{G_0, G_1, \ldots, G_N\}$—a finite subcollection of the original open cover C_0.

Example 8: The interval $(0, 1)$ does not have the Heine-Borel property. To prove this assertion we must show there exists an open cover C of $(0, 1)$ with the property that *no finite* subcollection of C covers $(0, 1)$. Consider the collection C of all open intervals of the form $I_x = (0, x)$, where $0 < x < 1$:

$$C = \{I_x : 0 < x < 1\}$$

Clearly, every real number in $(0, 1)$ is in one of these intervals (in fact, in many). Thus, C is an open cover of $(0, 1)$. But no finite subcollection of C covers $(0, 1)$. If we choose any finite number of elements of C, say,

$I_{x_1}, I_{x_2}, \ldots, I_{x_n}$, then their union is simply I_x, where

$$x = \max \{x_1, x_2, \ldots, x_n\}.$$

The point $(x + 1)/2$, for example, belongs to $(0, 1)$ but is not covered by this finite subcollection.

Example 9: The set Z of integers does not have the Heine-Borel property. Consider the collection of open intervals $C = \{I_n : n \in Z\}$, where

$$I_n = (n - 1, n + 1).$$

Clearly, C is an open cover of Z. However, any finite subcollection C_0 of C contains some I_N with maximal N; it follows that no integer greater than N is covered by the intervals in such a subcollection.

The preceding examples illustrate the following theorem:

Theorem 3.11.2 Heine-Borel Theorem *A set of real numbers is compact if and only if it has the Heine-Borel property.*

PROOF:

(1) Let S have the Heine-Borel property. We show that S is bounded and closed. Consider the collection $C = \{I_x : x \in S\}$, where I_x is the open interval $(x - 1, x + 1)$. Clearly, C is an open cover of S; by the Heine-Borel property, a finite subcollection of C covers S. Let this finite subcollection be $I_{x_1}, I_{x_2}, \ldots, I_{x_N}$ where the subscripts are chosen so that $x_1 < x_2 < \cdots < x_N$. Then S is contained in the interval $(x_1 - 1, x_N + 1)$, so that S is bounded. To complete the proof of compactness we show that S is closed. Suppose by way of contradiction that there exists an $x \in R$ such that $x \notin S$ but is a cluster point of S. Now every neighborhood of x contains elements of S. Let $\delta_n = \frac{1}{n}$ and choose $x_n \in S$ such that $x - \frac{1}{n} \leq x_n \leq x + \frac{1}{n}$. Clearly the sequence $\{x_n\}_{n=1}^{\infty}$ converges to x. Now define

$$G_n = \left\{y : y \in R \text{ and } |y - x| > \tfrac{1}{n}\right\} = \left[x - \tfrac{1}{n}, x + \tfrac{1}{n}\right]^c.$$

It is easy to see that the collection $C_1 = \{G_n : n \in Z^+\}$ is an open cover of S. By applying the Heine-Borel property, we can say that some finite subcollection of C_1 covers S. We denote the elements of this subcollection by $G_{n_1}, G_{n_2}, \ldots, G_{n_k}$ and choose the notation so that $n_1 < n_2 < \cdots < n_k$. Then the union of these open sets is just G_{n_k} and therefore does not contain any terms of the sequence $\{x_n\}_{n=1}^{\infty}$ with $n > n_k$. But these terms are elements of S, contradicting the assumption that $\{G_{n_1}, \ldots, G_{n_k}\}$ is an open cover of S. Thus, a set with Heine-Borel property must be compact.

(2) To prove the converse, let S be compact and suppose we are given an open cover C of S. Let $a = \inf S$, $b = \sup S$, and note that $a, b \in S$ (work Exericse 3.11.2). Define a set A as follows:

$$A = \{x : x \in [a, b] \text{ and a finite subcollection of } C \text{ covers } [a, x] \cap S\}$$

Now A is nonempty ($a \in A$) and bounded above (by b), so we can define $c = \sup A$. Certainly $a < c \leq b$.

It is sufficient to prove that $c = b$. For then we can let G_b be that member of the open cover C that contains b. For some $\delta > 0$, $N_\delta(b) \subseteq G_b$. Now $b - \delta$ is not an upper bound of A, so there exists an $x > b - \delta$ such that $x \in A$; this means that $[a, x] \cap S$ is covered by a finite number of members of C. Together with G_b, these sets form an open cover of S.

It remains to show that $c = b$. Suppose $c < b$. Now S is closed, so S^c is open; if $c \in S^c$, there is an open interval $(c - \delta, c + \delta)$ contained in $[a, b]$ and having no points in common with S. This is inconsistent with the definition of c as the least upper bound of A (work Exercise 3.11.5). Therefore, $c \in S$, and there is a member G_c of the open cover C such that $c \in G_c$. Choose y and z in G_c such that $y < c < z$. Arguing as in the last paragraph, there exists a finite subcollection of C whose members cover $[a, y] \cap S$. Together with G_c, these sets cover $[a, z] \cap S$, and it follows that $z \in A$, contradicting the fact that c is an upper bound of A. ■

You may well wonder how the Heine-Borel property can be applied. The next theorem is an illustration. It extends the result of Theorem 3.9.1 from closed bounded intervals to compact sets in general.

Theorem 3.11.3 *If a real function f is continuous on a compact set $S \subseteq R$, then f is uniformly continuous on S.*

PROOF: Let $a \in S$ and let $\varepsilon > 0$ be given. By the continuity of f at a, we can determine a positive number δ_a such that for all $x \in S$ with $|x - a| < \delta_a$ we have $|f(x) - f(a)| < \frac{\varepsilon}{2}$. Now associate with each $a \in S$ the neighborhood of a with half-width $\frac{1}{2}\delta_a$; denote this neighborhood N_a.

Clearly the collection $C = \{N_a : a \in S\}$ is an open cover of S. Since S is compact, some finite subcollection of C covers S. Let the members of this subcollection be $N_{a_1}, N_{a_2}, \ldots, N_{a_n}$. Choose δ to be the minimum of the numbers $\frac{1}{2}\delta_{a_1}, \frac{1}{2}\delta_{a_2}, \ldots, \frac{1}{2}\delta_{a_n}$, so that $\delta > 0$. We prove that, for any pair $a, x \in S$ with $|x - a| < \delta$, we have $|f(x) - f(a)| < \varepsilon$. Since $a \in S$, a must belong to one of the special neighborhoods $N_{a_1}, N_{a_2}, \ldots, N_{a_n}$. Suppose $a \in N_{a_k}$, then $|a - a_k| < \frac{1}{2}\delta_{a_k}$. If $|x - a| < \delta$, we have

$$\begin{aligned} |x - a_k| &= |x - a + a - a_k| \\ &\leq |x - a| + |a - a_k| \end{aligned}$$

$$< \delta + \frac{1}{2}\delta_{a_k}$$
$$\leq \frac{1}{2}\delta_{a_k} + \frac{1}{2}\delta_{a_k} = \delta_{a_k}$$

Because of the way the δ_{a_k} were chosen, $|f(x) - f(a_k)| < \frac{\varepsilon}{2}$. Finally, we have for $|x - a| < \delta$,

$$|f(x) - f(a)| \leq |f(x) - f(a_k)| + |f(a_k) - f(a)|$$
$$< \frac{\varepsilon}{2} + \frac{\varepsilon}{2} = \varepsilon \quad \blacksquare$$

EXERCISES 3.11

1. Determine which of the following sets are compact subsets of R, and explain your conclusions.
   (a) $(0, \infty)$
   (b) $[0, \infty)$
   (c) $(0, 1]$
   (d) $[0, 1] \cup \{2\}$
   (e) $\left\{\frac{1}{n} : n \in Z^+\right\}$
   (f) $\left\{\frac{1}{n} : n \in Z\right\} \cup \{0\}$
   (g) $\cup\left\{\left(\frac{1}{n}, 2\right] : n \in Z^+\right\}$
   (h) $\cap\left\{\left(-\frac{1}{n}, 2\right] : n \in Z^+\right\}$
   (i) $\cap\{[n, \infty) : n \in Z^+\}$
   (j) The Cantor set (as defined in the project following Exercises 3.10).
2. Show that if S is a nonempty compact subset of R, then sup S and inf S exist and belong to S.
3. Show that if S is bounded, then $\overline{S}$ is compact.
4. Prove or disprove the following statements, for subsets of R.
   (a) The union of a finite collection of compact sets is compact.
   (b) The union of any collection of compact sets is compact.
   (c) The intersection of any collection of compact sets is compact.
5. Refer to the last paragraph of the proof of the Heine-Borel Theorem and justify the claim that if $c < b$, then we must have $c \in S$.
6. Illustrate the Heine-Borel property in the following ways:
   (a) Show that the property fails for the set $\left\{1 - \frac{1}{n} : n \in Z^+\right\}$.
   (b) Show that the property holds for the set $\left\{1 - \frac{1}{n} : n \in Z^+\right\} \cup \{1\}$.
   (c) Show that the property fails for the set $(-\infty, 0]$.
7. (a) Let S be a compact subset of R. Suppose that $\{x_n\}_{n=1}^{\infty}$ is a sequence of points in S with the property that all convergent subsequences converge to the same number L. Show that $\{x_n\}_{n=1}^{\infty}$ converges to L.
   (b) Show by an example that the conclusion in part (a) may fail if S is not compact.
   (c) Let f be a real function whose domain is a compact set S. Show that if f is continuous on S and invertible, then f^{-1} is continuous.

★ 3.12 *Continuity and Open Sets*

In more advanced analysis and topology, the property of continuity can be studied by means of open sets. We will see how in this section; first, however, we introduce some relevant terminology concerning mappings.

Definition 3.12.1 Let $f: A \to B$ be a mapping. If S is a subset of B, then the ***inverse image*** of S under f is the set

$$f^{-1}(S) = \{x : x \in A \text{ and } f(x) \in S\}.$$

Example 1: If $f: R \to R$ is given by $f(x) = x^2$ then $f^{-1}([0, 4]) = [-2, 2]$; $f^{-1}((1, 4]) = [-2, -1) \cup (1, 2]$; $f^{-1}([-1, 0]\} = \{0\}$.

Theorem 3.12.1 *Let f be a real function with domain R. Then f is continuous on R if and only if for every open set G, the set $f^{-1}(G)$ is also open.*

PROOF: Let f be continuous, let G be open, and let $x_0 \in f^{-1}(G)$, so that $f(x_0) \in G$. Then there is an $\varepsilon > 0$ such that $N_\varepsilon(f(x_0)) \subseteq G$. Corresponding to this ε, there is a $\delta > 0$ such that $f(x) \in N_\varepsilon(f(x_0)) \subseteq G$ whenever $x \in N_\delta(x_0)$ (see Corollary 3.4.3). Therefore, $N_\delta(x_0) \subseteq f^{-1}(G)$. We conclude that $f^{-1}(G)$ is open.

For the converse, suppose that f has the property that $f^{-1}(G)$ is open whenever G is open. Let x_0 be any real number; we show that f is continuous at x_0. Let $\varepsilon > 0$ be given. Choose for G the open set $N_\varepsilon(f(x_0))$; then $f^{-1}(N_\varepsilon(f(x_0)))$ is an open set containing x_0; hence, there is a $\delta > 0$ such that $N_\delta(x_0) \subseteq f^{-1}(N_\varepsilon(f(x_0)))$. Then $f(x) \in N_\varepsilon(f(x_0))$ whenever $x \in N_\delta(x_0)$, so f is continuous at x_0. ■

In order to obtain a result like Theorem 3.12.1 for a real function whose domain is a subset D of R, we must generalize the notion of openness. Specifically, we need to consider "openness in D" as well as "openness in R".

Definition 3.12.2 Let $D \subseteq R$. A subset S of D is ***open in D*** provided that S is the intersection of D with a set G that is open in R; that is:

$$S = D \cap G \qquad G \text{ open in } R$$

Example 2: Let $D = [0, 1]$. Then the intervals

$$\left[0, \frac{1}{2}\right) = D \cap \left(-\frac{1}{2}, \frac{1}{2}\right)$$
$$\left(\frac{1}{2}, 1\right] = D \cap \left(\frac{1}{2}, \frac{3}{2}\right)$$

are open in D, although they are not open in R. The interval $\left(\frac{1}{2}, \frac{2}{3}\right)$ is open in D as well as in R.

Example 3: Let $D = Z$, the set of integers. Then

$$Z^+ = Z \cap (0, \infty)$$

is an open subset of Z.

Example 4: Let $D = Q$, the set of rationals. Then Z is not an open subset of Q. If it were, then we would have $Z = Q \cap G$, where G is open in R. Since $1 \in G$, there would be a neighborhood of 1 contained in G. By the Rational Density Theorem (Theorem 1.2.3), this neighborhood contains rationals that are not integers, and so the intersection $Q \cap G$ cannot be Z.

It is interesting to note that the properties stated in Theorem 3.10.1 remain valid if "open" is understood as "open in D", where D is some subset of R. For example, let us prove property (i): If S is a union of sets, all of which are open in D, then S is open in D. Accordingly, let $S = \cup\{H : H \in H\}$ where each member H of H is open in D; then for each $H \in H$, we have

$$H = D \cap G_H$$

where G_H is open in R. Now the collection of all these open sets G_H corresponding to the sets H is a collection of open sets in R:

$$\{G_H : H \in H \text{ and } H = D \cap G_H\}$$

and so its union is open in R. Thus,

$$\begin{aligned} S = \cup\{H : H \in H\} &= \cup\{D \cap G_H : H \in H\} \\ &= D \cap (\cup\{G_H : H \in H\}) \end{aligned}$$

is the intersection of D with an open set in R, showing that S is open in D. (In Exercise 3.12.6 you are asked to prove part (ii) of Theorem 3.10.1 where, again, "open" is interpreted as "open in D".)

This new understanding about openness permits a nice generalization of Theorem 3.12.1:

Theorem 3.12.2 *A real function f with domain D is continuous on D if and only if $f^{-1}(G)$ is open in D whenever G is open in R.*

PROOF: Assume that f is continuous on D. Let G be open in R, and let $S = f^{-1}(G)$. If $x_0 \in S$, then $f(x_0) \in G$. Because G is open, there is an $\varepsilon > 0$ such that $N_\varepsilon(f(x_0))$ is contained in G. Corresponding to this ε there is a $\delta > 0$ such that if $x \in N_\delta(x_0)$ and $x \in D$, then $f(x) \in N_\varepsilon(f(x_0)) \subseteq G$. That is, $N_\delta(x_0) \cap D$ is contained in $f^{-1}(G)$; by definition, $N_\delta(x_0) \cap D$ is open in D. Now as x_0 varies over S, the union of all these sets $N_\delta(x_0) \cap D$ is just the set S, and so $f^{-1}(G) = S$ is open in D. (Exercise 3.12.7 asks you to prove the converse.) ■

By applying Theorem 3.12.2, we can see the results of Section 3.6 in a broader context.

Theorem 3.12.3 *If a real function f is continuous on a compact set D, then $f(D)$ is compact.*

PROOF: Let $\mathcal{G}$ be a collection of open sets in $\mathcal{R}$ that covers $f(D)$. For each $G \in \mathcal{G}$, let $H_G = f^{-1}(G)$. By Theorem 3.12.2, H_G is open in D, so that H_G is the intersection of D with a set that is open in $\mathcal{R}$; we denote this set by K_G because it depends upon G. Now the collection $\{K_G : G \in \mathcal{G}\}$ is an open cover of D, and by the Heine-Borel Theorem (Theorem 3.11.2), some finite subcollection $\{K_{G_1}, K_{G_2}, \ldots, K_{G_n}\}$ covers D. But then $\{G_1, G_2, \ldots, G_n\}$ covers $f(D)$. ■

Corollary 3.12.1 *If a real function f is continuous on a compact set D, then f assumes an absolute minimum and an absolute maximum there.*

Theorem 3.12.3 says that continuous functions preserve compactness. It follows from the Intermediate Value Theorem that they also preserve intervals. Consequently, closed bounded intervals are carried into closed bounded intervals by continuous functions.

EXERCISES 3.12

1. Use Theorem 3.12.1 to show that if a real function f is continuous on $\mathcal{R}$, then the following sets are open.
   (a) $\{x : f(x) > 0\}$
   (b) $\{x : c < f(x) < d \text{ where } c < d\}$
   (c) $\{x : f(x) \neq 0\}$
   (d) $\{x : f(x) \neq c \text{ where } c \in \mathcal{R}\}$

2. Determine in each of the following cases whether S is open in D.
   (a) $D = (-\infty, 1]$; $S = \left(\frac{1}{2}, 1\right)$
   (b) $D = (-\infty, 1]$; $S = \left(\frac{1}{2}, 1\right]$
   (c) $D = \mathcal{Q}$; $S = \mathcal{Q}^+$ (the positive rationals)
   (d) $D = [0, \infty)$; $S = \mathcal{Z}^+$
   (e) $D = \mathcal{Q}$; $S = \mathcal{Z}^+$
   (f) $D = \mathcal{Q}$; $S = \{r : r \text{ is rational and } 0 < r < 1\}$
   (g) $D = \mathcal{Q}$; $S = \{r : r \text{ is rational and } 0 \leq r \leq 1\}$
   (h) $D = \mathcal{Q}$; $S = \{r : r \text{ is rational and } -\sqrt{2} < r < \sqrt{2}\}$

3. Show that if D is an open subset of $\mathcal{R}$, then a subset S of D is open in D if and only if S is open in $\mathcal{R}$.

4. Let D be a subset of $\mathcal{R}$ and let S be a subset of D. Then we define S to be *closed in D* provided that $S = D \cap F$, where F is some closed subset of $\mathcal{R}$. Determine in each of the following cases whether S is closed in D.
   (a) $D = (0, 1]$; $S = \left(0, \frac{1}{2}\right]$
   (b) $D = (0, 1]$; $S = \left[\frac{1}{2}, 1\right]$
   (c) $D = (0, 1]$; $S = \left(\frac{1}{2}, 1\right]$
   (d) $D = (0, \infty)$; $S = \mathcal{Z}^+$
   (e) $D = (0, \infty)$; $S = \left\{\frac{1}{n} : n \in \mathcal{Z}^+\right\}$
   (f) $D = \mathcal{Q}$; $S = \mathcal{Z}$

5. Show that a subset S of D is closed in D (see Exercise 4) if and only if $D \dashv S$ is open in D.

6. (a) Show that if S is the intersection of a finite number of sets each of which is open in D, then S is open in D.
   (b) Give an example showing that the intersection of an infinite number of sets, each of which is open in D, is not necessarily open in D.

7. Prove the converse part of Theorem 3.12.2.

8. Prove Corollary 3.12.1.

9. Let S be a compact set of real numbers and let a be a fixed real number. Show that there exists a point x in S that is closest to a, and a point y in S that is farthest from a.

10. Show that if the real functions f and g are defined and continuous on a set $D \subseteq \mathcal{R}$, then
$$\{x : x \in D, f(x) > g(x)\}$$
is open in D and
$$\{x : x \in D, f(x) = g(x)\}$$
is closed in D.

CHAPTER FOUR

CONCEPTS OF CALCULUS

4.1 *The Derivative*

Elementary calculus deals extensively with the concept of the derivative, with its interpretations as slope or rate of change, and with methods of calculating derivatives. Here we concentrate on the theory of derivatives.

Definition 4.1.1

(1) Let f be a real function whose domain D contains a neighborhood of the real number a. Then f is said to be ***differentiable*** at a provided that

$$\lim_{x \to a} \frac{f(x) - f(a)}{x - a} \qquad \textbf{(1)}$$

exists and is finite.

(2) Let D_1 be the set of all points a of D for which the limit (1) exists. Then the function f' defined for $a \in D_1$ by

$$f'(a) = \lim_{x \to a} \frac{f(x) - f(a)}{x - a}$$

is called the ***derivative*** of f.

Note: If the domain of f is a closed interval $[a, b]$ with $a < b$, then $f'(a)$ and $f'(b)$ are one-sided limits; that is,

$$f'(a) = \lim_{x \to a^+} \frac{f(x) - f(a)}{x - a} \quad \text{and} \quad f'(b) = \lim_{x \to b^-} \frac{f(x) - f(b)}{x - b}$$

The process of finding the derivative of a function is called *differentiation*.

Example 1: Let $f(x) = 3x - x^3$ for $x \in R$. Then for any $a \in R$, we have

$$\begin{aligned}
f'(a) &= \lim_{x \to a} \frac{f(x) - f(a)}{x - a} \\
&= \lim_{x \to a} \frac{3x - x^3 - 3a + a^3}{x - a} \\
&= \lim_{x \to a} \frac{3(x - a) - (x^3 - a^3)}{x - a} \\
&= \lim_{x \to a} \frac{3(x - a) - (x - a)(x^2 + ax + a^2)}{x - a} \\
&= \lim_{x \to a} \frac{[3 - (x^2 + ax + a^2)](x - a)}{x - a} \\
&= 3 - 3a^2
\end{aligned}$$

By changing the independent variable in this formula to x, we obtain the result in a more familiar looking form: $f'(x) = 3 - 3x^2$.

Example 2: Let $f(x) = \sqrt{25 - x^2}$ for $x \in [-5, 5]$. If $a \in (-5, 5)$, we obtain

$$\begin{aligned}
f'(a) &= \lim_{x \to a} \frac{\sqrt{25 - x^2} - \sqrt{25 - a^2}}{x - a} \\
&= \lim_{x \to a} \frac{a^2 - x^2}{(x - a)(\sqrt{25 - x^2} + \sqrt{25 - a^2})}
\end{aligned}$$

$$= \lim_{x \to a} \frac{-(a + x)}{\sqrt{25 - x^2} + \sqrt{25 - a^2}} = \frac{-2a}{2\sqrt{25 - a^2}} = \frac{-a}{\sqrt{25 - a^2}}$$

At the endpoints -5 and 5 we find that the limits defining $f'(-5)$ and $f'(5)$ are infinite; for example,

$$\lim_{x \to 5^-} \frac{\sqrt{25 - x^2} - 0}{x - 5} = \lim_{x \to 5^-} \left(-\frac{\sqrt{25 - x^2}}{5 - x} \right)$$

$$= \lim_{x \to 5^-} \left(-\sqrt{\frac{5 + x}{5 - x}} \right) = -\infty$$

Thus, the function defined by $f(x) = \sqrt{25 - x^2}$ for $x \in [-5, 5]$ is differentiable for $x \in (-5, 5)$ and $f'(x) = x/\sqrt{25 - x^2}$.

Note: You are no doubt aware that there are many notations for derivatives. If $y = f(x)$, then $f'(x)$ is denoted variously by y', $D_x f(x)$, $D_x y$, and so on. A particularly interesting notation, attributed to Leibniz,* is dy/dx. This symbol reflects the fact that the derivative is the limit of a quotient; that is

$$\frac{dy}{dx} = \lim_{\Delta x \to 0} \frac{\Delta y}{\Delta x}$$

where Δy is the change induced in the function by a given change Δx in the independent variable.

Differentiability and Continuity

A fundamental relationship between differentiability and continuity is conveyed by the following theorem.

Theorem 4.1.1 *Let f be a real function whose domain contains a neighborhood of the point a. If f is differentiable at a, then f is continuous at a.*

PROOF: Our assumption is that the limit

$$\lim_{x \to a} \frac{f(x) - f(a)}{x - a} = f'(a)$$

exists. For all $x \neq a$ in the domain of f, we have

$$f(x) - f(a) = \frac{f(x) - f(a)}{x - a} \cdot (x - a)$$

so that

$$f(x) = f(a) + \frac{f(x) - f(a)}{x - a} \cdot (x - a)$$

*A brief biography of Leibniz appears on page 299.

By letting $x \to a$, we obtain from Theorem 3.2.7, the limit theorem,

$$\lim_{x\to a} f(x) = \lim_{x\to a} f(a) + f'(a) \cdot 0$$
$$= f(a)$$

By Theorem 3.4.5, f is continuous at a. ■

The result of Theorem 4.1.1 is valid also if a is an endpoint of the domain of f; the proof is then carried out with appropriate one-sided limits. However, the converse of Theorem 4.1.1 is false, as Example 3 shows.

Example 3: Let $f(x) = |x|$ for $x \in R$. Then f is continuous everywhere, but it is not differentiable at zero. The required limit

$$\lim_{x\to 0} \frac{|x|}{x}$$

fails to exist because the right- and left-hand limits are 1 and -1, respectively. However, f is differentiable at every other point. For example, if $a > 0$, then $f(x) = x$ for all x in the neighborhood $(0, 2a)$ of a. By Theorem 3.2.3,

$$\lim_{x\to a} \frac{f(x) - f(a)}{x - a} = \lim_{x\to a} \frac{x - a}{x - a} = 1$$

If $a < 0$, then $f(x) = -x$ for all x in the neighborhood $(2a, 0)$ of a, and

$$\lim_{x\to a} \frac{f(x) - f(a)}{x - a} = \lim_{x\to a} \frac{-x + a}{x - a} = -1$$

Thus, the derivative of the absolute value function is

$$f'(x) = \begin{cases} 1 & \text{if } x > 0 \\ -1 & \text{if } x < 0 \end{cases}$$

Techniques for the differentiation of elementary functions are studied extensively in calculus. You are asked to develop these techniques in the exercises. The following theorem will be useful in obtaining derivatives of piecewise-defined functions.

Theorem 4.1.2 *Let f and g be real functions whose domains both contain an open interval I, and supose that $f(x) = g(x)$ for $x \in I$. If f is differentiable on I, then so is g, and $f'(x) = g'(x)$ for $x \in I$.*

PROOF: This theorem is a straightforward consequence of Theorem 3.2.3. ■

Example 4: Let

$$f(x) = \begin{cases} x & \text{if } x < 0 \\ x^2 & \text{if } x \geq 0 \end{cases}$$

On the open interval $(-\infty, 0)$, $f(x) = x$, so $f'(x) = 1$ for $x < 0$. And on the open interval $(0, \infty)$, $f(x) = x^2$, so $f'(x) = 2x$ for $x > 0$. *Caution:* The fact that $f(x) = x^2$ on the interval $[0, \infty)$ does not permit us to conclude that $f'(x) = 2x$ at $x = 0$. The endpoint 0 must be checked specifically; indeed, after calculating limits from the right and left, we conclude that $f'(0)$ does not exist. So the derivative of f is

$$f'(x) = \begin{cases} 1 & \text{if } x < 0 \\ 2x & \text{if } x > 0 \end{cases}$$

⋆ *Periodic Extension of a Function*

The following theorem will be useful when we develop the trigonometric functions in Section 4.12. It says that if we start with a function differentiable on a closed bounded interval, and if the values of the function, as well as the values of its derivative, agree at the endpoints of this interval, then we can extend the function* to one that is periodic and differentiable on all of R.

Theorem 4.1.3 **Periodic Extension Theorem** *Let a and p be real numbers with $p > 0$, and let f be a real function differentiable on the interval $[a, a + p]$ with the properties $f(a + p) = f(a)$ and $f'(a + p) = f'(a)$. Then there exists an extension F of f that is defined and differentiable for all real numbers and has period p; the derivative F' of F also has period p.*

Proof: For convenience of notation we assume that $a = 0$. Thus, we start with a function f differentiable on $[0, p]$ with the properties that $f(p) = f(0)$ and $f'(p) = f'(0)$. [Keep in mind that $f'(p)$ and $f'(0)$ are one-sided limits.]

Each real number lies in one and only one interval I_n of the form

$$I_n = [np, (n + 1)p) \qquad n \in Z$$

For $x \in I_n$, the number $x - np$ lies in the interval $[0, p]$ where f is defined, and we define $F(x)$ by copying the value of f at $x - np$; that is,

$$F(x) = f(x - np) \qquad x \in I_n$$

The result is to make the graph of F over each interval I_n a copy of the graph of f on $[0, p]$, translated horizontally a distance np. It is clear that F is an extension of f and that F has period p.

It is also easy to see that F is differentiable on the interior† of each I_n. The following proof carries out the translation idea: Let $x_0 \in I_n^0$, where I_n^0 denotes

*Periodic functions and extensions of functions are discussed in Section R.3.

†The interior of an interval I is the set of points of I that are not endpoints of I.

the interior of the interval I_n. Then, by definition,

$$F'(x_0) = \lim_{x \to x_0} \frac{F(x) - F(x_0)}{x - x_0}$$

Since $x_0 \in I_n^0$ and since $x \to x_0$, we can restrict x to lie in I_n^0. Since I_n^0 is open, we have, by Theorem 3.2.3,

$$F'(x_0) = \lim_{x \to x_0} \frac{f(x - np) - f(x_0 - np)}{x - x_0}$$

Now, by letting $y = x - np$ and $y_0 = x_0 - np$, this limit can be written

$$F'(x_0) = \lim_{y \to y_0} \frac{f(y) - f(y_0)}{y - y_0} = f'(y_0) = f'(x_0 - np)$$

As expected, the value of F' at each such point exists and is equal to the value of f' at the corresponding point in the interval $(0, p)$. To complete the proof that F' exists and is periodic, we consider the endpoints of the intervals I_n; these are the points np with $n \in Z$. We must examine two one-sided limits:

$$\begin{aligned}\lim_{x \to np^+} \frac{F(x) - F(np)}{x - np} &= \lim_{x \to np^+} \frac{f(x - np) - f(0)}{x - np} \\ &= \lim_{y \to 0^+} \frac{f(y) - f(0)}{y - 0} \qquad \text{where } y = x - np \\ &= f'(0)\end{aligned}$$

$$\begin{aligned}\lim_{x \to np^-} \frac{F(x) - F(np)}{x - np} &= \lim_{x \to np^-} \frac{f(x - (n - 1)p) - f(0)}{x - np} \\ &= \lim_{y \to p^-} \frac{f(y) - f(0)}{y - p} \qquad \text{where } y = x - (n - 1)p \\ &= \lim_{y \to p^-} \frac{f(y) - f(p)}{y - p} \qquad \text{since } f(p) = f(0) \\ &= f'(p)\end{aligned}$$

Now, since $f'(0)$ and $f'(p)$ are equal, the two one-sided limits agree, and $F'(np)$ exists and is equal to $f'(0)$. You are asked to prove that F' has period p in Exercise 4.1.8. ∎

EXERCISES 4.1

1. Show that if f and g are defined in a neighborhood of a and are differentiable at a, and if c is a constant, then $f + g$, $f - g$, and cf are also differentiable at a and

$$\begin{aligned}(f + g)'(a) &= f'(a) + g'(a) \\ (f - g)'(a) &= f'(a) - g'(a) \\ (cf)'(a) &= cf'(a)\end{aligned}$$

2. Show that for $n \in Z^+$, the function defined for $x \in R$ by

$$f(x) = x^n$$

is differentiable at every real number a and its derivative is given by $f'(a) = na^{n-1}$.

3. Use the results of Exercises 1 and 2 to obtain the derivative of the polynomial

$$p(x) = \sum_{k=0}^{n} a_k x^k$$

4. Show that if f and g are defined in a neighborhood of a and are differentiable at a, then the function h defined by

$$h(x) = f(x)g(x)$$

is also differentiable at a with

$$h'(a) = f(a)g'(a) + g(a)f'(a)$$

(*Hint:* Start with the equation

$$\frac{h(x) - h(a)}{x - a} = f(x)\frac{g(x) - g(a)}{x - a} + g(a)\frac{f(x) - f(a)}{x - a}$$

and use the fact that f is continuous at a by Theorem 4.1.1.)

5. (a) Show that if f and g are defined in a neighborhood of a and are differentiable at a, and if $g(a) \neq 0$, then the function h defined by

$$h(x) = \frac{f(x)}{g(x)}$$

is differentiable at a with

$$h'(a) = \frac{g(a)f'(a) - f(a)g'(a)}{[g(a)]^2}$$

(*Hint:* Show that the quotient is defined in a neighborhood of a. Then obtain a formula for $[h(x) - h(a)]/(x - a)$ like the one in Exercise 4.)

(b) Show that if n is a negative integer and if $f(x) = x^n$ $(x \neq 0)$, then

$$f'(x) = nx^{n-1} \qquad x \neq 0$$

6. Determine whether the given function is differentiable at the given value(s) of a, and if it is, find $f'(a)$.

(a) $f(x) = \begin{cases} 2x + 3 & \text{if } x \geq 1 \\ 4x + 1 & \text{if } x < 1 \end{cases} \qquad a = 0, 1, 2$

(b) $f(x) = \begin{cases} x^2 & \text{if } x \geq 0 \\ x^3 & \text{if } x < 0 \end{cases} \qquad a = -1, 0, 1$

(c) $f(x) = \begin{cases} x^2 - x & \text{if } x \leq 1 \\ x - 1 & \text{if } x > 1 \end{cases} \qquad a = 0, 1, 2$

(d) $f(x) = \begin{cases} x^2 - 4x & \text{if } x \geq 2 \\ x^3 - 12 & \text{if } x < 2 \end{cases} \qquad a = 2$

(e) $f(x) = \begin{cases} x^2 - 4x - 12 & \text{if } x \geq 2 \\ x^3 - 12x & \text{if } x < 2 \end{cases} \qquad a = 2$

(f) $f(x) = \begin{cases} x^2 - 4x & \text{if } x \geq 2 \\ x^3 - 12x & \text{if } x < 2 \end{cases} \qquad a = 2$

7. Prove or disprove: If a real function f is defined in a neighborhood of a and is differentiable at a, then there exists a neighborhood of a in which f is bounded.

8. Show that the function F' in the proof of Theorem 4.1.3 has period p.

9. Let f be a real function with the properties

(i) $f(x + y) = f(x)f(y)$ for all $x, y \in R$
(ii) f is differentiable at 0 with $f'(0) = 1$
(iii) $f(x) \neq 0$ for all $x \in R$

Show that f is differentiable everywhere, with $f'(x) = f(x)$. (*Note:* This exercise extends Exercise 3.4.11.)

4.2 *Rolle's Theorem*

Relative Extrema

In Section 3.6 we discussed absolute maxima and minima. Recall from calculus that absolute extrema were sometimes found among relative* extrema.

Definition 4.2.1 Let f be a real function with domain D and let $c \in D$. If there exists a $\delta > 0$ such that $f(x) \geq f(c)$ $(f(x) \leq f(c))$ whenever $x \in N_\delta(c) \cap D$, then f is said to assume a ***relative minimum (maximum)*** at c. The

*Relative extrema are sometimes called *local* extrema.

point $(c, f(c))$ is said to be a ***relative minimum (maximum) point*** on the graph of f and $f(c)$ is a ***relative minimum (maximum) value*** of f.

Clearly any absolute extremum of f on D is also a relative extremum, although the converse does not hold.

Example 1: Let $f(x) = x^2 + 1$ $(-2 \le x \le 1)$. It is clear that the graph of f has an absolute minimum point at $(0, 1)$. There is a relative maximum at 1 because if $x \in N_2(1) \cap [-2, 1]$, then $x^2 + 1 \le 2 = f(1)$; however, this relative maximum is not an absolute maximum because f assumes values greater than 2 in the interval $[-2, -1)$. There is a relative and absolute maximum at -2.

The following theorem gives a useful condition that must hold when a differentiable function assumes a relative extremum at an interior point of the domain.

Theorem 4.2.1 *Let f be a real function whose domain contains a neighborhood of the real number c. If f assumes a relative extremum at c and if f is differentiable at c, then $f'(c) = 0$.*

PROOF: We consider the case in which f assumes a relative minimum at c. Then there is a $\delta > 0$ such that $f(x) \ge f(c)$ for all $x \in N_\delta(c)$. Since

$$f'(c) = \lim_{x \to c} \frac{f(x) - f(c)}{x - c}$$

exists by hypothesis, the two one-sided limits

$$\lim_{x \to c^+} \frac{f(x) - f(c)}{x - c} \quad \text{and} \quad \lim_{x \to c^-} \frac{f(x) - f(c)}{x - c}$$

exist and are both equal to $f'(c)$. By the minimum property of $f(c)$, we have, for $x \in (c, c + \delta)$,

$$\frac{f(x) - f(c)}{x - c} \ge 0$$

Consequently (see the Note following Theorem 3.3.3),

$$f'(c) = \lim_{x \to c^+} \frac{f(x) - f(c)}{x - c} \ge 0$$

On the other hand, by considering $x \in (c - \delta, c)$, we find that

$$f'(c) = \lim_{x \to c^-} \frac{f(x) - f(c)}{x - c} \le 0$$

The only possible conclusion is that $f'(c) = 0$. ∎

"Quick Test" for Extrema

The following method gives a quick test for *absolute* extrema of a function continuous on a closed bounded interval. The Extreme Value Theorem (Theorem 3.6.5) guarantees that such a function assumes an absolute maximum and an absolute minimum there. If an absolute extremum occurs at an interior point of the interval, and if the derivative exists at that point, then by Theorem 4.2.1, the derivative is zero there. Thus, the extrema may occur only at the following kinds of points:

(i) The endpoints of the interval
(ii) Points where the function is not differentiable (if any)
(iii) Points where the derivative is zero (if any)

If the list of points in (i), (ii), (iii) is finite, we can determine the absolute extrema by evaluating the function at each point on the list and selecting the least and greatest values.

Example 2: Let $f(x) = (x^2 - 1)^2$ $(x \in [-2, 2])$. The derivative $f'(x) = 4x(x^2 - 1)$ exists throughout $[-2, 2]$ and so there are no points of type (ii). In addition to the endpoints -2 and 2, we check the value of f at $x = -1$, 0, and 1, where $f'(x) = 0$:

x	$f(x)$
−2	9
−1	0
0	1
1	0
2	9

Absolute maxima occur at -2 and 2, absolute minima occur at -1 and 1. At 0, f assumes a relative maximum because $f(x) \le f(0)$ for all x in a suitably small neighborhood of zero.

Example 3: The function

$$f(x) = |\sin x| \qquad -\pi \le x \le \pi$$

has no derivative at $x = 0$ since $\lim_{x \to 0^+} |\sin x|/x = 1$ while $\lim_{x \to 0^-} |\sin x|/x = -1$. Clearly, $x = 0$ gives an absolute minimum. Other absolute minima occur at the endpoints $x = -\pi, \pi$; absolute maxima occur at $x = \pi/2, -\pi/2$, where $f'(x) = 0$.*

*Assume without proof for the time being that the derivative of the sine is the cosine.

Theorem 4.2.2 **Rolle's Theorem** *If a real function f is continuous on a closed interval $[a, b]$, where $a < b$, and differentiable on the open interval (a, b), and if $f(a) = f(b) = 0$, then there exists a number $\xi \in (a, b)$ such that $f'(\xi) = 0$.*

PROOF: If f is the constant zero, the result is obvious. Otherwise f must assume at least one extreme value that is either positive or negative, and since $f(a)$ and $f(b)$ are zero, this extremum must occur at a point $\xi \in (a, b)$. Thus, f is differentiable at ξ, and so $f'(\xi) = 0$ by Theorem 4.2.1. ∎

Example 4: The function $f(x) = x^3 - 4x$ satisfies the hypothesis of Rolle's Theorem on the interval $[0, 2]$. Since $f'(x) = 3x^2 - 4$, the number ξ must satisfy $3\xi^2 - 4 = 0$, and so $\xi = 2/\sqrt{3}$.

Example 5: We saw in Section 4.1 that the function $f(x) = \sqrt{25 - x^2}$ is differentiable on the interval $(-5, 5)$; it is continuous (but not differentiable) at -5 and 5. In any event, the function satisfies the hypothesis of Rolle's Theorem, and we find that the number ξ in this case is zero.

Rolle's Theorem has a variety of consequences, the most important of which is the Mean Value Theorem discussed in Section 4.3. The next example is an application of Rolle's Theorem to the theory of equations.

Example 6: We show that the equation $x^2 + 5x - 17 = 0$ has a unique solution. Let $f(x) = x^3 + 5x - 17$ and observe that $f(1)$ and $f(2)$ are of opposite sign. By the Intermediate Value Theorem (Theorem 3.6.2), a solution of $f(x) = 0$ exists between 1 and 2. Rolle's Theorem shows the uniqueness. If $f(a) = f(b) = 0$ with $a \neq b$, there would exist a number ξ between a and b with $f'(\xi) = 0$. This cannot happen because $f'(\xi) = 3\xi^2 + 5 > 0$ for all $\xi \in R$.

Theorem 4.2.3 *If a real function f is differentiable on an interval containing a and b, and if $f'(a)$ and $f'(b)$ are of opposite sign, then there exists a number ξ between a and b such that $f'(\xi) = 0$.*

PROOF: Suppose that $a < b$ and that $f'(a) < 0 < f'(b)$. By the Extreme Value Theorem (Theorem 3.6.5), f assumes an absolute minimum on the interval $[a, b]$. It does not assume an absolute minimum at a. For $f'(a) < 0$ and by Theorem 3.3.3, there is an interval $(a, a + \delta)$ $(\delta > 0)$ in which

$$\frac{f(x) - f(a)}{x - a} < 0$$

Thus, for $x \in (a, a + \delta)$, we have $f(x) < f(a)$, so $f(a)$ is not a minimum. Similarly, the condition $f'(b) > 0$ implies that f does not assume a minimum at b. Consequently, f assumes its minimum on $[a, b]$ at a point $\xi \in (a, b)$ and by Theorem 4.2.1, $f'(\xi) = 0$. The case $f'(a) > 0 > f'(b)$ is proved similarly. ∎

The following corollary is an immediate consequence of Theorem 4.2.3.

Corollary 4.2.1 *If f is differentiable on an interval I and if f' is nonzero on I, then f' is of constant sign on I.*

The Intermediate Value Theorem says that a continuous function assumes all values between any two of its values. It is interesting that the derivative of a function differentiable on an interval has this property too, even though it may not be continuous. The proof is a straightforward application of Theorem 4.2.3 (work Exercise 5).

Corollary 4.2.2 **Darboux Property** *If a real function f is differentiable on an interval containing the numbers a and b and if μ is between $f'(a)$ and $f'(b)$, then there is a ξ between a and b such that $f'(\xi) = \mu$.*

EXERCISES 4.2

1. Locate any relative extrema for the following functions on the indicated intervals.
   (a) $f(x) = 2x + 1 \qquad -1 \leq x \leq 2$
   (b) $f(x) = 2x + 1 \qquad 1 < x < 2$
   (c) $f(x) = 2x + 1 \qquad x \in R$
   (d) $f(x) = 4 - x^2 \qquad -1 \leq x \leq 2$
   (e) $f(x) = x(1 - x) \qquad 0 \leq x \leq 2$
   (f) $f(x) = \sin x \qquad 0 \leq x \leq 2\pi$
   (g) $f(x) = \cos x + 2 \qquad 0 \leq x \leq 2\pi$
2. Use the quick test to find the absolute maxima and minima of the following functions on the indicated intervals.
   (a) $f(x) = 4x + 3 \qquad -2 \leq x \leq 3$
   (b) $f(x) = x^3 - 3x \qquad -2 \leq x \leq 1$
   (c) $f(x) = x^4 - 2x^2 \qquad -2 \leq x \leq 2$
   (d) $f(x) = 2 \sin x - x \qquad 0 \leq x \leq 2\pi$
   (e) $f(x) = \sin x + \cos x \qquad 0 \leq x \leq 2\pi$
   (f) $f(x) = |2x - 3| \qquad 0 \leq x \leq 2$
   (g) $f(x) = |x| + |x - 1| \qquad -1 \leq x \leq 2$
3. Show that if $p > 0$ and the equation
$$x^3 + px + q = 0$$
has a real solution, then this solution is unique.
4. Give a proof of Theorem 4.2.1 for the case in which the function assumes a relative maximum at c.
5. Prove Corollary 4.2.2. [*Hint:* Apply Theorem 4.2.3 to the function $g(x) = f(x) - \mu x$.]
6. Show that the greatest integer function is not the derivative of a differentiable function on R.

4.3 *Mean Value Theorem*

The Mean Value Theorem is a straightforward consequence of Rolle's Theorem. It has many important applications in the theory of calculus, several of which are included in this section.

Theorem 4.3.1 **Mean Value Theorem** *If a real function f is continuous on a closed interval $[a, b]$, where $a < b$, and differentiable on the open interval (a, b), then there exists a number $\xi \in (a, b)$ for which*

$$f'(\xi) = \frac{f(b) - f(a)}{b - a}$$

PROOF: We define a function F on $[a, b]$ as follows:

$$F(x) = f(x) - f(a) - \frac{f(b) - f(a)}{b - a}(x - a)$$

It is easy to verify that F satisfies the hypothesis of Rolle's Theorem on the interval $[a, b]$, with derivative

$$F'(x) = f'(x) - \frac{f(b) - f(a)}{b - a}$$

Thus, there exists a $\xi \in (a, b)$ such that $F'(\xi) = 0$; for this ξ, it follows that

$$f'(\xi) - \frac{f(b) - f(a)}{b - a} = 0$$

You are asked to investigate a geometric interpretation of the function F in Exercise 4.3.9. ■

Remark: The Mean Value Theorem is used so frequently that it is important to fix it firmly in mind. The following interpretation based on slopes may help. By recalling that $f'(\xi)$ gives the slope of the line tangent to the graph of f at the point $(\xi, f(\xi))$, and that $[f(b) - f(a)]/(b - a)$ is the slope of the chord connecting A and B in Figure 4.1, we can paraphrase the conclusion of the theorem as follows. There is at least one number $\xi \in (a, b)$ such that the tangent line at $(\xi, f(\xi))$ is parallel to the chord AB.

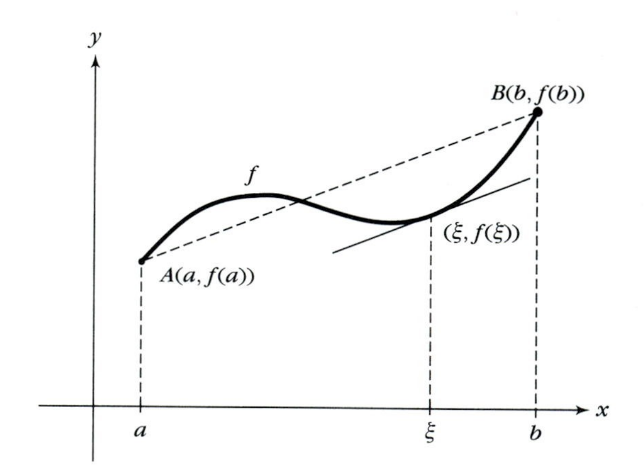

Figure 4.1
Mean Value Theorem

Example 1: Let $f(x) = 6x - x^3$ $(x \in [-2, 2])$. Since $f'(x) = 6 - 3x^2$, the Mean Value Theorem asserts the existence of $\xi \in (-2, 2)$ satisfying

$$f'(\xi) = \frac{f(2) - f(-2)}{4} = 2$$

By solving the equation $6 - 3\xi^2 = 2$, we obtain two values in the interval $(-2, 2)$, namely, $\xi = \pm 2\sqrt{3}/3$.

The derivative of a function can be used to determine whether the function is increasing or decreasing on an interval. In the following theorem, the notation I^0 refers to the *interior* of the interval I, the set of points of I that are not endpoints of I.

Theorem 4.3.2 **Sign of the Derivative** *Let f be a real function continuous on the interval I and differentiable on its interior I^0. Then*

(i) If $f'(x) \geq 0$ for all $x \in I^0$, then f is increasing on I.
(ii) If $f'(x) > 0$ for all $x \in I^0$, then f is strictly increasing on I.
(iii) If $f'(x) \leq 0$ for all $x \in I^0$, then f is decreasing on I.
(iv) If $f'(x) < 0$ for all $x \in I^0$, then f is strictly decreasing on I.

PROOF: Let x_1 and x_2 be two points of I for which $x_1 < x_2$. By applying the Mean Value Theorem to f on the interval $[x_1, x_2]$, we have

$$f(x_2) - f(x_1) = f'(\xi)(x_2 - x_1)$$

where $\xi \in (x_1, x_2) \subseteq I^0$. In case (i), $f'(\xi) \geq 0$, so that $f'(\xi)(x_2 - x_1) \geq 0$, and $f(x_1) \leq f(x_2)$; this proves that f is increasing on I. The remaining cases follow similarly. ∎

Example 2: Consider $f(x) = x^3$ on the inverval $[-1, 1]$. The derivative of f is positive on $[-1, 0)$ and on $(0, 1]$, and since f is continuous at zero, it is strictly increasing on each of the closed intervals $[-1, 0]$ and $[0, 1]$. It follows that f is also strictly increasing on their union $[-1, 1]$.

The following theorem and its corollary are fundamental to the theory of integration and differential equations.

Theorem 4.3.3 *If a real function f is continuous on an interval I and differentiable with derivative zero on I^0, then f is constant on I.*

PROOF: The proof is similar to the proof of Theorem 4.3.2. ∎

Corollary 4.3.1 *If two real functions are continuous on an interval I and have the same derivative on I^0, then they differ by a constant on I.*

Example 3: The function $f(x) = x^3/3$ has the derivative x^2; consequently, any other function with derivative x^2 must have the form $g(x) = x^3/3 + C$ for some constant C.

Tests for Maxima and Minima

The familiar First and Second Derivative Tests concerning relative extrema are stated below (Theorems 4.3.5 and 4.3.6); their proofs are left as exercises. We begin by giving a very elementary condition for an extremum to occur.

Theorem 4.3.4 *Let f be a real function defined on an open interval (a, b) and suppose that $a < c < b$. Then:*

(i) If f is increasing on $(a, c]$ and decreasing on $[c, b)$, then f assumes a relative maximum at c.
(ii) If f is decreasing on $(a, c]$ and increasing on $[c, b)$ then f assumes a relative minimum at c.

Theorem 4.3.5 **First Derivative Test** *Let f be a real function continuous on a set that contains an open interval (a, b) and suppose that $a < c < b$. If $f'(x)$ exists and is positive (negative) for $x \in (a, c)$, and if $f'(x)$ exists and is negative (positive) for $x \in (c, b)$, then f assumes a relative maximum (minimum) at c.*

Example 4: Consider the function $f(x) = x^3 - 3x^2$. Its derivative, $f'(x) = 3x^2 - 6x$, changes sign as x increases through zero because f' is positive on $(-1, 0)$ and negative on $(0, 1)$. According to the First Derivative Test, f assumes a relative maximum at 0. And f' also changes sign as x increases through 2, this time from negative to positive, so that f assumes a relative minimum at 2.

Example 5: Consider the function $f(x) = |1 - x^2|$. Its derivative

$$f'(x) = \begin{cases} -2x & \text{if } -1 < x < 1 \\ 2x & \text{if } x < -1 \text{ or } x > 1 \end{cases}$$

is not defined at 1, but f is continuous there and the change in sign of the derivative signals a relative minimum. There is also a relative minimum at -1 and a relative maximum at 0.

Theorem 4.3.6 **Second Derivative Test** *Let f be a real function differentiable on a set that contains an open interval (a, b) and suppose that $a < c < b$. Suppose also that $f'(c) = 0$ and that $f''(c)$ exists.* If $f''(c) < 0$, then f assumes a relative maximum at c; if $f''(c) > 0$, then f assumes a relative minimum at c.*

Derivative of the Inverse Function

Let us now look at a theorem that relates the derivative of a function and the derivative of its inverse.

*The *second derivative* f'' is the derivative of f'; the *third derivative* f''' is the derivative of f'', and so on.

Theorem 4.3.7 *Let f be a real function whose domain is an interval I, and suppose that f is differentiable on I with f' nonzero there. Then f is invertible and its inverse g is differentiable on the interval $f(I)$, with*

$$g'(y) = \frac{1}{f'(g(y))}$$

PROOF: You are asked to prove the theorem in Exercise 4.3.11. ∎

Example 6: Let $n = Z^+$ and consider $f(x) = x^n$ $(x > 0)$. The inverse of f is given by

$$g(y) = \sqrt[n]{y} \qquad y > 0$$

By using Theorem 4.3.7 and familiar rules of operation with radicals and exponents, we get

$$g'(y) = \frac{1}{n(\sqrt[n]{y})^{n-1}} = \frac{\sqrt[n]{y}}{ny} = \frac{1}{n} y^{\frac{1}{n}-1}$$

Note: It is interesting to state the conclusion of Theorem 4.3.7 in Leibniz's notation. If $y = f(x)$, then $f'(x) = dy/dx$; for the inverse function $x = g(y)$, the derivative is expressed as $g'(y) = dx/dy$. In this notation, then, the conclusion is

$$\frac{dx}{dy} = \frac{1}{dy/dx}$$

Thus, the symbolism seems to suggest what is true.

EXERCISES 4.3

1. Find all numbers ξ satisfying the Mean Value Theorem for the following functions and intervals. If no such number ξ exists, explain why.
   (a) $f(x) = x^2 + 5x$ $[0, 4]$
   (b) $f(x) = 3x - x^3$ $[0, \sqrt{3}]$
   (c) $f(x) = \sin x$ $[\pi/2, 3\pi/2]$
   (d) $f(x) = |x|$ $[-1, 2]$
   (e) $f(x) = |\sin x|$ $[-\pi/2, \pi/2]$

2. Find the intervals on which the following functions are increasing.
   (a) $f(x) = 3x - x^3$
   (b) $f(x) = x/(x^2 + 1)$
   (c) $f(x) = \sqrt{25 - x^2}$
   (d) $f(x) = |x - x^2|$
   (e) $f(x) = 1/(x^2 - 4)$

3. Show that $f(x) = x + \sin x$ is strictly increasing on the interval $[0, 4\pi]$ even though its derivative takes the value zero at two points in the interior of the interval.

4. Show that for a quadratic function

$$f(x) = px^2 + qx + r,$$

   the mean value number ξ in Theorem 4.3.1 occurs at the midpoint of the interval $[a, b]$.

5. Let p be a polynomial function of degree n:

$$p(x) = a_0x^n + a_1x^{n-1} + \cdots + a_n \qquad a_0 \neq 0$$

   (a) Show that if r and s are different roots of the equation $p(x) = 0$, then there is a root of the equation $p'(x) = 0$ between r and s.
   (b) If $p(x) = 0$ has n different roots, how many different roots does $p'(x) = 0$ have?

6. (a) Prove the Corollary 4.3.1.
(b) Show that the functions

$$f(x) = \begin{cases} 0 & \text{if } 0 < x < 1 \\ 1 & \text{if } 1 < x < 2 \end{cases}$$

and

$$g(x) = \begin{cases} 0 & \text{if } 0 < x < 1 \\ 2 & \text{if } 1 < x < 2 \end{cases}$$

have identical derivatives but do not differ by a constant. Why does this fact not contradict the corollary?

7. (a) Prove Theorem 4.3.4.
(b) Use Theorem 4.3.4 to prove Theorem 4.3.5.
(c) Use Theorem 4.3.5 to prove Theorem 4.3.6.

[*Hint:* If $f''(c) < 0$, there exists an $h > 0$ such that $[f'(x) - f'(c)]/(x - c) < 0$ for $x \in N_h(c)$.]

8. Use the facts (to be proved in Section 4.12) that the derivative of sine is cosine and $|\cos x| \leq 1$ for all $x \in R$, to show that $|\sin \alpha| \leq |\alpha|$ for all $\alpha \in R$.

9. Show that the function F defined in the proof of the Mean Value Theorem represents the directed vertical distance between the point $(x, f(x))$ on the graph of f and the point on the chord AB corresponding to that value of x.

10. Show that if a real function has a bounded derivative on an interval, then it is uniformly continuous there.

11. Prove Theorem 4.3.7 by justifying each of the following statements under the hypotheses of that theorem:

(a) f' is of constant sign on I
(b) f is strictly monotone on I
(c) f is invertible on I
(d) The inverse g of f is continuous on $f(I)$
(e) $g'(y) = \dfrac{1}{f'(g(y))}$ for each $y \in f(I)$

12. (a) Show that Theorem 4.3.7 applies to the function $f(x) = \sin x$ $(-\pi/2 < x < \pi/2)$.
(b) Show that the inverse g of f has the derivative

$$g'(y) = \frac{1}{\sqrt{1 - y^2}} \qquad -1 < y < 1$$

4.4 *Chain Rule*

The derivative of f at a has been defined as the limit of a quotient:

$$f'(a) = \lim_{x \to a} Q(x) \qquad \text{where } Q(x) = \frac{f(x) - f(a)}{x - a}$$

The following theorem expresses the differentiability of f at a in terms of Q and leads us to one of the most important rules for differentiation—the Chain Rule.

Theorem 4.4.1 *Let f be a real function defined on an open interval I and let $a \in I$. Then f is differentiable at a if and only if there exists a real function Q defined on I and continuous at a such that*

$$f(x) = f(a) + Q(x)(x - a) \tag{1}$$

for $x \in I$.

PROOF: If f is differentiable at a, then Q may be defined on I as follows:

$$Q(x) = \begin{cases} \dfrac{f(x) - f(a)}{x - a} & \text{if } x \in I, x \neq a \\ f'(a) & \text{if } x = a \end{cases}$$

Clearly Q satisfies Equation (1) for $x \in I$ and is continuous at a. Now for the converse, suppose a function Q exists, satisfying (1) for $x \in I$ and continuous at a. Then, for $x \in I$, $x \neq a$, we have

$$\frac{f(x) - f(a)}{x - a} = Q(x)$$

and by the continuity of Q at a, we see that $f'(a)$ exists:

$$f'(a) = \lim_{x \to a} \frac{f(x) - f(a)}{x - a} = \lim_{x \to a} Q(x) = Q(a) \qquad \blacksquare$$

Example 1: The function $f(x) = x^2$ is differentiable on the open interval R of all real numbers. We examine this function at the real number $a = 3$, where Equation (1) becomes

$$x^2 = 9 + Q(x)(x - 3)$$

For $x \neq 3$, we see that

$$Q(x) = \frac{x^2 - 9}{x - 3} = x + 3$$

Since Q is continuous at 3, we have $Q(3) = \lim_{x \to 3}(x + 3) = 6$, and so $Q(x) = x + 3$ for all $x \in R$.

Example 2: Consider $f(x) = \sin x$, and take for granted once more that f is differentiable everywhere with $f'(x) = \cos x$. The differentiability of f at zero implies that there is a function Q defined on R and continuous at zero such that

$$\sin x = 0 + Q(x)(x - 0)$$

and so, for $x \neq 0$,

$$Q(x) = \frac{\sin x}{x}$$

Moreover,

$$\lim_{x \to 0} Q(x) = f'(0) = \cos 0 = 1$$

and so there arises a limit familiar from calculus:

$$\lim_{x \to 0} \frac{\sin x}{x} = 1$$

(This is not a proof of that limit statement because the latter was used in calculus to prove that the derivative of sine is cosine.)

Theorem 4.4.2 **The Chain Rule** *Let g be defined in a neighborhood of a and be differentiable at a. Let f be defined in a neighborhood of $g(a)$ and be differentiable* *at $g(a)$. Then the composite function $h = f \circ g$ is defined in a neighborhood of a and*

$$h'(a) = f'(g(a))\, g'(a)$$

PROOF: Since f is defined in a neighborhood of $g(a)$, we can choose $\varepsilon > 0$ with the property that $N_\varepsilon(g(a))$ is contained in the domain of f. By hypothesis, g is defined in a neighborhood of a and is continuous at a (Theorem 4.1.1), so there exists a $\delta > 0$ such that $N_\delta(a)$ is contained in the domain of g and has the property that $g(x) \in N_\varepsilon(g(a))$ whenever $x \in N_\delta(a)$. It follows that $h(x)$ is defined for all $x \in N_\delta(a)$ (see Figure 4.2).

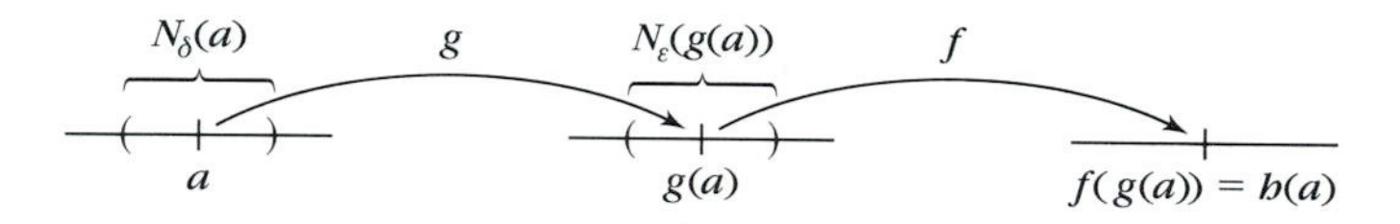

Figure 4.2
Composition of f on g

By Theorem 4.4.1, there exist functions Q and R such that Q is continuous at $g(a)$, R is continuous at a, and the following equations hold:

$$f(u) = f(g(a)) + Q(u)[u - g(a)] \qquad \text{for } u \in N_\varepsilon(g(a)) \tag{2}$$

$$g(x) = g(a) + R(x)(x - a) \qquad \text{for } x \in N_\delta(a) \tag{3}$$

Now consider the limit defining $h'(a)$:

$$\lim_{x \to a} \frac{h(x) - h(a)}{x - a}$$

By using Equations (2) and (3), we can calculate $h(x) - h(a)$ for $x \in N_\delta(a)$:

$$\begin{aligned} h(x) - h(a) &= f(g(x)) - f(g(a)) \\ &= \{f(g(a)) + Q(g(x))[g(x) - g(a)]\} - f(g(a)) \\ &= Q(g(x))[g(x) - g(a)] \\ &= Q(g(a) + R(x)(x - a))[g(x) - g(a)] \\ &= Q(g(a) + R(x)(x - a))R(x)(x - a) \end{aligned}$$

So for $x \in N_\delta(a)$ with $x \neq a$, we have

$$\frac{h(x) - h(a)}{x - a} = Q(g(a) + R(x)(x - a))R(x)$$

As $x \to a$, $R(x) \to g'(a)$; also, $g(a) + R(x)(x - a) \to g(a)$, so that $Q(g(a) + R(x)(x - a)) \to Q(g(a)) = f'(g(a))$. Finally,

$$\lim_{x \to a} \frac{h(x) - h(a)}{x - a} = f'(g(a))g'(a) \qquad \blacksquare$$

The following examples typify the use of the Chain Rule in calculus.

Example 3: If $h(x) = (x^2 + 5)^{10}$, then h is the composite function $f \circ g$, where $g(x) = x^2 + 5$ and $f(u) = u^{10}$. By using the Chain Rule together with the basic rules for differentiating powers, we obtain

$$\begin{aligned} h'(x) = f'(g(x))g'(x) &= 10[g(x)]^9 g'(x) \\ &= 10(x^2 + 5)^9(2x) \\ &= 20x(x^2 + 5)^9 \end{aligned}$$

Example 4: Let $h(x) = \cos\left(\frac{\pi}{3} - 2x\right)$. By using the fact that the derivative of cosine is the negative of the sine, we have

$$\begin{aligned} h'(x) &= -\sin\left(\frac{\pi}{3} - 2x\right)(-2) \\ &= 2\sin\left(\frac{\pi}{3} - 2x\right) \end{aligned}$$

Note: In Leibniz's notation, the Chain Rule says that if $y = f(u)$ and $u = g(x)$ are differentiable functions, then

$$\frac{dy}{dx} = \frac{dy}{du} \cdot \frac{du}{dx}$$

We have another situation in which the Leibniz notation suggests what is true.

The Functions $f_n(x) = x^n \sin \frac{1}{x}$

Example 5: The function $f_0(x) = \sin \frac{1}{x}$ is undefined at 0; moreover, this discontinuity is not removable because $\lim_{x \to 0} f(x)$ does not exist (work Exercise 4.4.4). By Theorem 4.1.1, f_0 is not differentiable at 0. However, f_0 is differentiable at all other real numbers; the Chain Rule gives us

$$f_0'(x) = \frac{-\cos(1/x)}{x^2} \qquad x \neq 0$$

Example 6: The function

$$f_1(x) = \begin{cases} x \sin \dfrac{1}{x} & \text{if } x \neq 0 \\ 0 & \text{if } x = 0 \end{cases}$$

is continuous everywhere. (Check the continuity of f at 0 in Exercise 4.4.5.) It is not differentiable at 0 because

$$\lim_{x \to 0} \frac{x \sin \frac{1}{x} - 0}{x} = \lim_{x \to 0} \sin \frac{1}{x}$$

does not exist. Elsewhere, f_1 is differentiable with its derivative given by

$$f_1'(x) = \frac{-1}{x} \cos \frac{1}{x} + \sin \frac{1}{x} \qquad x \neq 0$$

Example 7: The function

$$f_2(x) = \begin{cases} x^2 \sin \dfrac{1}{x} & \text{if } x \neq 0 \\ 0 & \text{if } x = 0 \end{cases}$$

is differentiable at 0:

$$f_2'(0) = \lim_{x \to 0} \frac{f_2(x) - f_2(0)}{x - 0} = \lim_{x \to 0} x \sin \frac{1}{x} = 0$$

However, $f_2'(x)$ is not continuous at 0, as may be seen by calculating $f_2'(x)$ for $x \neq 0$ and examining the limit as $x \to 0$ (work Exercise 4.4.6). Thus, f_2 is an example of a function differentiable everywhere, whose derivative has a point of discontinuity.

EXERCISES 4.4

1. Find the function Q satisfying Equation (1) for the given real function f and real number a.
   (a) $f(x) = x^2 + 2x, \quad a = 0$
   (b) $f(x) = \cos x, \quad a = \pi$
   (c) $f(x) = \sqrt{x + 4}, \quad a = 0$
   (d) $f(x) = \frac{1}{x}, \quad a = 2$
2. Use the Chain Rule to find the derivative of $h = f \circ g$ at $x = a$, where
   (a) $f(u) = \sin u; \quad g(x) = \pi\left(x + \frac{1}{x}\right); \quad a = 1$
   (b) $f(u) = \cos u; \quad g(x) = \pi + 2x; \quad a = \pi/4$
   (c) $f(u) = \sqrt{u}; \quad g(x) = x^2 + 6x; \quad a = 2$
3. Let

$$\operatorname{sgn} u = \begin{cases} 1 & \text{if } u > 0 \\ -1 & \text{if } u < 0 \end{cases}$$

   Show that if $f(u) = |u|$, then $f'(u) = \operatorname{sgn} u$. Then use the Chain Rule to find $h'(x)$ where $g(x)$ is differentiable everywhere and $h(x) = |g(x)|$. At what points might h fail to be differentiable?
4. Show that $f(x) = \sin\frac{1}{x}$ has no limit as $x \to 0$. [*Hint:* Consider the sequences $\{x_n\}_{n=1}^{\infty}$ and $\{y_n\}_{n=1}^{\infty}$ defined by $x_n = 1/2n\pi$ and $y_n = 1/\left(2n + \frac{1}{2}\right)\pi$.]
5. Show that f_1 in Example 6 is continuous at 0.
6. Determine $f_2'(x)$ for $x \neq 0$, where f_2 is the function in Example 7. Then show that f_2' is discontinuous at zero.
7. Show that the function f_2 in Example 7 has infinitely many critical points of type (iii).
8. Let

$$f_3(x) = \begin{cases} x^3 \sin \dfrac{1}{x} & \text{if } x \neq 0 \\ 0 & \text{if } x = 0 \end{cases}$$

   Show that f_3 is differentiable at 0 and determine whether its derivative is continuous at 0.
9. Let $r = p/q$ be a rational number in lowest terms, so that $p \in Z$ and $q \in Z^+$. For $x > 0$, x^r is defined to be $\sqrt[q]{x^p}$. Use the result of Example 6 in Section 4.3 and the Chain Rule to verify that the function $f(x) = x^r$ is differentiable for $x > 0$, with $f'(x) = rx^{r-1}$.

4.5 *The Riemann Integral*

Introduction

Many applications of calculus involve the process of integration. Literally, *integration* means "making a whole by putting together all of its parts".

A familiar example is the calculation of the area of the region under the graph of a nonnegative real function f over an interval $[a, b]$ (see Figure 4.3).

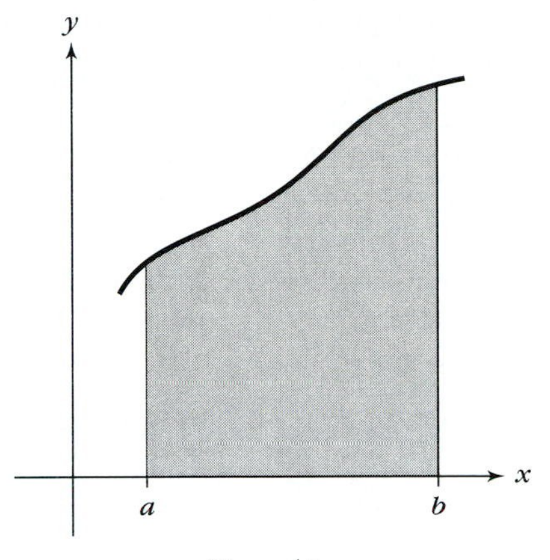

Figure 4.3
Ordinate set of f

This region is called the *ordinate set* of f over $[a, b]$; it is defined by

$$S_f[a, b] = \{(x, y) : a \le x \le b \quad \text{and} \quad 0 \le y \le f(x)\}$$

We can approximate the area of $S_f[a, b]$ by adding together the areas of vertical rectangular strips like those in Figure 4.4. Of course, there are many ways to choose the heights of the approximating rectangles. If the function is continuous, it assumes a minimum and a maximum on each of the subintervals of $[a, b]$ shown in Figure 4.4. By taking the height to be the minimum value of the function on each subinterval, we get a lower approximation to the desired area, as shown in Figure 4.4(a); using the maximum gives an upper approximation as in 4.4(b). We expect the exact area (if such a thing exists)

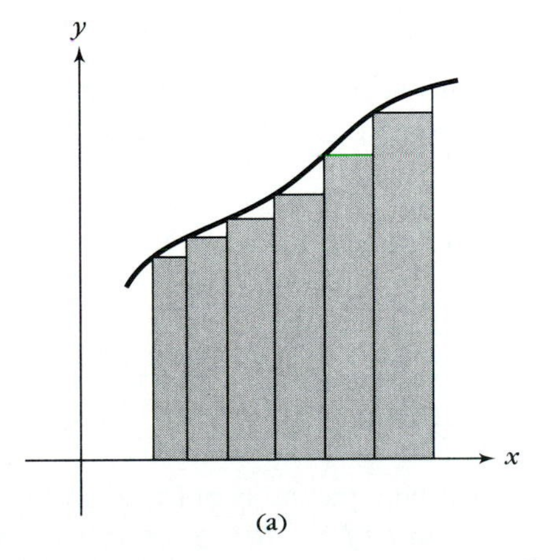

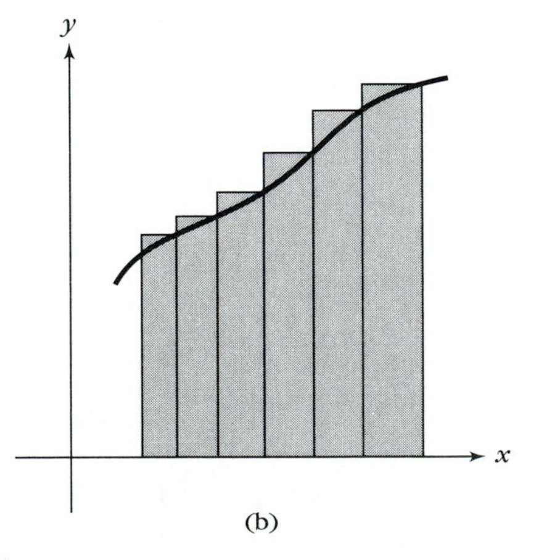

Figure 4.4
Approximations to the area of the ordinate set

to lie between these extremes. Mathematically, the exact area is defined as the integral of the function from a to b, and conditions for the existence of the integral are a major concern in the study of analysis.

Our analytic development of the integral will follow closely the ideas outlined above: the lower and upper approximations will be represented analytically by lower and upper sums. We will use the concepts of infimum and supremum to try to squeeze down on a number that lies between *every* pair of numbers consisting of a lower sum and an upper sum. In Section 4.6 we will find that a function is integrable over the given interval if and only if such a number exists and is unique.

There are two ways in which our formal development will differ from our informal description:

(1) We will not assume that the given function is continuous on $[a, b]$—only that it is bounded there. Consequently, instead of using minima and maxima to form lower and upper sums, we will rely on infima and suprema.

(2) We will not assume that the function is nonnegative on $[a, b]$; thus, it is not to be expected that the lower and upper sums (or the integral itself) represent areas in the ordinary geometric sense. (In spite of this fact, many properties of the integral can be interpreted nicely in terms of area, and we will feel free to take advantage of these interpretations. Of course, we will not use intuitive area arguments as *proofs* of our theorems.)

Lower and Upper Sums

We assume throughout the remainder of this section that the real function f is bounded on the interval $[a, b]$ and that $a < b$.

Definition 4.5.1 A ***partition*** π of $[a, b]$ is a set of points $\{x_0, x_1, \ldots, x_n\}$ where

$$a = x_0 < x_1 < \cdots < x_n = b$$

If π_1 and π_2 are partitions of $[a, b]$ such that every point of π_1 belongs to π_2, then π_2 is a ***refinement*** of π_1. The set of all partitions of $[a, b]$ will be denoted $P[a, b]$.

Example 1: The following are partitions of $[1, 3]$:

$$\pi_1 = \left\{1, \tfrac{3}{2}, 2, \tfrac{9}{4}, 3\right\}$$
$$\pi_2 = \left\{1, \tfrac{3}{2}, 2, \tfrac{9}{4}, \tfrac{5}{2}, 3\right\}$$

Notice that π_2 is a refinement of π_1.

Definition 4.5.2 Let $\pi = \{x_0, x_1, \ldots, x_n\}$ be a partition of $[a, b]$ and let f be bounded on $[a, b]$. Then the ***lower sum of f relative to π*** is defined to be

$$L_\pi(f) = m_1(x_1 - x_0) + m_2(x_2 - x_1) + \cdots + m_n(x_n - x_{n-1})$$
$$= \sum_{k=1}^{n} m_k(x_k - x_{k-1})$$

where $m_k = \inf\{f(x) : x \in [x_{k-1}, x_k]\}$. The ***upper sum of f relative to π*** is defined to be

$$U_\pi(f) = M_1(x_1 - x_0) + M_2(x_2 - x_1) + \cdots + M_n(x_n - x_{n-1})$$
$$= \sum_{k=1}^{n} M_k(x_k - x_{k-1})$$

where $M_k = \sup\{f(x) : x \in [x_{k-1}, x_k]\}$.

It is the boundedness of the real function f that assures us of the existence of the numbers m_k and M_k—hence, of the upper and lower sums themselves. Moreover, it is clear from the definition that for every $\pi \in P[a, b]$,

$$L_\pi(f) \leq U_\pi(f).$$

If $f(x) \geq 0$ for $a \leq x \leq b$, then any lower sum of f relative to a partition of $[a, b]$ may be interpreted geometrically as the area of a polygonal region [such as Figure 4.4(a)] that is contained in the ordinate set $S_f[a, b]$. Any upper sum of f relative to a partition of $[a, b]$ represents the area of a polygonal region that contains $S_f[a, b]$.

Example 2: Let $f(x) = [x]$ $(1 \leq x \leq 3)$. If π is the partition $\left\{1, \frac{3}{2}, 2, 3\right\}$, we have

$$L_\pi(f) = m_1\left(\frac{3}{2} - 1\right) + m_2\left(2 - \frac{3}{2}\right) + m_3(3 - 2)$$
$$= 1 \cdot \frac{1}{2} + 1 \cdot \frac{1}{2} + 2 \cdot 1 = 3$$
$$U_\pi(f) = M_1\left(\frac{3}{2} - 1\right) + M_2\left(2 - \frac{3}{2}\right) + M_3(3 - 2)$$
$$= 1 \cdot \frac{1}{2} + 2 \cdot \frac{1}{2} + 3 \cdot 1 = 4\frac{1}{2}$$

We have already noted that for a given function and a given partition of $[a, b]$ the lower sum is less than or equal to the upper sum. We will prove a stronger result, that *every* lower sum of f on $[a, b]$ is less than or equal to *every* upper sum (Theorem 4.5.1 below). The following lemma will simplify that proof.

Lemma 4.5.1 *Let f be bounded on $[a, b]$ and let π_1 and π_2 be partitions of $[a, b]$ such that π_2 is a refinement of π_1. Then $L_{\pi_1}(f) \leq L_{\pi_2}(f)$* and $U_{\pi_2}(f) \leq U_{\pi_1}(f)$.

PROOF: We prove the theorem in the case of upper sums. Because of mathematical induction, there is no harm in assuming that π_2 is the result of adding *one* more point of subdivision to the partition π_1. Suppose then that π_1 and

π_2 are partitions of $[a, b]$ given by

$$\pi_1 = \{x_0, x_1, \ldots, x_{k-1}, x_k, \ldots, x_n\}$$

and

$$\pi_2 = \{x_0, x_1, \ldots, x_{k-1}, y, x_k, \ldots, x_n\},$$

where $a = x_0 < x_1 < \cdots < x_{k-1} < y < x_k < \cdots < x_n = b$; thus, π_2 includes the point y between successive points x_{k-1} and x_k of π_1. The sums $U_{\pi_1}(f)$ and $U_{\pi_2}(f)$ differ only insofar as, in $U_{\pi_1}(f)$, the kth term, $M_k(x_k - x_{k-1})$, in which

$$M_k = \sup\{f(x) : x \in [x_{k-1}, x_k]\}$$

is replaced by two terms in $U_{\pi_2}(f)$:

$$M_k'(y - x_{k-1}) + M_k''(x_k - y)$$

in which

$$M_k' = \sup\{f(x) : x \in [x_{k-1}, y]\} \qquad \text{and} \qquad M_k'' = \sup\{f(x) : x \in [y, x_k]\}$$

But, clearly, both M_k' and M_k'' are less than or equal to M_k, and so

$$\begin{aligned} U_{\pi_1}(f) - U_{\pi_2}(f) &= M_k(x_k - x_{k-1}) - [M_k'(y - x_{k-1}) + M_k''(x_k - y)] \\ &\geq M_k(x_k - x_{k-1}) - [M_k(y - x_{k-1}) + M_k(x_k - y)] \\ &= M_k(x_k - x_{k-1}) - M_k(x_k - x_{k-1}) = 0 \end{aligned}$$

Thus, $U_{\pi_1}(f) - U_{\pi_2}(f) \geq 0$, and so $U_{\pi_2}(f) \leq U_{\pi_1}(f)$. ■

Note: The lemma says that refinement tends to increase lower sums and decrease upper sums.

Theorem 4.5.1 *Let f be bounded on $[a, b]$ and let π and σ be partitions of $[a, b]$. Then*

$$L_\pi(f) \leq U_\sigma(f)$$

PROOF: Let $\tau = \pi \cup \sigma$, so that τ is a refinement of each of the partitions π and σ. By Lemma 4.5.1, $L_\pi(f) \leq L_\tau(f)$ and $U_\tau(f) \leq U_\sigma(f)$, and the conclusion of the theorem follows from the fact that $L_\tau(f) \leq U_\tau(f)$. ■

Lower and Upper Integrals; the Riemann Integral

We consider the set $\mathcal{L}$ of all lower sums of f for all partitions of $[a, b]$:

$$\mathcal{L} = \{L_\pi(f) : \pi \in \mathcal{P}[a, b]\}$$

Similarly, we let $\mathcal{U}$ consist of all upper sums of f on $[a, b]$:

$$\mathcal{U} = \{U_\pi(f) : \pi \in \mathcal{P}[a, b]\}$$

Theorem 4.5.1 tells us that $\mathcal{L}$ is bounded above (by every element of $\mathcal{U}$) and that $\mathcal{U}$ is bounded below (by every element of $\mathcal{L}$). Consequently, $\sup \mathcal{L}$ and $\inf \mathcal{U}$ exist, and we can make the following definition.

Definition 4.5.3 Let f be bounded on $[a, b]$. The ***lower integral*** of f from a to b, denoted $\underline{\int_a^b} f$, is $\sup L = \sup \{L_\pi(f) : \pi \in P[a, b]\}$. The ***upper integral*** of f from a to b, denoted $\overline{\int_a^b} f$, is $\inf U = \inf \{U_\pi(f) : \pi \in P[a, b]\}$.

The lower and upper integrals of a function bounded on $[a, b]$ always exist, and it is clear that

$$\underline{\int_a^b} f \leq \overline{\int_a^b} f$$

Definition 4.5.4 Let f be bounded on $[a, b]$. Then f is ***Riemann integrable*** on $[a, b]$ provided that $\underline{\int_a^b} f = \overline{\int_a^b} f$; in case of integrability, the common value of the lower and upper integrals is called the ***Riemann integral*** of f from a to b and is denoted $\int_a^b f$ or $\int_a^b f(x)\, dx$.

Note: The Riemann integral defined above is the most elementary formulation of the integral concept. A more powerful theory of integration due to Henri Lebesgue (1875–1941) is usually based on the theory of measure. (An elementary approach to the Lebesgue integral is found in the text by Mikusiński and Mikusiński.*)

Georg Friedrich Bernhard Riemann

(b. Sept. 17, 1826; d. July 20, 1866)

Although he died at age 39, Riemann's contributions to mathematics were profound and far-reaching. His doctoral dissertation "Foundations for a General Theory of Functions of a Complex Variable" (1851) met with the enthusiastic approval of Karl Friedrich Gauss (1777–1855). He further impressed that famous mathematician in 1854 with his lecture, "On the Hypotheses Which Lie at the Foundations of Geometry." Riemann extended Cauchy's work with the definite integral, allowing certain discontinuous functions to be integrated. And he introduced the surfaces over the complex domain that bear his name.

*Jan Mikusiński and Piotr Mikusiński, *An Introduction to Analysis* (New York: Wiley, 1993), chapter 7.

Example 3: Let f be a constant function, say $f(x) = K$ $(a \le x \le b)$. For every partition $\pi = \{x_0, x_1, \ldots, x_n\}$ where $a = x_0 < x_1 < \cdots < x_n = b$, we have

$$m_k = K \qquad k = 1, 2, \ldots, n$$

so that

$$L_\pi(f) = \sum_{k=1}^{n} K(x_k - x_{k-1}) = K(x_n - x_0) = K(b - a)$$

It follows that L contains just the single real number $K(b - a)$, and so $\underline{\int_a^b} f = K(b - a)$. But for $k = 1, 2, \ldots, n$ we also have $M_k = K$, and it follows likewise that $\overline{\int_a^b} f = K(b - a)$. Thus, f is Riemann integrable on $[a, b]$, and $\int_a^b f = \int_a^b K\,dx = K(b - a)$.

Example 4: Consider the function

$$f(x) = \begin{cases} 1 & \text{if } x \text{ rational}, \quad a \le x \le b \\ 0 & \text{if } x \text{ irrational}, \quad a \le x \le b \end{cases}$$

Let π be any partition of $[a, b]$, say $\pi = \{x_0, x_1, \ldots, x_n\}$ where $a = x_0 < x_1 < \cdots < x_n = b$. Then, by the Irrational Density Theorem (Theorem 1.2.4), we have $m_k = 0$ for $k = 1, 2, \ldots, n$. Consequently, the lower sum of f relative to an arbitrary partition π of $[a, b]$ is 0, and so $\underline{\int_a^b} f = 0$. However, by the Rational Density Theorem (Theorem 1.2.3), each $M_k = 1$, and we see by a calculation similar to that in Example 3 that $\overline{\int_a^b} f = b - a > 0$. Thus, f is not Riemann integrable on $[a, b]$.

Regular Partitions

Let $[a, b]$ be a fixed closed interval with $a < b$. For each $n \in Z^+$, we denote by π_n the partition of $[a, b]$ with n subintervals of equal length:

$$\pi_n = \{x_0, x_1, \ldots, x_n\},$$

where

$$x_k = a + \frac{k}{n}(b - a) \qquad k = 0, 1, 2, \ldots, n$$

We call π_n a *regular* partition of $[a, b]$. Regular partitions are often useful in calculating integrals.

Example 5: Consider $f(x) = x^2$ on the interval $[0, b]$, where $b > 0$. For $n \in Z^+$, let π_n be the regular partition of $[0, b]$ with n subintervals; then $\pi_n = \{x_0, x_1, \ldots, x_n\}$, where $x_k = kb/n$ for $k = 0, 1, \ldots, n$. Because f

is increasing, we see that, for $k = 1, 2, \ldots, n$, $m_k = f(x_{k-1})$ and $M_k = f(x_k)$. Using the formula given in Exercise 8a, section P.2, it is easy to calculate the lower and upper sums of f relative to π_n:

$$\begin{aligned}
L_{\pi_n}(f) = \sum_{k=1}^{n} f(x_{k-1})\frac{b}{n} &= \sum_{k=1}^{n}\left[\frac{(k-1)b}{n}\right]^2 \frac{b}{n} \\
&= \frac{b^3}{n^3}\sum_{k=1}^{n}(k-1)^2 = \frac{b^3}{n^3}\frac{n(n-1)(2n-1)}{6} \\
&= \frac{b^3}{6}\left(1-\frac{1}{n}\right)\left(2-\frac{1}{n}\right) \\
U_{\pi_n}(f) = \sum_{k=1}^{n} f(x_k)\frac{b}{n} &= \sum_{k=1}^{n}\left(\frac{kb}{n}\right)^2\frac{b}{n} \\
&= \frac{b^3}{n^3}\sum_{k=1}^{n} k^2 = \frac{b^3}{n^3}\frac{n(n+1)(2n+1)}{6} \\
&= \frac{b^3}{6}\left(1+\frac{1}{n}\right)\left(2+\frac{1}{n}\right)
\end{aligned}$$

It is clear that the sequences $\{L_{\pi_n}(f)\}_{n=1}^{\infty}$ and $\{U_{\pi_n}(f)\}_{n=1}^{\infty}$ both converge to $b^3/3$. We claim that f is integrable on $[0, b]$ and that its integral is $b^3/3$.

The sequence $\{L_{\pi_n}(f)\}_{n=1}^{\infty}$ is increasing and bounded above; and by part (i) of Theorem 2.6.1, $\{L_{\pi_n}(f)\}_{n=1}^{\infty}$ converges to its least upper bound. Therefore,

$$\begin{aligned}
\frac{b^3}{3} &= \lim_{n\to\infty} L_{\pi_n}(f) \\
&= \sup\{L_{\pi_n}(f) : n \in Z^+\} \\
&\le \sup\{L_{\pi}(f) : \pi \in P[a, b]\} \qquad \text{(Why?)} \\
&= \underline{\int_0^b} f
\end{aligned}$$

But a parallel argument shows that $\{U_{\pi_n}(f)\}_{n=1}^{\infty}$ converges to its greatest lower bound:

$$\begin{aligned}
\frac{b^3}{3} &= \lim_{n\to\infty} U_{\pi_n}(f) \\
&= \inf\{U_{\pi_n}(f) : n \in Z^+\} \\
&\ge \inf\{U_{\pi}(f) : \pi \in P[a, b]\} \\
&= \overline{\int_0^b} f
\end{aligned}$$

Since $\overline{\int_0^b} f \le b^3/3 \le \underline{\int_0^b} f$, we conclude that the upper and lower integrals are equal; thus, f is integrable on $[0, b]$ and

$$\int_0^b f = \int_0^b x^2\,dx = \frac{b^3}{3}$$

EXERCISES 4.5

1. Prove Lemma 4.5.1 for lower sums.

2. Show that if $b > 0$, then $f(x) = x$ is integrable on $[0, b]$ and $\int_0^b f = b^2/2$.

3. Show that if $b > 0$, then $f(x) = x^3$ is integrable on $[0, b]$ and $\int_0^b f = b^4/4$.

4. Show that if there is a constant M such that
$$|f(x)| \le M \text{ for } a \le x \le b,$$
and if f is integrable on $[a, b]$, then
$$\left|\int_a^b f\right| \le M(b - a)$$

5. Let f be bounded on $[a, b]$ where $a < b$ and let K be a constant. Prove that if $K > 0$, then
$$\overline{\int_a^b} Kf = K\overline{\int_a^b} f$$
What if $K < 0$?

6. Let f and g be bounded on $[a, b]$ where $a < b$. Prove the inequalities
$$\underline{\int_a^b}(f + g) \ge \underline{\int_a^b} f + \underline{\int_a^b} g$$
$$\overline{\int_a^b}(f + g) \le \overline{\int_a^b} f + \overline{\int_a^b} g$$
In each case give examples of functions f and g for which equality does not hold.

4.6 *Conditions for Integrability*

The class of real functions that are integrable on an interval is very large indeed. We will see in this section that any function that is either monotone on $[a, b]$ or continuous on $[a, b]$ is also integrable there. These special results are consequences of a theorem giving a simple condition (Riemann's Condition) that is both necessary and sufficient for integrability.

We continue to assume in this section that the real function f is bounded on the interval $[a, b]$ and that $a < b$.

Riemann's Condition

Theorem 4.6.1 **Riemann's Condition** *A bounded real function f is integrable on $[a, b]$ if and only if for every $\varepsilon > 0$ there exists a partition π of $[a, b]$ such that*
$$U_\pi(f) - L_\pi(f) < \varepsilon$$

PROOF: We assume first that for every $\varepsilon > 0$ such a partition π exists. It follows that, for every $\varepsilon > 0$, we have
$$\overline{\int_a^b} f \le U_\pi(f) < L_\pi(f) + \varepsilon \le \underline{\int_a^b} f + \varepsilon$$
Accordingly, $\overline{\int_a^b} f \le \underline{\int_a^b} f$, and so f is integrable on $[a, b]$.

Conversely, suppose that f is integrable on $[a, b]$ and let $I = \int_a^b f$. Using the notation of the previous section, we have

$$I = \overline{\int_a^b} f = \inf \mathcal{U}$$

It follows that $I + \frac{\varepsilon}{2}$ is not a lower bound of $\mathcal{U}$ and so there exists a partition π_1 of $[a, b]$ such that

$$U_{\pi_1}(f) < I + \tfrac{\varepsilon}{2}$$

On the other hand, $I = \underline{\int_a^b} f = \sup \mathcal{L}$, so that $I - \frac{\varepsilon}{2}$ is not an upper bound of $\mathcal{L}$ and there exists a partition π_2 of $[a, b]$ such that

$$L_{\pi_2}(f) > I - \tfrac{\varepsilon}{2}$$

Let $\pi = \pi_1 \cup \pi_2$. Then $U_\pi(f) \leq U_{\pi_1}(f) < I + \frac{\varepsilon}{2}$, and $L_\pi(f) \geq L_{\pi_2}(f) > I - \frac{\varepsilon}{2}$, and it follows that

$$U_\pi(f) < I + \frac{\varepsilon}{2} < L_\pi(f) + \varepsilon$$

$$U_\pi(f) - L_\pi(f) < \varepsilon \qquad \blacksquare$$

Note: The difference $U_\pi(f) - L_\pi(f)$ can be interpreted geometrically as the shaded area in Figure 4.5. Riemann's Condition says simply that f is integrable if and only if this shaded area can be made arbitrarily small by taking a fine enough partition.

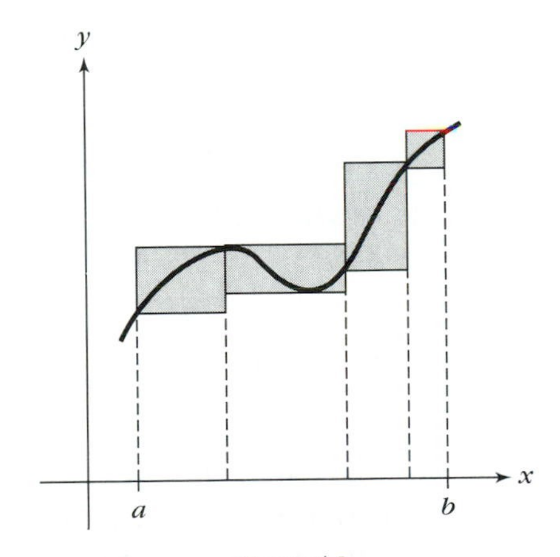

Figure 4.5
Riemann's Condition

Corollary 4.6.1 **Integrability of Monotone Functions** *A real function that is monotone on a closed bounded interval is integrable there.*

PROOF: Let f be monotone on $[a, b]$ with $a < b$, and suppose for definiteness that f is increasing. If f is not constant, then $f(a) < f(b)$. Let $\varepsilon > 0$ and choose $n \in Z^+$ such that $[(b - a)/n][f(b) - f(a)] < \varepsilon$. Form the regular partition π_n of $[a, b]$, and observe that because f is increasing, the numbers m_k and M_k are $f(x_{k-1})$ and $f(x_k)$, respectively. Thus,

$$L_{\pi_n}(f) = \sum_{k=1}^{n} f(x_{k-1})(x_k - x_{k-1}) = \sum_{k=1}^{n} f(x_{k-1}) \frac{b-a}{n}$$

and

$$U_{\pi_n}(f) = \sum_{k=1}^{n} f(x_k)(x_k - x_{k-1}) = \sum_{k=1}^{n} f(x_k) \frac{b-a}{n}$$

Therefore,

$$\begin{aligned}
U_{\pi_n}(f) - L_{\pi_n}(f) &= \sum_{k=1}^{n} [f(x_k) - f(x_{k-1})] \frac{b-a}{n} \\
&= \frac{b-a}{n} \sum_{k=1}^{n} [f(x_k) - f(x_{k-1})] \\
&= \frac{b-a}{n} \{[f(x_1) - f(x_0)] + [f(x_2) - f(x_1)] + \cdots + [f(x_n) - f(x_{n-1})]\} \\
&= \frac{b-a}{n} [f(x_n) - f(x_0)] = \frac{b-a}{n} [f(b) - f(a)] < \varepsilon
\end{aligned}$$

Because of the way n was chosen, we have $U_{\pi_n}(f) - L_{\pi_n}(f) < \varepsilon$, and so f is integrable on $[a, b]$ by Theorem 4.6.1. ■

Note: It is often helpful to express the condition $U_\pi(f) - L_\pi(f) < \varepsilon$ in Riemann's Condition as follows:

$$\sum_{k=1}^{n} (M_k - m_k)(x_k - x_{k-1}) < \varepsilon$$

(where the notation is the same as in Section 4.5).

Corollary 4.6.2 **Integrability of Continuous Functions** *A real function that is continuous on a closed bounded interval is integrable there.*

PROOF: Let f be continuous on $[a, b]$ with $a < b$ and let $\varepsilon > 0$ be given. By Theorem 3.9.1, f is uniformly continuous on $[a, b]$, and corresponding to $\varepsilon/(b - a)$ there is a $\delta > 0$ such that $|f(u) - f(v)| < \varepsilon/(b - a)$ whenever u and v are points in $[a, b]$ with $|u - v| < \delta$. Let π be a partition of $[a, b]$ in which all subintervals have lengths less than δ. Then

$$\begin{aligned}
U_\pi(f) - L_\pi(f) &= \sum_{k=1}^{n} (M_k - m_k)(x_k - x_{k-1}) \\
&= \sum_{k=1}^{n} [f(x_k') - f(x_k'')](x_k - x_{k-1})
\end{aligned}$$

where x_k' is a point at which f assumes its maximum on $[x_{k-1}, x_k]$, and x_k'' is a point at which f assumes its minimum on that subinterval. Clearly, $|x_k' - x_k''| < \delta$ for $k = 1, 2, \ldots, n$, so that $f(x_k') - f(x_k'') < \varepsilon/(b - a)$ and

$$\begin{aligned} U_\pi(f) - L_\pi(f) &< \sum_{k=1}^{n} \frac{\varepsilon}{b - a}(x_k - x_{k-1}) \\ &= \frac{\varepsilon}{b - a} \sum_{k=1}^{n} (x_k - x_{k-1}) \\ &= \frac{\varepsilon}{b - a}(b - a) = \varepsilon \end{aligned}$$ ∎

Example 1: We see immediately from Corollaries 4.6.1 and 4.6.2 that the following functions are integrable on the indicated intervals:

$$S(x) = [x] \quad \text{on } [a, b]\ (a < b)$$

$$g(x) = \begin{cases} x & \text{if } 0 \le x < 1 \\ 2x + 1 & \text{if } 1 \le x \le 2 \end{cases} \quad \text{on } [0, 2]$$

$$f(x) = |x| \quad \text{on } [a, b]\ (a < b)$$

Next we use Riemann's Condition to prove a result promised in the introduction of Section 4.5.

Theorem 4.6.2 *A bounded real function f is integrable on $[a, b]$ if and only if there exists a unique real number I such that*

$$L_\pi(f) \le I \le U_\pi(f) \tag{1}$$

for all partitions π of $[a, b]$.

PROOF: If f is integrable on $[a, b]$, then $I = \int_a^b f$ clearly satisfies (1); if some other real number $J \ne I$ (say $J > I$) should also satisfy (1), we would have, for all partitions π of $[a, b]$,

$$L_\pi(f) \le I < J \le U_\pi(f)$$

It follows then that $\underline{\int_a^b} f < \overline{\int_a^b} f$, contradicting the integrability of f. Thus, I is unique.

If f is not integrable on $[a, b]$, then for every partition π of $[a, b]$,

$$L_\pi(f) \le \underline{\int_a^b} f < \overline{\int_a^b} f \le U_\pi(f)$$

and in that case every number beween $\underline{\int_a^b} f$ and $\overline{\int_a^b} f$ satisfies (1). ∎

The area problem in calculus that we discussed in Section 4.5 can be expressed as follows: If f is a nonnegative function on $[a, b]$, is it possible to assign an area to the ordinate set of f on $[a, b]$ in a unique way? Theorem 4.6.2 suggests that the answer is yes if and only if f is integrable on $[a, b]$.

Some Consequences of Riemann's Condition

Let us conclude this section by stating several more interesting consequences of Riemann's Condition. Proofs of these theorems are left for you to work out in the exercises.

Theorem 4.6.3 *Let f be bounded and integrable on $[a, b]$. Then the function g defined by $g(x) = [f(x)]^2$ is also integrable on $[a, b]$.*

PROOF: The proof is outlined in Exercise 4.6.2. ∎

Theorem 4.6.4 *Let f be bounded and integrable on $[a, b]$. If g is defined on $[a, b]$ with the property that $g(x) = f(x)$ for all except a finite number of points in $[a, b]$, then g is also integrable on $[a, b]$ and $\int_a^b g = \int_a^b f$.*

PROOF: Exercise 4.6.5 asks you to prove the integrability of g. The proof that the two integrals are equal is left for you to work out in Exercise 4.7.6. ∎

Note: Theorem 4.6.4 says that we can change the values of an integrable function at a finite number of points in the interval of integration without changing the fact of integrability or the value of the integral.

Example 2: The function

$$S_1(x) = [x] \qquad 0 \le x \le 2$$

is monotone on $[0, 2]$ and therefore integrable there. The function

$$S_2(x) = \begin{cases} 0 & \text{if } 0 \le x \le 1 \text{ or } x = 2 \\ 1 & \text{if } 1 < x < 2 \end{cases}$$

has the same values as $S_1(x)$ at all points of $[0, 2]$ except 1 and 2. By Theorem 4.6.4, S_2 is integrable on $[0, 2]$ and $\int_0^2 S_2 = \int_0^2 S_1$.

Theorem 4.6.5 *Let f be bounded and integrable on $[a, b]$ and suppose that c and d are real numbers with $a \le c < d \le b$. Then f is also integrable on $[c, d]$.*

PROOF: You are asked to prove this theorem in Exercise 4.6.6. ∎

Theorem 4.6.6 *Let f be bounded on $[a, b]$. If f is integrable on $[c, d]$ for every pair c, d of real numbers with $a < c < d < b$, then f is integrable on $[a, b]$.*

PROOF: You are asked to prove this theorem in Exercise 4.6.7.

Theorem 4.6.7 *Let f be bounded on $[a, b]$. If f is integrable on $[a, b]$, then so is $|f|$, and*

$$\left|\int_a^b f\right| \le \int_a^b |f|$$

PROOF: The proof is outlined in Exercise 4.6.8. ■

EXERCISES 4.6

1. For each of the following functions prove integrability by finding for an arbitrary $\varepsilon > 0$, a partition π of the given interval with the property that $U_\pi(f) - L_\pi(f) < \varepsilon$.
   (a) $f(x) = x$; $[0, 1]$
   (b) $f(x) = |x|$; $[-1, 1]$
   (c) $f(x) = [x]$; $[0, 10]$ ([ ] denotes the greatest integer function)
   (d) $f(x) = \begin{cases} 1 & \text{if } x = \frac{1}{2} \\ 2 & \text{if } x \neq \frac{1}{2} \end{cases}$; $[0, 1]$

   [*Note:* Parts (b), (c), and (d) show that neither monotonicity nor continuity is necessary for integrability.]

2. (a) Let f be bounded on a nonempty interval I. The *oscillation* of f on I is defined to be

   $$\omega_f(I) = \sup\{f(x) : x \in I\} - \inf\{f(x) : x \in I\}$$

   Show that

   $$\omega_f(I) = \sup\{f(x_1) - f(x_2) : x_1, x_2 \in I\}.$$

   (b) Let f be bounded on a nonempty interval I, with $|f(x)| \le M$ for $x \in I$. Let $g(x) = [f(x)]^2$. Show that

   $$\omega_g(I) \le 2M\omega_f(I).$$

   (c) Prove Theorem 4.6.3. [*Hint:* Let s_k and S_k denote, respectively, the infimum and supremum of g on $[x_{k-1}, x_k]$ and let m_k and M_k denote the infimum and supremum of f on $[x_{k-1}, x_k]$. By parts (a) and (b),

   $$S_k - s_k \le 2M\,(M_k - m_k)$$

   for $k = 1, 2, \ldots, n$, where M is the supremum of f on $[a, b]$.]

3. Show that the integrability of $g(x) = [f(x)]^2$ on $[a, b]$ is not a sufficient condition for integrability of f there.

4. Prove or disprove: If f and g are bounded and integrable on $[a, b]$, then their product $h(x) = f(x)g(x)$ is integrable on $[a, b]$.

5. Prove the integrability of g in Theorem 4.6.4. (Because of mathematical induction, it is sufficient to do this in the case in which f and g differ at only one point.)

6. Prove Theorem 4.6.5.

7. Prove Theorem 4.6.6.

8. (a) Let f be bounded on $[a, b]$. Define f^+ and f^-, the *positive* and *negative* parts of f, respectively, as follows:

   $$f^+(x) = \begin{cases} f(x) & \text{if } f(x) \ge 0 \\ 0 & \text{if } f(x) < 0 \end{cases}$$

   $$f^-(x) = \begin{cases} 0 & \text{if } f(x) \ge 0 \\ -f(x) & \text{if } f(x) < 0 \end{cases}$$

   Show that if f is integrable on $[a, b]$, then so are f^+ and f^-.

   (b) Use the identity $|f| = f^+ + f^-$ to show that if f is integrable on $[a, b]$, then so is $|f|$.

   (c) Show that if f is integrable on $[a, b]$, then

   $$\left|\int_a^b f\right| \le \int_a^b |f|$$

4.7 *Properties of the Integral*

This section presents a number of the most important properties of the definite integral. The key results of linearity and additivity (Theorems 4.7.1 and 4.7.2) are proved in detail. Others are left as exercises, as they are fairly straightforward consequences of Theorems 4.7.1 and 4.7.2 together with the results of Section 4.6. You should familiarize yourself with all the theorems of this section, whether or not you are asked to prove them, because they are often called upon in later work. In most cases, the geometric content makes them easy to remember.

Linearity

The integral, like the derivative, is a "linear" operator in the sense that it preserves sums and constant multipliers.

Theorem 4.7.1 **Linearity of the Integral** *Let f and g be bounded and integrable on the interval $[a, b]$, where $a < b$, and let K be a real constant. Then $f + g$ and Kf are integrable on $[a, b]$ and*

$$\int_a^b (f + g) = \int_a^b f + \int_a^b g \tag{1}$$

$$\int_a^b Kf = K \int_a^b f \tag{2}$$

PROOF: We begin with the proof of Equation (1). Observe that on any subset S of $[a, b]$,

$$\sup \{f(x) + g(x) : x \in S\} \leq \sup \{f(x) : x \in S\} + \sup \{g(x) : x \in S\}$$

(work Exercise 4.7.1). It follows that if π is a partition of $[a, b]$, then $U_\pi(f + g) \leq U_\pi(f) + U_\pi(g)$. Similarly, it can be shown that $L_\pi(f + g) \geq L_\pi(f) + L_\pi(g)$.

Now let $\varepsilon > 0$. Because of the integrability of f and g, there exist partitions π_1 and π_2 of $[a, b]$ such that

$$U_{\pi_1}(f) < L_{\pi_1}(f) + \frac{\varepsilon}{2} \qquad \text{and} \qquad U_{\pi_2}(g) < L_{\pi_2}(g) + \frac{\varepsilon}{2}$$

Let $\pi = \pi_1 \cup \pi_2$. Then, by noting that π is a refinement of both π_1 and π_2 and recalling the effect of refinement on upper and lower sums, we get the following chain of inequalities:

$$\begin{aligned} U_\pi(f + g) \leq U_\pi(f) + U_\pi(g) \leq U_{\pi_1}(f) + U_{\pi_2}(g) &< \left[L_{\pi_1}(f) + \frac{\varepsilon}{2}\right] + \left[L_{\pi_2}(g) + \frac{\varepsilon}{2}\right] \\ &\leq \left[L_\pi(f) + \frac{\varepsilon}{2}\right] + \left[L_\pi(g) + \frac{\varepsilon}{2}\right] \\ &\leq L_\pi(f + g) + \varepsilon \end{aligned}$$

Since $f + g$ satisfies Riemann's Condition on $[a, b]$, it is integrable there. Moreover, we see that

$$\overline{\int_a^b}(f+g) \leq U_\pi(f+g) < L_\pi(f) + L_\pi(g) + \varepsilon \leq \underline{\int_a^b} f + \underline{\int_a^b} g + \varepsilon$$

and because f, g, and $f + g$ are integrable, we have

$$\int_a^b (f+g) < \int_a^b f + \int_a^b g + \varepsilon$$

Since this inequality holds for every $\varepsilon > 0$, we have established one half of the desired result:

$$\int_a^b (f+g) \leq \int_a^b f + \int_a^b g$$

The other half is obtained by a parallel argument. For $\varepsilon > 0$,

$$\underline{\int_a^b}(f+g) + \varepsilon \geq L_\pi(f+g) + \varepsilon > U_\pi(f) + U_\pi(g) \geq \overline{\int_a^b} f + \overline{\int_a^b} g$$

so that

$$\int_a^b (f+g) \geq \int_a^b f + \int_a^b g$$

and Equation (1) follows.

For Equation (2), if $K = 0$, the result is trivial. If $K > 0$, then it is easy to see that for any partition π of $[a, b]$, $U_\pi(Kf) = KU_\pi(f)$ and $L_\pi(Kf) = KL_\pi(f)$. For $\varepsilon > 0$, a partition π exists such that $U_\pi(f) < L_\pi(f) + \frac{\varepsilon}{K}$; hence, $U_\pi(Kf) < L_\pi(Kf) + \varepsilon$, and we see that Kf is integrable. The proof that $\int_a^b Kf = K\int_a^b f$ when $K > 0$ is left for you to work out in Exercise 4.7.2a. For $K < 0$, it is now sufficient to consider the case in which $K = -1$. Here the pertinent relations are $U_\pi(-f) = -L_\pi(f)$, and $L_\pi(-f) = -U_\pi(f)$. (Work Exercise 4.7.2b.) ■

Example 1: In earlier examples we saw that for $b > 0$ and K constant,

$$\int_0^b x^2\,dx = \frac{b^3}{3} \qquad \text{and} \qquad \int_0^b K\,dx = Kb$$

It follows from Theorem 4.7.1 that

$$\int_0^2 (5x^2 - 4)\,dx = 5 \cdot \frac{8}{3} - 4 \cdot 2 = \frac{16}{3}$$

With the aid of mathematical induction, Theorem 4.7.1 can be generalized to sums of more than two terms:

Corollary 4.7.1 *If $n \in Z^+$, if $f_k (k = 1, 2, \ldots, n)$ are bounded and integrable on $[a, b]$, and if c_k $(k = 1, 2, \ldots, n)$ are real constants, then the function*

$$\sum_{k=1}^{n} c_k f_k$$

is also integrable on $[a, b]$ and

$$\int_a^b \sum_{k=1}^{n} c_k f_k = \sum_{k=1}^{n} c_k \int_a^b f_k$$

Additivity

Next we consider the additive property of the integral. For nonnegative, integrable functions on $[a, c]$, this property says that if b is between a and c, then the area of the ordinate set over $[a, c]$ is equal to the sum of the areas of the ordinate sets over $[a, b]$ and $[b, c]$ (see Figure 4.6).

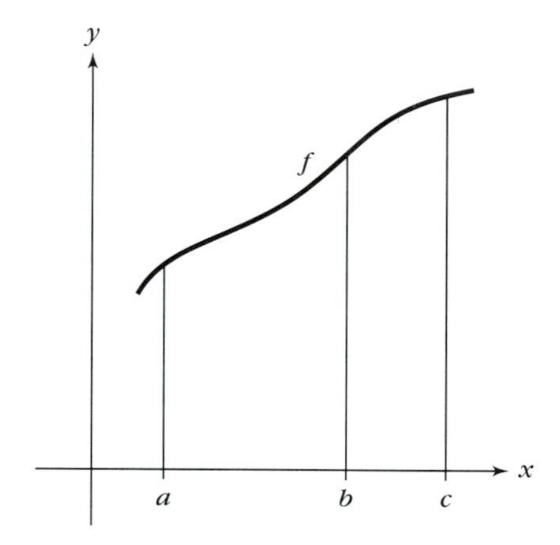

Figure 4.6
The additive property

Theorem 4.7.2 **Additivity of the Integral** *Let f be bounded and integrable on $[a, b]$ and $[b, c]$, where $a < b < c$. Then f is also integrable on $[a, c]$ and*

$$\int_a^c f = \int_a^b f + \int_b^c f$$

PROOF: To prove the integrability of f on $[a, c]$, we need to show that Riemann's Condition holds. Let $\varepsilon > 0$. Because of the integrability of f on the intervals $[a, b]$ and $[b, c]$, there exist partitions $\pi_1 \in P[a, b]$, $\pi_2 \in P[b, c]$ such that

$$U_{\pi_1}(f) < L_{\pi_1}(f) + \frac{\varepsilon}{2} \quad \text{and} \quad U_{\pi_2}(f) < L_{\pi_2}(f) + \frac{\varepsilon}{2}$$

Let $\pi = \pi_1 \cup \pi_2$; then $\pi \in P[a, c]$. We see that $U_\pi(f) = U_{\pi_1}(f) + U_{\pi_2}(f)$ and $L_\pi(f) = L_{\pi_1}(f) + L_{\pi_2}(f)$. Therefore,

$$U_\pi(f) - L_\pi(f) = U_{\pi_1}(f) - L_{\pi_1}(f) + U_{\pi_2}(f) - L_{\pi_2}(f) < \varepsilon$$

and f is integrable on $[a, c]$.

Furthermore, since any lower sum is less than or equal to the integral, we obtain (for the same partitions π_1, π_2, and π above):

$$U_{\pi_1}(f) < \int_a^b f + \frac{\varepsilon}{2}$$

$$U_{\pi_2}(f) < \int_b^c f + \frac{\varepsilon}{2}$$

$$U_\pi(f) < \int_a^b f + \int_b^c f + \varepsilon$$

Accordingly,

$$\int_a^c f = \overline{\int_a^c} f < \int_a^b f + \int_b^c f + \varepsilon$$

and since ε is an arbitrary positive number, it follows that

$$\int_a^c f \le \int_a^b f + \int_b^c f$$

Now by Equation (2), Theorem 4.7.1, with $c = -1$, we can apply the result just obtained to the function $-f$ on $[a, c]$:

$$\int_a^c (-f) \le \int_a^b (-f) + \int_b^c (-f)$$

so that

$$-\int_a^c f \le -\int_a^b f - \int_b^c f$$

and

$$\int_a^c f \ge \int_a^b f + \int_b^c f$$

The theorem follows. ■

It is often useful to apply the additive property in the following form:

$$\int_b^c f = \int_a^c f - \int_a^b f$$

where $a < b < c$ and f is integrable on $[a, c]$.

Example 2: To calculate $\int_1^2 (5x^2 - 4)\,dx$, we write

$$\int_1^2 (5x^2 - 4)\,dx = \int_0^2 (5x^2 - 4)\,dx - \int_0^1 (5x^2 - 4)\,dx$$
$$= \frac{16}{3} - \left(\frac{5}{3} - 4\right) = \frac{23}{3}$$

With the aid of mathematical induction, Theorem 4.7.2 can be generalized to any finite number of intervals:

Corollary 4.7.2 *If $n \in Z^+$ and $a_0, a_1, \ldots, a_n$ are real numbers such that $a_0 < a_1 < \cdots < a_n$, and if f is bounded and integrable on each interval $[a_{k-1}, a_k]$ ($k = 1, 2, \ldots, n$), then f is integrable on $[a_0, a_n]$ and*

$$\int_{a_0}^{a_n} f = \sum_{k=1}^{n} \int_{a_{k-1}}^{a_k} f.$$

PROOF: You are asked to prove this corollary in Exercise 4.7.3. ■

Positivity

Theorem 4.7.3 **Positivity of the Integral** *If f is bounded and integrable on $[a, b]$ where $a < b$ and if $f(x) \geq 0$ there, then $\int_a^b f \geq 0$.*

PROOF: You are asked to prove this theorem in Exercise 4.7.4a. ■

Corollary 4.7.3 *If f and g are bounded and integrable on $[a, b]$ where $a < b$, and if $f(x) \geq g(x)$ for $a \leq x \leq b$, then $\int_a^b f \geq \int_a^b g$.*

PROOF: You are asked to prove this corollary in Exercise 4.7.4b. ■

The function

$$f(x) = \begin{cases} 0 & \text{if } 0 \leq x < \frac{1}{2} \text{ or } \frac{1}{2} < x \leq 1 \\ 1 & \text{if } x = \frac{1}{2} \end{cases}$$

is an example of a nonnegative function, not identically zero, whose integral on $[0, 1]$ is zero by Theorem 4.6.4. Among *continuous* nonnegative functions, however, the only one whose integral on an interval is zero is the one that is zero everywhere on that interval. To put it another way, a continuous nonnegative function that is not identically zero has a positive integral. You are asked for a proof of this fact in Exercise 4.7.5a.

Theorem 4.7.4 *If f is continuous and nonnegative on [a, b] where $a < b$, and if $f(c) > 0$ for some $c \in [a, b]$, then $\int_a^b f > 0$.*

Corollary 4.7.4 *If f and g are continuous on [a, b] where $a < b$, if $f(x) \geq g(x)$ for $a \leq x \leq b$, and if $f(c) > g(c)$ for some $c \in [a, b]$, then*

$$\int_a^b f > \int_a^b g$$

PROOF: You are asked to prove this corollary in Exercise 4.7.5b. ■

Step Functions and the Integral

Definition 4.7.1 Let $[a, b]$ be a closed interval with $a < b$. A ***step function*** on $[a, b]$ is a real function S with the property that S is constant on the interior of each subinterval of some partition of $[a, b]$.

Example 3: The greatest integer function S defined for $x \in [0, 10]$ by $S(x) = [x]$ is a step function on $[0, 10]$.

As a consequence of Theorem 4.6.4 and Corollary 4.7.2, it can be shown that any step function on $[a, b]$ is necessarily integrable there.

Theorem 4.7.5 *Let $a = x_0 < x_1 < \cdots < x_n = b$, so that $\pi = \{x_0, x_1, \ldots, x_n\}$ is a partition of [a, b]. Let S be a step function whose values on the subintervals of π are given by*

$$S(x) = S_k \qquad \text{if } x \in (x_{k-1}, x_k)$$

for $k = 1, 2, \ldots, n$. Then S is integrable on [a, b] and

$$\int_a^b S = \sum_{k=1}^{n} S_k(x_k - x_{k-1})$$

PROOF: You are asked to prove this theorem in Exercise 4.7.8. ■

Step functions give us another very useful condition for integrability: A function is integrable if and only if it can be squeezed between two step functions whose integrals differ by less than a prescribed $\varepsilon > 0$.

Theorem 4.7.6 *Let f be bounded on [a, b] where $a < b$. Then f is integrable on [a, b] if and only if, for every $\varepsilon > 0$, there exist step functions S and T such that*

$$S(x) \leq f(x) \leq T(x) \text{ for } x \in [a, b] \qquad \text{and} \qquad \int_a^b (T - S) < \varepsilon$$

PROOF: You are asked to prove this theorem in Exercise 4.7.9. ■

EXERCISES 4.7

1. Show that if f and g are real functions, defined and bounded on a set S, then

$$\sup\{f(x) + g(x) : x \in S\} \le \sup\{f(x) : x \in S\} + \sup\{g(x) : x \in S\}$$

and

$$\inf\{f(x) + g(x) : x \in S\} \ge \inf\{f(x) : x \in S\} + \inf\{g(x) : x \in S\}$$

Give examples in each case showing that strict inequality can hold.

2. Complete the proof of Theorem 4.7.1:
   (a) For the case in which $K > 0$
   (b) For the case in which $K = -1$

3. Prove Corollary 4.7.2.

4. (a) Prove Theorem 4.7.3.
   (b) Prove Corollary 4.7.3.

5. (a) Prove Theorem 4.7.4.
   (b) Prove Corollary 4.7.4.

6. This exercise completes the proof of Theorem 4.6.4.
   (a) Let $h(x)$ be zero at all except a finite number of points of the closed bounded interval $[a, b]$. (By Exercise 4.6.3, it follows that h is integrable on $[a, b]$.) Show that $\int_a^b h = 0$.
   (b) Let f and g be bounded and integrable on $[a, b]$ where $a < b$, and suppose that these functions have equal values except at a finite number of points of $[a, b]$. Show that

$$\int_a^{b,} f = \int_a^b g.$$

7. Let f and g be bounded and integrable on $[a, b]$ where $a < b$. Show that fg is also integrable there.

8. Prove Theorem 4.7.5.

9. Prove Theorem 4.7.6. [*Hint:* If $\pi \in P[a, b]$ then $U_\pi(f)$ and $L_\pi(f)$ are integrals of step functions.]

10. Give a proof of Equation (1) of Theorem 4.7.1 based on Exercise 4.5.6.

4.8 *The Relationship Between Derivatives and Integrals*

At first encounter, it seems that the concepts of derivative and integral have nothing to do with one another. The derivative is defined to be the limit of a quotient and represents slope or rate of change. The integral is obtained by applying a limiting process (infimum or supremum) to a special kind of sum and represents such things as area, volume, mass, and so on.

Of the two concepts, the integral is much the older. Integration as a process of summation goes back at least as far as the ancient Greeks. Archimedes, for example, calculated areas of regions bounded by curves, including a parabolic region*. The summation process is difficult and lengthy, at best, as Example 5 in Section 4.5 indicates. Differentiation, on the other hand, arose much later in connection with the mechanics of moving bodies, tangents, and velocities. It is a much easier process to execute.

Seventeenth century mathematicians, most notably Newton and Leibniz, discovered a profound relationship between differentiation and integration that greatly simplified the calculation of many integrals. This result, known as the Fundamental Theorem of Calculus, has two aspects: (1) it shows that

*See George F. Simmons, "Archimedes' Quadrature of the Parabola", in *Calculus Gems* (New York: McGraw Hill, 1992) pp. 233–236.

differentiation and integration are inverse processes; and (2) it shows specifically how an integral can be calculated using a process of "reverse differentiation". The discovery of the Fundamental Theorem marks the emergence of calculus as one subject rather than two, and its impact on the development of mathematics and science would be hard to exaggerate.

Isaac Newton

(b. January 4, 1643; d. March 31, 1727)

In 1661 Isaac Newton entered Trinity College, Cambridge, where he was a solitary and serious student. During the years of the plague in England, the university was closed, and Newton dedicated himself to studying and advancing the mathematics of the day. In 1665–1666 he developed a method of approximating functions by series, the general binomial theorem, the method of tangents, the method of "fluxions", and the "inverse method" (differential and integral calculus). A delay in publishing his work on calculus caused a dispute with Leibniz and others over who invented the subject. His work, *Philosophiae Naturales Principia Mathematica*, developed the theory of gravitation. Many honors were bestowed on Newton, among them the presidency of the Royal Society, and knighthood.

Gottfried Wilhelm Leibniz

(b. July 1, 1646; d. November 14, 1716)

A University of Leipzig student at age 15, Leibniz was interested in history, theology, diplomacy, linguistics, geology, and biology as well as mathematics. He was anxious to obtain knowledge and make inventions. He spent the years 1672–1676 in Paris where he developed calculus from geometric concepts. He published his work in 1684, using symbols for differentiating and integrating that are still in use today. Both Newton and Leibniz presented the foundations of the calculus with some vagueness, but their geometric ideas were sound. An active controversy over the honor for discovery began during their lifetimes and continued long after the deaths of both men.

A More General Integral; Mean Value Theorem for Integrals

In defining the integral of a function from a to b, we have so far stipulated that $a < b$. Now we extend the definition to include cases in which $b \leq a$.

Definition 4.8.1 Let f be a real function that is bounded on an interval containing the points a and b. Then:

$$\int_a^a f = 0 \tag{1}$$

If $b < a$, and f is integrable on $[b, a]$, then

$$\int_a^b f = -\int_b^a f \tag{2}$$

It is easy to see that the linearity properties of the integral (Theorem 4.7.1) hold under this more general understanding. Moreover, the additivity property (Theorem 4.7.2),

$$\int_a^c f = \int_a^b f + \int_b^c f$$

holds whenever these integrals exist, regardless of the order of a, b, and c. Consider the case in which $c < a < b$ and suppose that f is integrable on $[c, a]$ and on $[a, b]$. Then, by Theorem 4.7.2, we know that f is integrable on $[c, b]$ and that

$$\int_c^b f = \int_c^a f + \int_a^b f$$

Now $\int_c^b f = -\int_b^c f$ and $\int_c^a f = -\int_a^c f$, so the equation becomes

$$-\int_b^c f = -\int_a^c f + \int_a^b f \qquad \text{or} \qquad \int_a^c f = \int_a^b f + \int_b^c f$$

The other cases follow similarly.

Theorem 4.8.1 **Mean Value Theorem for Integrals** *If f is continuous on an interval containing a and b, then there exists a number ξ between a and b such that*

$$f(\xi)(b - a) = \int_a^b f$$

PROOF: The proof is trivial if $a = b$. So assume that $a < b$. Since f is continuous, it assumes a minimum m and a maximum M on the interval $[a, b]$. From the inequalities

$$m \leq f(x) \leq M \qquad a \leq x \leq b$$

and Corollary 4.7.3, we get

$$m(b-a) \leq \int_a^b f \leq M(b-a)$$

In fact, by Corollary 4.7.4, it follows that $m(b-a)$ is *strictly* less than $\int_a^b f$ unless f is the constant m. Similarly, $M(b-a)$ is strictly greater than $\int_a^b f$ unless f is the constant M. Therefore, unless f is constant, we have

$$m(b-a) < \int_a^b f < M(b-a)$$

and the number

$$\mu = \frac{\int_a^b f}{b-a}$$

is strictly between m and M. By the Intermediate Value Theorem (Theorem 3.6.2), there is a $\xi \in (a, b)$ such that $f(\xi) = \mu$; for this ξ, we have

$$f(\xi)(b-a) = \int_a^b f$$

If f is constant, say $f(x) = m$, then the equation holds because both sides are equal to $m(b-a)$.

If $b < a$, then by what we have already proved, there exists a number ξ between b and a such that

$$f(\xi)(a-b) = \int_b^a f = -\int_a^b f$$

so that

$$f(\xi)(b-a) = \int_a^b f \qquad \blacksquare$$

Remark: The number $f(\xi)$ in the Mean Value Theorem is called the *mean value*, or *average value*, of f over $[a, b]$. A geometric interpretation is suggested in Figure 4.7.

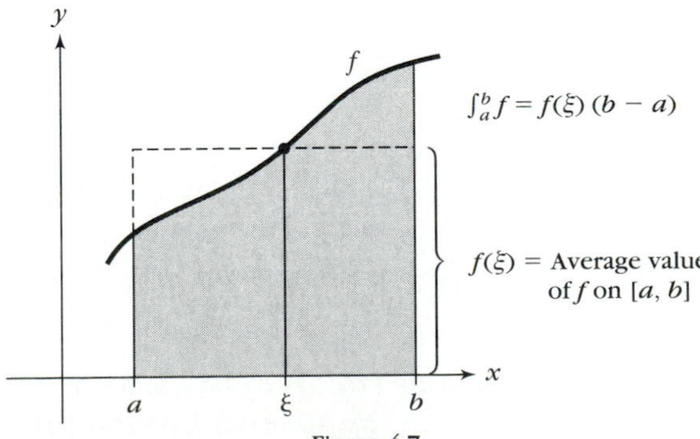

Figure 4.7
Average value

Fundamental Theorem of Calculus

Let f be a real function continuous on an interval I, let $a \in I$, and define a function G as follows:

$$G(x) = \int_a^x f \qquad x \in I$$

Thus, $G(x)$ is an integral with a variable upper limit. If f is nonnegative and if $x > a$, then $G(x)$ represents the area of the ordinate set of f over $[a, x]$ (see Figure 4.8).

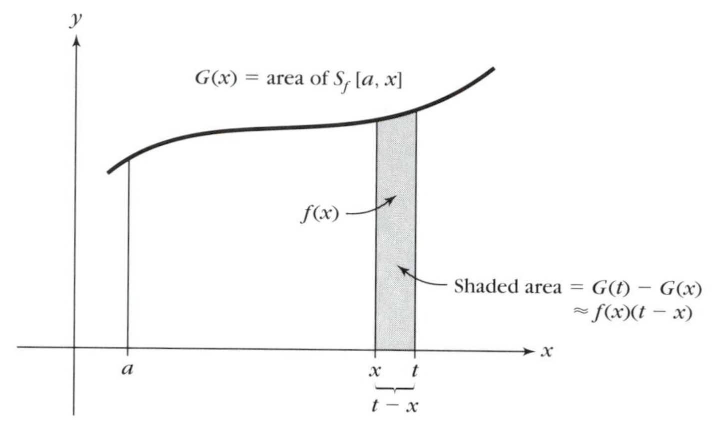

Figure 4.8
Area of ordinate set

As x changes, the value of $G(x)$ changes, and we can ask at what rate this area is changing with respect to x—in other words, what is $G'(x)$? The change induced in the area when the upper limit of the integral changes from x to t ($t > x$) is equal to $G(t) - G(x)$ (shaded in Figure 4.8). The continuity of f suggests that for t close to x, this shaded area is approximately $f(x)(t - x)$, and so

$$G(t) - G(x) \approx f(x)(t - x)$$

or

$$\frac{G(t) - G(x)}{t - x} \approx f(x)$$

We expect, therefore, that

$$G'(x) = \lim_{t \to x} \frac{G(t) - G(x)}{t - x} = f(x)$$

Theorem 4.8.2 **Fundamental Theorem of Calculus, First Form** *Let f be a real function, continuous on an interval I containing the number a. Let $G(x)$ be defined* *for $x \in I$ by*

$$G(x) = \int_a^x f$$

Then G is differentiable on I and

$$G'(x) = f(x) \qquad x \in I$$

PROOF: The proof follows closely the geometric ideas preceding the statement of the theorem. Observe first that the integral defining G exists because of the continuity of f.

For $x \in I$, we calculate $G'(x)$ starting from the definition:

$$\begin{aligned} G'(x) &= \lim_{t \to x} \frac{G(t) - G(x)}{t - x} \\ &= \lim_{t \to x} \frac{\int_a^t f - \int_a^x f}{t - x} \\ &= \lim_{t \to x} \frac{\int_x^t f}{t - x} \qquad \text{by the additive property} \end{aligned}$$

To evaluate this limit, we investigate the sequential limit

$$\lim_{n \to \infty} \frac{\int_x^{t_n} f}{t_n - x}$$

where $\{t_n\}_{n=1}^{\infty}$ is an arbitrary sequence in $I \dashv \{x\}$ converging to x. The Mean Value Theorem for Integrals states that for each $n \in Z^+$ there is a number ξ_n between x and t_n such that

$$\int_x^{t_n} f = f(\xi_n)(t_n - x)$$

Therefore,

$$\lim_{n \to \infty} \frac{\int_x^{t_n} f}{t_n - x} = \lim_{n \to \infty} f(\xi_n)$$

Since ξ_n is between x and t_n ($n \in Z^+$) and since $t_n \to x$, the Squeeze Theorem (Theorem 2.3.8) tells us that $\xi_n \to x$, and by the continuity of f at x, we have

$$G'(x) = \lim_{n \to \infty} f(\xi_n) = f(x) \qquad \blacksquare$$

Theorem 4.8.2 means that the derivative of the integral is the original function—that is, that differentiation and integration are inverse processes.

Example 1: Let $f(t) = t/(t^2 + 1)$ and let $G(x) = \int_0^x f$. Then

$$G'(x) = \frac{x}{x^2 + 1}$$

We now consider the second aspect of the Fundamental Theorem, which leads to the familiar calculus method of evaluating integrals.

Theorem 4.8.3 **Fundamental Theorem of Calculus, Second Form** *Let f be a real function, continuous on an interval I containing the numbers a and b. Let F be any function differentiable on I such that $F'(x) = f(x)$ for $x \in I$. Then*

$$\int_a^b f = F(b) - F(a)$$

PROOF: We know from the first form of the Fundamental Theorem that there is at least one function defined on I whose derivative is f. Indeed,

$$G(x) = \int_a^x f$$

has the property that $G'(x) = f(x)$ for $x \in I$. Now if F is any other function whose derivative on I is also f, then $G'(x) = F'(x)$ for $x \in I$, and by Corollary 4.3.1, F and G differ by a constant there; that is,

$$G(x) = F(x) + C \qquad x \in I$$

Since $G(a) = 0$, we see that $C = -F(a)$ and so

$$G(x) = F(x) - F(a) \qquad x \in I$$

By letting $x = b$, we get

$$\int_a^b f = G(b) = F(b) - F(a) \qquad \blacksquare$$

Example 2: To evaluate $\int_0^{\pi/2} \cos x \, dx$, we look first for a function F whose derivative is the cosine function; the function $F(x) = \sin x$ has this property, so that

$$\int_0^{\pi/2} \cos x \, dx = \sin \frac{\pi}{2} - \sin 0 = 1$$

EXERCISES 4.8

1. Let f be continuous on $[a, b]$ where $a < b$. Show that if m is the minimum value of f on $[a, b]$, and if $\int_a^b f = m(b - a)$, then $f(x) = m$ for all $x \in [a, b]$.

2. Find the average value of the function on the indicated interval.
   (a) $f(x) = x^2$; $[0, b]$ $(b > 0)$
   (b) $f(x) = x^2$; $[-b, b]$ $(b > 0)$

(c) $F(x) = Ax + B$; $[a, b]$ $(a < b)$
(d) $f(x) = x^3$; $[a, b]$ $(a < b)$

3. Give an example of a real function bounded and integrable on a closed interval such that the conclusion of the Mean Value Theorem for Integrals does not hold.

4. Determine those values of x for which the given function G is differentiable and find its derivative.

(a) $G(x) = \int_1^x (t^2 + 1)\,dt$

(b) $G(x) = \int_0^x \cos(t^2)\,dt$

(c) $G(x) = \int_1^{x^2} \cos(t^2)\,dt$

(d) $G(x) = \int_{-x^2}^{x^2} \cos t\,dt$

(e) $G(x) = \int_0^x [t]\,dt$
([ ] denotes the greatest integer function)

(f) $G(x) = \int_{-1}^x |t|\,dt$

(g) $G(x) = \int_0^x \sqrt{1 - t^2}\,dt$

5. Verify that if f is continuous everywhere and if a and b are constants, then the functions

$$\int_a^x f(t)\,dt \qquad \text{and} \qquad \int_b^x f(t)\,dt$$

have the same derivative. By what constant do they differ?

6. Let f be bounded on an interval I and integrable on every closed subinterval of I, and let a be a fixed point in I. Define

$$G(x) = \int_a^x f \qquad x \in I$$

(a) Show that G is uniformly continuous on I.
(b) Show that the relation $G'(x) = f(x)$ holds at every point of continuity of f in I.

★ 4.9 *Antiderivatives*

The Second Form of the Fundamental Theorem immediately calls for investigation of "antidifferentiation".

Definition 4.9.1 If F and f are real functions defined on an interval I and if F is differentiable with derivative given by

$$F'(x) = f(x) \qquad x \in I$$

then F is an ***antiderivative*** of f on I.

It is customary to denote $F(x)$ by $\int f(x)\,dx$, which is also called an *indefinite integral* or *integral** of f. Because of Theorem 4.8.3, once an antiderivative of f has been found, the definite integral is obtained by an easy substitution and subtraction:

$$\int_a^b f = F(b) - F(a) \tag{1}$$

The right-hand side of Equation (1) is frequently written

$$F(x)\Big|_a^b \quad \text{or} \quad \int f(x)\,dx\Big|_a^b.$$

The vertical bar means "evaluate and subtract".

*By far the most descriptive terminology for F is *antiderivative*, but we will adhere to tradition and often use the term *integral* as well.

Example 1: Since $F(x) = x^3 + 2x$ is an antiderivative of $f(x) = 3x^2 + 2$, the integration of f on $[0, 2]$ is carried out as follows:

$$\int_0^1 (3x^2 + 2)\, dx = \int (3x^2 + 2)\, dx \Big|_0^2$$
$$= (x^3 + 2x)\Big|_0^2 = 12 - 0 = 12$$

Recall from elementary calculus that a "constant of integration" is added to an antiderivative in order to obtain the most general antiderivative of a function. The situation is summarized precisely in the next theorem, which follows immediately from Corollary 4.3.1 and the fact that the derivative of a constant is zero.

Theorem 4.9.1 *If a real function $f(x)$ has an antiderivative $F(x)$ on an interval I, then, for an arbitrary constant $C \in R$, the function*

$$G(x) = F(x) + C \tag{2}$$

is also an antiderivative of $f(x)$ on I; moreover, every antiderivative of $f(x)$ on I is given by Equation (2) for some $C \in R$.

Techniques of integration (more precisely, techniques of antidifferentiation) are studied extensively in elementary calculus. We will not dwell on these procedures here, except to mention the two most important: (1) integration by parts and (2) substitution. It should come as no surprise that each of these methods follows from a corresponding derivative property.

Integration by Parts

The product rule for derivatives (see Exercise 4 in Section 4.1) says that, for functions f and g differentiable on an interval I, the derivative of the product

$$h(x) = f(x)g(x)$$

is given by

$$h'(x) = f(x)g'(x) + f'(x)g(x)$$

Thus,

$$f(x)g'(x) = h'(x) - f'(x)g(x)$$

and so

$$\int f(x)g'(x)\, dx = \int h'(x)\, dx - \int f'(x)g(x)\, dx$$
$$= h(x) - \int f'(x)g(x)\, dx$$

giving us the "integration by parts" formula

$$\int f(x)g'(x)\,dx = f(x)g(x) - \int f'(x)g(x)\,dx \tag{3}$$

Equation (3) allows us to "exchange" the integral of fg' for the integral of $f'g$, which may be simpler. Here is a familiar example:

Example 2: Consider $\int x \cos x\,dx$, which is of the form $\int f(x)g'(x)\,dx$, where $f(x) = x$ and $g(x) = \sin x$. We obtain*

$$\begin{aligned}\int x \cos x\,dx &= x \sin x - \int \sin x\,dx \\ &= x \sin x + \cos x + C\end{aligned}$$

Note: Equation (3) is often written in the form

$$\int u\,dv = uv - \int v\,du \tag{4}$$

in which the *differential* of a function $u(x)$ is defined to be $du = u'(x)\,dx$.

Integration by Substitution (Change of Variables)

The method of substitution, or change of variables, is familiar to any calculus student.

Example 3: To evaluate

$$\int_0^4 \frac{2x}{\sqrt{9 + x^2}}\,dx$$

we let $u = 9 + x^2$, so that $du = 2x\,dx$; then $u = 9$ when $x = 0$, and $u = 25$ when $x = 4$. Making these substitutions in the given integral gives us a new one, which we evaluate by the second form of the Fundamental Theorem:

$$\int_9^{25} \frac{du}{\sqrt{u}} = 2\sqrt{u}\,\Big|_9^{25} = 10 - 6 = 4$$

Justification of the procedure in Example 3 is given by Theorem 4.9.2 below. It is a straightforward application of the Chain Rule (Theorem 4.4.2) and the Fundamental Theorem.

Definition 4.9.2 A real function is ***continuously differentiable*** on an interval provided that it has a continuous derivative there.

*In the examples and exercises of this section we make free use of certain familiar identities and facts relating to the trigonometric and inverse trigonometric functions. These will be discussed in Section 4.12.

Theorem 4.9.2 **Substitution** *Let f and u be real functions with the following properties:*

(i) *u is continuously differentiable on the interval* $I = [a, b]$ *where* $a < b$

(ii) *f is continuous on the interval* $J = u(I)$

Then the following integrals exist and are equal:

$$\int_a^b f(u(x))\, u'(x)\, dx = \int_{u(a)}^{u(b)} f(u)\, du$$

PROOF: By Theorem 3.6.6, $J = u(I)$ is a closed bounded interval. The definite integrals in the equation exist because the integrands are continuous on the required intervals. Now let G be an antiderivative of f on J, so that $G'(u) = f(u)$. Then, by the Fundamental Theorem,

$$\int_{u(a)}^{u(b)} f(u)\, du = G(u(b)) - G(u(a))$$

Next, define $g(x) = G(u(x))$ for all $x \in [a, b]$; the Chain Rule applies to this composite function, so that

$$\begin{aligned} g'(x) &= G'(u(x))u'(x) \\ &= f(u(x))u'(x) \qquad x \in [a, b] \end{aligned}$$

Thus,

$$\begin{aligned} \int_a^b f(u(x))u'(x)\, dx &= g(b) - g(a) \\ &= G(u(b)) - G(u(a)) \\ &= \int_{u(a)}^{u(b)} f(u)\, du \qquad \blacksquare \end{aligned}$$

In Example 3 we made a "direct" substitution into the integral—that is, the integrand was immediately expressible as a function of the new variable $u = 9 + x^2$ and its derivative. Sometimes we wish to make an "indirect" substitution by setting x equal to a function of some other variable. For instance, consider the trigonometric substitution in the next example.

Example 4: To evaluate

$$\int_0^3 \frac{dx}{9 + x^2}$$

we let $x = 3 \tan \theta$, so that $dx = 3 \sec^2\theta\, d\theta$. Then $\theta = 0$ when $x = 0$, and $\theta = \frac{\pi}{4}$ when $x = 3$. By using the identity $\sec^2\theta = 1 + \tan^2\theta$, the integral is transformed as follows:

$$\int_0^{\pi/4} \frac{3 \sec^2\theta\, d\theta}{9 + 9 \tan^2\theta} = \int_0^{\pi/4} \frac{1}{3}\, d\theta = \frac{\pi}{12}$$

Setting $x = 3 \tan \theta$ in Example 4 is justified because for each $x \in [0, 3]$, there is a θ satisfying $x = 3 \tan \theta$; in fact, θ can be determined by means of the equation $\theta = \tan^{-1}\left(\frac{x}{3}\right)$. The values of θ lie in the interval $\left[0, \frac{\pi}{4}\right]$. The point to observe here is that the original substitution function ($x = 3 \tan \theta$) has an inverse.

Theorem 4.9.3 *Let f be continuous on the closed interval $I = [a, b]$ where $a < b$. Let J be a closed bounded interval, and let ψ be a continuously differentiable function mapping J onto I with $\psi'(\theta) \neq 0$ for $\theta \in J$. Then*

$$\int_a^b f(x)\, dx = \int_\alpha^\beta f(\psi(\theta))\psi'(\theta)\, d\theta$$

where $\alpha = \psi^{-1}(a)$ and $\beta = \psi^{-1}(b)$.

PROOF: The fact that $\psi'(\theta) \neq 0$ for $\theta \in J$ tells us that ψ is strictly monotone and hence has an inverse. Thus, $\alpha = \psi^{-1}(a)$ and $\beta = \psi^{-1}(b)$ are well defined. There are two cases:

(i) If ψ is increasing, then so is ψ^{-1}, and therefore $\alpha < \beta$. Then, by Theorem 4.9.2,

$$\int_\alpha^\beta f(\psi(\theta))\psi'(\theta)\, d\theta = \int_{\psi(\alpha)}^{\psi(\beta)} f(x)\, dx = \int_a^b f(x)\, dx$$

(ii) If ψ is decreasing, then $\beta < \alpha$, and by Theorem 4.9.2,

$$\int_\beta^\alpha f(\psi(\theta))\psi'(\theta)\, d\theta = \int_{\psi(\beta)}^{\psi(\alpha)} f(x)\, dx = \int_b^a f(x)\, dx$$

so that $\displaystyle\int_\alpha^\beta f(\psi(\theta))\psi'(\theta)\, d\theta = \int_a^b f(x)\, dx$ ■

EXERCISES 4.9

1. Use integration by parts to find the following antiderivatives.

(a) $\int x\, e^x\, dx$

(b) $\int x \sin x\, dx$

(c) $\int \ln x\, dx$

(d) $\int \cos mx \sin nx\, dx$

(e) $\int \sec^3 x\, dx$

2. Verify the following "reduction" formulas, in which n is a positive integer and a is a nonzero real number.

(a) $\displaystyle\int x^n\, e^{ax}\, dx = \frac{1}{a} x^n e^{ax} - \frac{n}{a}\int x^{n-1} e^{ax}\, dx$

(b) $\displaystyle\int \sin^n x\, dx = -\frac{1}{n}\sin^{n-1} x \cos x + \frac{n-1}{n}\int \sin^{n-2} x\, dx \qquad n \geq 2$

(c) $\displaystyle\int \sec^n x\, dx = \frac{\sec^{n-2} x \tan x}{n-1} + \frac{n-2}{n-1}\int \sec^{n-2} x\, dx \qquad n \geq 2$

3. Show that if f is continuous everywhere and has period p, then, for every $a \in R$,

$$\int_a^{a+p} f = \int_0^p f$$

4. Let f be continuous on $[-a, a]$ where $a \geq 0$. Use a simple substitution to verify the following statements.

(a) If f is even, then $\int_{-a}^0 f = \int_0^a f$, so that for an even function f, $\int_{-a}^a f = 2\int_0^a f$.

(b) If f is odd, then $\int_{-a}^0 f = -\int_0^a f$, so that for an odd function f, $\int_{-a}^a f = 0$.

5. Let f be continuously differentiable and invertible on the interval $[a, b]$, and suppose that $f(a) = \alpha$ and $f(b) = \beta$.

(a) Show that

$$\int_\alpha^\beta f^{-1}(y)\,dy = \int_a^b xf'(x)\,dx$$

(b) Integrate the right member of part (a) by parts. Obtain a geometric interpretation of your result for the case in which $0 < a < b$ and $0 < \alpha < \beta$.

(c) Part (a) indicates that

$$\int_0^1 \tan^{-1}y\,dy = \int_0^{\pi/4} x\sec^2 x\,dx$$

and

$$\int_1^e \ln y\,dy = \int_0^1 x\,e^x\,dx$$

Verify these equations by integrating both sides.

★ 4.10 *Elementary Functions: I. The Natural Logarithmic and Exponential Functions*

Virtually all branches of mathematics that depend on calculus use the natural exponential function e^x and its inverse function $\ln x$, the natural logarithm. In elementary calculus it is typical to introduce the exponential function first and then to define the logarithm as its inverse. The basic rules for differentiation

$$\frac{d}{dx}(\ln x) = \frac{1}{x} \qquad x > 0 \tag{1}$$

$$\frac{d}{dx}(e^x) = e^x \tag{2}$$

are subsequently developed; in the process it is usually assumed that these functions are not only well defined but also continuous.

A theoretical development such as ours requires precise definitions of these functions, and derivations of their properties that do not use unproved assumptions.

The number e was introduced in Section 2.6, and in Section 3.7 we defined e^x for *rational* x; namely,

$$e^x = e^{p/q} = \sqrt[q]{e^p}$$

where $x = \frac{p}{q}$ is in lowest terms. This definition is fine for rational x, but clearly it is of no use when x is irrational. We are now going to construct an extension of the rational exponential function that is defined and differentiable for all real numbers with derivative given by Equation (2).

We approach the construction of the exponential function "through the back door", so to speak, since we begin by obtaining its inverse, the natural logarithm. Equation (1) provides one clue as to how we can define the natural logarithm: its derivative should be $\frac{1}{x}$. Is there a function, definable in terms of the analytic concepts at our disposal, which has the property that its derivative, for $x > 0$, is $\frac{1}{x}$? Yes, the first form of the Fundamental Theorem of Calculus tells us that every function continuous on an interval has an antiderivative there—specifically, if f is continuous on I, then for a fixed $a \in I$, the function defined for $x \in I$ by

$$F(x) = \int_a^x f(t)\, dt$$

is an antiderivative of f. As a special case, the integral $\int_1^x \frac{1}{t}\, dt$ is an antiderivative of $\frac{1}{x}$ for $x > 0$, and we take it as our definition of the natural logarithm. (The choice $a = 1$ has the desirable effect of giving the logarithm the value 0 when $x = 1$.) We denote this function temporarily by $L(x)$, to avoid presuming it has the very properties we wish to establish.

The Natural Logarithm

Definition 4.10.1 The ***natural logarithm*** L is the real function defined for $x > 0$ by

$$L(x) = \int_1^x \frac{1}{t}\, dt$$

The following theorem indicates that L behaves in the manner expected of a logarithm.

Theorem 4.10.1 *The natural logarithm $L : (0, \infty) \to R$ has the following properties:*

(i) $L(1) = 0$
(ii) $L'(x) = 1/x \qquad x > 0$
(iii) L *is strictly increasing and continuous on* $(0, \infty)$
(iv) $L(ab) = L(a) + L(b) \qquad a, b > 0$
(v) $L(1/a) = -L(a) \qquad a > 0$
(vi) $L(b/a) = L(b) - L(a) \qquad a, b > 0$
(vii) $L(a^r) = rL(a) \qquad r \in Q, a > 0.$

PROOF: As noted, the choice of 1 as lower limit ensures that $L(1) = 0$, and the first form of the Fundamental Theorem of Calculus tells us that $L'(x) = \frac{1}{x} (x > 0)$. Then property (iii) follows immediately from (ii) by part (ii) of Theorem 4.3.2. To obtain property (iv), we apply the Chain Rule (Theorem 4.4.2) to differentiate the composite function

$$H(x) = L(ax) \qquad a, x > 0$$

We obtain

$$H'(x) = L'(ax) \cdot a = \frac{1}{ax} \cdot a = \frac{1}{x} = L'(x)$$

Because H and L have the same derivative on $(0, \infty)$, they differ by a constant (Corollary 4.3.1):

$$H(x) = L(x) + C$$

Since $H(1) = L(a)$ and $L(1) = 0$, we see that

$$L(a) = 0 + C$$

Thus, for all $x \in (0, \infty)$,

$$L(ax) = L(a) + L(x)$$

and property (iv) follows by letting $x = b$. The next two properties are established as follows: since $L\left(a \cdot \frac{1}{a}\right) = L(1) = 0$, we have $L(a) + L\left(\frac{1}{a}\right) = 0$, so that $L\left(\frac{1}{a}\right) = -L(a)$; furthermore,

$$L\left(\frac{b}{a}\right) = L\left(b \cdot \frac{1}{a}\right) = L(b) + L\left(\frac{1}{a}\right) = L(b) - L(a)$$

The proof of property (vii) is outlined in Exercise 4.10.1. ∎

The next theorem makes an important connection with Euler's number e.

Theorem 4.10.2 *(i)* $L(e) = 1$
(ii) $L(e^r) = r \qquad r \in \mathbb{Q}$

PROOF: Recall the definition of e given in Section 2.6:

$$e = \lim_{n\to\infty} \left(1 + \frac{1}{n}\right)^n$$

Thus,

$$L(e) = L\left[\lim_{n\to\infty} \left(1 + \frac{1}{n}\right)^n\right]$$

and because L is continuous on $(0, \infty)$,

$$L(e) = \lim_{n\to\infty} L\left[\left(1 + \frac{1}{n}\right)^n\right]$$

By property (vii) of Theorem 4.10.1,

$$L(e) = \lim_{n\to\infty} \left[nL\left(1 + \frac{1}{n}\right)\right]$$

Now the sequence $\left\{nL\left(1 + \frac{1}{n}\right)\right\}_{n=1}^{\infty}$ has a limit we can recognize by rewriting it:

$$nL\left(1 + \frac{1}{n}\right) = \frac{L\left(1 + \frac{1}{n}\right)}{\frac{1}{n}} = \frac{L\left(1 + \frac{1}{n}\right) - L(1)}{\left(1 + \frac{1}{n}\right) - 1}$$

The terms of this sequence are values of the difference quotient

$$\frac{L(x_n) - L(1)}{x_n - 1}$$

corresponding to the sequence $\{x_n\}_{n=1}^{\infty} = \left\{1 + \frac{1}{n}\right\}_{n=1}^{\infty}$, which converges to 1. Thus, $\left\{nL\left(1 + \frac{1}{n}\right)\right\}_{n=1}^{\infty}$ converges to $L'(1) = 1$, and we have established property (i). Now, property (ii) is a consequence of this result and property (vii) of Theorem 4.10.1. ∎

Since L is strictly increasing, it maps $(0, \infty)$ one-to-one into R. That the range of L includes arbitrarily large positive and negative values follows from the fact that $L(e^n) = nL(e) = n$ and $L(e^{-n}) = -nL(e) = -n$ for $n \in Z^+$. By the Intermediate Value Theorem (Theorem 3.6.2), we conclude that the range of L is all of R, and so $L : (0, \infty) \rightarrow R$ is a bijection. Moreover, it is easy to show that L has the limit properties stated in the following theorem. The graph of L appears in Figure 4.9.

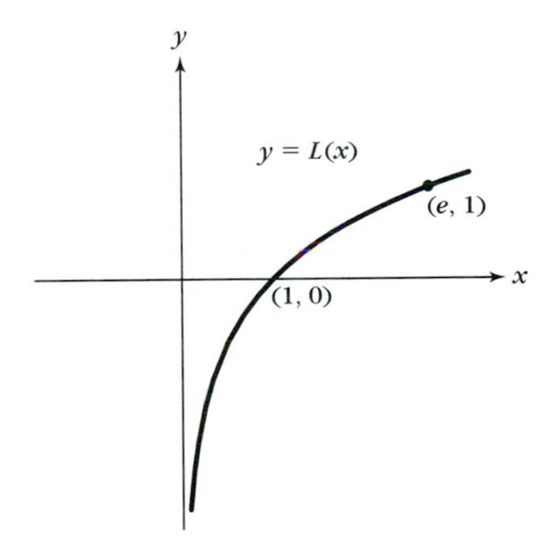

Figure 4.9
Natural logarithm function

Theorem 4.10.3 *(i)* $\lim_{x \to \infty} L(x) = \infty$

(ii) $\lim_{x \to 0^+} L(x) = -\infty$

PROOF: You are asked to prove this theorem in Exercise 4.10.2. ∎

The Natural Exponential Function

Since $L : (0, \infty) \to R$ is a bijection, L has an inverse, which we denote by E.

Definition 4.10.2 The ***natural exponential function*** E is the inverse of the natural logarithm; that is, $E = L^{-1}$.

Because L is strictly increasing, so also is E (Theorem 3.8.1); moreover, since L maps $(0, \infty)$ one-to-one onto R, E maps R one-to-one onto $(0, \infty)$. The graph of E can be obtained from the graph of L by reflection through the line $y = x$ (see Figure 4.10).

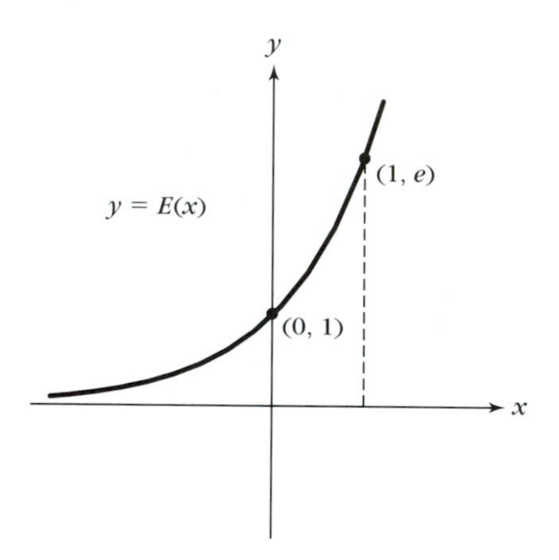

Figure 4.10
Natural exponential function

Theorem 4.10.4 *The natural exponential function* $E : R \to (0, \infty)$ *has the following properties:*

(i) $E(0) = 1$

(ii) $E'(x) = E(x) \qquad x \in R$

(iii) E *is strictly increasing and continuous on* R

(iv) $E(u + v) = E(u)E(v) \quad u, v \in R$

(v) $E(-v) = \dfrac{1}{E(v)} \qquad v \in R$

(vi) $E(u - v) = \dfrac{E(u)}{E(v)} \qquad u, v \in R$

PROOF: By definition, $y = E(x)$ if and only if $x = L(y)$. Thus, property (i) follows from the fact that $L(1) = 0$. Since L is differentiable with nonzero derivative on $(0, \infty)$, we can apply Theorem 4.3.7, concluding that E is differentiable on R with

$$E'(y) = \frac{1}{L'[E(y)]} = \frac{1}{1/E(y)} = E(y) \qquad y \in R$$

If we replace y by x, we obtain property (ii), and (iii) follows from the fact that $E'(x)$ is positive for $x \in R$. Property (iv) is really just property (iv) of Theorem 4.10.1 in exponential notation: Let $a = E(u)$ and $b = E(v)$, so that $u = L(a)$ and $v = L(b)$. By property (iv) of Theorem 4.10.1, we have

$$u + v = L(ab)$$

If we write this equation in exponential form, we obtain

$$E(u + v) = ab = E(u)E(v)$$

Now properties (v) and (vi) follow easily from (iv). ∎

Because E is the inverse of L, and $L(e^r) = r$ for $r \in Q$, it follows that $E(r) = e^r$ for $r \in Q$. This fact is important because it tells us that the new function E is an extension of the "old" (rational) exponential function. Of course, E is an improvement because it is not only defined on all of R but is differentiable, hence also continuous, there. Indeed, E is the *only* continuous real function that extends the rational exponential function to R (answer Exercise 4.10.3).

The following limit properties follow from Theorem 4.10.3:

Theorem 4.10.5 *(i)* $\lim_{x \to \infty} E(x) = \infty$

(ii) $\lim_{x \to -\infty} E(x) = 0$

Henceforth, we will denote the natural exponential and logarithmic functions in the traditional way:

$$E(x) = e^x \quad \text{and} \quad L(x) = \ln x$$

Since they are inverses, we always have

$$e^{\ln x} = x \quad x > 0 \qquad \text{and} \qquad \ln(e^x) = x \quad (x \in R)$$

Other Bases

Having defined e^x for $x \in R$, and $\ln x$ for $x > 0$, we can introduce exponential and logarithmic functions with other bases. Suppose a is a positive real number not equal to 1. Keeping in mind that $a = e^{\ln a}$, we define a^x as follows:

Definition 4.10.3 If $a > 0$ and $a \neq 1$, the ***exponential function with base*** a is defined by

$$a^x = e^{x \ln a} \qquad x \in R$$

It is easy to see that a^x has many properties like those for e^x in Theorem 4.10.4. An exception is that in place of property (ii), we get, by the Chain Rule,

$$\frac{d}{dx}(a^x) = \frac{d}{dx}(e^{x \ln a}) = e^{x \ln a} \ln a = a^x \ln a \tag{3}$$

It follows that if $0 < a < 1$, then $\ln a < 0$ and a^x is strictly decreasing rather than increasing (work Exercises 4.10.4 and 4.10.5).

A familiar algebraic property of exponents can now be derived in general. (You are asked for the proof in Exercise 4.10.6.)

Theorem 4.10.6 *If a is a positive real number different from 1, then, for all x, y ∈ R,*

$$(a^x)^y = a^{xy}$$

Of course, the inverse of the exponential function with base a is known as the logarithmic function with base a; it is usually denoted $\log_a$. By definition, $y = \log_a x$ if and only if $a^y = x$.

It is seldom necessary or even desirable to use bases other than the natural base e. For a^x is immediately expressible in terms of the natural base by means of the definition, and its derivative is obtainable as in Equation (3). Similarly, general logarithms can be expressed in terms of natural logarithms by means of the equation

$$\log_a x = \frac{\ln x}{\ln a}$$

(Verify this in Exercise 4.10.7.) It follows that

$$\frac{d}{dx}(\log_a x) = \frac{d}{dx}\left(\frac{\ln x}{\ln a}\right) = \frac{1}{\ln a} \cdot \frac{1}{x} \tag{4}$$

The "nuisance" factor $\ln a$ that occurs in formulas (3) and (4) disappears when we take $a = e$, since $\ln e = 1$.

Let us now examine the limiting behavior of some typical real functions that involve exponentials and logarithms.

Example 1: The function $f(x) = e^{-x^2}$ is the basis of the normal distribution in statistics. For $x > 0$, this function is strictly decreasing. It approaches zero through positive values as x approaches infinity or negative infinity. Since f is even, its graph (see Figure 4.11) is symmetric with respect to the y-axis.

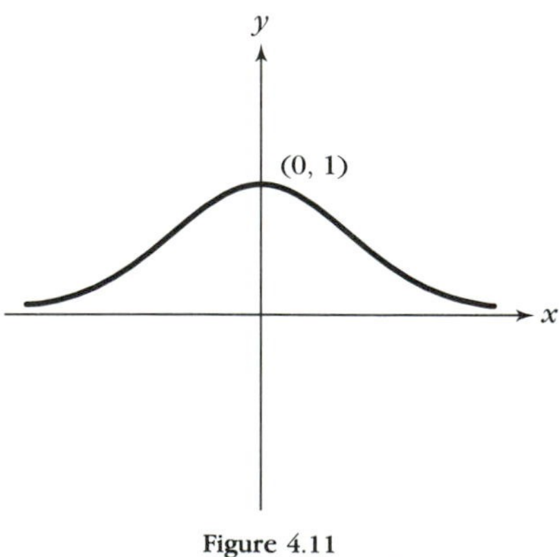

Figure 4.11
Graph of $f(x) = e^{-x^2}$

Example 2: The hyperbolic sine and cosine functions, sinh and cosh, are defined by the equations

$$\sinh x = \frac{e^x - e^{-x}}{2} \qquad \text{and} \qquad \cosh x = \frac{e^x + e^{-x}}{2}$$

and provide solutions for certain differential equations. Because $e^{-x} \to 0$ as $x \to \infty$, the graph of $\sinh x$ (see Figure 4.12a) approaches that of $y = \frac{1}{2}e^x$ from below on the right; the portion of the graph to the left of the y-axis can be obtained by symmetry because $\sinh x$ is odd. On the other hand, $\cosh x$ is even, and its graph approaches $y = \frac{1}{2}e^x$ from above on the right and approaches $y = \frac{1}{2}e^{-x}$ on the left. With suitable adjustment of scale, the graph of $\cosh x$ (Figure 4.12b) is the curve along which a cable would lie if suspended at two points and hanging under its own weight.

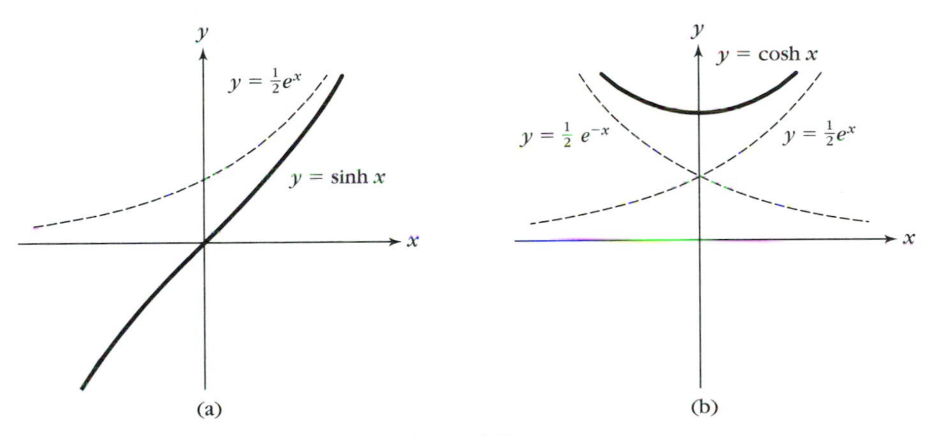

Figure 4.12
(a) The hyperbolic sine function; (b) the hyperbolic cosine function

Example 3: A graph of the function $f(x) = (\ln x)/x$ $(x > 0)$ is shown in Figure 4.13. In Exercise 4.10.10, you are asked to prove that

$$\lim_{x \to 0} f(x) = -\infty \qquad \text{and} \qquad \lim_{x \to \infty} f(x) = 0$$

An absolute maximum occurs when $x = e$, $f(e) = \frac{1}{e}$ (work Exercise 4.10.11).

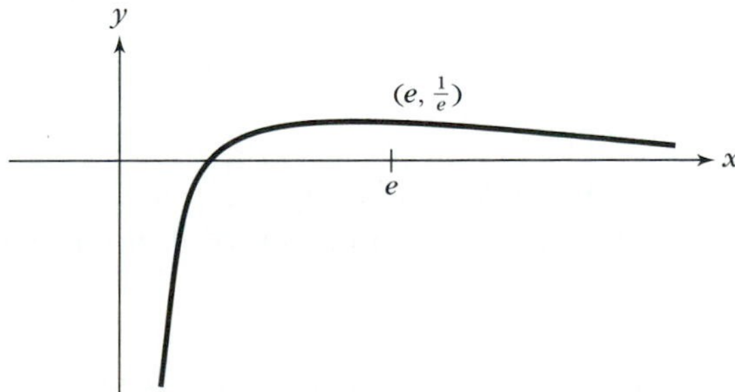

Figure 4.13
Graph of $f(x) = \dfrac{\ln x}{x}$ $\quad x > 0$

Example 4: A graph of $f(x) = e^{1/x}$ $(x \neq 0)$ exhibits the following limit properties (see Figure 4.14):

$$\lim_{x \to -\infty} e^{1/x} = \lim_{x \to \infty} e^{1/x} = 1$$
$$\lim_{x \to 0^+} e^{1/x} = \infty \qquad \text{and} \qquad \lim_{x \to 0^-} e^{1/x} = 0$$

You are asked to verify these properties in Exercise 4.10.14. This function is of interest because it has a discontinuity with an infinite limit on one side of zero and a finite limit on the other side.

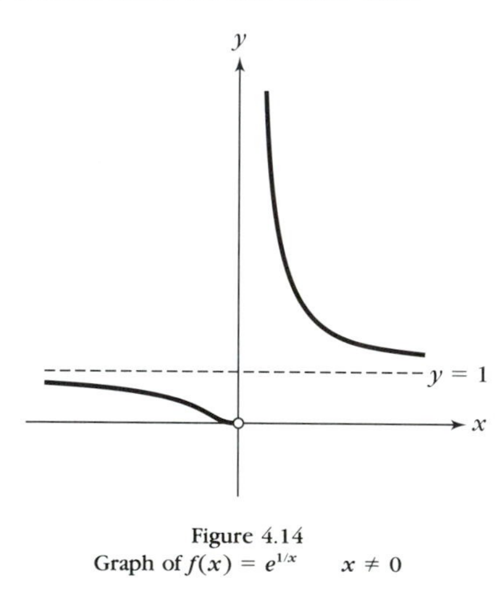

Figure 4.14
Graph of $f(x) = e^{1/x}$ $\quad x \neq 0$

Logarithmic Differentiation

Sometimes it is convenient to take the logarithm of a function prior to obtaining the derivative. The following is a well-known example:

Example 5: Let us find the derivative of $f(x) = x^x$ $(x > 0)$. Now

$$\ln f(x) = x \ln x$$

so that

$$\frac{f'(x)}{f(x)} = x \cdot \frac{1}{x} + \ln x$$
$$f'(x) = (1 + \ln x)f(x)$$
$$= (1 + \ln x)x^x$$

The result of Exercise 4.4.9 shows that the power rule $(d/dx)(x^p) = px^{p-1}$ holds whenever p is *rational*. Logarithmic differentiation enables us to extend this rule to arbitrary real powers.

Theorem 4.10.7 **Power Rule** *If p is any fixed real number, then the function*

$$f(x) = x^p \qquad x > 0$$

is differentiable and

$$f'(x) = px^{p-1}$$

PROOF: By taking the logarithm of both sides, we obtain

$$\ln f(x) = p \ln x$$
$$\frac{f'(x)}{f(x)} = \frac{p}{x}$$
$$f'(x) = \frac{p}{x} \cdot f(x) = \frac{p}{x} x^p = px^{p-1} \quad \blacksquare$$

EXERCISES 4.10

1. The following steps lead to a proof of property (vii) of Theorem 4.10.1. In all parts, a is assumed to be a fixed positive real number.
   (a) Show by mathematical induction that the equation in property (iv) can be extended to n terms. That is, show that for $n \in Z^+$ and $a_1, a_2, \ldots, a_n > 0$, we have
   $$L(a_1a_2 \cdots a_n) = L(a_1) + L(a_2) + \cdots + L(a_n)$$
   (b) Use the result in part (a) to conclude that if $n \in Z^+$, then $L(a^n) = nL(a)$.
   (c) If $q \in Z^+$, the product $a^{1/q}a^{1/q} \cdots a^{1/q}$ (consisting of q factors all equal to $a^{1/q}$) is equal to a, so that
   $$L(a) = L(a^{1/q}a^{1/q} \cdots a^{1/q}) = qL(a^{1/q})$$
   Conclude that $L(a^{1/q}) = \frac{1}{q}L(a)$, and proceed to show that for $p, q \in Z^+$,
   $$L(a^{p/q}) = \frac{p}{q}L(a)$$
   (d) The result of part (c) is that $L(a^r) = rL(a)$, where r is a positive rational number. Now apply property (v) of Theorem 4.10.1 to conclude that $L(a^r) = rL(a)$ for all $r \in Q$.

2. Prove Theorem 4.10.3.

3. Explain why E is the only continuous real function that extends the rational exponential function to R.

4. Show that $\ln a < 0$ for $0 < a < 1$; and that $\ln a > 0$ for $a > 1$.

5. Let a be a fixed positive real number with $a \neq 1$. Verify the following properties of the function $A(x) = a^x$ $(x \in R)$:
   (i) $a^0 = 1$
   (ii) $A'(x) = \dfrac{d}{dx}(a^x) = a^x \ln a \qquad x \in R$
   (iii) a^x is strictly increasing if $a > 1$ (decreasing if $a < 1$) and continuous on R
   (iv) $a^{u+v} = a^u a^v \qquad u, v \in R$
   (v) $a^{-u} = \dfrac{1}{a^u} \qquad u \in R$
   (vi) $a^{u-v} = \dfrac{a^u}{a^v} \qquad u, v \in R$

6. Prove Theorem 4.10.6. (*Hint:* Let $b = a^x$ and show that $b^y = e^{y \ln b} = e^{xy \ln a}$.)

7. Show that for $a > 0$, $a \neq 1$,
   $$\log_a x = \frac{\ln x}{\ln a} \qquad x > 0$$

8. Refer to the functions in Example 2 to verify the following relations.
   (a) $\dfrac{d}{dx}(\sinh x) = \cosh x \qquad x \in R$
   (b) $\dfrac{d}{dx}(\cosh x) = \sinh x \qquad x \in R$
   (c) $\dfrac{d}{dx}(\tanh x) = \text{sech}^2 x \qquad x \in R$

   where $\tanh x$ is defined to be $\sinh x/\cosh x$, and $\text{sech}\, x$ is defined to be $1/\cosh x$.

9. Compare the integral $\int_1^x dt/t$ with $\int_1^x dt/\sqrt{t}$, and show that $\ln x \leq 2(\sqrt{x} - 1)$ for $x > 1$.

10. Refer to the function $f(x) = (\ln x)/x$ $(x > 0)$ in Example 3 to verify the following limits.

(a) $\lim_{x\to 0^+} f(x) = -\infty$

(b) $\lim_{x\to\infty} f(x) = 0$ (*Hint:* Use the result of Exercise 9.)

11. Show that $f(x) = (\ln x)/x$ $(x > 0)$ assumes an absolute maximum at $x = e$.

12. Let $f(x) = x \ln x$ $(x > 0)$. Verify the following facts and draw the graph of f.

(a) $\lim_{x\to\infty} f(x) = \infty$

(b) $\lim_{x\to 0^+} f(x) = 0$ (*Hint:* Use Exercise 10b, substituting $x = 1/y$.)

(c) f assumes an absolute minimum at $x = 1/e$.

13. (a) Show that $e^x > x^2$ for $x \geq 1$. [*Hint:* Consider the function $h(x) = e^x - x^2$; show that $h(1)$ is positive and that $h''(x)$ and $h'(x)$ are positive for $x \geq 1$.]

(b) Show that $\lim_{x\to\infty} xe^{-x} = 0$ and $\lim_{x\to\infty} \frac{e^x}{x} = \infty$.

(c) Draw the graph of $f(x) = xe^{-x}$ $(x \in R)$

(d) Draw the graph of $g(x) = e^x/x$ $(x \neq 0)$.

14. Verify the limit properties of $e^{1/x}$ stated in Example 4.

15. **Logarithmic Differentiation** Find the derivative of each of the following functions by taking the logarithm before differentiating.

(a) $x^{\sin x}$ $\quad 0 < x < \pi$

(b) $x^{1/x}$ $\quad x > 0$

(c) x^{x^2} $\quad x > 0$

(d) x^{x^x} $\quad x > 0$

16. Show that the series $\sum_{n=1}^{\infty} \frac{(-1)^{n+1}}{n\sqrt[n]{n}}$ converges.

[*Hint:* To prove that $\left\{\frac{1}{n\sqrt[n]{n}}\right\}_{n=1}^{\infty}$ is decreasing, consider the function $f(x) = x^{1+1/x}$ and use logarithmic differentiation.]

★ 4.11 *Improper Integrals of the First Type*

The definition of the Riemann integral of f on $[a, b]$ requires that f be bounded on $[a, b]$. There are situations in which it is useful to relax this restriction, giving rise to what is called an "improper" integral of the first type. (A second type of improper integral will be discussed in Section 4.13.)

Example 1: The function

$$f(x) = \begin{cases} \dfrac{1}{\sqrt{1-x}} & \text{if } 0 \leq x < 1 \\ 0 & \text{if } x = 1 \end{cases}$$

is defined, but unbounded, on the interval $[0, 1]$. The upper sum of f relative to any partition of $[0, 1]$ is therefore undefined, and so is the Riemann integral of f on $[0, 1]$. However, if $0 < c < 1$, then f is integrable on $[0, c]$, and we can denote its integral there by $I(c)$:

$$I(c) = \int_0^c f = \int_0^c \frac{dx}{\sqrt{1-x}}$$

Geometrically, $I(c)$ represents the area of the region shaded in Figure 4.15.

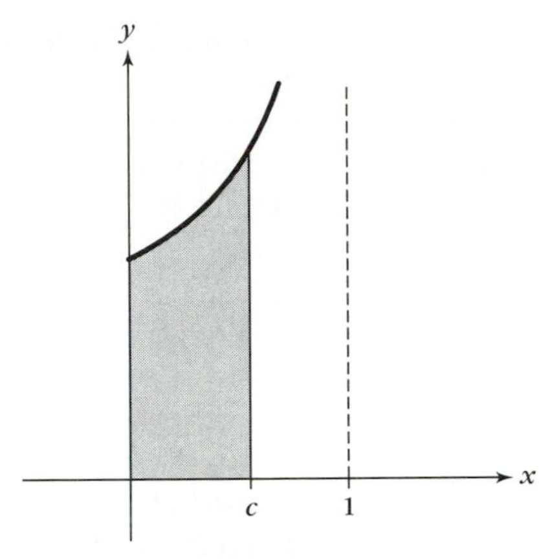

Figure 4.15
$I(c) = \int_0^c \frac{1}{\sqrt{1-x}}\,dx$

The second form of the Fundamental Theorem of Calculus (Theorem 4.8.3) gives

$$I(c) = 2 - 2\sqrt{1-c}$$

and if we let $c \to 1^-$, we find that

$$\lim_{c \to 1^-} I(c) = 2$$

It is natural to define $\int_0^1 f$ by this limit; that is,

$$\int_0^1 f = \lim_{c \to 1^-} I(c) = \lim_{c \to 1^-} \int_0^c f = 2$$

In so doing we have extended the definition of the integral to a situation in which it was previously undefined.

Definition 4.11.1 Let f be a real function that is bounded and integrable on every interval of the form $[a, c]$, where $a < c < b$, but unbounded on $[a, b]$ itself. If the limit

$$\lim_{c \to b^-} \int_a^c f \tag{1}$$

exists, then the ***improper integral***

$$\int_a^b f \tag{2}$$

is said to ***converge***, and we define its value to be the limit (1). If the limit (1) fails to exist, then the improper integral (2) is said to ***diverge***.

The result of Example 1 is that the improper integral from 0 to 1 of the function f defined there converges and its value is 2.

Note: Neither the convergence nor the value of the improper integral (2) depends on the value of the function at the endpoint in whose neighborhood it becomes unbounded. Therefore, when it is convenient to do so, we will simply leave the function undefined at such a point.

Example 2: Let f be given by

$$f(x) = \frac{1}{(x-2)^2}$$

so that f is undefined at 2. Consider the improper integral $\int_1^2 f$. By the Fundamental Theorem, we get, for $1 < c < 2$,

$$\int_1^c f = -\frac{1}{c-2} + 1$$

which approaches infinity as $c \to 2^-$. Thus, the limit required for convergence does not exist, and the improper integral $\int_1^2 f$ diverges.

Of course, it is possible to consider an improper integral whose integrand is unbounded in the neighborhood of the left endpoint of the interval; here we must examine

$$\lim_{c \to a^+} \int_c^b f$$

Example 3: Let $f(x) = 1/\sqrt{x}$ $(0 < x \le 1)$. Then, for $c > 0$,

$$\int_c^1 \frac{dx}{\sqrt{x}} dx = 2 - 2\sqrt{c}$$

so that

$$\lim_{c \to 0^+} \int_c^1 \frac{dx}{\sqrt{x}} dx = \lim_{c \to 0^+} (2 - 2\sqrt{c}) = 2$$

Thus, $\int_0^1 dx/\sqrt{x}$ converges to 2.

The following theorem is reminiscent of comparison tests studied in connection with infinite series. We state and prove this theorem for the case in which the function is unbounded at the right end of the interval; of course, a similar result holds for functions unbounded at the left end.

Theorem 4.11.1 *Let f and g be real functions that are bounded and integrable on every interval $[a, c]$, where $a < c < b$, but unbounded on $[a, b]$. Suppose that $0 \le g(x)$*

$\le f(x)$ for $a \le x < b$. If $\int_a^b f$ converges, then $\int_a^b g$ also converges, and we have $\int_a^b g \le \int_a^b f$; if $\int_a^b g$ diverges, then so does $\int_a^b f$.

PROOF: Let $a < c < b$ and define $I(c) = \int_a^c g$ and $J(c) = \int_a^c f$. Since $0 \le g(x) \le f(x)$ for $x \in [a, b)$, Corollary 4.7.3 tells us that

$$I(c) \le J(c) \qquad a < c < b \tag{3}$$

Because f and g are nonnegative on $[a, b)$, it follows that I and J are both increasing functions of c. (Why?) Assuming that $\int_a^b f$ converges, $J(c)$ is bounded above for $c \in (a, b)$, and we can apply Theorem 3.8.3; as $c \to b^-$, $J(c)$ increases to its least upper bound:

$$J(c) \to \sup_{c \in (a,\, b)} J(c) = \lim_{c \to b^-} J(c) = \int_a^b f$$

From Inequality (3) it follows that $I(c)$ is also bounded for $c \in (a, b)$:

$$I(c) \le J(c) \le \int_a^b f$$

and so, by Theorem 3.8.3, $I(c)$ has a limit as $c \to b^-$:

$$\lim_{c \to b^-} I(c) = \sup_{c \in (a,\, b)} I(c) \le \sup_{c \in (a,\, b)} J(c) = \int_a^b f$$

That is,

$$\int_a^b g \le \int_a^b f \qquad \blacksquare$$

Example 4: The function

$$f(x) = \frac{1}{\sqrt{1 - x^2}} \qquad 0 \le x < 1$$

has a convergent improper integral on $[0, 1]$ because, for $x \in [0, 1)$,

$$0 < \frac{1}{\sqrt{1 - x^2}} = \frac{1}{\sqrt{1 - x}\sqrt{1 + x}} \le \frac{1}{\sqrt{1 - x}}$$

and we have already seen that the integral $\int_0^1 dx/\sqrt{1 - x}$ converges to 2. Therefore, $\int_0^1 dx/\sqrt{1 - x^2}$ converges to a value less than or equal to 2.

If a function is unbounded at both ends of $[a, b]$ but is bounded and integrable on every interval $[c, d]$ with $a < c < d < b$, we define the improper integral from a to b to be the sum of *two* improper integrals:

$$\int_a^b f = \int_a^m f + \int_m^b f \tag{4}$$

where m is a real number such that $a < m < b$. Then the integral on the left is considered to be convergent if and only if both integrals on the right are convergent; its value is the sum of their values.

Note: It can be shown (work Exercise 4.11.21) that this definition does not depend on the particular choice of the number m in Equation (4).

Example 5: The integral

$$\int_{-1}^{1} \frac{x\,dx}{(x^2 - 1)^{2/3}}$$

is improper because the integrand is unbounded near both endpoints of the interval $[-1, 1]$. Its convergence can be tested by choosing $m = 0$ and writing

$$\int_{-1}^{1} \frac{x\,dx}{(x^2 - 1)^{2/3}} = \int_{-1}^{0} \frac{x\,dx}{(x^2 - 1)^{2/3}} + \int_{0}^{1} \frac{x\,dx}{(x^2 - 1)^{2/3}}$$

Now,

$$\int_{0}^{1} \frac{x\,dx}{(x^2 - 1)^{2/3}} = \lim_{c \to 1^-} \int_{0}^{c} \frac{x\,dx}{(x^2 - 1)^{2/3}} = \lim_{c \to 1^-} \left[\frac{3}{2}(x^2 - 1)^{1/3} \Big|_0^c \right] = \frac{3}{2}$$

and a similar calculation shows that

$$\int_{-1}^{0} \frac{x\,dx}{(x^2 - 1)^{2/3}} = -\frac{3}{2}$$

Therefore, the given integral converges, and its value is zero.

Example 6: The integral

$$\int_{-1}^{1} \frac{x\,dx}{(x^2 - 1)^2}$$

diverges because the integrals

$$\int_{-1}^{0} \frac{x\,dx}{(x^2 - 1)^2} \quad \text{and} \quad \int_{0}^{1} \frac{x\,dx}{(x^2 - 1)^2}$$

both diverge. (In Exercise 4.11.22, you are asked to show that the first of these integrals diverges to negative infinity, while the second diverges to positive infinity.)

Note: The fact that the integrand in Example 6 is odd means that, for $0 < c < 1$,

$$\int_{-c}^{c} \frac{x\,dx}{(x^2 - 1)^2} = 0$$

tempting some students to say that the positive and negative areas in Figure 4.16 cancel. Be aware that our definition of convergence does not permit this kind of "cancellation" for infinite regions.

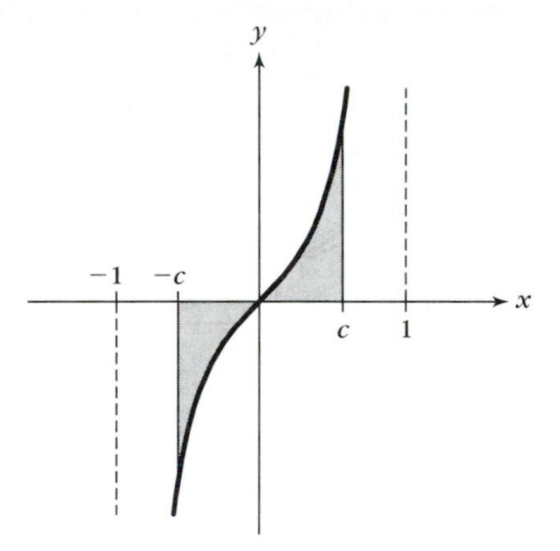

Figure 4.16
$$\int_{-c}^{c} \frac{x\,dx}{(x^2-1)^2} = 0 \text{ but } \int_{-1}^{1} \frac{x\,dx}{(x^2-1)^2} \text{ diverges}$$

If a function is unbounded in the neighborhood of a point d in the interior of $[a, b]$, but is bounded and integrable on every closed subinterval of $[a, b]$ that does not contain d, then we consider the improper integral of f on $[a, b]$ to be the sum

$$\int_a^b f = \int_a^d f + \int_d^b f$$

The integral on the left converges if and only if each integral on the right converges,* and its value is the sum of their values.

Example 7: The integral

$$\int_0^3 \frac{x\,dx}{(x^2-1)^{2/3}}$$

is improper because the integrand is unbounded near $x = 1$. Therefore, we must examine

$$\int_0^1 \frac{x\,dx}{(x^2-1)^{2/3}} \quad \text{and} \quad \int_1^3 \frac{x\,dx}{(x^2-1)^{2/3}}$$

We find

$$\lim_{c\to 1^-} \int_0^c \frac{x\,dx}{(x^2-1)^{2/3}} = \lim_{c\to 1^-} \left[\frac{3}{2}(x^2-1)^{1/3}\Big|_0^c \right] = \frac{3}{2}$$

*It is possible that one of the integrals on the right is proper; for example, if f is bounded on the left of d, then $\int_a^d f$ would be proper (work Exercise 4.11.20).

$$\lim_{c\to 1^+}\int_c^3 \frac{x\,dx}{(x^2-1)^{2/3}} = \lim_{c\to 1^+}\left[\frac{3}{2}(x^2-1)^{1/3}\Big|_c^3\right] = 3$$

so that the given integral converges to $\frac{9}{2}$.

EXERCISES 4.11

In Exercises 1–15 determine whether the improper integral converges. If it does converge, find its value.

1. $\int_0^8 \frac{dx}{\sqrt[3]{x}}$
2. $\int_0^1 \frac{dx}{\sqrt[3]{1-x}}$
3. $\int_{-1}^1 \frac{dx}{x^3}$
4. $\int_0^1 \frac{x^2+1}{x}\,dx$
5. $\int_0^1 \frac{dx}{(x-1)^2}$
6. $\int_1^4 \frac{x\,dx}{x^2-1}$
7. $\int_1^3 \frac{x\,dx}{\sqrt[3]{x^2-1}}$
8. $\int_0^1 \frac{dx}{\sqrt{x}+x}$
9. $\int_0^1 \frac{dx}{\sqrt{x}+\sqrt[3]{x}}$
10. $\int_0^1 \ln x\,dx$
11. $\int_0^1 \frac{\ln x}{\sqrt{x}}\,dx$
12. $\int_0^1 \frac{\ln x}{x}\,dx$
13. $\int_0^1 \frac{dx}{e^x - e^{-x}}$
14. $\int_0^1 \frac{dx}{x\ln x}$
15. $\int_0^1 \frac{dx}{e^x - 1}$

16. (a) Explain why the integral $\int_0^1 x\ln x\,dx$ is proper, and evaluate it.
    (b) Is the integral $\int_0^1 e^{-1/x}\,dx$ improper? Explain.

In Exercises 17–20 determine whether the given integral converges but do not try to evaluate it.

17. $\int_0^1 \frac{dx}{(x-1)^2+\sqrt{1-x}}$
18. $\int_0^1 \frac{dx}{\sqrt[3]{x^4+x}}$
19. $\int_0^1 \frac{e^x\,dx}{x^2+\sqrt{x}}$
20. $\int_{-1}^1 e^{1/x}\,dx$

21. Show that the choice of $m \in (a, b)$ in Equation (4) affects neither the question of convergence nor the value of the integral.

22. Show that the integrals

$$\int_{-1}^0 \frac{x\,dx}{(x^2-1)^2} \quad \text{and} \quad \int_0^1 \frac{x\,dx}{(x^2-1)^2}$$

diverge to negative infinity and infinity, respectively.

23. The integrals

$$I(\alpha, \beta) = \int_0^1 x^{\alpha-1}(1-x)^{\beta-1}\,dx \qquad \alpha, \beta > 0$$

arise in the study of the beta distribution in statistics.

(a) For which values of α and β are these integrals improper?
(b) Show that $I(\alpha, \beta)$ is convergent for $\alpha, \beta > 0$.
(c) Show that $I(\alpha, \beta) = I(\beta, \alpha)$ for $\alpha, \beta > 0$.

★ 4.12 *Elementary Functions: II. Trigonometric Functions*

In Section 4.10 we defined the natural exponential function in terms of its inverse, the natural logarithm. We are now going to apply a similar procedure to give an analytic definition of the sine function by means of its inverse, the arcsine. Once we have the sine function, we can define the cosine as its derivative, and the remaining four trigonometric functions can be defined in the usual way as reciprocals or quotients of sine and cosine.

In elementary calculus the derivative formula

$$\frac{d}{dx}(\arcsin x) = \frac{1}{\sqrt{1 - x^2}}$$

leads to an integral formula

$$\int_0^x \frac{1}{\sqrt{1 - t^2}}\,dt = \arcsin x \qquad -1 \le x \le 1 \tag{1}$$

Now, proceeding by analogy with our development of the logarithm, we use this integral formula to *define* the arcsine function, to establish its basic properties, and then to determine its inverse.

Notice first that the integral (1) is proper when $-1 < x < 1$ but is improper when $x = \pm 1$. However, we showed in Example 4 of Section 4.11 that

$$\int_0^1 \frac{1}{\sqrt{1 - t^2}}\,dt$$

converges to a real number L that is less than 2. Clearly, then, the integral

$$\int_0^{-1} \frac{1}{\sqrt{1 - t^2}}\,dt = -\int_{-1}^{0} \frac{1}{\sqrt{1 - t^2}}\,dt$$

converges to $-L$, and so (1) converges for $-1 \le x \le 1$.

Definition 4.12.1 The ***arcsine*** function A is defined for $x \in [-1, 1]$ by the formula

$$A(x) = \int_0^x \frac{dt}{\sqrt{1 - t^2}}$$

The real number π is defined to be $2A(1)$; that is,

$$\pi = 2\int_0^1 \frac{dt}{\sqrt{1 - t^2}}$$

Note: The number π just defined is the same number π encountered in geometry, as you will see in Exercise 4.12.4c.

Theorem 4.12.1 *The arcsine function A has the following properties:*

(i) *A is odd:* $A(-x) = -A(x) \qquad -1 \le x \le 1$

(ii) *A is continuous on $[-1, 1]$ and differentiable on $(-1, 1)$, with derivative given by*

$$A'(x) = \frac{1}{\sqrt{1 - x^2}} \qquad -1 < x < 1 \tag{2}$$

(iii) *A is strictly increasing on $[-1, 1]$ with range* $\left[-\frac{\pi}{2}, \frac{\pi}{2}\right]$

PROOF:

(i) That A is odd follows from the fact that the integrand in the definition is even (work Exercise 4.12.1).

(ii) The derivative formula (2) follows from the Fundamental Theorem of Calculus (First Form) because the integrand is continuous on $(-1, 1)$. Continuity of A at -1 and 1 follows immediately from the definition of the improper integral; for instance,

$$\lim_{x\to 1} A(x) = \lim_{x\to 1} \int_0^x \frac{dt}{\sqrt{1-t^2}} = \int_0^1 \frac{dt}{\sqrt{1-t^2}} = A(1)$$

(iii) The derivative of A is strictly positive on $(-1, 1)$, so that A is strictly increasing on $[-1, 1]$ by Theorem 4.3.2. ■

Since the function A is strictly increasing on $[-1, 1]$ (see Figure 4.17), it is invertible, and its inverse s is a strictly increasing odd function with domain $\left[-\frac{\pi}{2}, \frac{\pi}{2}\right]$. By Theorem 4.3.7, the derivative of s is given on the open interval $\left(-\frac{\pi}{2}, \frac{\pi}{2}\right)$ by

$$s'(x) = \frac{1}{A'[s(x)]} = \frac{1}{1/\sqrt{1-[s(x)]^2}} = \sqrt{1-[s(x)]^2}$$

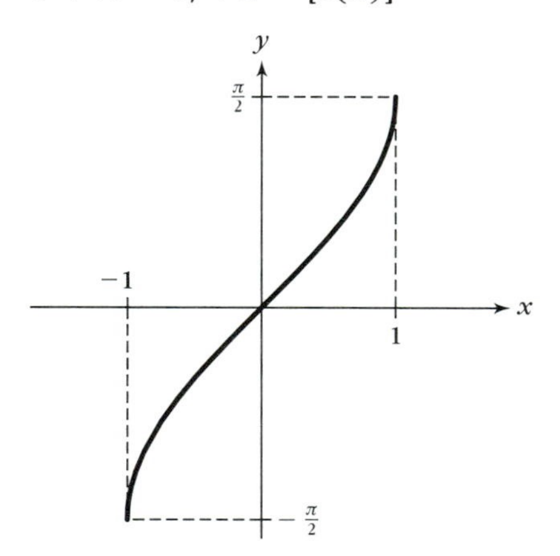

Figure 4.17
Graph of the arcsine function A

Note that Theorem 4.3.7 does not give information about s' at the endpoints $-\frac{\pi}{2}$ or $\frac{\pi}{2}$ because A' itself exists only on the open interval $(-1, 1)$. However, it can be shown (work Exercise 4.12.2) that $s'\left(\frac{\pi}{2}\right)$ and $s'\left(-\frac{\pi}{2}\right)$ both exist and are equal to zero. Thus, the formula above for s' actually holds throughout the interval $\left[-\frac{\pi}{2}, \frac{\pi}{2}\right]$:

$$s'(x) = \sqrt{1-[s(x)]^2} \qquad -\tfrac{\pi}{2} \le x \le \tfrac{\pi}{2}$$

The graph of the function s (shown in Figure 4.18) contains the essential part of the sine function.

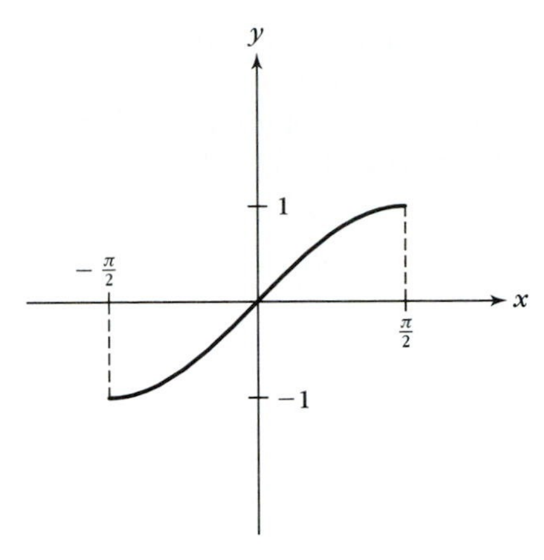

Figure 4.18
Graph of $s = A^{-1}$

Our next step is to extend the function s from the interval $\left[-\frac{\pi}{2}, \frac{\pi}{2}\right]$ to $[-\pi, \pi]$. We define an extension function S as follows:

$$S(x) = \begin{cases} s(x) & \text{if } -\frac{\pi}{2} \leq x \leq \frac{\pi}{2} \\ s(\pi - x) & \text{if } \frac{\pi}{2} < x \leq \pi \\ -s(\pi + x) & \text{if } -\pi \leq x < -\frac{\pi}{2} \end{cases}$$

There are several things to observe concerning this definition. To begin with, when $x \in \left(\frac{\pi}{2}, \pi\right]$, we have $\pi - x \in \left[0, \frac{\pi}{2}\right)$, so that $s(\pi - x)$ is defined. In defining $S(x)$ to be $s(\pi - x)$, we have been guided by a property (an identity) that we want sine to satisfy. Similarly, if $x \in \left[-\pi, -\frac{\pi}{2}\right)$, then $\pi + x \in \left[0, \frac{\pi}{2}\right)$, so that $s(\pi + x)$ is defined. The extended function is odd, for if $-x \in \left[-\pi, -\frac{\pi}{2}\right)$, then $x \in \left(\frac{\pi}{2}, \pi\right]$, and $S(-x) = -s[\pi + (-x)] = -s(\pi - x) \equiv -S(x)$. Also, we find that S assumes the expected values at special points—for example, it is easy to check that $S(-\pi) = S(0) = S(\pi) = 0$, $S\left(-\frac{\pi}{2}\right) = -1$, and $S\left(\frac{\pi}{2}\right) = 1$. Figure 4.19 shows the graph of the sine function over its basic period interval $[-\pi, \pi]$.

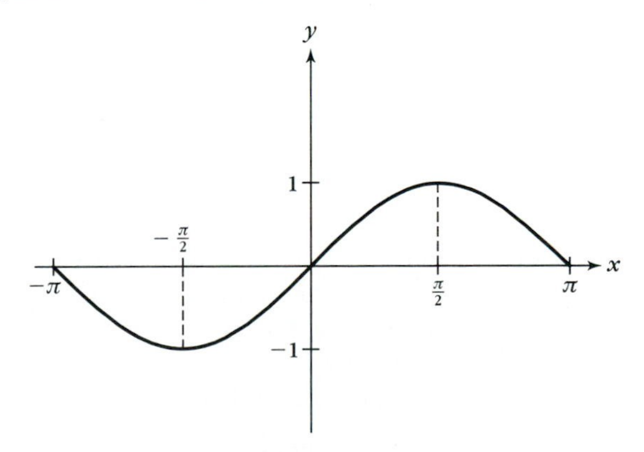

Figure 4.19
Extension of s to $[-\pi, \pi]$

It remains to extend S periodically to a function differentiable on all of R; for this purpose, of course, we will use the Periodic Extension Theorem (Theorem 4.1.3). To apply this theorem we need to verify that S is differentiable on $[-\pi, \pi]$ with $S'(-\pi) = S'(\pi)$.

On the open interval $\left(-\frac{\pi}{2}, \frac{\pi}{2}\right)$, S agrees with s, and so by Theorem 4.1.2,

$$S'(x) = s'(x) = \sqrt{1 - [s(x)]^2} \qquad \frac{-\pi}{2} < x < \frac{\pi}{2}$$

At the point $x = \frac{\pi}{2}$ we must take care to check the limit defining $S'\left(\frac{\pi}{2}\right)$ on both sides:

$$\begin{aligned}
\lim_{x \to \pi/2^-} \frac{S(x) - S(\pi/2)}{x - \pi/2} &= \lim_{x \to \pi/2^-} \frac{s(x) - s(\pi/2)}{x - \pi/2} = s'\left(\frac{\pi}{2}\right) = 0 \\
\lim_{x \to \pi/2^+} \frac{S(x) - S(\pi/2)}{x - \pi/2} &= \lim_{x \to \pi/2^+} \frac{s(\pi - x) - s(\pi/2)}{x - \pi/2} \\
&= \lim_{u \to \pi/2^-} \frac{s(u) - s(\pi/2)}{\pi/2 - u} = -s'\left(\frac{\pi}{2}\right) = 0
\end{aligned}$$

In the last step of this calculation we made a change of variable $u = \pi - x$. We conclude that $S'\left(\frac{\pi}{2}\right)$ exists and is zero.

Now we consider points in the interval $\left(\frac{\pi}{2}, \pi\right]$, where $S(x) = s(\pi - x)$. The Chain Rule applies, and (with $u = \pi - x$), we obtain

$$\begin{aligned}
S'(x) &= s'(\pi - x)(-1) \\
&= -\sqrt{1 - [s(\pi - x)]^2} \\
&= -\sqrt{1 - [S(x)]^2}
\end{aligned}$$

Similar results are obtained at $-\frac{\pi}{2}$ and on the interval $\left[-\pi, -\frac{\pi}{2}\right)$; we conclude that S is differentiable, hence continuous, on $[-\pi, \pi]$. We have thus proved the following theorem:

Theorem 4.12.2 *The function S is differentiable on the interval $[-\pi, \pi]$, and its derivative is given by*

$$S'(x) = \begin{cases} \sqrt{1 - [S(x)]^2} & \text{if } x \in \left[-\frac{\pi}{2}, \frac{\pi}{2}\right] \\ -\sqrt{1 - [S(x)]^2} & \text{if } x \in \left[-\pi, -\frac{\pi}{2}\right) \cup \left(\frac{\pi}{2}, \pi\right] \end{cases} \tag{3}$$

The derivative formula in Theorem 4.12.2 leads immediately to the identity

$$[S'(x)]^2 = 1 - [S(x)]^2 \qquad -\pi \le x \le \pi \tag{4}$$

Because $S(-\pi) = S(\pi) = 0$ and $S'(-\pi) = S'(\pi), = -1$, we can extend S to all real numbers in such a way that the extension is differentiable everywhere and has period 2π (Theorem 4.1.3). We define this extension to be the *sine function*, and its derivative to be the *cosine function*. By Theorem 4.1.3, the cosine also has period 2π. We denote the values of these functions at x in the traditional way, $\sin x$ and $\cos x$.

Theorem 4.12.3 *The sine and cosine functions have the following properties for all* $x, y \in R$:

(i) $\sin(x + 2n\pi) = \sin x$ and $\cos(x + 2n\pi) = \cos x$ for $n \in Z$
(ii) $\sin n\pi = 0$ and $\cos\left(n + \frac{1}{2}\right)\pi = 0$ for $n \in Z$
(iii) $\sin^2 x + \cos^2 x = 1$
(iv) $\dfrac{d}{dx}(\sin x) = \cos x$ and $\dfrac{d}{dx}(\cos x) = -\sin x$
(v) $\sin(x + y) = \sin x \cos y + \sin y \cos x$
(vi) $\cos(x + y) = \cos x \cos y - \sin x \sin y$

PROOF: Many of the properties stated in this theorem follow immediately from the discussion above. The fundamental identity, property (iii), follows from Equation (4) together with the fact that sine and cosine both have period 2π. The first part of property (iv) is just our definition of cosine, while the second follows upon differentiating both sides of the equation in property (iii), using the Chain Rule (work Exercise 4.12.7). The identity in (v) is obtained by letting a be a fixed real number and differentiating the function defined for $x \in R$ by

$$f(x) = \sin x \cos(a - x) + \sin(a - x) \cos x \tag{5}$$

The derivative of f is zero everywhere, hence f is constant by Theorem 4.3.3. The constant value of f can be determined by setting $x = 0$:

$$f(0) = 0 + \sin a$$

Now, by letting $y = a - x$, we get $a = x + y$, and so property (v) follows:

$$\sin(x + y) = f(0) = \sin x \cos y + \sin y \cos x$$

To prove property (vi), we simply differentiate (v) with respect to x, treating y as a constant. ∎

Other Trigonometric Functions

The remaining trigonometric functions are defined as usual:

tangent: $\tan x = \dfrac{\sin x}{\cos x}$ $\quad x \neq \left(n + \dfrac{1}{2}\right)\pi, \quad n \in Z$

secant: $\sec x = \dfrac{1}{\cos x}$ $\quad x \neq \left(n + \dfrac{1}{2}\right)\pi, \quad n \in Z$

cotangent: $\cot x = \dfrac{\cos x}{\sin x}$ $\quad x \neq n\pi, \quad n \in Z$

cosecant: $\sec x = \dfrac{1}{\sin x}$ $\quad x \neq n\pi, \quad n \in Z$

Identities relating these functions, as well as derivative properties, can be derived from those pertaining to sine and cosine.

Example 1: The identity $\tan^2 x + 1 = \sec^2 x$ follows from property (iii) of Theorem 4.12.3 by dividing both sides by $\cos^2 x$.

Example 2: The tangent function is differentiable for $x \neq \left(n + \frac{1}{2}\right)\pi$, $n \in Z$. Its derivative is obtained by using the quotient rule:

$$\begin{aligned}\frac{d}{dx}(\tan x) &= \frac{d}{dx}\left(\frac{\sin x}{\cos x}\right) \\ &= \frac{\cos^2 x + \sin^2 x}{\cos^2 x} \\ &= \frac{1}{\cos^2 x} = \sec^2 x \qquad x \neq \left(n + \frac{1}{2}\right)\pi, \quad n \in Z\end{aligned}$$

Example 3: It follows from Example 2 that the tangent function is strictly increasing on $\left(-\frac{\pi}{2}, \frac{\pi}{2}\right)$. Since

$$\lim_{x \to \pi/2^-} \tan x = \infty \qquad \text{and} \qquad \lim_{x \to \pi/2^+} \tan x = -\infty$$

we conclude that the tangent function maps $\left(-\frac{\pi}{2}, \frac{\pi}{2}\right)$ one-to-one onto R.

Limits of Trigonometric Functions

The following theorem is usually derived in elementary calculus by a geometric argument. In our discussion it arises as a special case of a derivative.

Theorem 4.12.4

$$\lim_{x \to 0} \frac{\sin x}{x} = 1$$

PROOF: Let $f(x) = \sin x$ and observe that the indicated limit is just $f'(0)$, hence its value is $\cos 0 = 1$. ■

Example 4: Note that

$$\lim_{x \to 0} \frac{\tan x}{x} = 1$$

can be derived either by writing $\tan x$ as $\sin x/\cos x$ or by recognizing the limit as the derivative of the tangent function at zero.

EXERCISES 4.12

1. Show that the function A is odd.
2. Show that $s'\left(-\frac{\pi}{2}\right)$ and $s'\left(\frac{\pi}{2}\right)$ exist and are equal to zero. [*Hint:* Use the definition of derivative and apply the Mean Value Theorem (Theorem 4.3.1).]

3. Differentiate the functions of Section 4.4:

$$f_n(x) = \begin{cases} x^n \sin \frac{1}{x} & \text{if } x \neq 0 \\ 0 & \text{if } x = 0 \end{cases}$$

where $n \in Z^+$. For what values of n is f_n differentiable at zero?

4. (a) Use properties (v) and (vi) of Theorem 4.12.3 to establish the following identities:

$$\sin 2x = 2 \sin x \cos x \qquad x \in R$$
$$\cos 2x = \cos^2 x - \sin^2 x \qquad x \in R$$

(b) Use the second identity in part (a) together with the identity in property (iii) of Theorem 4.12.3 to obtain

$$\sin^2 x = \tfrac{1}{2}(1 - \cos 2x)$$
$$\cos^2 x = \tfrac{1}{2}(1 + \cos 2x)$$

(c) Use the substitution $x = a \sin \theta$ to show that

$$\int_0^a \sqrt{a^2 - x^2}\, dx = \frac{\pi a^2}{4}$$

and thereby establish the geometric connection between our definition of π and the traditional one.

5. Show that $\sin^2 x$, $\cos^2 x$, $\sec^2 x$, $\csc^2 x$, $\tan x$, and $\cot x$ all have period π.

6. (a) Show that the function

$$f(x) = \tan x \left(-\tfrac{\pi}{2} < x < \tfrac{\pi}{2}\right)$$

is strictly increasing and differentiable on that interval.

(b) The inverse g of the function in part (a) is known as the *arctangent* function. Show that g is differentiable with

$$g'(x) = 1/(1 + x^2)\ (x \in R).$$

(c) Show that:

$$\lim_{x\to\infty}(\arctan x) = \frac{\pi}{2};\ \lim_{x\to-\infty}(\arctan x) = -\frac{\pi}{2}$$

7. (a) Show that differentiation of both sides of the equation $\sin^2 x + \cos^2 x = 1$ yields

$$\frac{d}{dx}(\cos x) = -\sin x$$

except at points where $\cos x = 0$.

(b) Show that the derivative formula also holds at the exceptional points, that is, at the points $\left(n + \frac{1}{2}\right)\pi$ for $n \in Z$. (*Hint:* Use the definition of the derivative and apply the Mean Value Theorem.)

8. Evaluate the following limits.

(a) $\displaystyle\lim_{x\to 0} \frac{\sin 2x}{x}$

(b) $\displaystyle\lim_{x\to 0} \frac{\tan x}{\sin x}$

(c) $\displaystyle\lim_{x\to 0} x \sin \frac{1}{x^2}$

(d) $\displaystyle\lim_{x\to 0^+} \frac{\sin x}{\sqrt{1 - \cos x}}$

(e) $\displaystyle\lim_{x\to 0^-} \frac{\sin x}{\sqrt{1 - \cos x}}$

9. Observe that $f(x) = \sin x/\sqrt{1 - \cos x}$ is discontinuous at $x = 0$. Sketch the graph of f over the interval $[-\pi, \pi]$, and then evaluate the following integrals.

(a) $\displaystyle\int_0^\pi \frac{\sin x}{\sqrt{1 - \cos x}}\, dx$

(b) $\displaystyle\int_{-\pi}^0 \frac{\sin x}{\sqrt{1 - \cos x}}\, dx$

(c) $\displaystyle\int_{-\pi}^\pi \frac{\sin x}{\sqrt{1 - \cos x}}\, dx$

Are any of these integrals improper?

10. Show that the functions sine and cosine are both bounded on R and that they are both uniformly continuous on R.

11. Determine whether the following functions are uniformly continuous on the given intervals. Justify your conclusions.

(a) $f(x) = (x + \sin x)^2$; $[-\pi, \pi]$
(b) $f(x) = \tan x$; $\left[-\frac{\pi}{4}, \frac{\pi}{4}\right]$
(c) $f(x) = \tan x$; $\left(-\frac{\pi}{2}, \frac{\pi}{2}\right)$
(d) $f(x) = \sin(x^2)$; $[-\sqrt{\pi}, \sqrt{\pi}]$
(e) $f(x) = \sin(x^2)$; R
(f) $f(x) = \sin(1/x)$; $(0, 1]$
(g) $f(x) = \sin x/x$; $(0, 1]$
(h) $f(x) = x \sin(1/x)$; $(0, 1]$
(i) The arctangent function defined as in part (b) Exercise 6; R.

12. (a) Let F be continuous on $[a, b]$ where $a < b$. Suppose that F is continuously differentiable on (a, b) with $F'(x) = f(x)$ there. Show that

$$\int_a^b f = F(b) - F(a)$$

regardless of whether $\int_a^b f$ is proper or improper.

(b) Use part (a) to evaluate $\displaystyle\int_0^1 \frac{dx}{\sqrt{x}}$.

★ 4.13 *Improper Integrals of the Second Type*

The definition of the Riemann integral of a real function requires not only the boundedness of the function but also the boundedness of the interval of integration. In this section we extend the integral concept to allow an infinite interval of integration (under appropriate conditions), giving rise to improper integrals of the second type.

Definition 4.13.1 Let a be a fixed real number and let f be bounded and integrable on every interval of the form $[a, b]$, where $b > a$. Let $I(b) = \int_a^b f$. If

$$\lim_{b \to \infty} I(b) \qquad (1)$$

exists, then the improper integral

$$\int_a^\infty f \qquad (2)$$

is said to ***converge***, and we define its value to be the limit (1). If the limit (1) fails to exist, then the improper integral (2) is said to ***diverge***.

Note: The "partial integral" $I(b) = \int_a^b f$, whose limit determines whether or not the integral converges, depends both on b and on the function f. For a nonnegative function, this partial integral increases with b and must either converge or diverge to infinity (by Theorem 3.8.4). The next two examples illustrate these possibilities.

Example 1: If $f(x) = \frac{1}{x^2}$, then the improper integral $\int_1^\infty f$ converges, and its value is 1:

$$\lim_{x \to \infty} \int_1^b \frac{dx}{x^2} = \lim_{b \to \infty} \left. \frac{-1}{x} \right|_1^b = 1$$

Example 2: If $g(x) = \frac{1}{x}$, then the improper integral $\int_1^\infty g$ diverges:

$$\lim_{b \to \infty} \int_1^b \frac{1}{x} = \lim_{b \to \infty} \ln x \Big|_1^b = \lim_{b \to \infty} \ln b = \infty$$

Figure 4.20 shows the graphs of the functions f and g in Examples 1 and 2. Convergence in Example 1 is due to the rapidity with which $f(x)$ approaches zero as x increases; even though $g(x)$ also approaches zero as x increases, it does so more slowly than $f(x)$, and the partial integral of g from 1 to b approaches infinity. The question then arises: How rapidly must a function approach zero in order for its integral to converge? A partial answer is given by the following theorem.

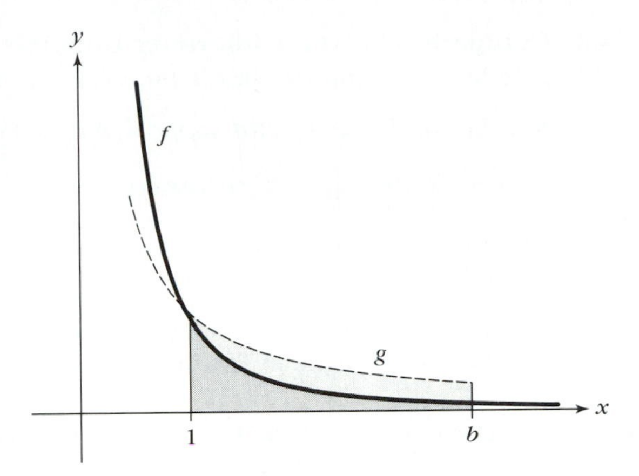

Figure 4.20
Shaded areas show partial integrals of f and g

Theorem 4.13.1 ***p* Integrals** *For $p > 1$, the integral $\int_1^\infty dx/x^p$ converges to $1/(p-1)$. For $p \leq 1$, the integral $\int_1^\infty dx/x^p$ diverges.*

PROOF: If $p \neq 1$, we have

$$\int_1^b \frac{dx}{x^p} = \left.\frac{x^{1-p}}{1-p}\right|_1^b = \frac{1-b^{1-p}}{p-1}$$

If $p > 1$, $b^{1-p} \to 0$ as $b \to \infty$, and the integral converges as stated. If $p < 1$, $b^{1-p} \to \infty$ as $b \to \infty$, and the integral diverges. The case $p = 1$ was discussed in Example 2. ■

Oscillation

Example 3: $\int_0^\infty \cos x\, dx$ diverges because

$$\lim_{b\to\infty} \int_0^b \cos x\, dx = \lim_{b\to\infty} \sin b\Big|_0^b = \lim_{b\to\infty} \sin b$$

which does not exist. This example shows that an integral can diverge without becoming infinite. The partial integral, like the integrand, oscillates between -1 and 1, without approaching any limit.

Comparisons

The following comparison test for integrals of nonnegative functions can be proved by the same method as Theorem 4.11.1. You are asked to prove it in Exercise 4.13.1a.

Theorem 4.13.2 **Comparison Test for Improper Integrals of the Second Type** *Let f and g be bounded and integrable on every interval $[a, b]$, where a is a fixed real number and $b > a$, and suppose that $0 \leq g(x) \leq f(x)$ for $x \geq a$. If $\int_a^\infty f$ converges, then $\int_a^\infty g$ also converges and $\int_a^\infty g \leq \int_a^\infty f$; if $\int_a^\infty g$ diverges, then so does $\int_a^\infty f$.*

Example 4: The improper integral $\int_1^\infty dx/(1 + x^2)$ converges by comparison with the integral in Example 1.

Theorem 4.13.2 can be also used to establish divergence.

Example 5: The improper integral $\int_2^\infty dx/\sqrt{x^2 - 1}$ diverges by comparison with $\int_2^\infty dx/x$.

Of course, it is possible to consider the convergence of an integral over an interval of the type $(-\infty, b]$ by means of $\lim_{a \to -\infty} \int_a^b f$. Theorem 4.13.2 can be modified in an obvious way to permit comparisons in this situation (work Exercise 4.13.1b).

Example 6: $\int_{-\infty}^{-1} e^x\, dx$ converges because

$$\lim_{a \to -\infty} \int_a^{-1} e^x\, dx = \lim_{a \to -\infty} (e^{-1} - e^a) = e^{-1} - 0 = e^{-1}$$

An improper integral over $(-\infty, \infty)$ converges if and only if the integrals over $(-\infty, 0]$ and $[0, \infty)$ both converge, in which case its value is defined by

$$\int_{-\infty}^{\infty} f = \int_{-\infty}^{0} f + \int_0^\infty f$$

Example 7: Consider $\int_{-\infty}^\infty dx/(1 + x^2)$. By Example 4, $\int_1^\infty dx/(1 + x^2)$ converges, and consequently so does $\int_0^\infty dx/\ (1 + x^2)$ (work Exercise 4.13.15). Arguing on the basis of symmetry, $\int_{-\infty}^0 dx/(1 + x^2)$ also converges, and so, finally, $\int_{-\infty}^\infty dx/(1 + x^2)$ converges. It can be shown that the value of this integral is π (work Exercise 4.13.16).

The following example concerns an integral that is important in probability and statistics.

Example 8: Consider $\int_{-\infty}^{\infty} e^{-x^2}\,dx$. On the interval $(-\infty, -1]$, $e^{-x^2} < e^x$ and so $\int_{-\infty}^{-1} e^{-x^2}\,dx$ converges by comparison with $\int_{-\infty}^{-1} e^x$ (Example 6). Thus,

$$\int_{-\infty}^{0} e^{-x^2}\,dx = \int_{-\infty}^{-1} e^{-x^2}\,dx + \int_{-1}^{0} e^{-x^2}\,dx$$

where the second integral is proper. On the interval $[0, \infty)$ we can appeal to symmetry because the function e^{-x^2} is even. Thus, the given integral converges, but it cannot be evaluated directly by means of the Fundamental Theorem of Calculus. By methods of multivariable calculus, its value can be shown to be $\sqrt{\pi}$. (An elementary proof is given in an article by Nicholas and Yates.*)

The Integral Test for Infinite Series

The analogy between improper integrals of the second type and infinite series is very strong indeed:

Theorem 4.13.3 **Integral Test** *Let f be nonnegative and decreasing on $[1, \infty)$ and let $\sum_{n=1}^{\infty} a_n$ be a series with the property that $a_n = f(n)$ for $n \in Z^+$. Then the improper integral $\int_1^{\infty} f$ converges if and only if the series $\sum_{n=1}^{\infty} a_n$ converges.*

Note: With minor adjustments in the statement and proof, the Integral Test applies when f is nonnegative and decreasing on $[a, \infty)$, where a is any real number. In such a case, the integral is taken over $[a, \infty)$ and the series is summed for $n \geq a$.

PROOF: Because f is nonnegative, the partial integral

$$I(b) = \int_1^b f$$

is an increasing function of b, and it follows from Corollary 3.8.3 that the improper integral $\int_1^{\infty} f$ converges if and only if the sequence $\{I(n)\}_{n=1}^{\infty}$ converges. It is therefore sufficient to show that the sequence $\{I(n)\}_{n=1}^{\infty}$ converges if and only if the infinite series $\sum_{n=1}^{\infty} a_n$ converges.

*C. P. Nicholas and R. C. Yates, "The Probability Integral", *American Mathematical Monthly* 59(1950): 412–413.

We consider two functions, σ and τ (Figures 4.21 and 4.22), defined for $x \geq 1$ as follows:

$$\left.\begin{aligned}\sigma(x) &= a_k \\ \tau(x) &= a_{k+1}\end{aligned}\right\} \quad \text{if } x \in [k, k+1), \quad \text{for } k \in Z^+$$

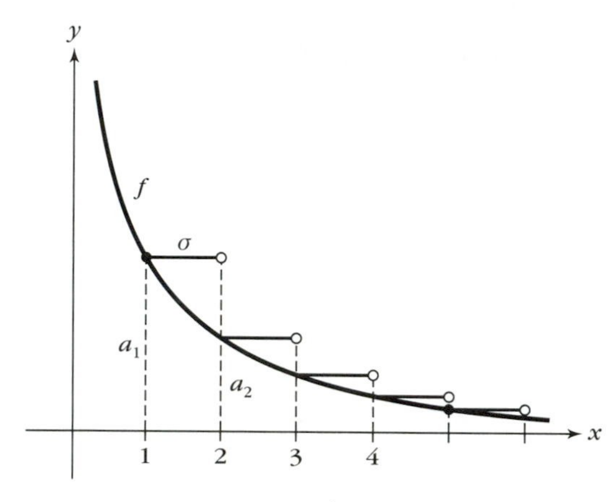

Figure 4.21
Graph of σ

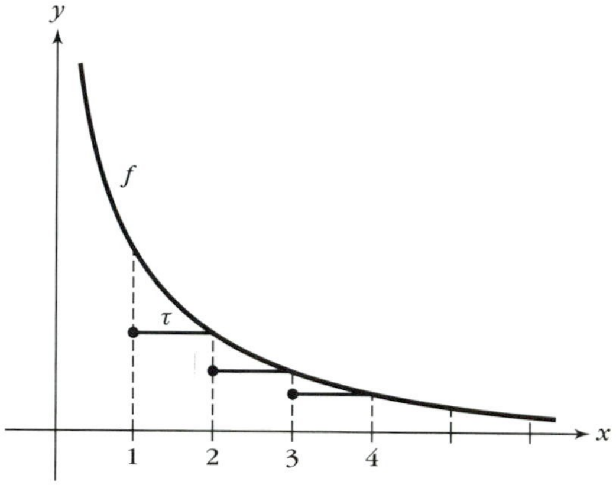

Figure 4.22
Graph of τ

On the interval $[1, n]$, where $n \in Z^+$, σ and τ are step functions with the property that

$$\tau(x) \leq f(x) \leq \sigma(x) \qquad 1 \leq x \leq n$$

Since $I(n) = \int_1^n f$, we have by Corollary 4.7.3,

$$\int_1^n \tau \leq I(n) \leq \int_1^n \sigma$$

Using Theorem 4.7.5, we can relate the integrals of τ and σ to the partial sums $\{S_n\}_{n=1}^{\infty}$ of the series $\sum_{n=1}^{\infty} a_n$:

$$\int_1^n \tau = a_2 + a_3 + \cdots + a_n = S_n - a_1; \int_1^n \sigma = a_1 + a_2 + \cdots + a_{n-1} = S_{n-1}$$

Therefore, for $n \in Z^+$, $n > 1$, we have

$$S_n - a_1 \leq I(n) \leq S_{n-1} \tag{3}$$

Since both $\{S_n\}_{n=1}^{\infty}$ and $\{I(n)\}_{n=1}^{\infty}$ are increasing sequences, the right inequality in (3) shows that the convergence of the series $\sum a_n$ implies the convergence of the sequence $\{I(n)\}_{n=1}^{\infty}$; the left inequality in (3) shows that the convergence of the sequence implies the convergence of the series. ∎

The following corollary is an immediate consequence of Theorems 4.13.1 and 4.13.3:

Corollary 4.13.1 *The p series* $\sum_{n=1}^{\infty} \frac{1}{n^p}$ *converges if and only if* $p > 1$.

Example 9: The series $\sum_{n=2}^{\infty} 1/(n \ln n)$ diverges because $\int_2^{\infty} dx/(x \ln x)$ diverges:

$$\lim_{b\to\infty} \int_2^b \frac{dx}{x \ln x} = \lim_{b\to\infty}[\ln(\ln b) - \ln(\ln 2)] = \infty$$

Example 10: The series $\sum_{n=2}^{\infty} \frac{1}{n(\ln n)^2}$ converges because $\int_2^{\infty} \frac{dx}{x(\ln x)^2}$ converges:

$$\begin{aligned} \lim_{b\to\infty} \int_a^b \frac{dx}{x(\ln x)^2} &= \lim_{b\to\infty} \left(-\frac{1}{\ln x}\Big|_2^b\right) \\ &= \lim_{b\to\infty} \left(\frac{1}{\ln 2} - \frac{1}{\ln b}\right) \\ &= \frac{1}{\ln 2} \end{aligned}$$

Example 11: The integral $\int_1^{\infty} xe^{-x}\, dx$ converges because $\sum_{n=1}^{\infty} n/e^n$ converges (by the Ratio Test).

Integrals that Are Improper Two Ways

Example 12: The integral $\int_0^{\infty} dx/x^p$ where $p > 0$ is improper in two ways. Because the integrand is unbounded near zero, and because the interval is unbounded, we must consider both terms of the sum

$$\int_0^1 \frac{dx}{x^p} + \int_1^{\infty} \frac{dx}{x^p}$$

The first integral converges only if $p < 1$ while the second converges only if $p > 1$; so $\int_0^{\infty} dx/x^p$ diverges for all $p > 0$.

EXERCISES 4.13

1. (a) Prove Theorem 4.13.2
   (b) State a theorem similar to Theorem 4.13.2 that applies to integrals over $(-\infty, b]$.

In Exercises 2–14 determine whether the integral converges; if it does, find its value.

2. $\int_0^{\infty} e^{-x}\, dx$
3. $\int_1^{\infty} \frac{dx}{x+4}$
4. $\int_0^{\infty} \frac{x\, dx}{x^2+4}$
5. $\int_2^{\infty} \frac{dx}{x^2-1}$
6. $\int_2^{\infty} \frac{dx}{x \ln x}$
7. $\int_1^{\infty} \frac{\ln x}{x}\, dx$
8. $\int_0^{\infty} xe^{-x}\, dx$
9. $\int_{-\infty}^{0} x^2e^x\, dx$
10. $\int_0^{\infty} \frac{e^x\, dx}{e^{2x}+1}$
11. $\int_0^{\infty} x \sin x\, dx$
12. $\int_{-\infty}^{\infty} \frac{(e^x - e^{-x})dx}{(e^x + e^{-x})^2}$
13. $\int_{-\infty}^{\infty} xe^{-x^2}\, dx$
14. $\int_{-\infty}^{\infty} \frac{x}{x^2+2}\, dx$

15. Show that if f is bounded and integrable on $[a, b]$ for every $b > a$, and if $c > a$, then $\int_c^\infty f$ is convergent if and only if $\int_a^\infty f$ is convergent, in which case $\int_a^\infty f = \int_a^c f + \int_c^\infty f$.

16. Show that $\int_{-\infty}^\infty dx/(1 + x^2) = \pi$. (*Hint:* See Exercise 4.12.6.)

17. Suppose that f is bounded and integrable on $[a, b]$ for every $b > a$. Then we say that $\int_a^\infty f$ *converges absolutely* provided $\int_a^\infty |f|$ converges. Show that an absolutely convergent integral is convergent and that in case of convergence $\left|\int_a^\infty f\right| \leq \int_a^\infty |f|$. (*Hint:* See Exercise 4.6.8.)

18.
(a) Show that $\int_1^\infty \frac{\cos x}{x^2}\, dx$ converges. (Use Exercise 17.)
(b) Use integration by parts together with part (a) to show that $\int_1^\infty \frac{\sin x}{x}\, dx$ converges.
(c) Does $\int_0^\infty \frac{\sin x}{x}\, dx$ converge?

In Exercises 19–26 use a comparison test to determine whether or not the integral is convergent. Do not attempt to evaluate.

19. $\int_0^\infty \frac{x\, dx}{x^3 + 8}$

20. $\int_0^\infty \frac{x^2\, dx}{x^3 + 8x + 1}$

21. $\int_1^\infty \frac{\ln x}{e^x}\, dx$

22. $\int_1^\infty \frac{\ln x}{x^2}\, dx$

23. $\int_1^\infty \frac{\ln x}{x^2 + 1}\, dx$

24. $\int_0^\infty \frac{dx}{x^2 + \sqrt{x}}$

25. $\int_0^\infty \frac{dx}{\sqrt{x^3 + x} + \sqrt{x}}$

26. $\int_0^\infty \frac{dx}{\sqrt{x^3 + x}}$

In Exercises 27–32 use the Integral Test to determine whether the given infinite series converges.

27. $\sum_{n=1}^\infty \frac{n}{\sqrt{n^2 + 1}}$

28. $\sum_{n=2}^\infty \frac{1}{n^2 - n}$

29. $\sum_{n=1}^\infty \frac{1}{n^2 + 1}$

30. $\sum_{n=2}^\infty \frac{\ln n}{n}$

31. $\sum_{n=2}^\infty \frac{\ln n}{n^2}$

32. $\sum_{n=1}^\infty \frac{1}{\sqrt{e^{2n} + 1}}$

33. Determine the values of p for which the series $\sum_{n=2}^\infty \frac{1}{n(\ln n)^p}$ converges.

★ 4.14 *L'Hospital's Rule*

The theorems known collectively as L'Hospital's Rule apply to the limiting behavior of a quotient $f(x)/g(x)$ when the limiting behavior of $f(x)$ and $g(x)$ is known. The limiting behavior of x itself may be any one of several types: $\lim_{x\to a^+}$, $\lim_{x\to a^-}$, $\lim_{x\to a}$, $\lim_{x\to\infty}$, and/or $\lim_{x\to-\infty}$. We will simply write "lim" until we need to be more specific.

Indeterminate Forms

Suppose f and g are real functions such that $\lim f(x) = L$ and $\lim g(x) = M$. If $M \neq 0$, then the limit theorems of Chapter 3 tell us that $\lim [f(x)/g(x)] = L/M$; while if $M = 0$ and $L \neq 0$, then $f(x)/g(x)$ becomes infinite and thus has no limit. But, if both L and M are zero, the limit theorems do not apply and the quotient may behave in a variety of ways. For example, by considering limits as $x \to 0$, we know that if $f(x) = x^2$ and $g(x) = x$, then $\lim [f(x)/g(x)] = 0$; but if $f(x) = x^2$ and $g(x) = x^2$, then $\lim [f(x)/g(x)] = 1$. Indeed, it is possible to find examples of functions for which $f(x)$ and $g(x)$ both approach zero and $f(x)/g(x)$ approaches an arbitrary preassigned number (or positive or negative infinity). For this reason, when $\lim f(x) = \lim g(x) = 0$, $f(x)/g(x)$ is said to be an *indeterminate form of type 0/0*. The meaning of indeterminacy is that various outcomes are possible, and in order to determine the limiting behavior of $f(x)/g(x)$ we must have more information than the mere fact that $f(x) \to 0$ and $g(x) \to 0$. The following are indeterminate forms of the 0/0 type where x assumes the indicated limiting behavior:

$$\frac{x^2 - 16}{\sqrt{x - 4}} \quad \text{as } x \to 4^+ \qquad \frac{\cos x - 1}{x} \quad \text{as } x \to 0$$

$$\frac{\ln x}{x - 1} \quad \text{as } x \to 1 \qquad x \sin \frac{1}{x} = \frac{\sin(1/x)}{1/x} \quad \text{as } x \to \infty$$

The 0/0 indeterminate form occurs very often. In fact, the existence of a derivative always arises from a 0/0 situation because, if $f'(a)$ exists, then, in the difference quotient

$$\frac{f(x) - f(a)}{x - a}$$

both numerator and denominator must approach zero as $x \to a$. (Why?)

Example 1: Taking the derivative of $f(x) = x^2$ at $a = 2$ requires the limit

$$\lim_{x \to 2} \frac{x^2 - 4}{x - 2}$$

The indeterminancy can be removed very simply by division:

$$\frac{x^2 - 4}{x - 2} = x + 2$$

showing that the difference quotient approaches 4 as x approaches 2.

Another common indeterminate form is ∞/∞. Here the problem is to evaluate the limit of $f(x)/g(x)$ when $f(x)$ and $g(x)$ both become infinite (positively or negatively) as x undergoes its limiting behavior. Again it is possible to construct examples showing that, literally, anything can happen: The quotient may approach positive or negative infinity or an arbitrary preassigned real number—or it may fail to do any of these things.

The following are indeterminate forms of the type ∞/∞:

$$\frac{e^x}{x} \quad \text{as } x \to \infty \qquad \frac{\tan x}{\sec^2 x} \quad \text{as } x \to \frac{\pi-}{2}$$

$$x \ln x = \frac{\ln x}{1/x} \quad \text{as } x \to 0^+$$

A device usually introduced in calculus for the purpose of evaluating indeterminate limits is L'Hospital's Rule. Before stating the rule formally, let us recall how it works.

To evaluate $\lim [f(x)/g(x)]$, where $f(x)/g(x)$ is indeterminate of type $0/0$ or ∞/∞, we take the derivative of the numerator and the derivative of the denominator (assuming that these exist) and attempt to evaluate the limit of $f'(x)/g'(x)$. If the latter limit exists, then so does the original limit, and they are equal. L'Hospital's Rule is remarkable because it works regardless of what kind of limiting behavior the independent variable undergoes ($x \to a$, $x \to a^+$, $x \to \infty$, and so on); and it works even if $f'(x)/g'(x)$ becomes infinite (positively or negatively). We illustrate the rule with two simple examples, representing different situations.

Example 2: Let $f(x) = \cos x - 1$, $g(x) = x$, and consider

$$\lim_{x\to 0^+} \frac{\cos x - 1}{x} = \lim_{x\to 0^+} \frac{f(x)}{g(x)}$$

which involves a $0/0$ indeterminate form. The quotient of derivatives

$$\frac{f'(x)}{g'(x)} = \frac{-\sin x}{1}$$

approaches zero as $x \to 0^+$. By L'Hospital's Rule,

$$\lim_{x\to 0^+} \frac{\cos x - 1}{x} = \lim_{x\to 0^+} \frac{-\sin x}{1} = 0$$

Example 3: Let $f(x) = e^x$, $g(x) = x$, and consider

$$\lim_{x\to\infty} \frac{e^x}{x} = \lim_{x\to\infty} \frac{f(x)}{g(x)}$$

which involves an ∞/∞ indeterminate form. The quotient of derivatives is $e^x/1 = e^x$, which approaches infinity as $x \to \infty$. By L'Hospital's Rule, $\lim_{x\to\infty} e^x/x = \infty$.

With certain understandings, L'Hospital's Rule can be stated briefly as follows:

L'Hospital's Rule: If $f(x)/g(x)$ is an indeterminate form of the type $0/0$ or ∞/∞, and if $\lim[f'(x)/g'(x)] = L$, then $\lim[f(x)/g(x)] = L$ also.

The understandings are:

(1) "lim" may indicate any type of limiting behavior of x
(2) f and g are differentiable with $g'(x) \neq 0$ on a set of numbers appropriate to the type of limiting behavior of x
(3) L may be any real number or infinity or negative infinity

We now state the rule in two cases: Theorem 4.14.1, involving the $0/0$ type as $x \to a^+$; Theorem 4.14.2, involving the ∞/∞ type as $x \to \infty$.

Theorem 4.14.1 L'Hospital's Rule for 0/0 Forms, as $x \to a^+$ *Let f and g be defined and differentiable on an open interval (a, b) with $g'(x) \neq 0$ for $a < x < b$. Suppose that $\lim_{x\to a^+} f(x) = \lim_{x\to a^+} g(x) = 0$. Then:*

(i) *If there is a real number L such that*

$$\lim_{x\to a^+} \frac{f'(x)}{g'(x)} = L$$

then also

$$\lim_{x\to a^+} \frac{f(x)}{g(x)} = L$$

(ii) *If*

$$\lim_{x\to a^+} \frac{f'(x)}{g'(x)} = \infty$$

then also

$$\lim_{x\to a^+} \frac{f(x)}{g(x)} = \infty$$

(iii) *If*

$$\lim_{x\to a^+} \frac{f'(x)}{g'(x)} = -\infty$$

then also

$$\lim_{x\to a^+} \frac{f(x)}{g(x)} = -\infty$$

Theorem 4.14.2 L'Hospital's Rule for ∞/∞ Forms as $x \to \infty$ *Let f and g be defined and differentiable on an open interval (a, ∞) with $g'(x) \neq 0$ for $x > a$. Suppose that f(x) and g(x) become infinite (positively or negatively) as $x \to \infty$. Then:*

(i) *If there is a real number L such that*

$$\lim_{x\to\infty} \frac{f'(x)}{g'(x)} = L$$

then also

$$\lim_{x\to\infty} \frac{f(x)}{g(x)} = L$$

(ii) *If*

$$\lim_{x\to\infty} \frac{f'(x)}{g'(x)} = \infty$$

then also

$$\lim_{x\to\infty} \frac{f(x)}{g(x)} = \infty$$

(iii) If

$$\lim_{x\to\infty} \frac{f'(x)}{g'(x)} = -\infty$$

then also

$$\lim_{x\to\infty} \frac{f(x)}{g(x)} = -\infty$$

Among the numerous cases remaining, note that the case involving ordinary (two-sided) limits ($x \to c$) requires the functions f and g to be differentiable with g' nonzero, on some deleted neighborhood of c.

Later in this section we will prove Theorem 4.14.1; a proof of Theorem 4.14.2 will be found in the text by Mikusiński and Mikusiński.*

In the following examples, the type of indeterminacy involved is indicated by the symbol $0/0$ or ∞/∞ in brackets beside the form. Keep in mind that each application of L'Hospital's Rule is conditional until the existence of $\lim\,[f'(x)/g'(x)]$ has been ascertained.

Example 4:

$$\begin{aligned}\lim_{x\to4^+} \frac{x^2-16}{\sqrt{x-4}} \quad \left[\frac{0}{0}\right] &= \lim_{x\to4^+} \frac{2x}{\frac{1}{2}(x-4)^{-1/2}} \\ &= \lim_{x\to4^+} 4x\sqrt{x-4} \\ &= 0\end{aligned}$$

Example 5:

$$\lim_{x\to1} \frac{\ln x}{x-1} \quad \left[\frac{0}{0}\right] = \lim_{x\to1} \frac{1/x}{1} = 1$$

Sometimes a function that is not given as a quotient can be rewritten.

Example 6:

$$\begin{aligned}\lim_{x\to\infty} x \sin\frac{1}{x} &= \lim_{x\to\infty} \frac{\sin(1/x)}{1/x} \quad \left[\frac{0}{0}\right] \\ &= \lim_{x\to\infty} \frac{(-1/x^2)\cos(1/x)}{-1/x^2} \\ &= \lim_{x\to\infty} \cos\frac{1}{x} = 1\end{aligned}$$

*Jan Mikusiński and Piotr Mikusiński, *An Introduction to Analysis* (New York: Wiley, 1993), pp. 167–169.

Example 7:

$$\lim_{x\to 0^+} x \ln x = \lim_{x\to 0^+} \frac{\ln x}{1/x} \quad \left[\frac{\infty}{\infty}\right]$$
$$= \lim_{x\to 0^+} \frac{1/x}{-1/x^2}$$
$$= \lim_{x\to 0^+} (-x) = 0$$

Example 8:

$$\lim_{x\to \pi/2^-} \frac{\tan x}{\sec^2 x} \quad \left[\frac{\infty}{\infty}\right] = \lim_{x\to \pi/2^-} \frac{\sec^2 x}{2 \sec^2 x \tan x}$$
$$= \lim_{x\to \pi/2^-} \frac{\cot x}{2} = 0$$

Sometimes several applications of the rule are required.

Example 9:

$$\lim_{x\to\infty} \frac{x^3}{e^x} \quad \left[\frac{\infty}{\infty}\right] = \lim_{x\to\infty} \frac{3x^2}{e^x} \quad \left[\frac{\infty}{\infty}\right]$$
$$= \lim_{x\to\infty} \frac{6x}{e^x} \quad \left[\frac{\infty}{\infty}\right]$$
$$= \lim_{x\to\infty} \frac{6}{e^x} = 0$$

Example 10:

$$\lim_{x\to 0} \frac{\cos x - 1 + (x^2/2)}{x^4} \quad \left[\frac{0}{0}\right] = \lim_{x\to 0} \frac{-\sin x + x}{4x^3} \quad \left[\frac{0}{0}\right]$$
$$= \lim_{x\to 0} \frac{-\cos x + 1}{12x^2} \quad \left[\frac{0}{0}\right]$$
$$= \lim_{x\to 0} \frac{\sin x}{24x} \quad \left[\frac{0}{0}\right]$$
$$= \lim_{x\to 0} \frac{\cos x}{24} = \frac{1}{24}$$

In order to prove Theorem 4.14.1, we first need to establish a generalization of the Mean Value Theorem, usually attributed to Cauchy:

Theorem 4.14.3 **Generalized Mean Value Theorem** *Let f and g be real functions continuous on $[a, b]$ and differentiable on (a, b), where $a < b$. If $g'(x) \neq 0$ for $x \in (a, b)$, then there exists a number $\xi \in (a, b)$ such that*

$$\frac{f(b) - f(a)}{g(b) - g(a)} = \frac{f'(\xi)}{g'(\xi)} \tag{1}$$

PROOF: Note that this theorem really is a generalization of the Mean Value Theorem in the sense that if we take $g(x) = x$, Equation (1) reduces to the conclusion of that theorem. Indeed, the proof of the new theorem can be obtained by "generalizing" the proof of the old one. Let $F(x)$ be defined on $[a, b]$ as follows:

$$F(x) = f(x) - f(a) - \frac{f(b) - f(a)}{g(b) - g(a)}[g(x) - g(a)]$$

It is easy to show that F satisfies the hypotheses of Rolle's Theorem (Theorem 4.2.2); therefore, there exists a number $\xi \in (a, b)$ such that $F'(\xi) = 0$. But

$$F'(\xi) = f'(\xi) - \frac{f(b) - f(a)}{g(b) - g(a)}g'(\xi)$$

and (1) follows immediately. ■

Note: In the study of parametric equations, the slope of the chord connecting the points of the curve $x = g(t)$, $y = f(t)$, corresponding to $t = a$ and $t = b$, is given by the quotient on the left side of Equation (1), while the right side gives the slope of the tangent line at $t = \xi$. So, once again we see that, under appropriate hypotheses, the tangent line corresponding to a suitable value ξ is parallel to the chord.

With Theorem 4.14.3 at our disposal, we can prove L'Hospital's Rule for the case described in Theorem 4.14.1.

PROOF OF THEOREM 4.14.1: Since the values of f and g at a itself have nothing to do with the values of any limits as $x \to a^+$, we can assume that $f(a) = g(a) = 0$. Then f and g are continuous from the right at a, and we can apply the Generalized Mean Value Theorem to f and g on any interval of the form $[a, x]$, where $x \in (a, b)$.

For case (i), let $\{x_n\}_{n=1}^{\infty}$ be an arbitrary sequence converging to a with $x_n \in (a, b)$. For each $n \in Z^+$, we can apply the Generalized Mean Value Theorem on $[a, x_n]$. Thus, there exists $\xi_n \in (a, x_n)$ such that

$$\frac{f'(\xi_n)}{g'(\xi_n)} = \frac{f(x_n) - f(a)}{g(x_n) - g(a)} = \frac{f(x_n)}{g(x_n)}$$

Since $\xi_n \to a$, $f'(\xi_n)/g'(\xi_n) \to L$, so that $f(x_n)/g(x_n) \to L$, and case (i) is proved. Cases (ii) and (iii) are established in the same way. ■

With obvious modifications, the statement and proof of Theorem 4.14.1 can be adapted to the situation in which $x \to a^-$. The case $x \to a$ then follows easily. The case in which $x \to \infty$ is left for Exercise 4.14.31.

Other Indeterminate Forms

There are several other indeterminate situations. For example, $0 \cdot \infty$ refers to a situation in which a function is the product of two functions, one of which approaches zero while the other approaches infinity. You should convince yourself by finding examples that this situation really is indeterminate. Another indeterminate situation is $\infty - \infty$. By means of algebra, it may be possible to reduce an indeterminate form of one of these types algebraically to one of the previous types.

Example 11:

$$\begin{aligned}\lim_{x\to 0^+} \sin x \ln x \quad [0 \cdot \infty] &= \lim_{x\to 0^+} \frac{\ln x}{\csc x} \quad \left[\frac{\infty}{\infty}\right] \\ &= -\lim_{x\to 0^+} \frac{1/x}{\csc x \cot x} \\ &= -\lim_{x\to 0^+} \frac{\sin x \tan x}{x} = -\lim_{x\to 0^+} \frac{\sin x}{x} \cdot \lim_{x\to 0^+} \tan x \\ &= -1 \cdot 0 = 0\end{aligned}$$

Example 12:

$$\begin{aligned}\lim_{x\to 0}\left(\frac{1}{x} - \frac{1}{\sin x}\right) \quad [\infty - \infty] &= \lim_{x\to 0} \frac{\sin x - x}{x \sin x} \quad \left[\frac{0}{0}\right] \\ &= \lim_{x\to 0} \frac{\cos x - 1}{x \cos x + \sin x} \quad \left[\frac{0}{0}\right] \\ &= \lim_{x\to 0} \frac{-\sin x}{-x \sin x + \cos x + \cos x} \\ &= \frac{0}{2} = 0\end{aligned}$$

The situations 1^∞, ∞^0, and 0^0 also represent indeterminant forms. However, the logarithm of such an indeterminate form is of the form $0 \cdot \infty$ and its limit may possibly be obtained by previous methods.

Example 13: We can obtain

$$\lim_{x\to\infty}\left(1 + \frac{1}{x}\right)^x \quad [1^\infty]$$

by using logarithms. Let $f(x) = \left(1 + \frac{1}{x}\right)^x$ and consider $g(x) = \ln f(x)$. Then

$$\begin{aligned}\lim_{x\to\infty} g(x) &= \lim_{x\to\infty} x \ln\left(1 + \tfrac{1}{x}\right) \quad [\infty \cdot 0] \\ &= \lim_{x\to\infty} \frac{\ln[1 + (1/x)]}{1/x} \quad \left[\frac{0}{0}\right] \\ &= \lim_{x\to\infty} \frac{\dfrac{1}{1 + (1/x)} \cdot \dfrac{-1}{x^2}}{-1/x^2} \\ &= \lim_{x\to\infty} \frac{1}{1 + (1/x)} = 1\end{aligned}$$

Since $\lim_{x\to\infty} \ln f(x) = \ln \lim_{x\to\infty} f(x) = 1$, we conclude that $\lim_{x\to\infty} f(x) = e$.

Example 14: We can evaluate $\lim_{x\to 0^+} x^{2/\ln x}$ $[0^0]$ (without L'Hospital's Rule) by first setting $y = x^{2/\ln x}$ and taking logarithms. We have

$$\ln y = \frac{2}{\ln x} \ln x$$

so that

$$\lim_{x\to 0^+} \ln y = 2$$

and it follows that

$$\lim_{x\to 0^+} y = e^2$$

This example shows why it is not a good idea to define 0^0 to be 1.

EXERCISES 4.14

1. Explain by means of examples why ∞/∞ is an indeterminate form. Do the same for $0 \cdot \infty$, $\infty - \infty$, and 1^∞.

In Exercises 2–27 find the limit, using L'Hospital's Rule (if it applies).

2. (a) $\lim_{x\to 3} \dfrac{x^3 - 27}{x^2 - 9}$
   (b) $\lim_{x\to 0} \dfrac{x^3 - 27}{x^2 - 9}$
3. $\lim_{x\to 0} \dfrac{e^x - x - 1}{x^2}$
4. $\lim_{x\to 0} \dfrac{\sqrt{x + 4} - 2}{x}$
5. $\lim_{x\to 0^+} \sqrt{x} \ln x$
6. $\lim_{x\to\infty} \dfrac{\ln x}{x}$
7. $\lim_{x\to 0^+} \dfrac{\ln(x + 1)}{x}$
8. $\lim_{x\to\infty} \dfrac{e^x}{e^{2x}}$
9. $\lim_{x\to 0} \dfrac{e^x - 1}{e^{3x} - 1}$
10. $\lim_{x\to -\infty} \dfrac{e^x}{e^x + e^{2x}}$
11. $\lim_{x\to 0} x \csc x$
12. $\lim_{x\to \pi/2^-} \left(x - \dfrac{\pi}{2}\right) \tan x$
13. $\lim_{x\to \pi/2} \dfrac{\cos x}{x - (\pi/2)}$
14. $\lim_{x\to\infty} \dfrac{\sqrt{x^2 + 1}}{x}$
15. $\lim_{x\to\infty} x \ln\left(\dfrac{x + 1}{x - 1}\right)$
16. $\lim_{x\to 0^+} x^x$
17. $\lim_{x\to 0^+} x^{1/x}$
18. $\lim_{x\to 0^+} x^{\sin x}$
19. $\lim_{x\to 0^+} x^{e^x}$
20. $\lim_{x\to 1} x^{\ln x}$
21. $\lim_{x\to\infty} \left(1 + \dfrac{3}{x}\right)^{2x}$
22. $\lim_{x\to \pi/2} (\sec x - \tan x)$
23. (a) $\lim_{x\to 0} \left(\dfrac{e^x + 1}{2}\right)^{1/x}$ (b) $\lim_{x\to 0^+} \left(\dfrac{e^x + 2}{2}\right)^{1/x}$
    (c) $\lim_{x\to 0^-} \left(\dfrac{e^x + 2}{2}\right)^{1/x}$
24. $\lim_{x\to 0} \left(\dfrac{x + 2}{2}\right)^{1/x}$

25. $\lim_{x \to 0} (2x^2 + 1)^{1/x^2}$

26. $\lim_{x \to 0} \left(\frac{1}{x} \int_0^x e^{-t^2}\, dt \right)$

27. $\lim_{x \to 0} \left(\frac{1}{x^3} \int_0^{2x} t \sin t\, dt \right)$

28. Show that

$$\lim_{x \to \infty} \frac{2x + 1}{x + \sin x}$$

exists, but that the quotient of derivatives $2/(1 + \cos x)$ has no limit as $x \to \infty$. (Thus, a "converse" of L'Hospital's Rule does not hold.)

29. It is said that the exponential function e^x dominates any positive power of x for large x, in the sense that

(i) $$\lim_{x \to \infty} \frac{x^p}{e^x} = 0$$

for every $p > 0$. Similarly, the logarithm is dominated by every such power; that is,

(ii) $$\lim_{x \to \infty} \frac{\ln x}{x^p} = 0$$

Prove these two limit statements.

30. (a) Write out L'Hospital's Rule for $0/0$ forms as $x \to a$.

(b) Write out L'Hospital's Rule for ∞/∞ forms as $x \to b^-$.

31. Prove L'Hospital's Rule for $0/0$ indeterminate forms in the case $x \to \infty$ by making use of the fact that

$$\lim_{x \to \infty} \phi(x) = \lim_{u \to 0^+} \phi\left(\frac{1}{u}\right)$$

where $u = 1/x$.

32. Let f be differentiable on a deleted neighborhood of a and denote by $f'_+(a)$ and $f'_-(a)$ the following limits:

$$f'_+(a) = \lim_{x \to a^+} f'(x) \qquad \text{and} \qquad f'_-(a) = \lim_{x \to a^-} f'(x)$$

Define

$$f^*(a) = \lim_{h \to 0^+} \frac{f(a + h) - f(a - h)}{2h}$$

(a) Find $f'_+(0)$, $f'_-(0)$, and $f^*(0)$ for $f(x) = |x|$.

(b) Show that if f is differentiable at a, then

$$f^*(a) = f'(a)$$

(c) Show that

$$f^*(a) = \frac{f'_+(a) + f'_-(a)}{2}$$

(provided that all three of these limits exist).

PROJECT: The Gamma Function

The following exercises are background for the development of the gamma function, a function important in statistics and other areas of applied mathematics.

(1) Show that if $0 < p < 1$ then the improper integral

$$\int_0^1 e^{-x} x^{p-1}\, dx$$

converges.

(2) Show that if $p > 0$, then the improper integral

$$\int_1^\infty e^{-x} x^{p-1}\, dx$$

converges. (*Hint:* By the first part of Exercise 4.14.29, $x^{p-1}/e^x \to 0$ as $x \to \infty$; so for sufficiently large x, we have $x^{p+1}/e^x < 1$ and

$$\frac{x^{p-1}}{e^x} = \frac{x^{p+1}}{e^x} \cdot \frac{1}{x^2} < \frac{1}{x^2}.)$$

For $p > 0$, the **gamma function** is defined by the integral

$$\Gamma(p) = \int_0^\infty e^{-x} x^{p-1}\, dx$$

Notice that this integral is an improper integral of the second type if $p \geq 1$, but is improper in two ways if $0 < p < 1$.

(3) Show that $\Gamma(p)$ converges for all $p > 0$.

(4) Derive the functional equation for the gamma function for $p > 0$:

$$\Gamma(p + 1) = p\,\Gamma(p) \tag{1}$$

(5) Show that $\Gamma(n + 1) = n!$ for $n \in Z^+$. (Thus, the gamma function generalizes the factorial function.)

(6) By changing variables in the integral for $\Gamma\left(\frac{1}{2}\right)$, show that

$$\Gamma\left(\tfrac{1}{2}\right) = 2\int_0^\infty e^{-u^2}\,du$$

It then follows from the remark at the end of Example 8 in Section 4.13 that

$$\Gamma\left(\tfrac{1}{2}\right) = \sqrt{\pi} \tag{2}$$

(7) Use Equations (1) and (2) to evaluate $\Gamma\left(\frac{3}{2}\right)$.

(8) By means of Equation (1), the domain of the gamma function can be extended to include negative numbers that are not integers. For instance, if $p \in (-1, 0)$, then $p + 1 \in (0, 1)$, and we can define

$$\Gamma(p) = \frac{\Gamma(p + 1)}{p} \tag{3}$$

Find the values of $\Gamma\left(-\frac{1}{2}\right)$ and $\Gamma\left(-\frac{3}{2}\right)$ using Equation (3).

★ 4.15 *Riemann Sums*

In calculus, the definition of the integral is often formulated as a limit of *Riemann sums*. Instead of using the infimum or supremum of the function on each subinterval of π, we can use the value of the function at an arbitrary point in the subinterval.

Definition 4.15.1 Let f be a real function, bounded on $[a, b]$ where $a < b$. Let

$$\pi = \{x_0, x_1, \ldots, x_n\}$$

be a partition of $[a, b]$, and for $k = 1, 2, \ldots, n$, let ξ_k be any point in $[x_{k-1}, x_k]$. Then the sum

$$\sum_{k=1}^{n} f(\xi_k)(x_k - x_{k-1})$$

is a ***Riemann sum*** for f on $[a, b]$ relative to π.

Notation: It will be convenient to denote the Riemann sum defined above by the symbol $R(f; \pi; \xi_1, \ldots, \xi_n)$. When we do this it is to be understood that $\pi = \{x_0, x_1, \ldots, x_n\}$ is a partition of $[a, b]$ and that the numbers ξ_k satisfy $x_{k-1} \leq \xi_k \leq x_k$.

Whereas the upper and lower sums for f are uniquely determined for a given partition π of $[a, b]$, there are infinitely many ways of forming a Riemann sum because each ξ_k can be chosen in infinitely many ways. Geometrically, a Riemann sum for a nonnegative function represents the area of a number of rectangles like those shown in Figure 4.23. The picture suggests that a Riemann sum may well give a closer approximation to the area of $S_f[a, b]$ than either the upper sum or the lower sum relative to the same partition of $[a, b]$.

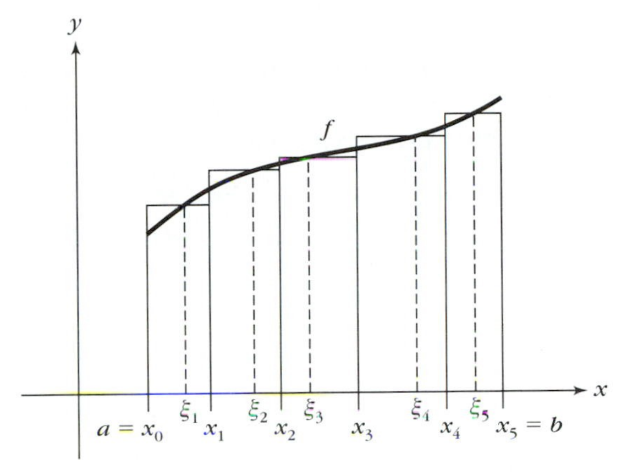

Figure 4.23
Riemann sum

Example 1: Let $f(x) = x^2$ for $0 \le x \le 1$, let $\pi = \left\{0, \frac{1}{2}, \frac{2}{3}, 1\right\}$, and $\xi_1 = \frac{1}{3}$, $\xi_2 = \frac{1}{2}$, $\xi_3 = \frac{3}{4}$. The corresponding Riemann sum is

$$R\left(f;\ \pi; \tfrac{1}{3}, \tfrac{1}{2}, \tfrac{3}{4}\right) = \sum_{k=1}^{3} f(\xi_k)(x_k - x_{k-1})$$

$$= \frac{1}{9} \cdot \frac{1}{2} + \frac{1}{4} \cdot \frac{1}{6} + \frac{9}{16} \cdot \frac{1}{3} = \frac{41}{144}$$

Example 2: The upper and lower sums for the last example are also Riemann sums. Because of the monotonicity of f, the m_k and M_k are values of f at x_{k-1} and x_k, respectively. Thus, $L_\pi(f) = R\left(f;\ \pi; 0, \frac{1}{2}, \frac{2}{3}\right)$ and $U_\pi(f) = R\left(f;\ \pi; \frac{1}{2}, \frac{2}{3}, 1\right)$.

Definition 4.15.2 The ***norm*** of a partition $\pi = \{x_0, x_1, \ldots, x_n\}$ is the length of the largest subinterval of π; that is, the maximum of the numbers $x_k - x_{k-1}$, for $k = 1, 2, \ldots, n$. The norm of π is denoted by $\|\pi\|$.

Example 3: In Example 1, the norm of π is $\|\pi\| = \frac{1}{2}$.

Definition 4.15.3 Let f be a real function bounded on $[a, b]$ and let I be real number. Then the limit statement

$$\lim_{\|\pi\| \to 0} R(f;\ \pi; \xi_1, \ldots, \xi_n) = I$$

means that for every $\varepsilon > 0$ there exists a $\delta > 0$ such that for all partitions π of $[a, b]$ with $\|\pi\| < \delta$, and for all choices of ξ_k with $\xi_k \in [x_{k-1}, x_k]$, we have

$$|R(f; \pi; \xi_1, \dots, \xi_n) - I| < \varepsilon$$

Our objective now is to prove that the condition in this definition is equivalent to integrability.

Theorem 4.15.1 *Let f be a real function, bounded on [a, b] where a < b. Then f is integrable on [a, b] if and only if there exists a real number I such that (with the notation described above)*

$$\lim_{\|\pi\| \to 0} R(f; \pi; \xi_1, \dots, \xi_n) = I$$

In case of integrability, $I = \int_a^b f$.

PROOF: Let us assume first that the limit statement

$$\lim_{\|\pi\| \to 0} R(f; \pi; \xi_1, \dots, \xi_n) = I$$

holds for some real number I. We prove that f is integrable on $[a, b]$. Let $\varepsilon > 0$ be given, and choose a fixed partition of $[a, b]$, $\pi = \{x_0\ x_1, \dots, x_n\}$ having sufficiently small norm that for all choices of $\xi_k \in [x_{k-1}, x_k]$ we have

$$I - \frac{\varepsilon}{3} < R(f; \pi; \xi_1, \dots, \xi_n) < I + \frac{\varepsilon}{3}$$

By taking the supremum of $R(f; \pi; \xi_1, \dots, \xi_n)$ over all choices of the ξ_k, we get

$$\begin{aligned} \sup R(f; \pi; \xi_1, \dots, \xi_n) &= \sup \sum_{k=1}^{n} f(\xi_k)(x_k - x_{k-1}) \\ &= \sum_{k=1}^{n} M_k(x_k - x_{k-1}) \end{aligned}$$

where

$$M_k = \sup\{f(x) : x_{k-1} \le x \le x_k\}$$

(work Exercise 4.15.3). But this sum is just $U_\pi(f)$, so that

$$U_\pi(f) = \sup R(f; \pi; \xi_1, \dots, \xi_n) \le I + \frac{\varepsilon}{3}$$

A parallel argument with infima and lower sums yields, for the same partition π,

$$I - \frac{\varepsilon}{3} \le L_\pi(f)$$

By combining these results, we obtain $U_\pi(f) - L_\pi(f) \leq 2\varepsilon/3 < \varepsilon$, and so (by Theorem 4.6.1), f is integrable on $[a, b]$. Now that f has been shown to be integrable, we can set $J = \int_a^b f$, and it is easy to show that $J = I$ (work Exercise 4.15.4).

To prove the converse, suppose that f is integrable on $[a, b]$ and that $\int_a^b f = I$. For $\varepsilon > 0$, we can choose (by Theorem 4.6.1) a partition π^* of $[a, b]$ such that

$$U_{\pi^*}(f) - L_{\pi^*}(f) < \frac{\varepsilon}{2}$$

Now let $M = \sup\{f(x) : a \leq x \leq b\}$, $m = \inf\{f(x) : a \leq x \leq b\}$, and $N =$ the number of subintervals of π^*. Define $\delta = \varepsilon/[2(M - m)N]$. We will show that if π is any partition of $[a, b]$ with norm less than δ, then

$$|R(f; \pi; \xi_1, \ldots, \xi_n) - I| < \varepsilon$$

for all choices of $\xi_k \in [x_{k-1}, x_k]$, for $k = 1, 2, \ldots, n$.

Accordingly, let π be a partition of $[a, b]$ with norm $\|\pi\| < \delta$; and let $\xi_k \in [x_{k-1}, x_k]$ for $k = 1, 2, \ldots, n$. Consider $U_\pi(f) - L_\pi(f)$, which can be expressed as a sum:

$$U_\pi(f) - L_\pi(f) = \sum_{k=1}^{n} (M_k - m_k)(x_k - x_{k-1})$$

where $M_k = \sup\{f(x) : x \in [x_{k-1}, x_k]\}$ and $m_k = \inf\{f(x) : x \in [x_{k-1}, x_k]\}$. We split this sum into two parts:

$$\sum_{k=1}^{n} (M_k - m_k)(x_k - x_{k-1}) = \sum{}' + \sum{}''$$

where Σ' contains all the terms corresponding to those subintervals of π that have a point of π^* in their interiors, and Σ'' contains all the remaining terms.

Now Σ' can have at most $N - 1$ terms because of the way N was defined; moreover, each term of Σ' is no larger than $(M - m) \cdot \delta$. Therefore,

$$\begin{aligned} \sum{}' \leq (N - 1)(M - m)\,\delta &= (N - 1)(M - m)\frac{\varepsilon}{2(M - m)N} \\ &< \frac{\varepsilon}{2} \end{aligned}$$

On the other hand, the sum of the terms in Σ'' cannot be greater than $U_{\pi^*}(f) - L_{\pi^*}(f)$, so that

$$\begin{aligned} U_\pi(f) - L_\pi(f) = \sum{}' + \sum{}'' &< \frac{\varepsilon}{2} + U_{\pi^*}(f) - L_{\pi^*}(f) \\ &< \frac{\varepsilon}{2} + \frac{\varepsilon}{2} = \varepsilon \end{aligned}$$

By working Exercise 4.15.2, you can show that any Riemann sum lies between the lower and upper sums. Thus,

$$I - \varepsilon \le U_\pi(f) - \varepsilon < L_\pi(f) \le R(f; \pi; \xi, \ldots, \xi_n) \le U_\pi(f) < L_\pi(f) + \varepsilon \le I + \varepsilon$$

and it follows that

$$|R(f; \pi; \xi_1, \xi_2, \ldots, \xi_n) - I| < \varepsilon \qquad \blacksquare$$

Theorem 4.15.1 permits a nice generalization of the Fundamental Theorem of Calculus in which we need not assume that the function f is continuous on $[a, b]$.

Theorem 4.15.2 *If f is a real function, bounded and integrable on $[a, b]$ and if there is a function F such that $F'(x) = f(x)$ for $a \le x \le b$, then*

$$\int_a^b f = F(b) - F(a)$$

PROOF: Let $\pi = \{x_0, x_1, \ldots, x_n\}$ be a partition of $[a, b]$, and on each subinterval $[x_{k-1}, x_k]$ select $\xi_k \in (x_{k-1}, x_k)$ by the Mean Value Theorem (Theorem 4.3.1) to be a number with the property that

$$f(\xi_k) = F'(\xi_k) = \frac{F(x_k) - F(x_{k-1})}{x_k - x_{k-1}}$$

Then the Riemann sum

$$R(f; \pi; \xi_1, \ldots, \xi_n) = \sum_{k=1}^{n} f(\xi_k)(x_k - x_{k-1}) = \sum_{k=1}^{n} \frac{F(x_k) - F(x_{k-1})}{x_k - x_{k-1}}(x_k - x_{k-1})$$
$$= \sum_{k=1}^{n} [F(x_k) - F(x_{k-1})] = F(x_n) - F(x_0) = F(b) - F(a)$$

We see that for *any* partition π, a Riemann sum can be constructed whose value is $F(b) - F(a)$. But the Riemann sums must approach $\int_a^b f$ by Theorem 4.15.1, and so we see that $\int_a^b f = F(b) - F(a)$. $\blacksquare$

Occasionally certain sequential limits can be recognized as limits of Riemann sums and can therefore be evaluated by integration.

Example 4: Let $f(x) = e^x$. The sum

$$S_n = \frac{e^{1/n} + e^{2/n} + \cdots + e^{n/n}}{n} = \sum_{k=1}^{n} f\left(\frac{k}{n}\right) \cdot \frac{1}{n}$$

is a Riemann sum for the function $f(x) = e^x$ corresponding to the regular partition π_n of $[0, 1]$. Since $\|\pi_n\| = 1/n \to 0$ as $n \to \infty$, we see that

$$\lim_{n\to\infty} S_n = \int_0^1 e^x\,dx = e - 1$$

Example 5: The sum

$$S_n = \sum_{k=1}^{n} \frac{1}{n+k}$$

is a Riemann sum for the function $f(x) = 1/x$ corresponding to the regular partition π_n of $[1, 2]$. Therefore,

$$\lim_{n\to\infty} S_n = \int_1^2 \frac{1}{x}\,dx = \ln 2$$

EXERCISES 4.15

1. Let $[a, b] = [0, 4]$, let $\pi = \{0, 1, 2, 3, 4\}$, and let $f(x) = 2x + 3$. Find $R(f; \pi; \xi_1, \ldots, \xi_4)$, where ξ_k is chosen to be (a) the left endpoint, (b) the right endpoint, (c) the midpoint, of $[k - 1, k]$ (for $k = 1, 2, 3, 4$).

2. Show that if π is a partition of $[a, b]$, then any Riemann sum for f on $[a, b]$ relative to π is between $L_\pi(f)$ and $U_\pi(f)$ inclusive.

3. Justify the claim in the proof of Theorem 4.15.1 that for a given partition the supremum of

$$R(f; \pi; \xi_1, \ldots, \xi_n)$$

over all choices of ξ_k for $k = 1, 2, \ldots, n$, is equal to $U_\pi(f)$.

4. Show that $J = I$ in the first part of the proof of Theorem 4.15.1.

5. (a) Verify the claim in Example 5 that

$$\sum_{k=1}^{n} 1/(n+k)$$

is a Riemann sum for $f(x) = 1/x$ corresponding to the regular partition π_n of $[1, 2]$.

(b) Identify $\sum_{k=1}^{2n} 1/(n+k)$ as an appropriate Riemann sum and evaluate its limit as $n \to \infty$.

(c) Evaluate $\lim_{n\to\infty} \sum_{k=1}^{mn} 1/(n+k)$, where m is a positive integer.

6. Evaluate the following limits by integration.

(a) $\displaystyle\lim_{n\to\infty} \sum_{k=1}^{n} \frac{1}{n} \sin \frac{k\pi}{n}$

(b) $\displaystyle\lim_{n\to\infty} \sum_{k=1}^{n} \frac{k}{n^2}$

(c) $\displaystyle\lim_{n\to\infty} \sum_{k=1}^{n} \frac{(k-1)^2}{n^3}$

(d) $\displaystyle\lim_{n\to\infty} \sum_{k=1}^{n} \frac{1}{n}\left(3 + \frac{2k}{n}\right)^2$

7. (a) Show that the function

$$g(x) = \begin{cases} -\cos\dfrac{1}{x} + 2x\sin\dfrac{1}{x} & x \neq 0 \\ 0 & x = 0 \end{cases}$$

is the derivative of the function f_2 in Example 4 in Section 4.7.

(b) Use Theorem 4.15.2 to evaluate

$$\int_0^{2/\pi} \left(2x \sin\frac{1}{x} - \cos\frac{1}{x}\right) dx.$$

4.16 *Taylor's Theorem*

Let f be differentiable on an interval I containing the real number a. Then the tangent line to the graph of f has the equation $y = f(a) + f'(a)(x - a)$. The linear function determined by this equation,

$$P_1(x) = f(a) + f'(a)(x - a)$$

is an approximation to the function f for x near a; P_1 has the property that $P_1(a) = f(a)$ and $P_1'(a) = f'(a)$.

If f is twice differentiable on I, the quadratic function

$$P_2(x) = f(a) + f'(a)(x - a) + \frac{f''(a)}{2}(x - a)^2$$

has the property that $P_2(a) = f(a)$, $P_2'(a) = f'(a)$, $P_2''(a) = f''(a)$. It is reasonable to expect that P_2 may be a better approximation to f near a than P_1 is because P_2 and f agree not only at a and in their first derivatives at a but also in their second derivatives at a.

If f has n derivatives on I for $n \in Z^+$, we can form the *nth Taylor polynomial for f at a*:

$$P_n(x) = f(a) + f'(a)(x - a) + \frac{f''(a)}{2!}(x - a)^2 + \cdots + \frac{f^{(n)}(a)}{n!}(x - a)^n \qquad \textbf{(1)}$$

It is easy to see that this polynomial and its first n derivatives agree at a, respectively, with f and its first n derivatives at a. Thus, we might hope that, as we take n larger, $P_n(x)$ becomes a better approximation to $f(x)$ for x near a.

Example 1: As a simple example we consider $f(x) = e^x$ at zero. For $n = 3$, we find $f(0) = f'(0) = f''(0) = f'''(0) = 1$. Therefore, the first few Taylor polynomials are given by

$$P_1(x) = 1 + x$$
$$P_2(x) = 1 + x + \frac{x^2}{2!}$$
$$P_3(x) = 1 + x + \frac{x^2}{2!} + \frac{x^3}{3!}$$

These polynomials are sketched in Figure 4.24.

Example 2: The derivatives of $f(x) = \sin x$ follow the scheme $\cos x$, $-\sin x$, $-\cos x$, $\sin x$, and so on. At $x = 0$, the values of these derivatives are 1, 0, -1, 0, 1, and so on, so that $f^{(2n)}(0) = 0$, and $f^{(2n-1)}(0) = (-1)^{n+1}$. Thus,

$$P_1(x) = x$$
$$P_3(x) = x - \frac{x^3}{3!}$$
$$P_5(x) = x - \frac{x^3}{3!} + \frac{x^5}{5!}$$

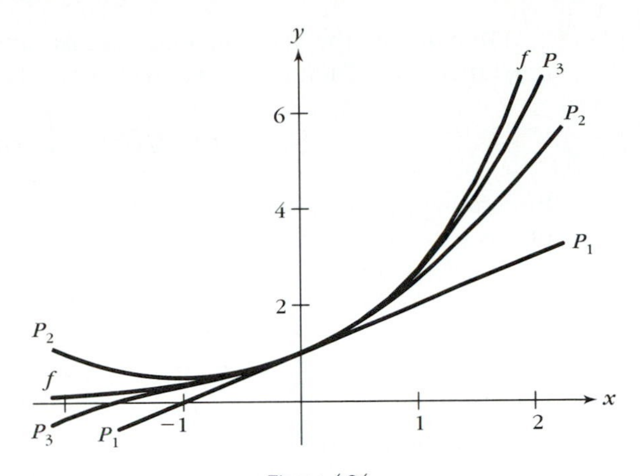

Figure 4.24
Taylor polynomials for e^x

and so on. Because $f^{(2n)}(0) = 0$, we see that $P_{2n}(x) = P_{2n-1}(x)$. Figure 4.25 shows P_1, P_3, and P_5. It is clear that the quality of the approximation of $f(x)$ by $P_n(x)$ depends upon x as well as n: For x near 0, the graph of P_5 appears to be a rather good fit to the sine curve, but for large x it wanders off.

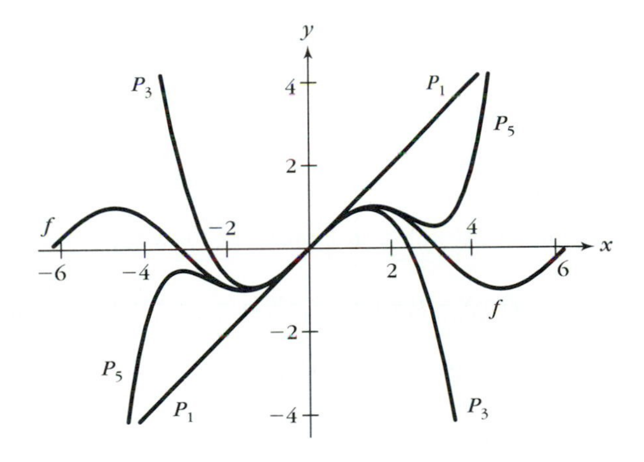

Figure 4.25
Taylor polynomials for sin x

Taylor's Formula with Remainder

In order to measure the accuracy with which $P_n(x)$ approximates its function, we define the *remainder* $R_n(x)$ as follows:

$$R_n(x) = f(x) - P_n(x) \tag{2}$$

The approximation of $f(x)$ by $P_n(x)$ is "good" when the remainder is "small"; keep in mind that these qualities depend on both x and n.

To compute the first remainder, we assume that f has a continuous second derivative on I. Then by the Fundamental Theorem of Calculus, we have

$$f(x) = f(a) + \int_a^x f'(t)\,dt$$

Integrating by parts, we take $u = f'(t)$ and $dv = dt$ so that a suitable choice of v is $v = -(x - t)$:

$$\begin{aligned} f(x) &= f(a) + \left[-f'(t)(x - t)\Big|_a^x + \int_a^x f''(t)(x - t)\,dt \right] \\ &= f(a) + f'(a)(x - a) + \int_a^x f''(t)(x - t)\,dt \end{aligned} \tag{3}$$

We have shown that

$$f(x) = P_1(x) + \int_a^x f''(t)(x - t)\,dt$$

Therefore,

$$R_1(x) = f(x) - P_1(x) = \int_a^x f''(t)(x - t)\,dt$$

This remainder gives the error when $f(x)$ is approximated by $P_1(x)$.

Next, we integrate the remainder $R_1(x)$ by parts, taking $u = f''(t)$ and $dv = (x - t)\,dt$ so that $v = -(x - t)^2/2$:

$$\begin{aligned} \int_a^x f''(t)(x - t)\,dt &= -f''(t)\frac{(x - t)^2}{2}\Big|_a^x + \int_a^x f'''(t)\frac{(x - t)^2}{2}\,dt \\ &= f''(a)\frac{(x - a)^2}{2} + \int_a^x f'''(t)\frac{(x - t)^2}{2}\,dt \end{aligned}$$

By substituting in Equation (3), we get

$$\begin{aligned} f(x) &= f(a) + f'(a)(x - a) + f''(a)\frac{(x - a)^2}{2} + \int_a^x f'''(t)\frac{(x - t)^2}{2}\,dt \\ &= P_2(x) + \int_a^x f'''(t)\frac{(x - t)^2}{2}\,dt \end{aligned}$$

so that

$$R_2(x) = f(x) - P_2(x) = \frac{1}{2}\int_a^x f'''(t)(x - t)^2\,dt$$

This calculation is valid if f has a continuous third derivative on I.

Continuing to integrate by parts in this way, and using mathematical induction, we obtain an integral formula for the remainder:

$$R_n(x) = \frac{1}{n!}\int_a^x f^{(n+1)}(t)(x - t)^n\,dt \tag{4}$$

provided $f^{(n+1)}$ is continuous on I. These results are summarized in the following theorem.

Theorem 4.16.1 **Taylor's Theorem with Integral Remainder** *If f has $n + 1$ continuous derivatives on an interval I containing the point a, then for $x \in I$ we have*

$$f(x) = P_n(x) + R_n(x)$$

where $P_n(x)$ is given by Equation (1) and $R_n(x)$ by Equation (4).

Taylor Series

If f has derivatives of all orders on I, we can continue approximating f by Taylor polynomials of higher degrees. If these polynomials do in fact give better and better approximations to the function f, it is reasonable to ask whether f can be expressed as an infinite series:

$$f(a) + f'(a)(x - a) + \frac{f''(a)}{2!}(x - a)^2 + \cdots + \frac{f^{(k)}(a)}{k!}(x - a)^k + \cdots = \sum_{k=0}^{\infty} \frac{f^{(k)}(a)}{k!}(x - a)^k \quad (5)$$

Clearly the sequence of partial sums of this series is just the sequence of Taylor polynomials at a:

$$S_1(x) = f(a) + f'(a)(x - a) = P_1(x)$$
$$S_2(x) = f(a) + f'(a)(x - a) + \frac{f''(a)}{2!}(x - a)^2 = P_2(x)$$
$$\vdots$$
$$S_n(x) = \sum_{k=0}^{n} \frac{f^{(k)}(a)}{k!}(x - a)^k = P_n(x)$$
$$\vdots$$

The infinite series (5) is known as the *Taylor series* for f at a.
The convergence of the series (5) depends not only on the function f but also on the value of x; it may converge for some values of x but diverge for others. And even if (5) converges at a certain value of x, it could conceivably converge to something other than $f(x)$.

By Taylor's Theorem, the difference between $f(x)$ and $S_n(x)$ is the remainder $R_n(x)$, given by

$$R_n(x) = f(x) - P_n(x) = f(x) - S_n(x) \quad (6)$$

By definition, the series (5) converges to $f(x)$ if and only if

$$\lim_{n \to \infty} S_n(x) = f(x)$$

hence, by Equation (6), if and only if

$$\lim_{n \to \infty} R_n(x) = 0$$

Thus, we have the following corollary of Taylor's Theorem:

Corollary 4.16.1 *Suppose f has derivatives of all orders on an interval I containing the real number a. Then the Taylor series for f at a converges to $f(x)$ at just those values of x for which*

$$\lim_{n\to\infty} R_n(x) = 0$$

where $R_n(x)$ is given by Equation (4).

Example 3: The exponential function $f(x) = e^x$ has derivatives of all orders on R; in fact, $f^{(n)}(x) = e^x$. In Example 1 we found the first three Taylor polynomials for this function at zero. The nth Taylor polynomial at zero is

$$P_n(x) = \sum_{k=0}^{n} \frac{f^{(k)}(0)}{k!} x^k = \sum_{k=0}^{n} \frac{x^k}{k!}$$

and the resulting Taylor series is

$$\sum_{k=0}^{\infty} \frac{x^k}{k!} \tag{7}$$

It can be shown by the Ratio Test (Theorem 2.8.7) that the series (7) is convergent for every $x \in R$ (work Exercise 4.16.2). We would like to show that this series actually converges to the function $f(x) = e^x$, and so we examine the remainder:

$$R_n(x) = \frac{1}{n!} \int_0^x f^{(n+1)}(t)(x - t)^n \, dt = \frac{1}{n!} \int_0^x e^t(x - t)^n \, dt$$

Let $x \in R$ be fixed. For each t in the interval of integration,* $e^t \leq e^{|x|}$ and $|x - t| \leq |x|$, so that

$$|R_n(x)| \leq \frac{1}{n!} e^{|x|} \left| \int_0^x (x - t)^n \, dt \right| \leq \frac{1}{n!} e^{|x|} \, |x|^{n+1}$$

(We have used Corollary 4.7.3 and Theorem 4.6.7.) Now $x^n/n!$ is the general term of the convergent series (7), and by Theorem 2.8.1, $x^n/n! \to 0$ as $n \to \infty$. It follows that $R_n(x) \to 0$ as $n \to \infty$. Since x was arbitrary, the series (7) converges to e^x for every $x \in R$.

Example 4: The sine function also has derivatives of all orders on R, whose values at zero follow the pattern $0, 1, 0, -1, 0, \ldots$. Therefore, the Taylor polynomials at zero for this function are

$$P_{2n}(x) = P_{2n-1}(x) = \sum_{k=1}^{n} (-1)^{k+1} \frac{x^{2k-1}}{(2k - 1)!}$$

and the corresponding Taylor series is

$$\sum_{k=1}^{\infty} (-1)^{k+1} \frac{x^{2k-1}}{(2k - 1)!}$$

*The interval of integration is $[0, x]$ or $[x, 0]$, depending upon the sign of x.

with remainder

$$R_{2n}(x) = \frac{1}{(2n)!}\int_0^x f^{(2n+1)}(t)(x - t)^{2n}\,dt$$

The values of $f^{(2n+1)}(t)$ lie in the interval $[-1, 1]$ so that

$$\left|R_{2n}(x)\right| \le \frac{1}{(2n)!}\left|\int_0^x (x - t)^{2n}\,dt\right| \le \frac{1}{(2n)!}|x|^{2n+1}$$

It follows as in the last example that $R_{2n}(x) \to 0$ as $n \to \infty$ for every $x \in R$ (work Exercise 4.16.3). We conclude that the Taylor series for $\sin x$ at zero converges to $\sin x$ everywhere:

$$\begin{aligned}\sin x &= x - \frac{x^3}{3!} + \frac{x^5}{5!} - \frac{x^7}{7!} + \cdots \\ &= \sum_{k=1}^{\infty}(-1)^{k+1}\frac{x^{2k-1}}{(2k - 1)!}\end{aligned}$$

The results of Example 3 and 4 are the best one could wish for: these Taylor series not only converge everywhere but also represent their functions everywhere. The next two examples show that less favorable outcomes are possible. In Example 5 we see that the logarithm function, which is defined for all positive real numbers, has a Taylor series at 1 that converges only in the interval $(0, 2]$. In Example 6 we find a function whose Taylor series at zero converges everywhere but converges to the function only at a single point.

Example 5: The natural logarithm function $f(x) = \ln x$ has derivatives of all orders for $x > 0$; these are given by

$$f^{(k)}(x) = (-1)^{k-1}(k - 1)!/x^k \qquad k \in Z^+$$

The coefficients of its Taylor polynomial at 1 are

$$\frac{f^{(k)}(1)}{k!} = \frac{(-1)^{k-1}}{k} \qquad k = 1, 2, \ldots, n$$

and by Theorem 4.16.1,

$$\ln x = \sum_{k=1}^{n}\frac{(-1)^{k-1}}{k}(x - 1)^k + R_n(x)$$

where

$$R_n(x) = \frac{1}{n!}\int_1^x \frac{(-1)^n n!}{t^{n+1}}(x - t)^n\,dt = (-1)^n\int_1^x \frac{(x - t)^n}{t^{n+1}}\,dt \qquad (8)$$

The methods of Sections 2.8 and 2.9 enable us to show that

$$\sum_{k=1}^{\infty}\frac{(-1)^{k-1}}{k}(x - 1)^k$$

converges for $x \in (0, 2]$ (work Exercise 4.16.4). We therefore turn to the question of whether the series converges to the function $\ln x$ on $(0, 2]$. To settle this question we must estimate the remainder. There are two cases:

(1) If $1 \leq x \leq 2$, then for any t in the interval of integration in Equation (8) we have $1 \leq t \leq x \leq 2$, so that $0 \leq x - t \leq 1$ and

$$|R_n(x)| = \int_1^x \frac{(x-t)^n}{t^{n+1}}\, dt$$
$$\leq \int_1^x \frac{dt}{t^{n+1}} = \frac{t^{-n}}{-n}\bigg|_1^x = \frac{1 - x^{-n}}{n} \leq \frac{1}{n}$$

Thus, $R_n(x) \to 0$ for $x \in [1, 2]$; it remains to show that $R_n(x) \to 0$ for $x \in (0, 1)$.

(2) If $0 < x < 1$, then in (8) we have $x \leq t \leq 1$, so that

$$|R_n(x)| = \left|\int_1^x \frac{(x-t)^n}{t^{n+1}}\, dt\right| \leq \int_x^1 \frac{(t-x)^n}{t^{n+1}}\, dt$$

Now since $0 < x \leq t \leq 1$, we have $1/x \geq 1/t$, and the last integral can be estimated as follows:

$$\int_x^1 \frac{(t-x)^n}{t^{n+1}}\, dt = \int_x^1 \frac{(t-x)^n}{t^n t}\, dt \leq \int_x^1 \frac{(t-x)^n}{t^n} \cdot \frac{1}{x}\, dt = \frac{1}{x}\int_x^1 \left(1 - \frac{x}{t}\right)^n dt$$

The maximum value of $[1 - (x/t)]^n$ on the interval $[x, 1]$ occurs when $t = 1$, and by Corollary 4.7.3,

$$\int_x^1 \left(1 - \frac{x}{t}\right)^n dt \leq \int_x^1 (1-x)^n\, dt = (1-x)^{n+1}$$

Therefore,

$$|R_n(x)| \leq \frac{1}{x}(1-x)^{n+1}$$

and since $|1 - x| < 1$, we see that $R_n(x) \to 0$ for $x \in (0, 1)$ (by Theorem 2.2.5). We have, finally,

$$\ln x = \sum_{k=1}^{\infty} \frac{(-1)^{k-1}}{k}(x-1)^k \qquad 0 < x \leq 2$$

An interesting consequence of the last example is the sum of the alternating harmonic series; taking $x = 2$ we get:

$$\sum_{k=1}^{\infty} \frac{(-1)^{k-1}}{k} = \ln 2$$

Example 6: Consider the function f defined by

$$f(x) = \begin{cases} e^{-1/x^2} & \text{if } x \neq 0 \\ 0 & \text{if } x = 0 \end{cases}$$

For $x \neq 0$, the derivative of f exists and can be found by the standard rules of differentiation. At $x = 0$, the derivative must be obtained directly from the definition of the derivative:

$$f'(0) = \lim_{x\to 0} \frac{e^{-1/x^2} - 0}{x - 0}$$

$$= \lim_{x\to 0} x^{-1}e^{-1/x^2} = \lim_{x\to 0} \frac{x^{-1}}{e^{1/x^2}}$$

By applying L'Hospital's Rule to this ∞/∞ indeterminate form, we obtain

$$f'(0) = \lim_{x\to 0} \frac{-x^{-2}}{e^{1/x^2}(-2x^{-3})}$$

$$= \lim_{x\to 0} \frac{x}{2e^{1/x^2}} = 0$$

Thus, f is differentiable on R and

$$f'(x) = \begin{cases} \dfrac{2e^{-1/x^2}}{x^3} & \text{if } x \neq 0 \\ 0 & \text{if } x = 0 \end{cases}$$

The higher derivatives of f can also be obtained for $x \neq 0$ by the rules of differentiation. Without trying to give a formula, note that each derivative is of the form

$$f^{(k)}(x) = g_k(x)\, e^{-1/x^2} \qquad x \neq 0$$

where $g_k(x)$ is a polynomial in $1/x$. [Verify this statement for $f''(x)$ and $f'''(x)$; a formal proof can be carried out by induction.]

Next we show by induction that the higher derivatives of f at zero are all equal to zero. By assuming that $f^{(k)}(0) = 0$, we can compute $f^{(k+1)}(0)$ from the definition

$$f^{(k+1)}(0) = \lim_{x\to 0} \frac{f^{(k)}(x) - f^{k}(0)}{x - 0} = \lim_{x\to 0} \frac{f^{(k)}(x)}{x}$$

$$= \lim_{x\to 0}[x^{-1}g_k(x)e^{-1/x^2}]$$

Now $x^{-1}g_k(x)$ is a polynomial in $1/x$, and it follows (work Exercise 4.16.5) that $\lim_{x\to 0}[x^{-1}g_k(x)e^{-1/x^2}] = 0$, showing that $f^{(k+1)}(0) = 0$. We conclude that f has derivatives of all orders on R and that $f^{(n)}(0) = 0$ for $n \in Z^+$. Thus, all terms in the Taylor series for f are zero, and since $f(x) = 0$ only when $x = 0$, the Taylor series for f represents f at only one point.

Lagrange's Remainder

The integral formula (4) for $R_n(x)$ is one of several ways to express the remainder in Taylor's Theorem. A useful alternative formula, attributed to Lagrange, can be developed using the following generalization of the Mean Value Theorem for Integrals.

Theorem 4.16.2 *If f and g are continuous on an interval containing a and b, and if $g(x) \neq 0$ for x between a and b, then there exists a number ξ between a and b such that*

$$\int_a^b f(x)g(x)\,dx = f(\xi)\int_a^b g(x)\,dx$$

PROOF: You are asked to prove this theorem in Exercise 4.16.6a. ∎

We can apply Theorem 4.16.2 to the integral in formula (4), obtaining, for some ξ between a and b,

$$\begin{aligned}\frac{1}{n!}\int_a^x f^{(n+1)}(t)(x-t)^n\,dt &= \frac{f^{(n+1)}(\xi)}{n!}\int_a^x (x-t)^n\,dt \\ &= \frac{f^{(n+1)}(\xi)}{n!}\left[-\frac{(x-t)^{n+1}}{n+1}\right]\Bigg|_a^x \\ &= \frac{f^{(n+1)}(\xi)}{(n+1)!}(x-a)^{n+1}\end{aligned}$$

Thus, the remainder can be expressed by the formula

$$R_n(x) = \frac{f^{(n+1)}(\xi)}{(n+1)!}(x-a)^{n+1} \tag{9}$$

where ξ is between a and x, and we can restate Taylor's Theorem:

Theorem 4.16.3 **Taylor's Theorem with Lagrange's Remainder** *If f has $n+1$ continuous derivatives on an interval I containing the point a, then, for $x \in I$, we have*

$$f(x) = P_n(x) + R_n(x)$$

where $P_n(x)$ is given by Equation (1) and $R_n(x)$ by Equation (9).

Theorem 4.16.3 has a useful application to the study of maxima and minima:

Theorem 4.16.4 *Suppose that f has $n+1$ continuous derivatives in an open interval containing a, and that $f'(a) = f''(a) = \cdots = f^{(n)}(a) = 0$ but $f^{(n+1)}(a) \neq 0$. Then:*

(i) If n is odd, f assumes a relative extremum at a; this extremum is a minimum if $f^{(n+1)}(a) > 0$ and a maximum if $f^{(n+1)}(a) < 0$.
(ii) If n is even, f does not assume a relative extremum at a.

PROOF: We begin by taking a neighborhood N of a with sufficiently small width that the sign of $f^{(n+1)}(x)$ for $x \in N$ is the same as that of $f^{(n+1)}(a)$;

such a neighborhood exists because $f^{(n+1)}$ is continuous at a. Then, if $x \in N$, Theorem 4.16.3 states that

$$f(x) = f(a) + R_n(x) = f(a) + \frac{f^{(n+1)}(\xi)}{(n+1)!}(x-a)^{n+1}$$

for some ξ between a and x.

(i) Now if n is odd, the factor $(x-a)^{n+1}$ is nonnegative in N, and the sign of $R_n(x)$ is the same as that of $f^{(n+1)}(a)$. (Why?) Thus, for $x \in N$, we have this situation: if $f^{(n+1)}(a) > 0$, then $f(x) \geq f(a)$ (so that f assumes a relative minimum at a), while if $f^{(n+1)}(a) < 0$, then $f(x) \leq f(a)$ (so that f assumes a relative maximum at a).

(ii) If n is even, the factor $(x-a)^{n+1}$ changes sign on either side of a, and therefore so does $f(x) - f(a)$, so that $f(x)$ does not assume an extremum of either kind at a. ■

Example 7: For the function

$$f(x) = x^2 + e^{-x^2}$$

we find that $f'(0) = f''(0) = f'''(0) = 0$ but $f^{(4)}(0) = 12 > 0$. (Verify.) Therefore, f assumes a relative minimum at zero.

EXERCISES 4.16

1. For each of the following functions, find the nth Taylor polynomial at a and write the corresponding integral formula for $R_n(x)$.
   (a) $f(x) = \cos x$; $a = 0$; $n = 6$
   (b) $f(x) = \sin x$; $a = \pi/2$; $n = 5$
   (c) $f(x) = e^{x^2}$; $a = 0$; $n = 4$
   (d) $f(x) = 1/x$; $a = 1$; $n = 4$
   (e) $f(x) = \ln(1 + x)$; $a = 0$; $n = 4$
   (f) $f(x) = \sqrt{1 + x}$; $a = 0$; $n = 4$

2. Show that the series $\sum_{k=0}^{\infty} x^k/k!$ is convergent for every $x \in R$.

3. Show that the remainder $R_{2n}(x)$ in Example 4 converges to zero as $n \to \infty$, for every $x \in R$.

4. Show that the series

$$\sum_{k=1}^{\infty} \frac{(-1)^{k-1}}{k}(x-1)^k$$

is convergent for every $x \in (0, 2]$.

5. (a) Show that if p is a polynomial, then

$$p(t)e^{-t^2} \to 0 \text{ as } t \to \infty$$

(also as $t \to -\infty$).
   (b) By replacing t by $1/x$ in part (a), conclude that if p is a polynomial, then

$$p(1/x)e^{-1/x^2} \to 0$$

as $x \to 0$.

6. (a) Prove Theorem 4.16.2.
   (b) Give an example to show that the conclusion of Theorem 4.16.2 may fail if the hypothesis that $g(x) \neq 0$ is not satisfied.

7. Show that if $f''(x)$ exists and is positive on an interval I, and if $c \in I$, then the tangent line to the graph of f at the point $(c, f(c))$ is never above the graph of f on I.

8. Show that $f(x) = \sin x - x + \frac{1}{6}x^3$ does not have a relative extremum at zero.

CHAPTER FIVE

SEQUENCES AND SERIES OF FUNCTIONS

This chapter extends the work of Sections 2.8 and 2.9 to include sequences and series whose terms are real functions.

The examples of Section 4.16 show that it is sometimes possible to approximate a given function by a sequence of functions that are "simpler" in some respects. The Taylor approximation allows us to represent certain transcendental functions such as e^x, $\sin x$, and $\ln x$ as infinite series that are actually limits of polynomial functions.*

We now take a slightly different approach. Instead of starting with a function and constructing a sequence, we start with a sequence of functions defined on a set, discuss what it means to say that the sequence converges on that set, and investigate properties of the limit function.

*It should be clear that polynomials are "simpler" in terms of their algebraic meaning; furthermore, they behave in a very simple way with respect to differentiation and integration.

5.1 *Pointwise Convergence*

Let D be a subset of R and suppose that for each $n \in Z^+$ there is a real function s_n defined on D. Suppose further that for each $x \in D$, the sequence $\{s_n(x)\}_{n=1}^{\infty}$ is convergent. Then it is possible to define a function s on D, as follows:

$$s(x) = \lim_{n\to\infty} s_n(x) \qquad x \in D$$

Definition 5.1.1 Let $\{s_n\}_{n=1}^{\infty}$ be a sequence of real functions defined on a subset D of R, let s be a function defined on D, and suppose that for every $x \in D$,

$$\lim_{n\to\infty} s_n(x) = s(x)$$

Then the sequence $\{s_n\}_{n=1}^{\infty}$ ***converges pointwise*** (or, simply, ***converges***) to s on D, and s is the ***limit function*** of the sequence $\{s_n\}_{n=1}^{\infty}$ We write

$$s_n(x) \to s(x) \qquad x \in D$$

Example 1: From our work in Section 4.16 we already have

$$1 + x + \frac{x^2}{2!} + \cdots + \frac{x^n}{n!} \to e^x \qquad x \in R$$

$$x - \frac{x^3}{3!} + \cdots + (-1)^{n+1}\frac{x^{2n-1}}{(2n-1)!} \to \sin x \qquad x \in R$$

$$(x-1) - \frac{(x-1)^2}{2} + \frac{(x-1)^3}{3} - \cdots + (-1)^{n+1}\frac{(x-1)^n}{n} \to \ln x \qquad x \in (0, 2]$$

Example 2: Let $D = R$ and let $s_n(x) = x/n$ for $x \in R$. If x is any real number, then we have

$$\lim_{n\to\infty} \frac{x}{n} = x \lim_{n\to\infty} \frac{1}{n} = 0$$

so that the sequence $\{x/n\}_{n=1}^{\infty}$ converges to the function $s(x) = 0$.

Example 3: Let $D = R$ and let $s_n(x) = (\sin nx)/n$ for $x \in R$. Because $0 \le |(\sin nx)/n| \le 1/n$ for each $x \in R$, it follows from the Squeeze Theorem (Theorem 2.2.4) that $\lim_{n\to\infty} (\sin nx)/n = 0$ for $x \in R$. Thus, $\{(\sin nx)/n\}_{n=1}^{\infty}$ converges to the function $s(x) = 0$.

Example 4: We know from Theorem 2.5.9 that the geometric sequence $\{x^n\}_{n=1}^{\infty}$ converges to 0 when x has any value between -1 and 1, converges to 1 when $x = 1$, and diverges for all other real values of x. So on the set $D = (-1, 1]$, the sequence

$$s_n(x) = x^n$$

converges to the function

$$s(x) = \begin{cases} 0 & \text{if } -1 < x < 1 \\ 1 & \text{if } x = 1 \end{cases}$$

Example 5: Consider the sequence of functions $s_n(x) = e^{-nx}$. For $x > 0$, $e^{-x} < 1$ and $e^{-nx} = (e^{-x})^n \to 0$ as $n \to \infty$. For $x = 0$, the sequence $\{s_n(0)\}_{n=1}^{\infty}$ is the constant 1. For $x < 0$, $e^{-x} > 1$ and $\{e^{-nx}\}_{n=1}^{\infty}$ diverges. The sequence $\{s_n(x)\}_{n=1}^{\infty}$ therefore converges only on the set $D = [0, \infty)$ and its limit is given by

$$s(x) = \begin{cases} 1 & \text{if } x = 0 \\ 0 & \text{if } x > 0 \end{cases}$$

Notice in Examples 4 and 5 that the limit of continuous functions may turn out to be discontinuous itself. Other peculiarities arise in Exercises 5.1.22 and 5.1.23.

Example 6: Let $s_n(x) = e^{-x/n}$. For each $x \in R$, we have $\lim\limits_{n\to\infty} (-x/n) = 0$, and by the continuity of the exponential function, $e^{-x/n} \to e^0 = 1$. Thus,

$$s_n(x) \to 1 \qquad x \in R$$

Example 7: Let $s_n(x) = n \ln[1 + (x/n)]$ $(x > 0)$. As $n \to \infty$, we have an indeterminate $(\infty \cdot 0)$ form. However,

$$\lim_{n\to\infty} s_n(x) = \lim_{n\to\infty} \frac{\ln\left(1 + \dfrac{x}{n}\right)}{1/n}$$

and by using L'Hospital's Rule,

$$\lim_{n\to\infty} s_n(x) = \lim_{n\to\infty} \frac{\dfrac{1}{\left(1 + \dfrac{x}{n}\right)}\left(-\dfrac{x}{n^2}\right)}{-1/n^2}$$

$$= \lim_{n\to\infty} \frac{x}{1 + (x/n)} = x$$

Thus,

$$s_n(x) \to x \qquad x > 0$$

EXERCISES 5.1

In Exercises 1–10 show that the sequence whose general term is given converges on the set D by finding its limit function.

1. $\dfrac{\cos nx}{n^2}$; $D = R$ **2.** e^{nx}; $D = (-\infty, 0]$

3. e^{-nx^2}; $D = R$

4. $\dfrac{x^n}{\sqrt{n}}$; $D = [-1, 1]$

5. $\dfrac{1}{n^2 + x^2}$; $D = R$

6. $\dfrac{x}{n + 2x}$; $D = [0, \infty)$

7. $\dfrac{n + x}{n + 2x}$; $D = [0, \infty)$

8. $\dfrac{x^n}{1 + x^n}$; $D = [0, 1]$

9. $\dfrac{x^n}{x^n + n^n}$; $D = [0, \infty)$

10. $\sin^n x$; $D = [0, \pi]$

In Exercises 11–20 find the set of all real numbers for which the indicated sequence converges and find the limit function.

11. $\dfrac{1 - x^n}{1 - x}$

12. ne^{-nx}

13. nxe^{-nx}

14. $\left(\dfrac{1}{x}\right)^n$

15. $\dfrac{nx^2 + \sin nx}{n}$

16. $\dfrac{x}{n}e^{-x/n}$

17. $xe^{-x/n}$

18. $n \sin \dfrac{x}{n}$

19. $\dfrac{\sin nx}{nx}$

20. $\left(\dfrac{\ln x}{x}\right)^n$

21. $\dfrac{\ln(1 + nx)}{n}$

22. Let $s_n(x) = e^{-nx}/n$. Show that, for $x \in [0, \infty)$, $\{s_n(x)\}_{n=1}^{\infty}$ converges to the function $s(x) = 0$. Find the limit of the sequence $\{s_n'(x)\}_{n=1}^{\infty}$ of derivatives and show that

$$\lim_{n\to\infty} s_n'(0) \neq s'(0)$$

23. Let $s_n(x) = nxe^{-nx^2}$. Show that, for $x \in [0, \infty)$, $\{s_n(x)\}_{n=1}^{\infty}$ converges to the function $s(x) = 0$. Find the limit of the sequence $\left\{\int_0^1 s_n\right\}_{n=1}^{\infty}$ and show that

$$\lim_{n\to\infty} \int_0^1 s_n \neq \int_0^1 s$$

5.2 *Uniform Convergence*

The examples and exercises relating to pointwise convergence in Section 5.1 illustrate certain peculiarities:

(1) Examples 4 and 5 show that the limit of a convergent sequence of continuous functions may fail to be continuous.

(2) Exercise 5.1.22 shows that the derivative of the limit of a convergent sequence of differentiable functions may not be the same as the limit of the derivatives.

(3) Exercise 5.1.23 shows that the integral of the limit of a convergent sequence of continuous functions may not be the same as the limit of the integrals.

Because pointwise convergence does not preserve continuity, derivatives, or integrals, it is desirable to introduce a stronger kind of convergence, known as "uniform" convergence. Before defining uniform convergence, let us take a closer look at pointwise convergence, defined in Section 5.1.

ε-N Formulation of Pointwise Convergence

According to Definition 5.1.1, the statement

$$s_n(x) \to s(x) \qquad x \in D$$

means that at each point $x \in D$, the sequence of real numbers $\{s_n(x)\}_{n=1}^{\infty}$ converges to the real number $s(x)$. If we apply the ε-N formulation of convergence given in Section 2.3, then we have, for each $x \in D$:

$$s_n(x) \to s(x)$$
$$\Leftrightarrow$$

for every $\varepsilon > 0$ there exists an $N \in Z^+$ such that $|s_n(x) - s(x)| < \varepsilon$ for $n \geq N$.

Example 1: In Example 2 of Section 5.1 we saw that the sequence $\{x/n\}_{n=1}^{\infty}$ converges to the zero function for each $x \in R$. Suppose $\varepsilon > 0$ has been given and we wish to find $N \in Z^+$ such that $|x/n| < \varepsilon$ for $n \geq N$. Since we want $n > |x|/\varepsilon$, we can take N to be the least positive integer greater than $|x|/\varepsilon$. Clearly this method of choosing N depends essentially on x as well as on ε. Suppose, for instance, that $\varepsilon = 1$. For $x = 10$, our choice would give $N = 11$; however, $N = 11$ is not an acceptable choice of N when $x = 100$. It is apparent that the convergence of this sequence to its limit becomes slower as $|x|$ increases. No matter how large an N we choose, there are values of x for which $|x|/\varepsilon > N$, so that $n > |x|/\varepsilon$ cannot hold for all $n \geq N$ and $x \in D$.

Uniform Convergence

Example 2: Suppose we now restrict the terms of $\{x/n\}_{n=1}^{\infty}$ to the interval $D = [0, 100]$. Then the condition $|x/n| < \varepsilon$ is satisfied whenever $100/n < \varepsilon$ because for $x \in D$ we have $|x| \leq 100$. Therefore, the choice of N as the least positive integer greater than $100/\varepsilon$ "works"; for $n \geq N$, we have

$$\left|\frac{x}{n}\right| < \epsilon \qquad \text{for all } x \in D$$

In Example 2 we found that for a given $\varepsilon > 0$ there is an N that depends only on ε and works simultaneously for all $x \in D$. This example exhibits what we call *uniform* convergence.

Definition 5.2.1 Let $\{s_n\}_{n=1}^{\infty}$ be a sequence of real functions defined on a subset D of R, and let s be a real function defined on D. We say that $\{s_n\}_{n=1}^{\infty}$ ***converges***

uniformly to s on D provided that for every $\varepsilon > 0$ there exists an $N \in Z^+$ such that whenever $n \geq N$, we have $|s_n(x) - s(x)| < \varepsilon$ for all $x \in D$. We write

$$s_n(x) \rightrightarrows s(x) \qquad x \in D$$

It is clear that if $s_n(x) \rightrightarrows s(x)$ $(x \in D)$, then $s_n(x) \to s(x)$ $(x \in D)$, so that uniform convergence is a stronger kind of convergence.

The distinction between pointwise convergence and uniform convergence is this: in the case of pointwise convergence, N may depend on both x and ε; in the case of uniform convergence, N must depend only on ε and must work for all x in the given set D. This distinction shows up in the order of the quantifiers when we display the two statements in logical notation:

POINTWISE CONVERGENCE

$$s_n(x) \to s(x) \qquad x \in D$$
$$\Leftrightarrow$$
$$(\forall x \in D)(\forall\ \varepsilon > 0)(\exists N \in Z^+)(n \geq N \Rightarrow |s_n(x) - s(x)| < \varepsilon)$$

UNIFORM CONVERGENCE

$$s_n(x) \rightrightarrows s(x) \qquad x \in D$$
$$\Leftrightarrow$$
$$(\forall \varepsilon > 0)(\exists N \in Z^+)(\forall\ x \in D)(n \geq N \Rightarrow |s_n(x) - s(x)| < \varepsilon)$$

Geometric Significance of Uniform Limit

If $s_n(x) \rightrightarrows s(x)$ $(x \in D)$, then corresponding to an $\varepsilon > 0$ there exists an $N \in Z^+$ with the property that

$$s(x) - \varepsilon < s_n(x) < s(x) + \varepsilon$$

for $n \geq N$ and all $x \in D$. The graph of each s_n with $n \geq N$ must therefore lie inside a band of width 2ε containing the graph of the limit function s (see Figure 5.1).

Example 3: Let $s_n(x) = (\sin x)/n$ for $x \in D = [0, 2\pi]$. Since for each x, the sequence $\{\sin x\}_{n=1}^{\infty}$ is bounded and $\{1/n\}_{n=1}^{\infty}$ is null, we have

$$\lim_{n\to\infty} \frac{\sin x}{n} = 0$$

by Theorem 2.2.2. Thus, $s_n(x) \to 0$ $(x \in [0, 2\pi])$. The convergence here is in fact uniform, as we can see from the inequality $|(\sin x)/n| \leq 1/n$. Given $\varepsilon > 0$, we can choose $N \in Z^+$ such that $1/n < \varepsilon$ for $n \geq N$; for

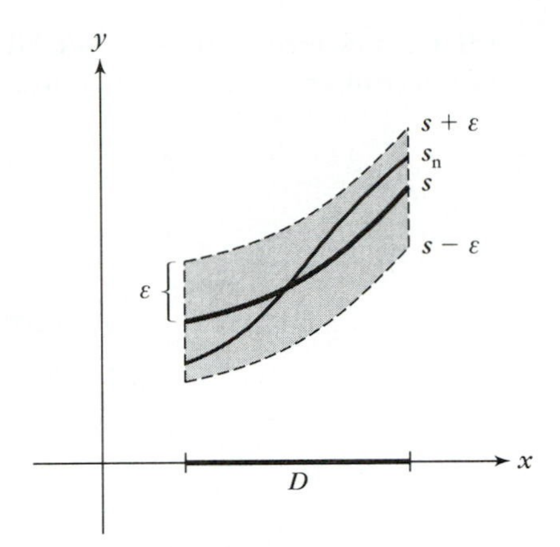

Figure 5.1
Geometric meaning of uniform convergence

that same N we have, for all $x \in [0, 2\pi]$, $|(\sin x)/n| < \varepsilon$ whenever $n \geq N$.

In Figure 5.2, we show the band of width 2ε around the limit function, where $\varepsilon = \frac{1}{4}$. All the s_n with $n > 4$ have the property that their graphs lie inside the shaded strip.

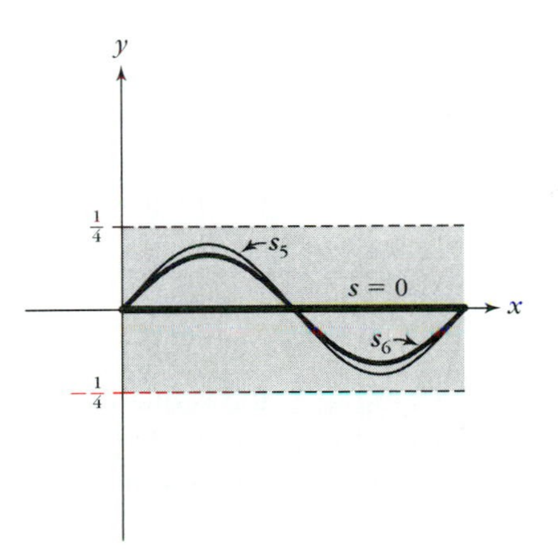

Figure 5.2
Uniform convergence of Example 3

Example 4: Let $s_n(x) = x/(n + x)$ for $x \geq 0$. Then

$$\lim_{n\to\infty} s_n(x) = \lim_{n\to\infty} \frac{x}{n + x} = 0$$

so that $\{s_n(x)\}_{n=1}^{\infty}$ converges pointwise to the zero function on $[0, \infty)$. But this convergence is not uniform. To see why, let $\varepsilon = \frac{1}{2}$ and consider the band of width 2ε about the graph of s (the x-axis). No matter how large n is chosen, the graph of s_n escapes from the band when x reaches the value n because $s_n(n) = \frac{1}{2}$ (see Figure 5.3). Thus, the geometric property of uniform convergence is contradicted.

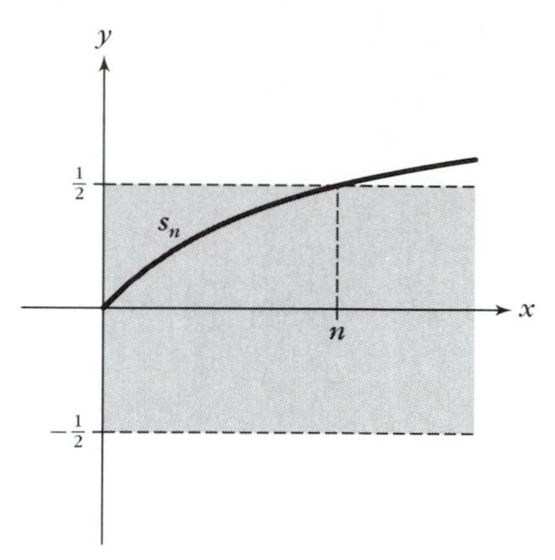

Figure 5.3

Dominance and the Little m Test

The discussion in Example 3 points the way to a very useful test for uniform convergence to the zero function.

Definition 5.2.2 Let $\{s_n\}_{n=1}^{\infty}$ and $\{t_n\}_{n=1}^{\infty}$ be sequences of real functions defined on a set $D \subseteq R$. Then $\{s_n\}_{n=1}^{\infty}$ is ***dominated*** by $\{t_n\}_{n=1}^{\infty}$ on D provided that for every $n \in Z^+$ the inequality

$$|s_n(x)| \leq t_n(x)$$

holds for $x \in D$.

In Example 3, the sequence $\{s_n(x)\}_{n=1}^{\infty}$ is dominated by the sequence of constants $\{1/n\}_{n=1}^{\infty}$ on $[0, 2\pi]$; our argument for the uniform convergence of $\{s_n(x)\}_{n=1}^{\infty}$ to zero is based on the fact that this sequence of constants is null.

Theorem 5.2.1 **Little *m* Test** *If $\{s_n(x)\}_{n=1}^{\infty}$ is a sequence of real functions defined on a set $D \subseteq R$ and if there is a null sequence $\{m_n\}_{n=1}^{\infty}$ of nonnegative numbers that dominates $\{s_n(x)\}_{n=1}^{\infty}$ on D, then $s_n(x) \rightrightarrows 0$ ($x \in D$).*

PROOF: Let $\{m_n\}_{n=1}^{\infty}$ be a null sequence and suppose that for each $n \in Z^+$,

$$|s_n(x)| \leq m_n \qquad x \in D \tag{1}$$

Given $\varepsilon > 0$, choose N so large that $n \geq N$ implies $m_n < \varepsilon$. Then, with that N, it follows from inequality (1) that $|s_n(x)| < \varepsilon$ for $n \geq N$ and all $x \in D$. Hence, $s_n(x) \rightrightarrows 0$ ($x \in D$). ■

Example 5: Let $s_n(x) = e^{-nx}$ for $x \in [1, \infty)$. Then $\{s_n(x)\}_{n=1}^{\infty}$ is dominated on $[1, \infty)$ by the null sequence $\{1/e^n\}_{n=1}^{\infty}$, and, according to Theorem 5.2.1, $s_n(x) \rightrightarrows 0$ $(x \in [1, \infty))$.

When applying Theorem 5.2.1, it may be helpful to use calculus to obtain the dominating null sequence.

Example 6: Let $s_n(x) = xe^{-nx}$ for $x \in [0, \infty)$. We let n be fixed and maximize s_n on $[0, \infty)$. By taking the derivative and simplifying, we obtain

$$s_n'(x) = \frac{1 - nx}{e^{nx}}$$

We find that a maximum occurs at $x = 1/n$, where s_n has the value $1/ne$. Thus, with $m_n = 1/ne$, we have

$$|s_n(x)| \leq m_n \qquad x \in [0, \infty)$$

and since $m_n \to 0$, we conclude by Theorem 5.2.1 that

$$s_n(x) \rightrightarrows 0 \qquad x \in [0, \infty)$$

Although Theorem 5.2.1 provides a test for uniform convergence to zero, it can be adapted to prove uniform convergence to other functions by observing that

$$s_n(x) \rightrightarrows s(x) \quad x \in D$$
$$\Leftrightarrow$$
$$s_n(x) - s(x) \rightrightarrows 0 \quad x \in D$$

Example 7: Let $s_n(x) = (x + 2n)/n$ $(x \in [-3, 2])$. Then, for each $x \in [-3, 2]$, we have $\lim_{n\to\infty} s_n(x) = 2$ so that the limit function is the constant 2. We have established *pointwise* convergence. To show that this convergence is in fact uniform on $[-3, 2]$, we consider

$$\begin{aligned} |s_n(x) - 2| &= \left|\frac{x}{n}\right| \\ &\leq \frac{3}{n} \qquad x \in [-3, 2] \end{aligned}$$

By applying Theorem 5.2.1 with $m_n = 3/n$, we have

$$s_n(x) - 2 \rightrightarrows 0 \qquad x \in [-3, 2]$$

hence,

$$s_n(x) \rightrightarrows 2 \qquad x \in [-3, 2]$$

Cauchy Criterion

The Cauchy Criterion (Theorem 2.7.2) provides a test for convergence of sequences of real numbers that does not require knowing the limit in advance. There is a similar theorem for uniform convergence of sequences of functions.

Theorem 5.2.2 **Cauchy Criterion for Uniform Convergence** *A sequence $\{s_n(x)\}_{n=1}^{\infty}$ of real functions, each of which is defined on a set $D \subseteq R$, is uniformly convergent on D if and only if, for each $\varepsilon > 0$, there exists an $N \in Z^+$ with the property that, whenever $m,n \geq N$, then*

$$|s_n(x) - s_m(x)| < \varepsilon \qquad x \in D$$

PROOF: Suppose that $s_n(x) \rightrightarrows s(x)$ $(x \in D)$. Let $\varepsilon > 0$ be given. Choose $N \in Z^+$ such that $|s_n(x) - s(x)| < \varepsilon/2$ whenever $n \geq N$ and $x \in D$. If m and n are both greater than or equal to N, then we have for all $x \in D$,

$$|s_n(x) - s_m(x)| \leq |s_n(x) - s(x)| + |s(x) - s_m(x)| < \frac{\varepsilon}{2} + \frac{\varepsilon}{2} = \varepsilon$$

showing that the Cauchy condition is necessary for uniform convergence.

To prove sufficiency, we begin by noting that for each $x \in D$, the sequence $\{s_n(x)\}_{n=1}^{\infty}$ is convergent by the "old" Cauchy Criterion (Theorem 2.7.2). Therefore, we can define a function s as follows

$$s(x) = \lim_{n\to\infty} s_n(x) \qquad x \in D$$

We must show that $s_n(x) \rightrightarrows s(x)$ $(x \in D)$. Let $\varepsilon > 0$ be given and choose $N \in Z^+$ with the property that, whenever $m,n \geq N$, we have

$$|s_n(x) - s_m(x)| < \frac{\varepsilon}{2} \qquad x \in D$$

Then, by letting $m \to \infty$, we obtain, for any fixed $n \geq N$ and all $x \in D$,

$$\lim_{m\to\infty} |s_n(x) - s_m(x)| \leq \frac{\varepsilon}{2}$$

$$\left|s_n(x) - \lim_{m\to\infty} s_m(x)\right| \leq \frac{\varepsilon}{2}$$

$$|s_n(x) - s(x)| \leq \frac{\varepsilon}{2}$$

Thus, $|s_n(x) - s(x)| < \varepsilon$ for $n \geq N$ and all $x \in D$. ∎

EXERCISES 5.2

1. Show that the sequence whose general term is given converges uniformly on the set D.

   (a) $\dfrac{\sin nx}{n}$; $D = R$

(b) $\dfrac{1}{n^2 + x^2}$; $D = R$
(c) $\dfrac{x^n}{\sqrt{n}}$; $D = [-1, 1]$
(d) e^{nx}; $D = (-\infty, -1]$
(e) $\ln \sqrt[n]{x}$; $D = [1, e]$
(f) $\tan^{-1} nx$; $D = [1, \infty)$
(g) $\sin^n x$; $D = \left[-\dfrac{\pi}{3}, \dfrac{\pi}{3}\right]$
(h) $1 - \cos^n x$; $D = \left[\dfrac{\pi}{4}, \dfrac{3\pi}{4}\right]$
(i) $e^{-x/n}$; $D = [0, 1]$

2. Show that the sequence whose general term is given converges pointwise but not uniformly on the given set D.

(a) e^{nx}; $D = (-\infty, 0]$
(b) $\sin^n x$; $D = [0, \pi]$
(c) $1 - \cos^n x$; $D = \left[-\dfrac{\pi}{2}, \dfrac{\pi}{2}\right]$
(d) $\ln \sqrt[n]{x}$; $D = [1, \infty)$
(e) $\tan^{-1} nx$; $D = (0, \infty)$
(f) $e^{-x/n}$; $D = [0, \infty)$
(g) $\dfrac{x}{n}e^{-x/n}$; $D = [0, \infty)$
(h) $\sqrt[n]{\sin x}$; $D = [0, \pi]$

3. Show that the sequence $\{e^{-nx}/n\}_{n=1}^{\infty}$ (Exercise 5.1.22) converges uniformly on $[0, \infty)$. Show that the sequence of derivatives converges pointwise but not uniformly there.

4. Show that the sequence $\{nxe^{-nx^2}\}_{n=1}^{\infty}$ (Exercise 5.1.23) converges pointwise but not uniformly on $[0, \infty)$.

5. Show that

$$n \sin \frac{x}{n} \to x \qquad x \in R$$

but that this convergence is not uniform.

6. Show that there exists a positive number c such that $\ln c = -c$. Show that $\{[(\ln x)/x]^n\}_{n=1}^{\infty}$ converges pointwise on (c, ∞). Show that if $d > c$, then $\{[(\ln x)/x]^n\}_{n=1}^{\infty}$ converges uniformly on $[d, \infty)$.

7. Show that if $\{s_n(x)\}_{n=1}^{\infty}$ is dominated on D by a sequence $\{t_n(x)\}_{n=1}^{\infty}$ that converges uniformly to zero on D, then $\{s_n(x)\}_{n=1}^{\infty}$ converges uniformly to zero on D.

5.3 *Consequences of Uniform Convergence*

In introducing uniform convergence we suggested that some of the peculiarities observed in earlier examples and exercises could be avoided by requiring this stronger kind of convergence. We now see that uniform convergence does indeed preserve continuity, integrals, and derivatives.

Preservation of Continuity

Theorem 5.3.1 *Let $\{s_n\}_{n=1}^{\infty}$ be a sequence of real functions converging uniformly on a set $D \subseteq R$ and let $s(x) = \lim_{n\to\infty} s_n(x)$ $(x \in D)$. If each s_n is continuous on D, then so is s.*

PROOF: We are given that

$$s_n(x) \rightrightarrows s(x) \qquad x \in D$$

Let x_0 be an arbitrary element of D; we show that s is continuous at x_0 by means of the ε-δ formulation of continuity (Theorem 3.4.6). For $\varepsilon > 0$, we must determine a $\delta > 0$ such that $|s(x) - s(x_0)| < \varepsilon$ whenever x is a point in D with $|x - x_0| < \delta$.

We begin by invoking uniform convergence by taking $N \in Z^+$ large enough that $|s_N(x) - s(x)| < \varepsilon/3$ for all $x \in D$. Since $s_N(x)$ is continuous at x_0, there is a $\delta > 0$ such that $|s_N(x) - s_N(x_0)| < \varepsilon/3$ whenever $x \in D$ and $|x - x_0| < \delta$. With this choice of δ we have, for all $x \in D$ with $|x - x_0| < \delta$,

$$\begin{aligned} |s(x) - s(x_0)| &= |s(x) - s_N(x) + s_N(x) - s_N(x_0) + s_N(x_0) - s(x_0)| \\ &\leq |s(x) - s_N(x)| + |s_N(x) - s_N(x_0)| + |s_N(x_0) - s(x_0)| \end{aligned}$$

The first and third terms are less than $\varepsilon/3$ because of the choice of N; the middle term is less than $\varepsilon/3$ because of the choice of δ. Thus, for $x \in D$ with $|x - x_0| < \delta$, we have

$$|s(x) - s(x_0)| < \frac{\varepsilon}{3} + \frac{\varepsilon}{3} + \frac{\varepsilon}{3} = \varepsilon \qquad \blacksquare$$

Example 1: Let $s_n(x) = (\sin x)/n$ $(x \in D = [0, 2\pi])$. We saw in Example 3 in Section 5.2 that the limit of this sequence is the zero function—which is, of course, continuous on D.

Example 2: Let $s_n(x) = \sin^n x$ $(x \in [0, \pi])$. The limit of this sequence of continuous functions is given by

$$s(x) = \begin{cases} 0 & \text{if } x \in [0, \pi],\ x \neq \dfrac{\pi}{2} \\ 1 & \text{if } x = \dfrac{\pi}{2} \end{cases}$$

The fact that the limit function is discontinuous tells us that the convergence is not uniform.

Integration

Exercise 5.1.23 shows that pointwise convergence does not necessarily preserve integration. We now consider another example of this phenomenon. It is described in an informal, geometric manner, but it is clear that the same conclusion can be reached by more formal methods.

Example 3: Let $n \in Z^+$. The graph of $s_n(x)$ consists of the line segment from $(0, 0)$ to $(1/n, n)$, the line segment from $(1/n, n)$ to $(2/n, 0)$, and the horizontal line segment from $(2/n, 0)$ to $(2, 0)$ (see Figure 5.4, where s_3 is sketched). For each n, the area of the ordinate set of s_n is clearly 1, hence, $\int_0^2 s_n = 1$ for $n \in Z^+$. On the other hand, it is easy to show that $s_n(x) \to 0$ $(x \in [0, 2])$ (work Exercise 5.3.2). Therefore, we have

$$\int_0^2 \lim_{n\to\infty} s_n = 0 \qquad \text{but} \qquad \lim_{n\to\infty} \int_0^2 s_n = 1$$

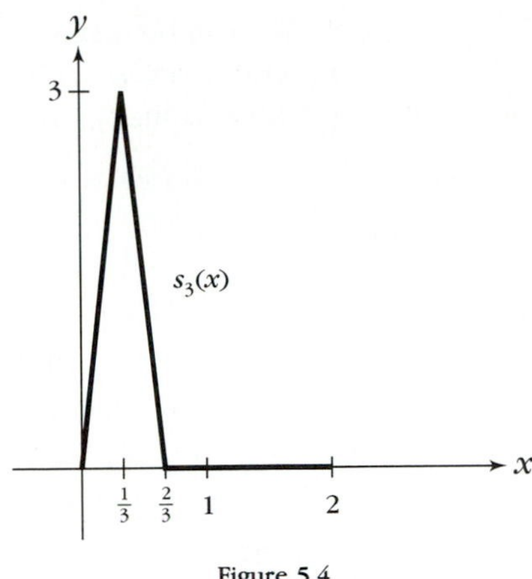

Figure 5.4
Graph of s_3

With uniform convergence, integrals are preserved:

Theorem 5.3.2 *Let $\{s_n\}_{n=1}^{\infty}$ be a sequence of real functions converging uniformly on the interval $[a, b]$ where $a < b$, and let $s(x) = \lim_{n\to\infty} s_n(x)$. If each s_n is integrable on $[a, b]$, then so is s and*

$$\int_a^b \lim_{n\to\infty} s_n = \int_a^b s = \lim_{n\to\infty} \int_a^b s_n$$

PROOF: There are two things to be proved: (1) that the limit function is integrable and (2) that the integral of the limit is the limit of the integrals.

(1) We use Theorem 4.7.6. It is sufficient to show that, for arbitrary $\varepsilon > 0$, there exist step functions S and T on $[a, b]$ such that

$$S(x) \leq s(x) \leq T(x) \qquad x \in [a, b]$$

and

$$\int_a^b (T - S) < \varepsilon$$

With ε given, we begin by choosing a fixed N large enough that the inequality

$$|s_N(x) - s(x)| < \frac{\varepsilon}{4(b - a)}$$

holds for all $x \in [a, b]$. Equivalently,

$$s_N(x) - \frac{\varepsilon}{4(b - a)} < s(x) < s_N(x) + \frac{\varepsilon}{4(b - a)}$$

This choice can be made because of the uniform convergence. Now s_N is integrable on $[a, b]$ by hypothesis; so, by Theorem 4.7.6, there exist step functions S_N and T_N such that

$$S_N(x) \leq s_N(x) \leq T_N(x) \qquad x \in [a, b]$$

and

$$\int_a^b (T_N - S_N) < \frac{\varepsilon}{2}$$

For $x \in [a, b]$, we define $S(x) = S_N(x) - \varepsilon/4(b - a)$ and $T(x) = T_N(x) + \varepsilon/4(b - a)$. Then S and T are the desired step functions, for s is between them, and

$$\int_a^b (T - S) = \int_a^b \left[T_N - S_N + \frac{\varepsilon}{2(b - a)}\right] = \int_a^b (T_N - S_N) + \int_a^b \frac{\varepsilon}{2(b - a)}$$
$$< \frac{\varepsilon}{2} + \frac{\varepsilon}{2} = \varepsilon$$

(2) To prove the limit statement, we let $\varepsilon > 0$ be given and choose N large enough that for all $n \geq N$, the inequality

$$|s_n(x) - s(x)| < \frac{\varepsilon}{b - a}$$

holds for all $x \in [a, b]$. By integrating the terms of the inequalities

$$s_n(x) - \frac{\varepsilon}{b - a} \leq s(x) \leq s_n(x) + \frac{\varepsilon}{b - a}$$

we get

$$\int_a^b s_n - \varepsilon \leq \int_a^b s \leq \int_a^b s_n + \varepsilon$$

Thus,

$$\left|\int_a^b s_n - \int_a^b s\right| < \varepsilon$$

whenever $n \geq N$. ■

Differentiation

A simple example shows that even though a sequence of functions converges uniformly on a set, the sequence of derivatives may fail to converge at a point in the set.

Example 4: Let $s_n(x) = x^n/n$ for $x \in [-1, 1]$. then $s_n(x) \rightrightarrows 0$ for $x \in [-1, 1]$ by the Little m Test with $m_n = 1/n$. However, the sequence of derivatives $\{x^{n-1}\}_{n-1}^{\infty}$ diverges at $x = -1$.

In order for differentiation to be preserved, we must require uniform convergence of the sequence of derivatives.

Theorem 5.3.3 *Let $\{s_n\}_{n=1}^{\infty}$ be a sequence of real functions defined on an interval I, converging pointwise to a function s on I:*

$$s_n(x) \to s(x) \qquad x \in I$$

If each s_n is continuously differentiable on I and if

$$s_n'(x) \rightrightarrows t(x) \qquad x \in I$$

where t is some function defined on I, then s is differentiable on I and

$$s'(x) = t(x) \qquad x \in I$$

PROOF: Let a be a fixed point in I. Because each s_n' is continuous, the Second Form of the Fundamental Theorem (Theorem 4.8.3) applies and

$$\int_a^x s_n' = s_n(x) - s_n(a)$$

Because $s_n'(x) \rightrightarrows t(x)$ $(x \in I)$, we can apply Theorem 5.3.2:

$$\begin{aligned} \lim_{n\to\infty} [s_n(x) - s_n(a)] &= \lim_{n\to\infty} \int_a^x s_n' \\ &= \int_a^x \lim s_n' \\ &= \int_a^x t \end{aligned}$$

But

$$\lim_{n\to\infty} [s_n(x) - s_n(a)] = s(x) - s(a)$$

and it follows that

$$s(x) = \int_a^x t + s(a)$$

Now t is continuous by Theorem 5.3.1, and, by the First Form of the Fundamental Theorem (Theorem 4.8.2), we see that s' is differentiable on I with

$$s'(x) = t(x) \qquad x \in I \qquad \blacksquare$$

Note: It is possible to prove a stronger version of this theorem: see the text by Goldberg.*

*Richard Goldberg, *Methods of Real Analysis*, 2nd ed. (New York: Wiley, 1976), p. 264.

Dini's Theorem on Monotone Convergence

Numerous examples can be found showing that even though a sequence of continuous functions converges to a continuous limit, the convergence may fail to be uniform. However, if the convergence takes place on a closed bounded interval, and if the functions converge *monotonically* to their limit, then the convergence must be uniform.

Theorem 5.3.4 **Dini's Theorem** *Let $\{s_n(x)\}_{n=1}^{\infty}$ be a sequence of continuous functions defined on $[a, b]$ and converging there to a continuous function s:*

$$s_n(x) \to s(x) \qquad x \in [a, b]$$

If the sequence $\{s_n(x)\}_{n=1}^{\infty}$ is decreasing for each $x \in [a, b]$ (or increasing for each $x \in [a, b]$), then

$$s_n(x) \rightrightarrows s(x) \qquad x \in [a, b]$$

PROOF: It is sufficient to consider the special case in which the limit is the zero function and $\{s_n(x)\}_{n=1}^{\infty}$ is decreasing for each $x \in [a, b]$ (work Exercise 5.3.10). Suppose by way of contradiction that the convergence is nonuniform. Then there exists an $\varepsilon > 0$ such that, no matter how large an $N \in Z^+$ is chosen, there exist positive integers $n > N$ with the property that $s_n(x) \geq \varepsilon$ for some $x \in [a, b]$. In the remainder of the proof, this ε is fixed.

Choose $n_1 \in Z^+$ and $x_1 \in [a, b]$ with the property that $s_{n_1}(x_1) \geq \varepsilon$. Then choose $n_2 \in Z^+$ and $x_2 \in [a, b]$ with the property that $n_2 > n_1$ and $s_{n_2}(x_2) \geq \varepsilon$. Continue in this manner, obtaining by induction a pair of sequences $\{n_k\}_{k=1}^{\infty}$ and $\{x_k\}_{k=1}^{\infty}$ such that $\{n_k\}_{k=1}^{\infty}$ is strictly increasing, $x_k \in [a, b]$, and $s_{n_k}(x_k) \geq \varepsilon$. Since the terms of $\{x_k\}_{k=1}^{\infty}$ lie in $[a, b]$, this sequence is bounded, and by the Bolzano-Weierstrass Theorem (Theorem 2.7.1), $\{x_k\}_{k=1}^{\infty}$ has a convergent subsequence. Suppose that $\{x_{k_i}\}_{i=1}^{\infty}$ is a subsequence converging to c. Then

$$x_{k_i} \to c = \lim_{i \to \infty} x_{k_i}$$

and it follows that $c \in [a, b]$. (Why?)

Now let m be an arbitrary positive integer. Consider the sequence of subscripts $\{n_{k_i}\}_{i=1}^{\infty}$ corresponding to $\{x_{k_i}\}_{i=1}^{\infty}$. Since this sequence approaches infinity, there is a value of i, say, i_0, such that $n_{k_i} > m$ whenever $i \geq i_0$. By the monotone convergence property,

$$s_m(x_{k_i}) \geq s_{n_{k_i}}(x_{k_i}) \qquad i \geq i_0$$

But $s_{n_{k_i}}(x_{k_i}) \geq \varepsilon$ and so

$$s_m(x_{k_i}) \geq \varepsilon \qquad i \geq i_0$$

We can now take the limit as $i \to \infty$, and because of the continuity of s_m at c, we obtain

$$s_m(c) = s_m\left(\lim_{i\to\infty} x_{k_i}\right) = \lim_{i\to\infty} s_m(x_{k_i}) \geq \varepsilon$$

Now m can be *any* positive integer, so the sequence $\{s_n(x)\}_{n=1}^{\infty}$ cannot converge to zero when $x = c$. This contradiction establishes the theorem. ■

Dini's Theorem can be generalized to compact subsets of R (work Exercise 5.3.11).

Example 5: Let $s_n(x) = x/(1 + nx)$ $(x \in [0, 1])$. Then each $s_n(x)$ is continuous on $[0, 1]$ and

$$s_{n+1}(x) \leq s_n(x) \qquad n \in Z^+$$

moreover, the limit function (zero) is continuous. Thus, the conditions of Dini's Theorem are fulfilled and we conclude that the convergence of $\{s_n(x)\}_{n=1}^{\infty}$ is uniform.

EXERCISES 5.3

1. Show that the convergence of each of the following sequences is not uniform on the set D by examining the limit function.
   (a) $s_n(x) = \sqrt[n]{x}$; $D = [0, 1]$
   (b) $s_n(x) = \cos^n x$; $D = \left[-\frac{\pi}{2}, \frac{\pi}{2}\right]$
   (c) $s_n(x) = x^n$; $D = [0, 1]$
   (d) $s_n(x) = e^{-nx}$; $D = [0, \infty)$
   (e) $s_n(x) = x^{1/(2n-1)}$; $D = [-1, 1]$
   (f) $s_n(x) = \dfrac{nx}{1 + nx}$; $D = [0, \infty)$
   (g) $s_n(x) = x^n e^{-x}$; $D = [0, 1]$
2. Show that the sequence of functions in Example 3 converges pointwise but not uniformly to zero on $[0, 2]$.
3. Let $s_n(x) = n^2 x^n (1 - x)$. Show that
   $$s_n(x) \to 0 \qquad x \in [0, 1].$$
   Show that the convergence is not uniform by examining the limit of the integrals $\int_0^1 s_n$.
4. Give an example of a sequence $\{s_n\}_{n=1}^{\infty}$ of real functions, integrable on an interval $[a, b]$ and converging to an integrable function there, but with the property that $\lim_{n\to 0} \int_a^b s_n$ does not exist.
5. Consider the set S of rational numbers in an interval $[a, b]$ where $a < b$. Because S is denumerable, its elements can be listed in a sequence $\{r_k\}_{k=1}^{\infty}$ such that every element of S occurs once and only once as a term of $\{r_k\}_{k=1}^{\infty}$. For $n \in Z^+$ and $x \in [a, b]$, define
   $$s_n(x) = \begin{cases} 1 & \text{if } x = r_k \text{ for some } k \leq n \\ 0 & \text{otherwise} \end{cases}$$
   Show that $\{s_n(x)\}_{n=1}^{\infty}$ converges on $[a, b]$ and that each s_n is integrable on $[a, b]$, but that the limit function is not integrable on $[a, b]$.
6. ★ Let $s_n(x) = n/(x^2 + n^2)$. Show that $\{s_n(x)\}_{n=1}^{\infty}$ is uniformly convergent on $[0, \infty)$ but that
   $$\lim_{n\to\infty} \int_0^{\infty} s_n \neq \int_0^{\infty} \lim_{n\to\infty} s_n$$
   (Thus, Theorem 5.3.2 cannot be applied to infinite intervals.)
7. Prove or disprove the following statements.
   (a) If $\{s_n(x)\}_{n=1}^{\infty}$ is a sequence of functions continuous on $[a, b]$, if $s_n(x) \to s(x)$ and if $s_{n+1}(x) \geq s_n(x)$ for each $x \in [a, b]$ and $n \in Z^+$, then $s_n(x) \rightrightarrows s(x)$.
   (b) If $\{s_n(x)\}_{n=1}^{\infty}$ is a sequence of functions integrable on $[a, b]$, if $s_n(x) \to s(x)$ where $s(x)$ is also integrable on $[a, b]$, and if $\int_a^b s = \lim_{n\to\infty} \int_a^b s_n$ then $s_n(x) \rightrightarrows s(x)$.

8. (a) Let $P_n(x) = 1 + x + \cdots + x_n/n!$, the nth Taylor polynomial for e^x at zero. Use Dini's Theorem to show that $\{P_n(x)\}_{n=1}^{\infty}$ converges uniformly to e^x on every closed bounded interval $[a, b]$ with $a > 0$.
(b) Is the convergence of $\{P_n(x)\}_{n=1}^{\infty}$ uniform on $[0, \infty)$?

9. Show that if $\{s_n(x)\}_{n=1}^{\infty}$ is a uniformly convergent sequence of continuous functions on a closed bounded interval $[a, b]$, then there exists a real number M such that $|s_n(x)| \leq M$ for $x \in [a, b]$ and $n \in Z^+$.

10. Explain why, in the proof of Dini's Theorem, it is sufficient to consider the special case described.

★11. Show that the conclusion of Dini's Theorem still holds if the monotone convergence property takes place on an arbitrary compact subset of R.

5.4 *Series of Functions*

In our discussion of Taylor series (Section 4.16) we found that some important functions can be represented as sums of infinite series (at least for certain values of the independent variable). Series representations of functions have many important uses in analysis, several of which we will be in a position to take up after we establish some basic facts.

We know that the convergence or divergence of an infinite series depends entirely on its sequence of partial sums. Once more, we distinguish between pointwise and uniform convergence:

Definition 5.4.1 Let $\{u_n(x)\}_{n=1}^{\infty}$ be a sequence of real functions, all defined on a set $D \subseteq R$. For $n \in Z^+$, let

$$S_n(x) = u_1(x) + u_2(x) + \cdots + u_n(x) = \sum_{k=1}^{n} u_n(x)$$

Then $\{S_n(x)\}_{n=1}^{\infty}$ is the sequence of ***partial sums*** of the infinite series

$$\sum_{k=1}^{\infty} u_k(x)$$

If there exists a real function $S(x)$, also defined on D, such that

$$S_n(x) \to S(x) \qquad x \in D$$

then $\sum_{k=1}^{\infty} u_k(x)$ ***converges pointwise*** to $S(x)$ on D and $S(x)$ is its ***sum***. If

$$S_n(x) \rightrightarrows S(x) \qquad x \in D$$

then $\sum_{k=1}^{\infty} u_k(x)$ ***converges uniformly*** to $S(x)$ on D.

Pointwise convergence of series is really nothing new—just think of x as fixed and proceed to determine convergence or divergence as in Chapter 2. For instance, the (pointwise) convergence of several of the Taylor series in

Section 4.16 was determined by means of the remainder formula in conjunction with the Ratio Test (Theorem 2.8.7) and Theorem 2.8.1.

Uniform convergence of a series is just a matter of the uniform convergence of its sequence of partial sums, and therefore the theorems about sequences in Section 5.3 have immediate consequences for series. Before we examine them, however, it is desirable to have some means of deciding whether or not a series converges uniformly. Our first result is similar to Theorem 2.8.1.

Theorem 5.4.1 *If the series* $\sum_{k=1}^{\infty} u_k(x)$ *converges uniformly on D, then* $u_k(x) \rightrightarrows 0$ $(x \in D)$.

PROOF: Let $S_n(x) = \sum_{k=1}^{n} u_k(x)$, so that, by hypothesis $\{S_n(x)\}_{n=1}^{\infty}$ converges uniformly on D to some real function $S(x)$. If $\varepsilon > 0$, we can choose N so that $|S_n(x) - S(x)| < \varepsilon/2$ for $n \geq N$. If $k \geq N$, then we have, for $x \in D$,

$$\begin{aligned} |u_{k+1}(x)| &= |S_{k+1}(x) - S_k(x)| \\ &\leq |S_{k+1}(x) - S(x)| + |S(x) - S_k(x)| \\ &< \frac{\varepsilon}{2} + \frac{\varepsilon}{2} = \varepsilon \end{aligned}$$

Therefore, $u_{k+1}(x) \rightrightarrows 0$ $(x \in D)$. ∎

Theorem 5.4.1 gives a necessary condition for uniform convergence; however, this condition is not sufficient. Like its predecessor, Theorem 5.4.1 is often used to rule out examples:

Example 1: The geometric series $\sum_{k=1}^{\infty} x^k$ is convergent for $x \in (-1, 1)$. However, $\{x^k\}_{k=1}^{\infty}$ is not uniformly convergent to zero there, so the series cannot converge uniformly.

Weierstrass M Test

The next theorem gives a *sufficent* condition for uniform convergence; it is a comparison test.

Theorem 5.4.2 **Weierstrass *M* Test** *Let* $\sum_{k=1}^{\infty} u_k(x)$ *be an infinite series whose terms are defined on a set* $D \subseteq R$, *and suppose that for* $k \in Z^+$ *the function* $|u_k(x)|$ *is bounded on D by a real number* M_k: *that is,*

$$|u_k(x)| \leq M_k \qquad x \in D$$

If the series $\sum_{k=1}^{\infty} M_k$ *is convergent, then* $\sum_{k=1}^{\infty} u_k(x)$ *is absolutely convergent at each point of D and uniformly convergent on D.*

KARL WEIERSTRASS

(b. October 31, 1815; d. February 19, 1897)

Weierstrass was in his forties before his abilities were recognized by the professional community and he was given a position at the University of Berlin. He became known as a great teacher and as the "father of modern analysis". Best known for his work in the theory of complex functions, power series, and uniform convergence, he is also known for his insistence on a greater rigor in proofs. He would highlight this need by constructing counterexamples to widely held beliefs; he was the first to devise a function that is continuous everywhere but has a derivative nowhere.

PROOF: For any given $x \in D$, the absolute convergence of $\sum_{k=1}^{\infty} u_k(x)$ is an immediate consequence of the Comparison Test for series of constants (Theorem 2.8.5).

To establish uniform convergence on D, we use the Cauchy Criterion (Theorem 5.2.2): Let $\varepsilon > 0$ be given. Denote the typical partial sum of the series $\sum_{k=1}^{\infty} M_k$ by T_n:

$$T_n = M_1 + M_2 + \cdots + M_n$$

Since $\sum_{k=1}^{\infty} M_k$ is assumed to be convergent, the sequence $\{T_n\}_{n=1}^{\infty}$ is convergent and must be Cauchy (Theorem 2.7.2). Therefore, there exists an $N \in Z^+$ such that

$$|T_n - T_m| < \varepsilon \qquad \text{for } n \geq m \geq N$$

It follows that for $n > m \geq N$,

$$T_n - T_m = M_{m+1} + \cdots + M_n < \varepsilon$$

Now consider the partial sums

$$S_n(x) = \sum_{k=1}^{n} u_k(x)$$

If $n > m \geq N$, then for $x \in D$,

$$\begin{aligned} |S_n(x) - S_m(x)| &= |u_{m+1} + u_{m+2}(x) + \cdots + u_n(x)| \\ &\leq |u_{m+1}(x)| + |u_{m+2}(x)| + \cdots + |u_n(x)| \\ &< M_{m+1} + M_{m+2} + \cdots + M_n \\ &< \varepsilon \end{aligned}$$

Therefore, by Theorem 5.2.2, the sequence $\{S_n(x)\}_{n=1}^{\infty}$ is uniformly convergent on D, and so $\sum_{k=1}^{\infty} u_k(x)$ is uniformly convergent on D. ■

Example 2: The series $\sum_{k=1}^{\infty} (\sin kx)/k^2$ is uniformly convergent on R because, for $M_k = 1/k^2$,

$$\left|\frac{\sin kx}{k^2}\right| \leq \frac{1}{k^2} = M_k \qquad x \in R$$

and the series of constants $\sum_{k=1}^{\infty} M_k$ converges.

Example 3: The series $\sum_{k=1}^{\infty} e^{-kx}$ is uniformly convergent on $[1, \infty)$ because for $x \geq 1$,

$$e^{-kx} \leq e^{-k}$$

and the series $\sum_{k=1}^{\infty} e^{-k}$ is a convergent geometric series.

Example 4: Let $u_k(x) = x^{2k}e^{-kx}$ $(x \geq 0)$. We will show that $\sum_{k=1}^{\infty} u_k(x)$ converges uniformly on $[0, \infty)$. For $x \geq 0$, the function x^2e^{-x} assumes its maximum value of $4/e^2$ when $x = 2$. Therefore,

$$|u_k(x)| = |(x^2e^{-x})^k| \leq \left(\frac{4}{e^2}\right)^k \qquad x \geq 0$$

With $M_k = (4/e^2)^k$, the geometric series $\sum_{k=1}^{\infty} M_k$ converges, and it follows that $\sum_{k=1}^{\infty} u_k(x)$ converges uniformly on $[0, \infty)$.

Example 5: Let $u_k(x) = x^{k^2} \ln x$ for $0 < x \leq 1$ and let $u_k(0) = 0$. On $[0, 1]$ the function $|u_k(x)|$ assumes a maximum of $1/k^2e$ when $x = e^{-1/k^2}$. (Verify.) With $M_k = 1/k^2e$, the series $\sum_{k=1}^{\infty} M_k$ converges, and so $\sum_{k=1}^{\infty} u_k(x)$ converges uniformly on $[0, 1]$.

Example 6: Let $u_k(x) = kxe^{-kx}$ $(x \geq 0)$. For a fixed $x > 0$, the Ratio Test (Theorem 2.8.7) tells us that the series $\sum_{k=1}^{\infty} u_k(x)$ is convergent because

$$\frac{u_{k+1}(x)}{u_k(x)} = \frac{(k+1)xe^{-(k+1)x}}{kxe^{-kx}} = \frac{k+1}{k}e^{-x} \to e^{-x} < 1$$

Clearly the series converges to zero when $x = 0$, so we have convergence on $[0, \infty)$. By Theorem 5.4.1, this convergence is not uniform—no matter how large k is chosen, the function $u_k(x)$ assumes the value e^{-1} at $x = 1/k$, so that $\{u_k(x)\}_{k=1}^{\infty}$ does not converge uniformly to zero. However, if a is a fixed positive number, the convergence is uniform on $[a, \infty)$. In that case we can first choose k such that $1/k < a$, and then for $k \geq K$ the terms of the series are decreasing on $[a, \infty)$ and the tail of the series is dominated by the convergent series

$$\sum_{k=K}^{\infty} kae^{-ka}$$

Continuity of the Sum

Now we turn to theorems about the continuity, integration, and differentiation of series.

Theorem 5.4.3 *If $\sum_{k=1}^{\infty} u_k(x)$ converges uniformly on a set $D \subseteq R$ and if each $u_k(x)$ is continuous on D, then the sum is continuous on D.*

PROOF: Let the nth partial sum of the given series be denoted by $S_n(x)$, and let the sum be $S(x)$. Then, by hypothesis,

$$S_n(x) = \sum_{k=1}^{n} u_k(x) \rightrightarrows S(x) \qquad x \in D$$

Since each $u_k(x)$ is continuous on D, so is $S_n(x)$. The continuity of $S(x)$ now follows by Theorem 5.3.1. ∎

Example 7: The sum of the series in Example 2,

$$S(x) = \sum_{k=1}^{\infty} \frac{\sin kx}{k^2}$$

is continuous on R.

Example 8: For $x \in [0, 2)$, the limit function of the series $\sum_{k=1}^{\infty} x(1 - x)^k$ is discontinuous:

$$S(x) = \begin{cases} 0 & \text{if } x = 0 \\ 1 & 0 < x \leq 2 \end{cases}$$

Thus, the convergence is not uniform.

Integral of the Sum

Theorem 5.4.4 *If $\sum_{k=1}^{\infty} u_k(x)$ converges uniformly on the interval $[a, b]$ and if each $u_k(x)$ is* *bounded and integrable on [a, b], then the sum S(x) is integrable on [a, b] and*

$$\int_a^b S = \sum_{k=1}^{\infty} \int_a^b u_k$$

(We say that the series can be integrated term by term.)

PROOF: Again, let the partial sums of the given series be denoted by $S_n(x)$ and the sum by $S(x)$, so that

$$S_n(x) = \sum_{k=1}^{n} u_k(x) \rightrightarrows S(x) \qquad x \in [a, b]$$

Each $S_n(x)$ is a sum of integrable functions on $[a, b]$, hence (by Corollary 4.7.1), integrable itself with

$$\int_a^b S_n = \sum_{k=1}^{n} \int_a^b u_k$$

By Theorem 5.3.2, $S(x)$ is integrable on $[a, b]$ and

$$\int_a^b S = \lim_{n\to\infty} \int_a^b S_n = \lim_{n\to\infty} \sum_{k=1}^{n} \int_a^b u_k = \sum_{k=1}^{\infty} \int_a^b u_k \qquad \blacksquare$$

Example 9: By Theorem 5.4.4, the series of Example 2 can be integrated term by term from zero to π to obtain

$$\begin{aligned}\int_0^{\pi} S &= \int_0^{\pi} \left(\sum_{k=1}^{\infty} \frac{\sin kx}{k^2} \right) dx = \sum_{k=1}^{\infty} \left(\int_0^{\pi} \frac{\sin kx}{k^2}\, dx \right) \\ &= \sum_{k=1}^{\infty} \left. \frac{-\cos kx}{k^3} \right|_0^{\pi} = \sum_{k=1}^{\infty} \left[\frac{(-1)^{k+1}}{k^3} + \frac{1}{k^3} \right] \\ &= \sum_{m=0}^{\infty} \frac{2}{(2m+1)^3}\end{aligned}$$

Example 10: We know that the geometric series $\sum_{k=1}^{\infty} x^k$ converges pointwise in the open interval $(-1, 1)$ to $S(x) = 1/(1 - x)$. By the M Test, the convergence is uniform on every closed interval of the type $[a, b]$, where $-1 < a < b < 1$. (To see this, let $M_k = c^k$, where c is the maximum of the two numbers $|a|$, $|b|$.)

It follows that Theorem 5.4.4 can be applied to this series on any interval whose endpoints lie in the open interval $(-1, 1)$. We choose $t \in (-1, 1)$ and integrate from 0 to t:

$$\begin{aligned}\int_0^t \frac{1}{1-x}\, dx &= \int_0^t \left(\sum_{k=0}^{\infty} x^k \right) dx = \sum_{k=0}^{\infty} \left(\int_0^t x^k\, dx \right) \\ &= \sum_{k=0}^{\infty} \frac{t^{k+1}}{k+1}\end{aligned}$$

On the other hand, we know that the integral on the left is equal to $-\ln(1 - t)$, giving us a representation of $\ln(1 - t)$ as a series:

$$\ln(1 - t) = -\sum_{k=0}^{\infty} \frac{t^{k+1}}{k + 1} \qquad -1 < t < 1$$

(Compare this result with Example 5 in Section 4.16.)

Derivative of the Sum

Next we consider differentiation of infinite series of functions. If $\sum_{k=1}^{\infty} u_k(x)$ is a series whose terms are all continuously differentiable, then the series $\sum_{k=1}^{\infty} u_k'(x)$ is known as the *derived* series. The next theorem states the conditions under which the derivative of the original series is equal to the derived series.

Theorem 5.4.5 *Let $\sum_{k=1}^{\infty} u_k(x)$ be an infinite series all of whose terms $u_k(x)$ are continuously differentiable on an interval I. Suppose that $\sum_{k=1}^{\infty} u_k(x)$ converges on I and that $\sum_{k=1}^{\infty} u_k'(x)$ converges uniformly on I. For $x \in I$, let $S(x)$ be the sum of the series $\sum_{k=1}^{\infty} u_k(x)$ and $T(x)$ be the sum of the series $\sum_{k=1}^{\infty} u_k'(x)$. Then $S(x)$ is differentiable on I, and $S'(x) = T(x)$ $(x \in I)$. (We say that the series can be differentiated term by term.)*

PROOF: This theorem follows from Theorem 5.3.3 in the same manner as the preceding theorems follow from their counterparts (work Exercise 5.4.5).

■

Example 11: For the geometric series

$$S(x) = \sum_{k=0}^{\infty} x^k$$

the derived series is

$$T(x) = \sum_{k=1}^{\infty} kx^{k-1}$$

By the Ratio Test (Theorem 2.8.7), the derived series converges pointwise on $(-1, 1)$, and an application of the M Test like that in Example 10 shows that uniform convergence occurs on every closed interval $[a, b]$ with $-1 < a < b < 1$ (work Exercise 5.4.7). Given any $x \in (-1, 1)$, we can choose a and b such that $-1 < a < x < b < 1$ and then apply Theorem 5.4.5 on $[a, b]$:

$$\begin{aligned} T(x) &= \sum_{k=1}^{\infty} kx^{k-1} \\ &= S'(x) \\ &= \frac{1}{(1-x)^2} \end{aligned}$$

Since x was an arbitrary element of $(-1, 1)$, the equation holds throughout the open interval.

Example 12: We cannot apply Theorem 5.4.5 to the series of Example 2 on the interval $[0, \pi]$ because the derived series

$$\sum_{k=1}^{\infty} \frac{\cos kx}{k}$$

is not uniformly convergent there. In fact, the derived series is not even pointwise convergent on $[0, \pi]$.

EXERCISES 5.4

1. Show that the following series are uniformly convergent on the indicated sets.
   (a) $\sum_{k=1}^{\infty} \left(\frac{\sin x}{2}\right)^k; \quad x \in R$
   (b) $\sum_{k=1}^{\infty} \sin^k x; \quad -\frac{\pi}{3} \le x \le \frac{\pi}{4}$
   (c) $\sum_{k=1}^{\infty} \frac{\cos kx}{k^2}; \quad x \in R$
   (d) $\sum_{k=1}^{\infty} \frac{x}{x^2 + k^2}; \quad -1 \le x \le 1$
   (e) $\sum_{k=1}^{\infty} \frac{1}{k^x}; \quad x \ge 2$
   (f) $\sum_{k=1}^{\infty} \frac{1}{k^2 x^2}; \quad |x| \ge 1$
   (g) $\sum_{k=1}^{\infty} xe^{-k^2 x}; \quad x \ge 0$
   (h) $\sum_{k=1}^{\infty} x^k e^{-kx}; \quad x \ge 0$
2. Find an infinite series for $\int_0^{\pi/2} S(x)\, dx$, where $S(x) = \sum_{k=1}^{\infty} \cos kx/k^2$. Justify.
3. Find an infinite series for $S'(x)$ where $S(x) = \sum_{k=1}^{\infty} \sin kx/k^3$. Where is this representation valid?
4. Use the geometric series
   $$\frac{1}{1+x^2} = \sum_{k=0}^{\infty} (-1)^k x^{2k} \qquad -1 < x < 1$$
   and the fact that
   $$\arctan t = \int_0^t \frac{1}{1+x^2}\, dx$$
   to obtain an infinite series for arctan t. Explain why your result is valid for $t \in (-1, 1)$.
5. Prove Theorem 5.4.5.
6. Let $S(x) = \sum_{k=1}^{\infty} x^k (1-x)^k$. Use Theorem 5.4.5 to show that S is differentiable on $[0, 1]$ and find an infinite series for its derivative.
7. Show that the derived series $\sum_{k=0}^{\infty} kx^{k-1}$ in Example 10 is uniformly convergent on every closed subinterval of $(-1, 1)$.
8. (a) Formulate a Cauchy Criterion for uniform convergence of series of functions.
   (b) Show that $x^k/k \rightrightarrows 0$ for $x \in [-1, 1]$.
   (c) Show that $\sum_{k=1}^{\infty} x^k/k$ converges pointwise but not uniformly on $[-1, 1)$.

9. Prove or disprove the following statements.

(a) If $\sum_{k=1}^{\infty} u_k(x)$ and $\sum_{k=1}^{\infty} v_k(x)$ are both uniformly convergent on D, then so is $\sum_{k=1}^{\infty} [u_k(x) + v_k(x)]$.

(b) If $u_k(x) \rightrightarrows 0$ $(x \in D)$, then $\sum_{k=1}^{\infty} u_k(x)$ converges on D.

(c) If $\sum_{k=1}^{\infty} u_k(x)$ converges uniformly on D and $g(x)$ is bounded on D, then $\sum_{k=1}^{\infty} u_k(x)g(x)$ converges uniformly on D. (*Hint:* Use Exercise 8a.)

10. (a) Prove the following Alternating Series Test for series of functions: Suppose that $\sum_{k=1}^{\infty} (-1)^{k+1}u_k(x)$ converges for $x \in [a, b]$ where $a < b$. If (1) $u_k(x) \geq u_{k+1}(x) \geq 0$ for each $x \in [a, b]$ and $k \in Z^+$ and if (2) each $u_k(x)$ is decreasing (as a function of $x \in [a, b]$) then the convergence of the series is uniform on $[a, b]$. [*Hint:* By the Estimate of Error in Section 2.9,

$$|S_n(x) - S(x)| \leq u_{n+1}(x)$$

and the Little m Test can be applied with $m_n = u_{n+1}(a)$.]
Note: In condition 2, *decreasing* can be replaced by *increasing*.

(b) Show that $\sum_{k=1}^{\infty} (-1)^{k+1}x^k/k$ is uniformly convergent on $[0, 1]$. (Thus, a series that converges uniformly on an interval can converge conditionally at an endpoint of the interval.)

5.5 *Power Series*

The Taylor series encountered in Section 4.16 is an important example of a type of series of functions in which the terms involve powers of $x - c$.

Definition 5.5.1 Let $c \in R$ and let $\{a_k\}_{k=1}^{\infty}$ be a sequence of real numbers. An infinite series of the form

$$\sum_{k=0}^{\infty} a_k(x - c)^k \tag{1}$$

is a ***power series*** in $x - c$; the numbers a_k are the ***coefficients*** of the series (1).

Several questions come to mind at once: Given a particular power series (1), what is the set D of real numbers x for which it converges? What can we say about uniform convergence? If f is a function defined for $x \in D$ by such a series; that is, if

$$f(x) = \sum_{k=0}^{\infty} a_k(x - c)^k$$

then what properties does f have? Where is f continuous? Where is it differentiable? And on what intervals is it integrable? Of course, we must begin by investigating the fundamental questions of convergence and uniform convergence. To simplify the discussion we will usually deal with the case in which $c = 0$; that is, we will consider power series in x:

$$\sum_{k=0}^{\infty} a_k x^k \tag{2}$$

Results for the more general series (1) can be obtained by simply replacing x by $x - c$ in (2).

Interval of Convergence

As we will soon see, the set of real numbers for which a power series converges is an interval. This interval may be open, closed, or half-open; it may be $(-\infty, \infty)$, or it may degenerate to a single point.

Theorem 5.5.1 *If the power series (2) converges for $x = x_1$ and if $|x_0| < |x_1|$, then the series converges absolutely for $x = x_0$.*

PROOF: Suppose that $\sum_{k=0}^{\infty} a_k x_1^k$ is convergent. Then the terms $a_k x_1^k \to 0$ (by Theorem 2.8.1) and are therefore bounded. Now suppose that $|a_k x_1^k| \le M$ for $k \in Z^+$. Then if $|x_0| < |x_1|$, we can write

$$|a_k x_0^k| = \left| a_k \cdot x_1^k \cdot \frac{x_0^k}{x_1^k} \right| \le M \cdot \left| \frac{x_0}{x_1} \right|^k$$

Since $|x_0/x_1| < 1$, the series $\sum_{k=0}^{\infty} |a_k x_0^k|$ is dominated by the convergent geometric series $\sum_{k=0}^{\infty} M|x_0/x_1|^k$ and must therefore converge itself; thus, $\sum_{k=0}^{\infty} a_k x_0^k$ converges absolutely. ∎

Theorem 5.5.2 *Let D be the set of real numbers x for which the power series (2),*

$$\sum_{k=0}^{\infty} a_k x^k$$

converges. Then just one of the following conditions holds:

(i) $D = \{0\}$
(ii) $D = R$
(iii) There exists a positive real number r such that the series (2) converges absolutely for $|x| < r$ and diverges for $|x| > r$.

PROOF: Clearly $0 \in D$. If D contains numbers other than 0, we consider two cases: (I) D is unbounded, and (II) D is bounded.

(I) If D is unbounded, then for any $x \in R$ there exists some $x_1 \in D$ such that $|x_1| > |x|$. Since $x_1 \in D$, $\sum_{k=0}^{\infty} a_k x^k$ converges absolutely by Theorem 5.5.1. But x was arbitrary, so $D = R$.

(II) If D is bounded and $D \ne \{0\}$, then D contains some positive numbers. (Why?) Let $r = \sup D > 0$. We will show that condition (iii) holds for this choice of r. Suppose x is a fixed real number with $|x| < r$. Because r is defined as the supremum of D and $|x| < r$, there exists an element x_1 of D such that $|x| < x_1$. Now $\sum_{k=0}^{\infty} a_k x_1^k$ converges, and by Theorem 5.5.1, $\sum_{k=0}^{\infty} a_k x^k$ converges absolutely. Next suppose $|x| > r$. If the series $\sum_{k=0}^{\infty} a_k x^k$ were to converge, there would exist real numbers greater than r but less than $|x|$ for which the series converges [for example, $x_2 = \frac{1}{2}(|x| + r)$]. But then we have contradicted the choice of r. Thus, the series diverges for $|x| > r$. ∎

Example 1: The geometric series $\sum_{k=0}^{\infty} x^k$ converges if and only if $|x| < 1$. Condition (iii) applies here, with $r = 1$.

Example 2: The series $\sum_{k=1}^{\infty} x^k/k$ converges by the Ratio Test whenever $|x| < 1$. It also converges when $x = -1$ by the Alternating Series Test (Theorem 2.9.1). It diverges at all other points. Again, $r = 1$.

Example 3: The series $\sum_{k=0}^{\infty} x^k/k!$ satisfies condition (ii) of the theorem because it converges by the Ratio Test for all real x.

Example 4: The series $\sum_{k=0}^{\infty} k!x^k$ converges for $x = 0$ but nowhere else, illustrating condition (i).

We now restate Theorem 5.5.2 as it applies to power series of the form (1).

Theorem 5.5.2′ *Let D be the set of real numbers for which the power series (1),*

$$\sum_{k=0}^{\infty} a_k(x - c)^k$$

converges. Then just one of the following conditions holds:

(i) $D = \{c\}$
(ii) $D = R$
(iii) *There exists a positive real number r such that the series (1) converges absolutely for $|x - c| < r$ and diverges for $|x - c| > r$.*

Example 5: The series

$$\sum_{k=1}^{\infty} \frac{(x + 3)^k}{k^2}$$

converges whenever $|x + 3| < 1$; it also converges when $x + 3 = \pm 1$, but it diverges whenever $|x + 3| > 1$. Thus, the series (1) converges if and only if $x \in [-4, -2]$.

In case (iii) of Theorem 5.5.2′, the number r is known as the *radius of convergence* of the series, and the *interval of convergence* is the open interval $(c - r, c + r)$. In case (ii) we say the radius of convergence is infinity, and in case (i) it is zero. The radius of convergence (hence, the interval of convergence) can frequently be determined by means of the Ratio Test. However, it is necessary to check the endpoints of the interval of convergence specifically (as in Examples 2 and 5).

The term *interval of convergence* is applied to an open interval; in Example 2, for instance, the interval of convergence is $(-1, 1)$, and in Example 5 it is $(-4, -2)$. In the interval of convergence, the convergence is absolute by Theorem 5.5.2. The examples above show that, at the endpoints of the interval of convergence, it is possible for the series to converge absolutely, converge conditionally, or diverge.

The next theorem gives us information about the uniform convergence of power series.

Theorem 5.5.3 *If a power series (2) has a positive radius of convergence, then it converges uniformly on every closed bounded interval whose endpoints are contained in the interval of convergence.*

PROOF: Let a, b $(a < b)$ belong to the interval of convergence of the series (2); since this interval is open, it is possible to find a number d, also in the interval of convergence, such that $|a| < d$ and $|b| < d$. Then $\sum_{k=1}^{\infty} a_k d^k$ is absolutely convergent (by Theorem 5.5.2), and for $x \in [a, b]$ we have

$$|a_k x^k| \leq |a_k| d^k$$

The theorem now follows by an application of the M Test, with $M_k = |a_k| d^k$. ■

Continuity and Integration

Of course, the conclusion of Theorem 5.5.3 applies as well to the more general type of power series (1): the convergence is uniform on any closed bounded subinterval of the interval of convergence. We let I be the interval of convergence of (1) and let $x_0 \in I$. Then x_0 can be enclosed in a closed bounded interval $J \subseteq I$, and the convergence is uniform on J; now the terms of the series are continuous on J and so, by Theorem 5.4.3, the sum is continuous on J, hence at x_0.

In the same way, given two points a and $b \in I$, we can apply Theorem 5.4.4 to obtain

$$\int_a^b \sum_{k=0}^{\infty} a_k(x-c)^k\,dx = \sum_{k=0}^{\infty} \int_a^b a_k(x-c)^k\,dx \tag{3}$$

We then have:

Theorem 5.5.4 *A power series is continuous at each point in its interval of convergence I. Moreover, the sum of the series is integrable on every closed bounded interval whose endpoints are contained in I and the integral of the sum is the sum of the integrals, as expressed by Equation (3).*

Differentiation

Consider a function defined by a power series

$$S(x) = \sum_{k=0}^{\infty} a_k(x-c)^k$$

with positive radius of convergence (possibly $r = \infty$). Its derived series is given by

$$T(x) = \sum_{k=1}^{\infty} ka_k(x-c)^{k-1}$$

We wish to show that $T(x) = S'(x)$ for each x in the interval of convergence of S. The key to this result is the following theorem, which we prove in the special case with $c = 0$.

Theorem 5.5.5 *If a power series has a positive radius of convergence, then its derived series has the same radius of convergence.*

PROOF: Let $S(x) = \sum_{k=0}^{\infty} a_k x^k$ have radius of convergence $r > 0$, so that $T(x) = \sum_{k=1}^{\infty} ka_k x^{k-1}$ is the derived series. Since

$$|a_k x^k| \le |ka_k x^k| = |x|\,|ka_k x^{k-1}|$$

it is clear by a comparison that the radius of convergence of the derived series is not greater than r. We show it is not less than r. Let x_0 be a fixed point in the interval of convergence of $S(x)$. Now we choose x_1 to be a point, also in the interval of convergence of $S(x)$, with $|x_1| > |x_0|$. Then

$$ka_k x_0^{k-1} = k(a_k x_1^{k-1})\left(\frac{x_0}{x_1}\right)^{k-1}$$

Since $\sum_{k=1}^{\infty} a_k x_1^k$ is convergent, the sequence $\{a_k x_1^k\}_{k=1}^{\infty}$ is null, hence bounded, say, by B: $|a_k x_1^k| \leq B$. So

$$\begin{aligned} |ka_k x_0^{k-1}| &\leq k \cdot \frac{B}{|x_1|} \cdot \left|\frac{x_0}{x_1}\right|^{k-1} \\ &= \frac{B}{|x_1|} k\rho^{k-1} \end{aligned}$$

where $\rho = |x_0/x_1| < 1$. The series $\sum_{k=1}^{\infty} |ka_k x_0^{k-1}|$ is therefore dominated by the series

$$\frac{B}{|x_1|} \sum_{k=1}^{\infty} k\rho^{k-1}$$

whose convergence follows by the Ratio Test. Thus, the derived series converges absolutely at every point in the interval of convergence of $S(x)$. ■

It is clear that the proof of Theorem 5.5.5 can be modified to take into account the possibility that $r = \infty$. It follows (work Exercise 5.5.3) that a power series can be differentiated term by term within its interval of convergence:

Corollary 5.5.1 *If the power series (1) has a positive radius of convergence, then it is differentiable within its interval of convergence and*

$$\frac{d}{dx} \sum_{k=0}^{\infty} a_k(x - c)^k = \sum_{k=0}^{\infty} ka_k(x - c)^{k-1}$$

Example 6: The series $\sum_{k=1}^{\infty} x^k/k^2$ and its derived series $\sum_{k=1}^{\infty} x^{k-1}/k$ have the interval of convergence $(-1, 1)$. Note, however, that they behave differently at the endpoints: The original series converges absolutely at $x = \pm 1$, but the derived series converges conditionally at $x = -1$ and diverges at $x = 1$.

Since the derivative of a function defined by a power series having interval of convergence I is another power series with the same interval of convergence, we can differentiate again to obtain the second derivative. By mathematical induction, the original function has derivatives of all orders on I.

Corollary 5.5.2 *The function defined by a power series on its interval of convergence has derivatives of all orders there.*

Applications

In its interval of convergence, a power series behaves in many ways like a polynomial. Limits, differentiation, and integration are relatively easy for polynomials, and because power series can be differentiated and integrated term by term under fairly relaxed conditions, these series are extremely useful tools in analysis.

Example 7: There is no elementary antiderivative of the function $f(x) = e^{-x^2}$, but f is continuous everywhere and Riemann integrable on every closed bounded interval. We will approximate $\int_0^1 e^{-x^2} dx$.

Since e^u is represented by the power series $\sum_{k=0}^{\infty} u^k/k!$ for all values of u, we may replace u by $-x^2$ to obtain

$$f(x) = \sum_{k=0}^{\infty} \frac{(-1)^k x^{2k}}{k!} = 1 - x^2 + \frac{x^4}{2!} - \cdots + \frac{(-1)^k x^{2k}}{k!} + \cdots.$$

By integrating term by term, we get

$$\int_0^1 f(x)\, dx = \sum_{k=0}^{\infty} \int_0^1 \frac{(-1)^k x^{2k}}{k!}\, dx = \sum_{k=0}^{\infty} \frac{(-1)^k}{(2k+1)k!}$$

This alternating series converges rapidly. The sum of the first five terms

$$S_4 = 1 - \frac{1}{3} + \frac{1}{10} - \frac{1}{42} + \frac{1}{216}$$

is found to be 0.747486. . . . By the estimate of error for the Alternating Series Test (Theorem 2.9.1), this approximation is within $\frac{1}{1320}$ of the exact value.

Example 8: We evaluate the limit

$$\lim_{x \to 0} \frac{\sin x - x}{x^3}$$

The numerator is given by the power series

$$(x - \frac{x^3}{3!} + \frac{x^5}{5!} - \frac{x^7}{7!} + \cdots) - x$$

or

$$-\frac{x^3}{3!} + \frac{x^5}{5!} - \frac{x^7}{7!} + \cdots$$

For $x \neq 0$, we can divide by x^3 to obtain

$$\frac{\sin x - x}{x^3} = -\frac{1}{3!} + \frac{x^2}{5!} - \frac{x^4}{7!} + \cdots$$

The power series on the right converges everywhere and therefore is continuous everywhere, so we may take the desired limit by substituting $x = 0$:

$$\lim_{x \to 0} \frac{\sin x - x}{x^3} = -\frac{1}{3!} = -\frac{1}{6}$$

Example 9: The power series for $\sin x$ can be differentiated term by term to obtain a series for $\cos x$:

$$\cos x = \frac{d}{dx}(\sin x) = \frac{d}{dx}\sum_{k=0}^{\infty}(-1)^k\frac{x^{2k+1}}{(2k+1)!} = \sum_{k=0}^{\infty}\frac{(-1)^k x^{2k}}{(2k)!}$$

This series is the same one we would have obtained in Section 4.16 had we applied Taylor's Theorem to $\cos x$ at zero.

Uniqueness

Example 9 raises a theoretical question of interest: If there are several ways to obtain a power series for a function at a point c, do they all lead to the same result? The answer is affirmative, and the proof is simple.

Theorem 5.5.6 *Suppose that for every x in an open interval $(c - r, c + r)$, where $r > 0$, a function $f(x)$ is represented by a power series*

$$f(x) = \sum_{k=0}^{\infty} a_k(x - c)^k \tag{4}$$

and also by a power series

$$f(x) = \sum_{k=0}^{\infty} b_k(x - c)^k \tag{5}$$

Then these series are identical:

$$a_k = b_k \qquad \text{for } k = 0, 1, 2, \ldots \tag{6}$$

PROOF: Since we can substitute $x = c$ in series (4) and (5), we have $a_0 = b_0$. Then we can differentiate term by term to obtain

$$\sum_{k=1}^{\infty} k a_k(x - c)^{k-1} = \sum_{k=1}^{\infty} k b_k(x - c)^{k-1}$$

It follows, by letting $x = c$ again, that $a_1 = b_1$. A second differentiation shows that $2a_2 = 2b_2$, so $a_2 = b_2$. By induction we can show that $a_k = b_k$ for $k \in Z^+$ (work Exercise 5.5.9). ■

Another way of stating Theorem 5.5.6 is that if two power series in $x - c$ converge to the same sum on an interval $(c - r, c + r)$ with $r > 0$, then they are identical in the sense of Equation (6). Consequently, the Taylor series for a function at c is unique (if it exists).

Series Solutions of Differential Equations

Power series can be applied effectively to solve certain differential equations. We assume that the equation to be solved has solutions that can be expressed by power series, and attempt to determine the coefficients by requiring them to satisfy the equation. Here is a very simple example of this approach:

Example 10: To determine a function y satisfying the differential equation

$$y' = 2y \tag{7}$$

on some interval centered at zero, we assume that there is a solution of the form $y = \sum_{k=0}^{\infty} a_k x^k$. Then we compute $2y$ and y' and require the resulting series to be identical. It is helpful to write out the series in extended form:

$$y = a_0 + a_1 x + a_2 x^2 + a_3 x^3 + \cdots + a_k x^k + \cdots$$

Then

$$2y = 2a_0 + 2a_1 x + 2a_2 x^2 + 2a_3 x^3 + \cdots + 2a_k x^k + \cdots$$

and

$$y' = a_1 + 2a_2 x + 3a_3 x^2 + 4a_4 x^3 + \cdots + (k+1)a_{k+1} x^k + \cdots$$

In order for y' and $2y$ to be identical, the corresponding coefficients must be equal by Theorem 5.5.6. Thus, for $k \geq 0$,

$$(k+1)a_{k+1} = 2a_k$$

If a_0 is chosen arbitrarily, the remaining coefficients can be expressed in terms of a_0 by means of the relation

$$a_{k+1} = \frac{2a_k}{k+1} \qquad k \in Z^+$$

So we get

$$a_1 = 2a_0 \qquad a_2 = \frac{2a_1}{2} = \frac{4a_0}{2} \qquad a_3 = \frac{2a_2}{3} = \frac{8a_0}{3 \cdot 2}$$

and so on. By induction

$$a_k = \frac{2^k a_0}{k!}$$

and the power series determined by these coefficients is

$$\sum_{k=0}^{\infty} \frac{2^k a_0}{k!} x^k = a_0 \sum_{k=0}^{\infty} \frac{(2x)^k}{k!}$$

In this form we can recognize the sum on the right as the series for e^{2x}, which converges for all $x \in R$. Since a_0 was an arbitrary constant, we write our result in the form

$$y = Ce^{2x} \tag{8}$$

It can be shown that this function does satisfy the differential equation (7) and that all solutions of (7) have the form given in (8).

Example 10 gives just a glimpse into the use of power series in solving differential equations. Those who would like to see more of this important application can consult the text by Kreyszig*, which contains a great deal of information at an elementary level.

Convergence at Endpoints

Our experience with the convergence of power series shows that the series may behave in various ways at the endpoints of the interval of convergence: it may converge absolutely, converge conditionally, or diverge. Now suppose a function is represented by its Taylor series in its interval of convergence:

$$f(x) = \sum_{k=0}^{\infty} a_k x^k \qquad -r < x < r$$

What if the series should converge at $x = r$? If $f(r)$ exists, must the series converge to $f(r)$? The answer is affirmative if f is continuous from the left at r.

Theorem 5.5.7 *If a power series* $\sum_{k=0}^{\infty} a_k x^k$ *converges in* $(-r, r)$ *to a function* $f(x)$*, and if it converges at* r*, then the sum at* r *is given by*

$$\sum_{k=0}^{\infty} a_k r^k = \lim_{x \to r^-} f(x)$$

A similar statement holds if the series converges at $-r$. Thus, a power series represents a continuous function wherever it converges. This result will not be proved here; it is due to the work of Niels Abel. For a proof, see the text by Taylor and Mann.†

Arithmetic of Power Series

The sum and difference of two infinite series were defined in Chapter 2. For power series in x these definitions mean that

$$\sum_{k=0}^{\infty} a_k x^k \pm \sum_{k=0}^{\infty} b_k x^k = \sum_{k=0}^{\infty} (a_k \pm b_k) x^k$$

Products and quotients were not defined in Chapter 2 since they are a little more complicated. We define these now for power series in x:

*E. Kreyszig, *Advanced Engineering Mathematics*, 6th ed. (New York: Wiley, 1988), pp. 180–241.
†A. E. Taylor and W. R. Mann, *Advanced Calculus*, 3rd ed. (New York: Wiley, 1983), pp. 643–646.

Definition 5.5.2 The ***product*** of $\sum_{k=0}^{\infty} a_k x^k$ and $\sum_{k=0}^{\infty} b_k x^k$ is the series $\sum_{k=0}^{\infty} c_k x^k$, where

$$c_k = a_0 b_k + a_1 b_{k-1} + \cdots + a_k b_0 = \sum_{j=0}^{k} a_j b_{k-j}$$

We write

$$\left(\sum_{k=0}^{\infty} a_k x^k\right)\left(\sum_{k=0}^{\infty} b_k x^k\right) = \sum_{k=0}^{\infty} c_k x^k$$

It is easy to show that this definition is consistent with the manner in which polynomials are multiplied in high school algebra: We multiply each term of the first series by each term of the second and then collect terms with like powers of x.

Example 11: The product of the series $\sum_{k=0}^{\infty} x^k$ and $\sum_{k=0}^{\infty} x^k/k!$,

$$(1 + x + x^2 + \cdots + x^k + \cdots)\left(1 + x + \frac{x^2}{2!} + \frac{x^3}{3!} + \cdots\right)$$

is given by $c_0 + c_1 x + c_2 x^2 + c_3 x^3 + \cdots + c_k x^k + \cdots$, where

$$c_0 = 1$$
$$c_1 = 1 + 1 = 2$$
$$c_2 = \frac{1}{2!} + 1 + 1 = \frac{5}{2}$$
$$c_3 = \frac{1}{3!} + \frac{1}{2!} + 1 + 1 = \frac{8}{3}$$
$$\vdots$$
$$c_k = \frac{1}{k!} + \frac{1}{(k-1)!} + \cdots + 1 = \sum_{j=0}^{k} 1 \cdot \frac{1}{(k-j)!}$$

Thus, the first few terms of the product are

$$1 + 2x + \frac{5}{2}x^2 + \frac{8}{3}x^3 + \cdots$$

The quotient of two power series in x is defined in a natural way in terms of the product:

Definition 5.5.3 The ***quotient*** of $\sum_{k=0}^{\infty} a_k x^k$ divided by $\sum_{k=0}^{\infty} b_k x^k$ where $b_0 \neq 0$ is the series $\sum_{k=0}^{\infty} d_k x^k$ (if one exists) such that

$$\left(\sum_{k=0}^{\infty} b_k x^k\right)\left(\sum_{k=0}^{\infty} d_k x^k\right) = \sum_{k=0}^{\infty} a_k x^k$$

It is not easy to give a formula for the coefficients $\{d_k\}_{k=1}^{\infty}$ in the quotient series; however, they must satisfy the equations

$$\begin{aligned} b_0 d_0 &= a_0 \\ b_1 d_0 + b_0 d_1 &= a_1 \\ &\vdots \\ \sum_{j=0}^{k} b_j d_{k-j} &= a_k \\ &\vdots \end{aligned}$$

and these may be computed as far as one wishes. Starting with the first equation, and recalling that $b_0 \neq 0$, we obtain $d_0 = a_0/b_0$. Using this value for d_0 in the second equation, we can solve for d_1, and so on. Observe that the condition $b_0 \neq 0$ is necessary to solve the equation at each step.

Example 12: The Taylor series for $\sin x$ and $\cos x$ at $x = 0$ are

$$\sin x = 0 + x + 0 - \frac{x^3}{6} + 0 + \frac{x^5}{120} + \cdots = \sum_{k=0}^{\infty} a_k x^k$$

and

$$\cos x = 1 + 0 - \frac{x^2}{2} + 0 + \frac{x^4}{24} + 0 + \cdots = \sum_{k=0}^{\infty} b_k x^k$$

Let us compute the quotient of the sine series by the cosine series. If we let the coefficients in the quotient be $d_0, d_1, d_2, \ldots$, we obtain

$$\begin{aligned} 1 \cdot d_0 \cdot &= 0 & &\text{so } d_0 = 0 \\ 1 \cdot d_1 + 0 \cdot d_0 &= 1 & &\text{so } d_1 = 1 \\ 1 \cdot d_2 + 0 \cdot d_1 - \frac{1}{2} \cdot d_0 &= 0 & &\text{so } d_2 = 0 \\ 1 \cdot d_3 + 0 \cdot d_2 - \frac{1}{2} \cdot d_1 + 0 \cdot d_0 &= -\frac{1}{6} & & \\ d_3 - \frac{1}{2} &= -\frac{1}{6} & &\text{so } d_3 = \frac{1}{3} \\ 1 \cdot d_4 + 0 \cdot d_3 + \left(-\frac{1}{2}\right) \cdot d_2 + 0 \cdot d_1 + \frac{1}{24} \cdot d_0 &= 0 & &\text{so } d_4 = 0 \\ 1 \cdot d_5 + 0 \cdot d_4 - \frac{1}{2} \cdot d_3 + 0 \cdot d_2 + \frac{1}{24} \cdot d_1 + 0 \cdot d_0 &= \frac{1}{120} & & \\ d_5 - \frac{1}{2} \cdot \frac{1}{3} + \frac{1}{24} &= \frac{1}{120} & &\text{so } d_5 = \frac{2}{15} \end{aligned}$$

Thus, the quotient series through terms of fifth degree is

$$x + \tfrac{1}{3}x^3 + \tfrac{2}{15}x^5 + \cdots.$$

These are the terms of the fifth Taylor polynomial for the tangent function at zero.

Naturally, questions of convergence and representation arise. What are the intervals of convergence of the product series and the quotient series? If the given series represent certain functions, do the product series and quotient series represent the product and quotient of the corresponding functions? Let us assume that both series have positive radii of convergence. Then, for the product series, there are favorable results: the product series converges and represents the product of the original functions at least in the intersection of their intervals of convergence. Although the quotient has a positive radius of convergence, this radius may be smaller than both of the radii of convergence of the given functions. (For a discussion of these matters, see the text by Taylor and Mann.*)

The project following the exercises asks you to discuss the product of series of constants.

EXERCISES 5.5

1. For each of the following series, use the Ratio Test (Theorem 2.8.7) to determine the radius of convergence and the interval of convergence. Check for convergence at any endpoints.

(a) $\sum_{k=1}^{\infty} \frac{x^k}{\sqrt{k}}$

(b) $\sum_{k=1}^{\infty} kx^k$

(c) $\sum_{k=0}^{\infty} \frac{x^k}{2^k}$

(d) $\sum_{k=1}^{\infty} \frac{(3x)^k}{k2^k}$

(e) $\sum_{k=1}^{\infty} \frac{(x-3)^k}{k^2}$

(f) $\sum_{k=1}^{\infty} \frac{(x+1)^k}{k^{3/2}}$

(g) $\sum_{k=1}^{\infty} \frac{2^k(x-2)^k}{3^{k+1}}$

(h) $\sum_{k=1}^{\infty} \frac{kx^k}{k^2+4}$

(i) $\sum_{k=1}^{\infty} \frac{x^{k-1}}{(k-1)!}$

2. Determine the radius of convergence of each of the following series.

(a) $\sum_{k=0}^{\infty} \frac{(2k)!}{(k!)^2}x^k$

(b) $\sum_{k=0}^{\infty} \frac{(k!)^3}{(3k)!}x^k$

3. Prove Corollary 5.5.1.

4. Establish the continuity of the sum of the following series on the indicated set. Does Theorem 5.10.4 apply?

(a) $\sum_{k=1}^{\infty} \frac{x^k}{k^2}$; $(-1, 1)$

(b) $\sum_{k=1}^{\infty} \frac{x^k}{k^2}$; $[-1, 1]$

(c) $\sum_{k=1}^{\infty} \frac{(x+4)^k}{k^2}$; $[-5, -3]$

(d) $\sum_{k=1}^{\infty} \frac{x^k}{k}$; $[-1, 1)$

(e) $\sum_{k=1}^{\infty} 3^k kx^k$; $\left(-\frac{1}{3}, \frac{1}{3}\right)$

*A.E. Taylor and W.R. Mann, *Advanced Calculus*, 3rd. ed. (New York: Wiley, 1983), pp. 600–604, 639–643.

5. Find power series for each of the following functions on the indicated interval.

   (a) $f(x) = x \sin x \qquad x \in R$

   (b) $f(x) = \begin{cases} \dfrac{\sin x}{x} & \text{if } x \neq 0, x \in R \\ 1 & \text{if } x = 0 \end{cases}$

   (c) $g(x) = \sin 2x \qquad x \in R$

   (d) $f(x) = (x - 1)\ln x \qquad 0 < x < 2$

   (e) $f(x) = \cosh x \qquad x \in R$
   $\left(\textit{Note:}\ \cosh x = \frac{1}{2}(e^x + e^{-x})\right.$.

   (f) $\displaystyle\int_0^x \frac{\sin t}{t}\,dt \qquad x \in R$

   (g) $\displaystyle\int_1^x (t - 1)\ln t\,dt \qquad 0 < x < 2$

   (h) $\displaystyle\int_0^x (e^{t^2} - 1)\,dt \qquad 0 \leq x \leq 1$

6. Find an explicit formula for the sum of the given series. Indicate the interval where your formula is valid.

   (a) $\displaystyle\sum_{k=1}^{\infty} kx^{k-1}$

   (b) $\displaystyle\sum_{k=1}^{\infty} kx^k$

   (c) $\displaystyle\sum_{k=2}^{\infty} k(k - 1)x^{k-2}$

   (d) $\displaystyle\sum_{k=0}^{\infty} \frac{x^{k+1}}{k!}$

   (e) $\displaystyle\sum_{k=0}^{\infty} \frac{3^k}{k!}x^k$

   (f) $\displaystyle\sum_{k=2}^{\infty} \frac{x^k}{k!}$

   (g) $\displaystyle\sum_{k=1}^{\infty} \frac{(-1)^{k+1}x^{2k-2}}{(2k - 1)!}$

   (h) $\displaystyle\sum_{k=0}^{\infty} \frac{(-1)^k}{(2k + 1)!}x^{2k}$

7. (a) Find the Taylor series for $f(u) = (1 + u)^{-1/2}$ about $u = 0$. Show that the radius of convergence is 1.

   (b) By substitution, obtain a power series on the interval $(-1, 1)$ for the function $1/\sqrt{1 - t^2}$.

   (c) Find a power series on the interval $(-1, 1)$ for the function $\arcsin x$ given that $\arcsin x = \displaystyle\int_0^x dt/\sqrt{1 - t^2}$.

   (d) Show that the series in part (c) converges for $x = \pm 1$. [Hint: Show that if
   $$a_k = \frac{1 \cdot 3 \cdot 5 \cdot \cdots \cdot (2k - 1)}{2 \cdot 4 \cdot 6 \cdot \cdots \cdot (2k)} \cdot \frac{1}{2k + 1}$$
   then $a_k^2 < \dfrac{1}{(2k + 1)^3}$.]

   (e) Explain why the series obtained in part (c) represents $\arcsin x$ on $[-1, 1]$. Find an infinite series for π.

8. Find series solutions of the form $\sum_{k=0}^{\infty} a_k x^k$ for the following differential equations.

   (a) $y' - 3y = 0$

   (b) $y'' + 4y = 0$

   (c) $y'' - y' = 0$

   (d) $y'' - y = 0$

   (e) $y' = 2xy$

   (f) $y + xy' = x^2$

9. In the proof of Theorem 5.5.6 we saw that $a_k = b_k$ for $k \in Z^+$. Show, in fact, that $a_k = b_k = f^{(k)}(c)/k!$, so that each series is just the Taylor series for the function f at c.

PROJECT: Cauchy Product of Series

If convergent, the series $\sum_{k=0}^{\infty} a_k$ and $\sum_{k=0}^{\infty} b_k$ can be regarded as the values, when $x = 1$, of the power series $\sum_{k=0}^{\infty} a_k x^k$ and $\sum_{k=0}^{\infty} b_k x^k$, respectively. This observation suggests the definition

$$\left(\sum_{k=0}^{\infty} a_k\right)\left(\sum_{k=0}^{\infty} b_k\right) = \sum_{k=0}^{\infty} c_k$$

where

$$c_k = \sum_{j=0}^{k} a_j b_{k-j} \qquad k \in Z^+$$

The product so defined is known as the *Cauchy product* of $\sum_{k=0}^{\infty} a_k$ and $\sum_{k=0}^{\infty} b_k$. In extended form,

$$\sum_{k=0}^{\infty} c_k = a_0b_0 + (a_0b_1 + a_1b_0) + (a_0b_2 + a_1b_1 + a_2b_0) + \cdots + (a_0b_k + a_1b_{k-1} + \cdots + a_kb_0) + \cdots$$

It is important to note the parentheses.

(1) Show that if $\sum_{k=0}^{\infty} a_k$ and $\sum_{k=0}^{\infty} b_k$ are absolutely convergent with $\sum_{k=0}^{\infty} a_k = A$ and $\sum_{k=0}^{\infty} b_k = B$, then their product converges absolutely to AB.

(2) Show that the Cauchy product of conditionally convergent series may diverge. [*Hint:* Let $a_k = b_k = (-1)^k/\sqrt{k+1}$.]

Note: The text by Olmsted is a good reference for this project.*

★ 5.6 *An Everywhere Continuous, Nowhere Differentiable Function*

We found early in the study of derivatives that if a real function is differentiable at a point c, then it is also continuous at c, but that the converse does not hold. Most examples given to show the failure of the converse are nondifferentiable at one or perhaps several points; for example, $f(x) = |x|$ is not differentiable at $x = 0$ but is differentiable elsewhere. We devote this section to an example, due to John McCarthy,† which illustrates the most extreme situation—a function that is continuous everywhere but differentiable nowhere.

Consider the function g with period 4 whose values on $[-2, 2]$ are given by

$$g(x) = \begin{cases} 1 + x & \text{if } -2 \le x \le 0 \\ 1 - x & \text{if } 0 \le x \le 2 \end{cases}$$

It is easy to see that g is continuous everywhere but fails to be differentiable at the even integers (work Exercise 5.6.1). The graph of g has a sawtooth appearance (see Figure 5.5).

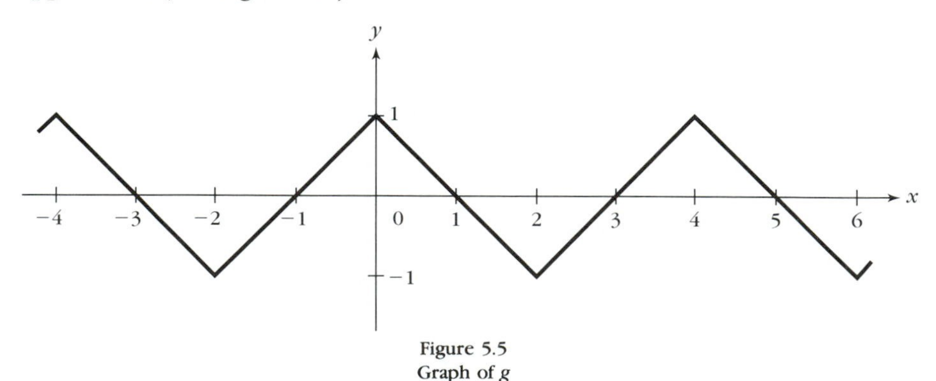

Figure 5.5
Graph of g

*John M.H. Olmsted, *Advanced Calculus*, (Englewood Cliffs, NJ, Prentice Hall, 1961), pp. 405–410.

†John McCarthy, "An Everywhere Continuous Nowhere Differentiable Function," *American Mathematical Monthly*, 60 (1953): 709.

The function f, whose derivative will be shown to exist nowhere, is an infinite sum of continuous functions:

$$f(x) = \sum_{n=1}^{\infty} 2^{-n} g_n(x) \tag{1}$$

where

$$g_n(x) = g(2^{2^n} x) \qquad x \in R, n \in Z^+$$

Since g has period 4, g_n has period $4 \cdot 2^{-2^n}$. The graph of g_n is also a sawtooth affair, whose "teeth" occur with greater frequency as n increases; indeed, g_n has 2^{2^n} teeth in the interval $[-2, 2]$. For $n = 1$, we have $g_1(x) = g(4x)$. A graph of g_1 appears in Figure 5.6.

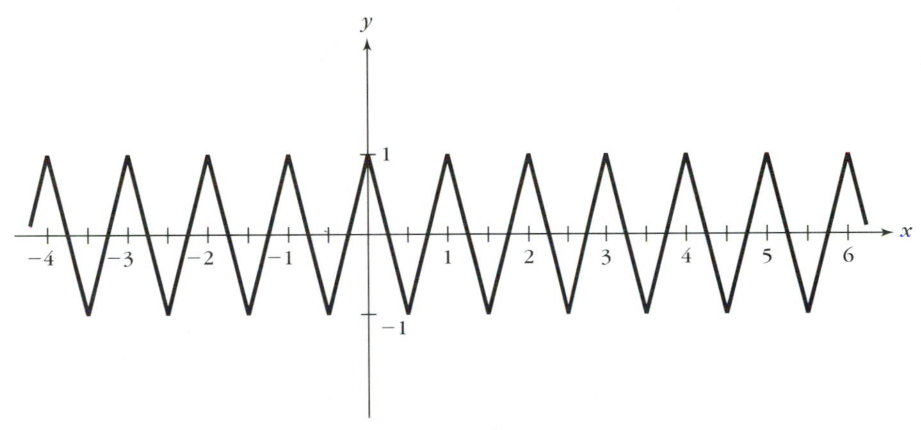

Figure 5.6
Graph of g_1

We claim that f is continuous everywhere. We note that since the values of g lie in the interval $[-1, 1]$, the same is true of each g_n. Thus, $|2^{-n} g_n(x)| \leq 2^{-n}$ for $n \in Z^+$, and so the series (1) converges uniformly on R by the Weierstrass M Test (Theorem 5.4.2). Each $g_n(x)$ is continuous for $x \in R$, and so also is $2^{-n} g_n(x)$. The uniform convergence of (1) thus guarantees that $f(x)$ is continuous everywhere (Theorem 5.4.3).

Now let x be an arbitrary real number. We will show that f is not differentiable at x by constructing a sequence $\{x_k\}_{k=1}^{\infty}$ converging to x such that the sequence of difference quotients

$$\left\{\frac{f(x_k) - f(x)}{x_k - x}\right\}_{k=1}^{\infty} \tag{2}$$

does not converge.

Before defining the sequence $\{x_k\}_{k=1}^{\infty}$, let us make a useful observation about the function g: If u is the x-coordinate of a point on a linear segment of the graph of g, then at least one of $u \pm 1$ is the x-coordinate of a point on the same segment and we have $|g(u \pm 1) - g(u)| = 1$. Noting that 1 is one-quarter of the period of g and that 2^{-2^k} is one-quarter of the period of g_k allows us to adapt the above observation to the function g_k: For the given $x \in R$, one of the two numbers $x \pm 2^{-2^k}$ corresponds to a point on the same linear segment of the graph of g_k. Let that number be x_k, and it follows that $|g_k(x_k) - g_k(x)| = 1$. In this manner we have defined a sequence $\{x_k\}_{k=1}^{\infty}$ that converges to x. It remains to show that the sequence (2) does not converge. For a given $k \in Z^+$ (fixed for the moment), we will abbreviate $x_k - x$ by Δx; observe that $|\Delta x| = 2^{-2^k}$. Likewise, we introduce the abbreviations

$$\Delta f = f(x_k) - f(x) \qquad \text{and} \qquad \Delta g_n = g_n(x_k) - g_n(x) \qquad n \in Z^+$$

It is easy to show (work Exercise 5.6.2) that for $n > k$, Δx is an integral multiple of the period of g_n, so that for $n > k$, $g_n(x_k) = g_n(x + \Delta x) = g_n(x)$ and, hence,

$$\Delta g_n = 0 \qquad n > k \tag{3}$$

Now consider the change induced in the function f as the independent variable changes from x to x_k:

$$\begin{aligned}\Delta f &= f(x_k) - f(x) \\ &= \sum_{n=1}^{\infty} 2^{-n} g_n(x_k) - \sum_{n=1}^{\infty} 2^{-n} g_n(x) \\ &= \sum_{n=1}^{\infty} 2^{-n} \Delta g_n\end{aligned}$$

Because of (3), this infinite series can be replaced by a finite sum; hence,

$$|\Delta f| = \left| \sum_{n=1}^{k} 2^{-n} \Delta g_n \right| \tag{4}$$

By separating the last term of the sum in (4) from the others, we have

$$\sum_{n=1}^{k} 2^{-n} \Delta g_n = 2^{-k} \Delta g_k + \sum_{n=1}^{k-1} 2^{-n} \Delta g_n \tag{5}$$

in which the second sum can be estimated as follows: since $2^{-n} < 1$ for all values of n concerned, the absolute value of the nth term is less than $|\Delta g_n|$; hence,

$$\left| \sum_{n=1}^{k-1} 2^{-n} \Delta g_n \right| < (k-1) \max \{|\Delta g_n| : n = 1, 2, \ldots, k-1\}$$

But the slope of each linear segment of g_n is $\pm 2^{2^n}$ (work Exercise 5.6.3), and so

$$|\Delta g_n| \le 2^{2^n} |\Delta x| = 2^{2^n} 2^{-2^k} \le 2^{2^{k-1}} 2^{-2^k}$$

Therefore,

$$\left| \sum_{n=1}^{k-1} 2^{-n} \Delta g_n \right| < (k-1) \cdot 2^{2^{k-1}} \cdot 2^{-2^k}$$

But $k - 1 < 2^k$ for $k \in Z^+$, and after some manipulation of exponents, we find

$$\left|\sum_{n=1}^{k-1} 2^{-n}\Delta g_n\right| < 2^k 2^{2^{k-1}-2^k} = 2^k 2^{-2^{k-1}} = 2^{k-2^{k-1}} \tag{6}$$

If we return to (5) and recall that $|\Delta g_k| = 1$, we see that

$$\sum_{n=1}^{k} 2^{-n}\Delta g_n = 2^{-k} + \sum_{n=1}^{k-1} 2^{-n}\Delta g_n$$

Hence,

$$\left|\frac{f(x_k) - f(x)}{x_k - x}\right| = \left|\frac{\Delta f}{\Delta x}\right| = \left|\frac{2^{-k} + \sum_{n=1}^{k-1} 2^{-n}\Delta g_n}{2^{-2^k}}\right| \geq \frac{2^{-k} - 2^{k-2^{k-1}}}{2^{-2^k}} = 2^{2^k-k} - 2^{2^{k-1}+k}$$

In this form it is easy to show (work Exercise 5.6.4) that the sequence (2) of difference quotients is unbounded as $k \to \infty$. Hence, f is not differentiable at x; since x was arbitrary, we have proved that f is differentiable nowhere.

EXERCISES 5.6

1. Verify that the function g defined in this section fails to be differentiable at the even integers.
2. Verify that the function g_n defines by $g_n(x) = g(2^{2^n}x)$ has period $4/2^{2^n}$, and show that for $n > k$, $\Delta x = 2^{-2^k}$ is an integral multiple of this period.
3. By using the fact that the slope of each linear segment of the graph of g_n is given by a derivative, show that this slope is $\pm\ 2^{2^n}$.
4. Show that the sequence $\{2^{2^k-k} - 2^{2^{k-1}+k}\}_{k=1}^{\infty}$ is unbounded.

CHAPTER SIX

A BRIEF INTRODUCTION TO METRIC SPACES

Many of the tools of analysis involve the notion of distance. For example, on the real number line, the δ-neighborhood of a, $N_\delta(a)$, consists of those points whose distance from a on the line is less than δ. A sequence $\{x_n\}_{n=1}^{\infty}$ of real numbers converges to L if and only if the distance of x_n from L becomes arbitrarily small as $n \to \infty$. A real function is continuous if the values of the function are close whenever the values of the independent variable are close.

In this chapter we consider a variety of situations in which the concept of distance can be introduced in an appropriate and useful way. Then we can investigate properties such as convergence, continuity, compactness, and completeness, which are fundamental to more advanced analysis.

6.1 *Metric Properties, Neighborhoods, Convergence*

Definition of Metric Space

Definition 6.1.1 Let A be a nonempty set. A ***metric*** on A is a function $\rho : A \times A \to R$ such that for all $x, y, z \in A$ the following properties hold:

(i) $\rho(x, y) \geq 0$ with equality holding if and only if $x = y$
(ii) $\rho(x, y) = \rho(y, x)$
(iii) **Triangle Inequality** $\rho(x, z) + \rho(z, y) \geq \rho(x, y)$

If ρ is a metric on A, then the ordered pair (A, ρ) is a ***metric space***. The metric ρ is often called a ***distance function*** on A, and $\rho(x, y)$ is the ***distance*** between x and y.

Example 1: For $x, y \in R$, let $\rho(x, y) = |x - y|$. Then ρ is a metric on R, known as the *absolute value* metric. Properties (i) and (ii) are obvious, and property (iii) is an immediate consequence of Theorem R.2.1.

Example 2: The coordinate plane $R \times R$ will be denoted R^2. If $P, Q \in R^2$, we can define the distance between P and Q by the familiar formula from coordinate geometry,

$$d(P, Q) = \sqrt{(x_2 - x_1)^2 + (y_2 - y_1)^2}$$

where $P = (x_1, y_1)$ and $Q = (x_2, y_2)$. The function d is a metric on R^2, known as the *Euclidean* metric. Property (iii) is illustrated in Figure 6.1; it is related to the familiar statement, "the shortest path between two points is along a straight line". Figure 6.1 also shows why property (iii) is named the Triangle Inequality. An analytic proof of this property is given in Section 6.2.

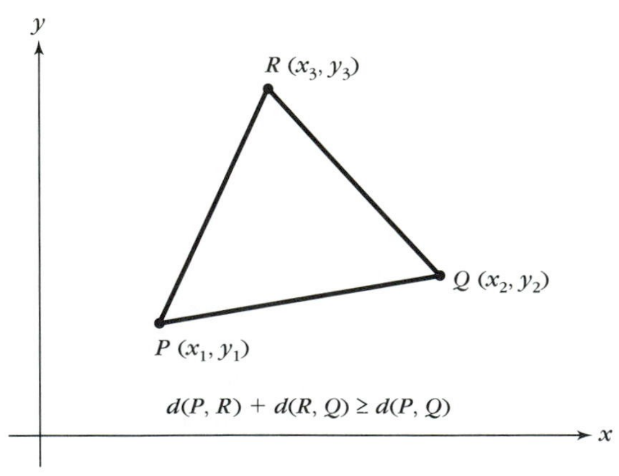

Figure 6.1
Triangle Inequality

Example 3: Here is a different metric for R^2. With P and Q as in Example 2, we define σ as follows

$$\sigma(P, Q) = |x_2 - x_1| + |y_2 - y_1|$$

You are asked to verify that σ satisfies properties (i), (ii), and (iii) in Exercise 6.1.1.

The last two examples show that different metrics can be assigned to the same set. The next example shows that every nonempty set can be made into a metric space.

Example 4: Let A be any nonempty set whatever, and for $x, y \in A$, define

$$\rho(x, y) = \begin{cases} 1 & \text{if } x \neq y \\ 0 & \text{if } x = y \end{cases}$$

Then ρ obviously satisfies the three properties required for a metric and is called the *discrete* metric for A.

A Space of Functions

Some of the most important metric spaces consist of functions.

Example 5: Let $C[a, b]$ be the set of all real functions defined and continuous on the interval $[a, b]$ where $a < b$. For $f, g \in C[a, b]$, define ρ as follows:

$$\rho(f, g) = \max_{x \in [a,b]} |f(x) - g(x)|$$

This definition is meaningful because $|f - g|$ is continuous on $[a, b]$ and its maximum therefore exists by Theorem 3.6.5. To verify that $(C[a, b], \rho)$ is a metric space, let $f, g, h \in C[a, b]$.

(1) Clearly, $\rho(f, g) \geq 0$. Now if $f = g$,* then certainly $\rho(f, g) = 0$. Conversely, if $\rho(f, g) = 0$, then $|f(x) - g(x)|$ must be zero for all $x \in [a, b]$, so that $f = g$.
(2) Property (ii) is obvious.
(3) To establish property (iii), we must prove that

$$\rho(f, h) + \rho(h, g) \geq \rho(f, g)$$

Now, for any functions $u, v \in C[a, b]$, we have (work Exercise 6.1.3a)

$$\max_{x \in [a,b]} u(x) + \max_{x \in [a,b]} v(x) \geq \max_{x \in [a,b]} [u(x) + v(x)]$$

Next, set $u(x) = |f(x) - h(x)|$ and $v(x) = |h(x) - g(x)|$ for $x \in [a, b]$. Then use the Triangle Inequality of R to obtain the desired conclusion (work Exercise 6.1.3c).

*Recall that $f = g$ means $f(x) = g(x)$ for all x in their common domain.

Neighborhoods

Definition 6.1.2 Let (A, ρ) be a metric space. If δ is a positive real number and $a \in A$, then the ***δ-neighborhood*** of a is the set

$$N_\delta(a) = \{x : x \in A \text{ and } \rho(a, x) < \delta\}$$

This definition of neighborhood coincides with the definition in Section R.2 when applied to R with the absolute value metric. Let us look at the neighborhood concept in some other metric spaces.

Example 6: In R^2 under the Euclidean metric, $N_\delta(P)$ consists of all points Q that lie inside a circle of radius δ centered at P (see Figure 6.2).

Example 7: In $C[a, b]$ with the metric described in Example 5, the δ neighborhood of a function f consists of all continuous functions on $[a, b]$ whose graphs lie within a band of width 2δ around the graph of f (see Figure 6.3).

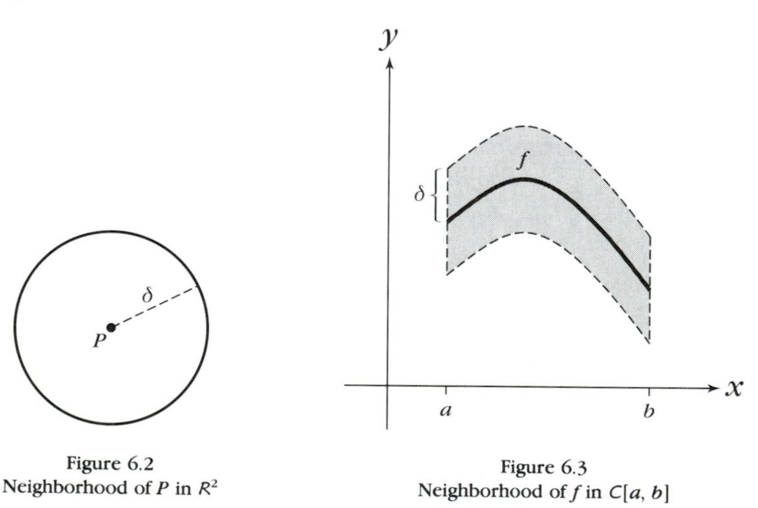

Figure 6.2
Neighborhood of P in R^2

Figure 6.3
Neighborhood of f in $C[a, b]$

In a metric space, it is always possible to separate two distinct points by enclosing them in disjoint neighborhoods:

Theorem 6.1.1 **Hausdorff Property** *If (A, ρ) is a metric space and if $x, y, \in A$ with $x \neq y$, then there exist neighborhoods of x and y, respectively, that are disjoint.*

PROOF: Since $x \neq y$, $\rho(x, y) > 0$, and we can let δ be any positive number less than or equal to $\frac{1}{2}\rho(x, y)$. Then $N_\delta(x)$ and $N_\delta(y)$ are disjoint (see Figure 6.4). For suppose $z \in N_\delta(x) \cap N_\delta(y)$. We would have both $\rho(x, z) < \delta$ and

$\rho(z, y) < \delta$, so that by the Triangle Inequality,

$$\rho(x, y) \leq \rho(x, z) + \rho(z, y) < 2\delta \leq \rho(x, y)$$

This contradiction establishes the theorem. ∎

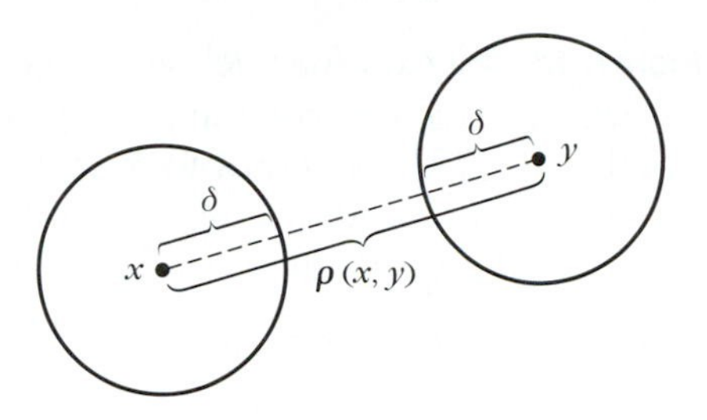

Figure 6.4
Hausdorff Property

Convergence of Sequences

Definition 6.1.3 Let (A, ρ) be a metric space, let $\{x_n\}_{n=1}^{\infty}$ be a sequence of elements of A, and let L be an element of A. Then $\{x_n\}_{n=1}^{\infty}$ ***converges*** to L provided that the sequence $\{\rho(x_n, L)\}_{n=1}^{\infty}$ is null. In that case, we write $x_n \to L$, and L is the ***limit*** of the sequence $\{x_n\}_{n=1}^{\infty} : L = \lim_{n\to\infty} x_n$.

Again observe that when applied to R with the absolute value metric, this definition is identical to that for the convergence of a sequence of real numbers. Other conditions for convergence in R generalize to metric spaces as well.

Theorem 6.1.2 *Let (A, ρ) be a metric space, let $\{x_n\}_{n=1}^{\infty}$ be a sequence of elements of A, and let $L \in A$. Then the following conditions are equivalent:*

(i) $x_n \to L$

(ii) For every $\varepsilon > 0$, there exists an $N \in Z^+$ such that $\rho(x_n, L) < \varepsilon$ whenever $n \geq N$

(iii) Every neighborhood of L contains x_n for all except a finite number of values of n

PROOF: If property (i) holds, then $\{\rho(x_n, L)\}_{n=1}^{\infty}$ is null and property (ii) follows immediately from the definition of null sequence. If (ii) holds and if $\varepsilon > 0$, then the neighborhood $N_\varepsilon(L)$ contains all x_n with $n \geq N$, that is, all except perhaps $x_1, \ldots, x_{N-1}$. If (iii) holds and $\varepsilon > 0$ is given, then $N_\varepsilon(L)$ contains x_n for all sufficiently large n, say, $n \geq N$; thus, $\rho(x_n, L) < \varepsilon$ for $n \geq N$ and $x_n \to L$. So (i) ⇒ (ii) ⇒ (iii) ⇒ (i), and the three conditions are equivalent. ∎

Theorem 6.1.3 *In any metric space, the limit of a convergent sequence is unique.*

PROOF: This theorem follows immediately from property (iii) of Theorem 6.1.2 and the Hausdorff Property. ■

Example 8: Consider R^2 with the Euclidean metric and suppose that $\{P_n\}_{n=1}^{\infty}$ is a sequence in R^2, where $P_n = (x_n, y_n)$ for $n \in Z^+$. We show that $\{P_n\}_{n=1}^{\infty}$ converges to a point $P = (x, y)$ if and only if $x_n \to x$ and $y_n \to y$ in R with the absolute value metric.

If $P_n \to P$, then, for $\varepsilon > 0$, there exists an $N \in Z^+$ such that $\sqrt{(x_n - x)^2 + (y_n - y)^2} < \varepsilon$ for $n \geq N$. But then, for $n \geq N$, we must have

$$|x_n - x| \leq \sqrt{(x_n - x)^2 + (y_n - y)^2} < \varepsilon$$

and

$$|y_n - y| \leq \sqrt{(x_n - x)^2 + (y_n - y)^2} < \varepsilon$$

Thus, $x_n \to x$ and $y_n \to y$. Conversely, let $x_n \to x$ and $y_n \to y$. If $\varepsilon > 0$, then there exist positive integers N_1 and N_2 such that $|x_n - x| < \varepsilon/\sqrt{2}$ for $n \geq N_1$, and $|y_n - y| < \varepsilon/\sqrt{2}$ for $n \geq N_2$. For $n \geq N = \max\{N_1, N_2\}$, the distance of P_n from P is

$$\sqrt{(x_n - x)^2 + (y_n - y)^2} < \sqrt{\frac{\varepsilon^2}{2} + \frac{\varepsilon^2}{2}} = \varepsilon$$

Thus, $P_n \to P$.

The next two examples concern $C[a, b]$ with the metric discussed in Example 5.

Example 9: For each $n \in Z^+$, let $f_n(x) = x/n$ $(0 \leq x \leq 1)$. Then $\{f_n\}_{n=1}^{\infty}$ is a sequence in $C[0, 1]$. We claim that $\{f_n\}_{n=1}^{\infty}$ is convergent to the zero function. For

$$0 \leq \rho(f_n, 0) = \max_{x \in [0,1]} \left|\frac{x}{n}\right| = \frac{1}{n}$$

and by the Squeeze Theorem, $\{\rho(f_n, 0)\}_{n=1}^{\infty}$ is null.

The sequence of functions in the last example converges uniformly to zero on $[0, 1]$. Indeed, it is easy to see that if a sequence of functions is convergent in $C[a, b]$, with the metric ρ of Example 5, then it is uniformly convergent on $[a, b]$ in the sense of Chapter 5. The converse of this proposition also holds (work Exercise 6.1.7). Thus, $f_n \to f$ in $C[a, b]$ if and only if $f_n(x) \rightrightarrows f(x)$ $(x \in [a, b])$.

Example 10: For each $n \in Z^+$, let $g_n(x) = x^n$ $(0 \leq x \leq 1)$. Then $\{g_n\}_{n=1}^{\infty}$ is a sequence in $C[0, 1]$. This sequence is not convergent in $C[0, 1]$ because $\{g_n(x)\}_{n=1}^{\infty}$ is not uniformly convergent on $[0, 1]$ (its limit function is discontinuous).

EXERCISES 6.1

1. (a) Refer to Example 3 and show that σ is a metric on R^2.
   (b) Sketch the δ-neighborhood of $(0, 0)$ with respect to the metric σ.

2. (a) Show that if ρ is a metric on the set A, then so is $k\rho$, where k is any fixed positive real number.
   (b) Show that if ρ_1 and ρ_2 are both metrics on the set A, then so is $\rho_1 + \rho_2$.

3. (a) Show that if the real functions u and v are continuous on $[a, b]$, then

$$\max_{x\in[a,b]} u(x) + \max_{x\in[a,b]} v(x) \geq \max_{x\in[a,b]} [u(x) + v(x)]$$

   (b) Give an example to show that strict inequality can occur in part (a).
   (c) Carry out the details of the proof of the Triangle Inequality for $C[a, b]$ in Example 5.

4. (a) Let $P = (x_1, y_1)$ and $Q = (x_2, y_2)$ be arbitrary points in R^2. Define $\tau(P, Q)$ to be zero if $P = Q$ and to be $\sqrt{x_1^2 + y_1^2} + \sqrt{x_2^2 + y_2^2}$ if $P \neq Q$. Show that τ is a metric on R^2.
   (b) Describe the δ-neighborhood of a point $P \in R^2$ relative to the metric τ in part (a).

5. Define two metrics ρ and σ on a set A to be *equivalent* provided that for each $a \in A$, every neighborhood of a under ρ contains a neighborhood of a under σ, and every neighborhood of a under σ contains a neighborhood of a under ρ.
   (a) Show geometrically that the metrics of Examples 2 and 3 are equivalent metrics on R^2.
   (b) Show that if ρ and σ are equivalent metrics on A, if $\{x_n\}_{n=1}^{\infty}$ is a sequence in A, and if $x \in A$, then $x_n \to x$ with respect to the metric ρ if and only if $x_n \to x$ with respect to σ.

6. (a) Let ρ be a metric on the set A and let f be a real function with the properties:
   (i) $f(u) \geq 0$ for $u \geq 0$; $f(0) = 0$
   (ii) f is strictly increasing on $[0, \infty)$; and
   (iii) $f(u + v) \leq f(u) + f(v)$ for $u, v \geq 0$.
   Show that $\sigma = f \circ \rho$ is a metric on A.
   (b) Show that if ρ is a metric on A, then the function $\sigma : A \times A \to R$ defined by

$$\sigma(x, y) = \frac{\rho(x, y)}{1 + \rho(x, y)}$$

   is also a metric on A.

7. Show that if a sequence of functions $\{f_n\}_{n=1}^{\infty}$, each of which is continuous on $[a, b]$, is uniformly convergent there (in the sense of Chapter 5), then $\{f_n\}_{n=1}^{\infty}$ is convergent in $C[a, b]$ with respect to the metric of Example 5.

8. Suppose that $\{x_n\}_{n=1}^{\infty}$ is a sequence of points of a set A and that $\{x_n\}_{n=1}^{\infty}$ is convergent in A with respect to the discrete metric. What can you conclude about the sequence $\{x_n\}_{n=1}^{\infty}$?

9. (a) For $f, g \in C[a, b]$, define

$$\tau(f, g) = \int_a^b |f - g|.$$

   Show that τ is a metric on $C[a, b]$.
   (b) Let $I[a\ b](a < b)$ denote the set of all Riemann integrable functions on $[a, b]$. Define

$$\sigma(f, g) = \int_a^b |f - g|.$$

   Is σ a metric on $I[a, b]$?

10. Prove or disprove: If $\{f_n\}_{n=1}^{\infty}$ is a sequence of functions in $C[a, b]$, then $\{f_n\}_{n=1}^{\infty}$ converges with respect to the metric of Example 5 if and only if it converges with respect to the metric of Exercise 9a.

6.2 R^n as a Metric Space

Two-dimensional space consists of all ordered pairs of real numbers; that is,

$$R^2 = \{(x, y) : x, y \in R\}$$

In the last section we saw several ways of introducing a metric on R^2; the most important of these is the Euclidean metric for which the distance between $P(x_1, y_1)$ and $Q(x_2, y_2)$ is given by

$$\sqrt{(x_2 - x_1)^2 + (y_2 - y_1)^2}$$

In three-dimensional coordinate geometry, points are given by ordered triples, and you will recall that the distance between two points (x_1, y_1, z_1) and (x_2, y_2, z_2) in R^3 is

$$\sqrt{(x_2 - x_1)^2 + (y_2 - y_1)^2 + (z_2 - z_1)^2}$$

We will soon define distance in n dimensions, but first we need to introduce some algebraic concepts.

Operations in R^n

If $n \in Z^+$, then n dimensional space is defined to be the set of all n-tuples* of real numbers. We denote an n-tuple by $(x_1, x_2, \ldots, x_n)$, now using subscripts to distinguish the different coordinates.† Thus, n dimensional space is

$$R^n = \{(x_1, x_2, \ldots, x_n) : x_1, x_2, \ldots, x_n \in R\}$$

We can, if we wish, regard the elements of R^n as vectors. Just like vectors in R^2 and R^3, elements of R^n can be added to one another and multiplied by real numbers (scalars). For $X = (x_1, x_2, \ldots, x_n)$, $Y = (y_1, y_2, \ldots, y_n)$, and $c \in R$, we define

$$X + Y = (x_1 + y_1, x_2 + y_2, \ldots, x_n + y_n)$$
$$cX = (cx_1, cx_2, \ldots, cx_n)$$

The *difference* is

$$X - Y = (x_1 - y_1, x_2 - y_2, \ldots, x_n - y_n)$$

We can also define an *inner product*

$$X \cdot Y = x_1y_1 + x_2y_2 + \cdots + x_ny_n = \sum_{k=1}^{n} x_ky_k$$

In two and three dimensions the inner product is often called the *dot product*. The *norm* of $X \in R^n$ is defined to be

$$\|X\| = \sqrt{\sum_{k=1}^{n} x_k^2}$$

*An *n-tuple* is an ordered set of n elements; more formally, it is a function defined on the set $\{1, 2, \ldots, n\}$.

†When we are dealing with examples and exercises involving $n = 2, 3$, we will use the familiar notation (x, y) for points of R^2 and (x, y, z) for points of R^3.

The norm of a vector in R^2 or R^3 is its magnitude or length.

Now you will recall that the dot product of two vectors in R^2 or R^3 satisfies

$$X \cdot Y = \|X\| \, \|Y\| \cos \theta$$

where θ is the angle between X and Y. Since $|\cos \theta| \leq 1$, we see that in R^2 and R^3,

$$|X \cdot Y| \leq \|X\| \, \|Y\|$$

The generalization of this property to n dimensions is known as Cauchy's Inequality.

Theorem 6.2.1 **Cauchy's Inequality** *For $X, Y, \in R^n$,*

$$|X \cdot Y| \leq \|X\| \, \|Y\| \tag{1}$$

PROOF: Let $X = (x_1, x_2, \ldots, x_n)$ and $Y = (y_1, y_2, \ldots, y_n)$. Consider the function

$$f(t) = \sum_{k=1}^{n} (x_k t + y_k)^2 \qquad t \in R$$

A little algebra shows that f is a quadratic function:

$$\begin{aligned} f(t) &= \left(\sum_{k=1}^{n} x_k^2\right) t^2 + 2\left(\sum_{k=1}^{n} x_k y_k\right) t + \sum_{k=1}^{n} y_k^2 \\ &= At^2 + Bt + C \end{aligned}$$

where $A = \sum_{n=1}^{n} x_k^2$, $B = 2 \sum_{k=1}^{n} x_k y_k$, and $C = \sum_{k=1}^{n} y_k^2$. Because it is a sum of squares, $f(t) \geq 0$ for all $t \in R$ and so the discriminant $B^2 - 4AC$ cannot be positive. So $B^2 \leq 4AC$, and we have

$$4\left(\sum_{k=1}^{n} x_k y_k\right)^2 \leq 4 \sum_{k=1}^{n} x_k^2 \sum_{k=1}^{n} y_k^2$$

Simplifying gives us

$$\left|\sum_{k=1}^{n} x_k y_k\right| \leq \sqrt{\sum_{k=1}^{n} x_k^2} \sqrt{\sum_{k=1}^{n} y_k^2} \tag{2}$$

Inequality (2) is just Inequality (1) expressed in terms of sums of real numbers. ■

Theorem 6.2.2 **Minkowski's Inequality** *For $X, Y \in R^n$,*

$$\|X + Y\| \leq \|X\| + \|Y\| \tag{3}$$

PROOF: Using the notation of the last proof,

$$\|X + Y\| = \sqrt{\sum_{k=1}^{n} (x_k + y_k)^2}$$

$$\begin{aligned}\|X + Y\|^2 &= \sum_{k=1}^{n} x_k^2 + 2 \sum_{k=1}^{n} x_k y_k + \sum_{k=1}^{n} y_k^2 \\ &= \|X\|^2 + 2X \cdot Y + \|Y\|^2\end{aligned}$$

and applying Theorem 6.2.1,

$$\begin{aligned}\|X + Y\|^2 &\leq \|X\|^2 + 2\|X\|\,\|Y\| + \|Y\|^2 \\ &= (\|X\| + \|Y\|)^2\end{aligned}$$

By taking square roots,

$$\|X + Y\| \leq \|X\| + \|Y\| \qquad \blacksquare$$

In terms of sums of real numbers, Minkowski's Inequality reads:

$$\sqrt{\sum_{k=1}^{n} (x_k + y_k)^2} \leq \sqrt{\sum_{k=1}^{n} x_k^2} + \sqrt{\sum_{k=1}^{n} y_k^2} \tag{4}$$

The Euclidean Metric on R^n

Now we are ready to introduce a metric on R^n. For $X, Y \in R^n$, we define $\rho_n(X, Y) = \|X - Y\|$. In terms of the coordinates of X and Y, the distance between X and Y is

$$\rho_n(X, Y) = \sqrt{\sum_{k=1}^{n} (x_k - y_k)^2}$$

That ρ_n is a metric on R^n is easily verified, the Triangle Inequality following from Minkowski's Inequality:

$$\begin{aligned}\rho_n(X, Y) + \rho_n(Y, Z) &= \|X - Y\| + \|Y - Z\| \\ &\geq \|(X - Y) + (Y - Z)\| = \|X - Z\| = \rho_n(X, Z)\end{aligned}$$

We will refer to ρ_n as the *Euclidean* metric for R^n; it clearly agrees with the absolute value metric when $n = 1$; and when $n = 2$ or 3, it agrees with the metrics of R^2 and R^3 mentioned at the beginning of the present section. When we speak of R^n as a metric space, it is to be assumed, unless otherwise specified, that the Euclidean metric is intended. Also, in the future we will not use the subscript n on ρ_n when it is clear from the context what n is.

Neighborhoods in R^n

According to the definition in the previous section, the δ-neighborhood of a point $Y \in R^n$ is the set of all points $X \in R^n$ such that $\rho(X, Y) < \delta$; that is,

$$\sqrt{(x_1 - y_1)^2 + (x_2 - y_2)^2 + \cdots + (x_n - y_n)^2} < \delta$$

where $X = (x_1, x_2, \ldots, x_n)$ and $Y = (y_1, y_2, \ldots, y_n)$. For $n = 3$, this set consists of the interior of a sphere of radius δ centered at Y (see Figure 6.5).

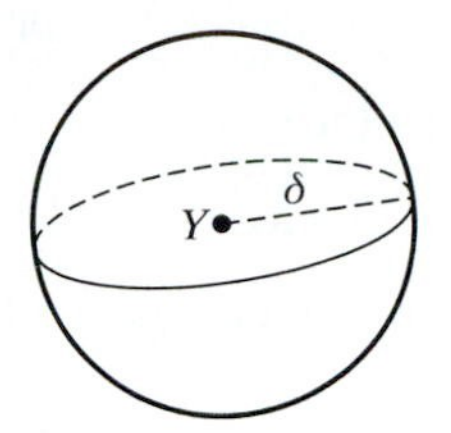

Figure 6.5
The δ-neighborhood of Y in R^3

Convergence in R^n

Let $\{X_k\}_{k=1}^{\infty}$ be a sequence in R^n. Since each X_k is an element of R^n, it has n coordinates; thus, we can write

$$X_k = (x_k^{(1)}, x_k^{(2)}, \ldots, x_k^{(j)}, \ldots, x_k^{(n)}),$$

where the superscripts now identify the various coordinates. Thus, $x_k^{(j)}$ is the jth coordinate of the kth term of the sequence.

Observe that to any such sequence of elements of R^n, there correspond n sequences of real numbers, one for each coordinate. These real sequences form the vertical columns to the right of the equality signs in the array below.

$$\begin{aligned} X_1 &= (x_1^{(1)}, x_1^{(2)}, \ldots, x_1^{(j)}, \ldots x_1^{(n)}) \\ X_2 &= (x_2^{(1)}, x_2^{(2)}, \ldots, x_2^{(j)}, \ldots x_2^{(n)}) \\ &\vdots \\ X_k &= (x_k^{(1)}, x_k^{(2)}, \ldots, x_k^{(j)}, \ldots x_k^{(n)}) \\ &\vdots \end{aligned}$$

The theorem we now prove says that the sequence of n-tuples on the left is convergent in R^n if and only if all of the real sequences on the right converge. In fact, if the jth sequence on the right converges to $x^{(j)}$, then the jth coordinate of the limit of X_k is $x^{(j)}$. You will see that the proof simply extends out the ideas of Example 8 of the last section.

Theorem 6.2.3 *A sequence $\{X_k\}_{k=1}^{\infty}$ converges to an element X in R^n if and only if for each $j = 1, 2, \ldots, n$,*

$$\lim_{k\to\infty} x_k^{(j)} = x^{(j)}$$

where $x_k^{(j)}$ is the jth coordinate of X_k and $x^{(j)}$ is the jth coordinate of X.

PROOF: Suppose that $X_k \to X$ in R^n, and let j be any one of the numbers $1, 2, \ldots, n$. If $\varepsilon > 0$ is given, there exists a positive integer K such that

$$\rho(X_k, X) < \varepsilon \qquad k \geq K$$

Remember that

$$\rho(X_k, X) = \sqrt{(x_k^{(1)} - x^{(1)})^2 + \cdots + (x_k^{(j)} - x^{(j)})^2 + \cdots + (x_k^{(n)} - x^{(n)})^2}$$

Clearly then, for our particular j, $|x_k^{(j)} - x^{(j)}| \leq \rho(X_k, X)$, so that

$$|x_k^{(j)} - x^{(j)}| < \varepsilon \qquad k \geq K$$

and we have

$$x_k^{(j)} \to x^{(j)}$$

But j was arbitrary and so the limit statement holds for $j = 1, 2, \ldots, n$.

Conversely, suppose that $x_k^{(j)} \to x^{(j)}$ for $j = 1, 2, \ldots, n$ and let $X = (x^{(1)}, x^{(2)}, \ldots, x^{(n)})$. We prove that $X_k \to X$. Let $\varepsilon > 0$ be given. For each j there exists a $K_j \in Z^+$ such that

$$|x_k^{(j)} - x^{(j)}| < \frac{\varepsilon}{\sqrt{n}} \qquad \text{for } k \geq K_j$$

If $K = \max\{K_1, K_2, \ldots, K_n\}$, then $|x_k^{(j)} - x^{(j)}| < \varepsilon/\sqrt{n}$ for $j = 1, 2, \ldots, n$ and $k \geq K$. Thus, for $k \geq K$,

$$\begin{aligned}\rho(X_k, X) &= \sqrt{(x_k^{(1)} - x^{(1)})^2 + \cdots + (x_k^{(j)} - x^{(j)}) + \cdots + (x_k^{(n)} - x^{(n)})^2} \\ &< \sqrt{\frac{\varepsilon^2}{n} + \cdots + \frac{\varepsilon^2}{n} + \cdots + \frac{\varepsilon^2}{n}} = \varepsilon \qquad \blacksquare\end{aligned}$$

EXERCISES 6.2

1. Show that the following inequality holds for all real numbers $x_1, x_2, \ldots, x_n$.

$$\frac{1}{n}\sum_{k=1}^{n} x_k \leq \sqrt{\sum_{k=1}^{n} \frac{x_k^2}{n}}$$

(The arithmetic mean of n real numbers is not greater than their quadratic mean.)

2. What can be concluded about X and Y if equality holds in Cauchy's Inequality?

3. (a) Show that the inner product of elements of R^n has the following properties:
   (i) $X \cdot Y = Y \cdot X$
   (ii) $(cX + dY) \cdot Z = c(X \cdot Z) + d(Y \cdot Z)$
   (iii) $X \cdot X = \|X\|^2$

   (b) Prove the Parallelogram Law, which says that for $X, Y \in R^n$,

$$\|X + Y\|^2 + \|X - Y\|^2 = 2\|X\|^2 + 2\|Y\|^2$$

4. In R^n, X and Y are *orthogonal* provided that $X \cdot Y = 0$. Given X and Y, the *projection* of X on Y is defined to be

$$P = \frac{X \cdot Y}{\|Y\|^2} Y$$

Show that P and $X - P$ are orthogonal. Interpret this result geometrically in R^3.

5. Show that if $\{X_k\}_{k=1}^{\infty}$ and $\{Y_k\}_{k=1}^{\infty}$ are sequences such that $X_k \to X$ and $Y_k \to Y$ in R^n and if $c, d \in R$, then $cX_k + dY_k \to cX + dY$.

6. Which of the following are metrics on R^n? Justify your conclusions.
   (a) $\sigma(X, Y) = \max\{|x_k - y_k| : k = 1, 2, \ldots, n\}$
   (b) $\tau(X, Y) = \min\{|x_k - y_k| : k = 1, 2, \ldots, n\}$

6.3 *Open and Closed Sets in Metric Spaces*

The concept of neighborhood allows us to define open and closed sets in any metric space. We will find that these sets have many of the same properties that were developed in Sections 3.10 through 3.12.

Open Sets

Definition 6.3.1 Let (A, ρ) be a metric space. A subset $G \subseteq A$ is ***open in*** A (briefly, ***open***) provided that for each $x \in G$ there exists a $\delta > 0$ such that $N_\delta(x) \subseteq G$.

Certainly in R with the absolute value metric, "openness" has the same meaning it had before. Thus, any open interval (a, b) is open in R, as are (a, ∞), $(-\infty, b)$, $\varnothing$, and R itself.

Example 1: If (A, ρ) is a metric space, then for any $x \in A$ and $\delta > 0$, the neighborhood $N_\delta(x)$ is open. For if $y \in N_\delta(x)$, then $\rho(x, y) < \delta$, and $\delta' = \delta - \rho(x, y)$ is a positive number. It follows by the Triangle Inequality that $N_{\delta'}(y) \subseteq N_\delta(x)$, for if $z \in N_{\delta'}(y)$, then

$$\rho(x, z) \leq \rho(x, y) + \rho(y, z) < \rho(x, y) + \delta - \rho(x, y) = \delta$$

Figure 6.6 shows the situation as it looks in R^2.

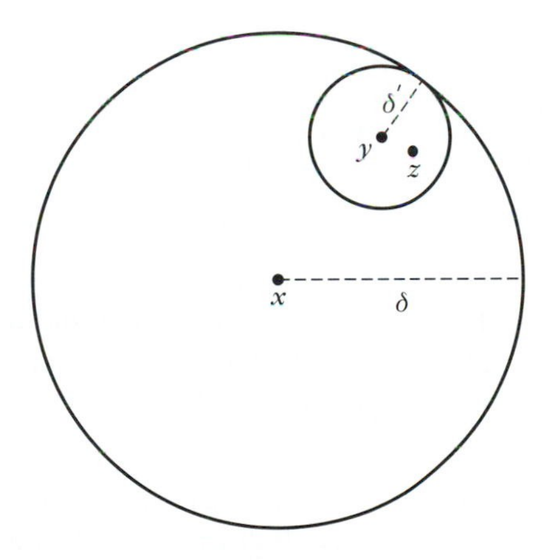

Figure 6.6
Neighborhood is open

Example 2: In R^2, the upper half-plane

$$H = \{(x, y) : y > 0\}$$

is open. Let (x_0, y_0) be a fixed point in H. If $\delta = y_0$, then the neighborhood $N_\delta((x_0, y_0))$ is contained in H. To prove this, let $(x, y) \in N_\delta((x_0, y_0))$, and note that

$$|y - y_0| \leq \sqrt{(x - x_0)^2 + (y - y_0)^2} < \delta = y_0$$

It follows that $-y_0 < y - y_0 < y_0$, so that

$$0 < y < 2y_0$$

Therefore, $(x, y) \in H$, so that $N_\delta((x_0, y_0)) \subseteq H$.

Example 3: In R^3, the first octant

$$S = \{(x, y, z): x > 0, y > 0, z > 0\}$$

is open. For if (x_0, y_0, z_0) is any fixed point in S, and if $\delta = \min\{x_0, y_0, z_0\}$, then $\delta > 0$ and $N_\delta((x_0, y_0, z_0)) \subseteq S$. (Verify this claim in Exercise 6.3.1.)

Theorem 6.3.1 *In a metric space:*

(i) The union of any collection of open sets is open
(ii) The intersection of a finite collection of open sets is open

PROOF: The proof of this theorem is exactly the same as the proof of Theorem 3.10.1. The only difference is that we now have a more general interpretation of neighborhoods. ■

Cluster Points

In a metric space (A, ρ) the *deleted δ-neighborhood* of a, $N_\delta^*(a)$, consists of all points in $N_\delta(a)$ except a itself. Thus, for $a \in A$ and $\delta > 0$,

$$N_\delta^*(a) = \{x : x \in A \text{ and } 0 < \rho(x, a) < \delta\}$$

Definition 6.3.2 Let (A, ρ) be a metric space, and let $S \subseteq A$. Then a point $a \in A$ is a ***cluster point*** of S provided that every deleted neighborhood of a contains one or more elements of S.

Definition 6.3.3 Let (A, ρ) be a metric space and let $S \subseteq A$. Then the ***derived*** set of S is the set S' of all cluster points of S.

The following theorem about cluster points, and its proof, are just like Theorem 3.1.1.

Theorem 6.3.2 *Let (A, ρ) be a metric space and let $S \subseteq A$ and $a \in A$. Then $a \in S'$ if and only if any of the following conditions holds:*

(i) For every $\delta > 0$, there exists an $x \in S$ such that $0 < \rho(x, a) < \delta$
(ii) There is a sequence $\{x_n\}_{n=1}^{\infty}$ in $S \dashv \{a\}$ such that $x_n \to a$

(iii) There is a sequence $\{y_k\}_{k=1}^{\infty}$ of distinct points of S such that $y_k \to a$

(iv) Every neighborhood of a contains infinitely many points of S

Corollary 6.3.1 *In any metric space, a finite set has no cluster points.*

Definition 6.3.4 Let (A, ρ) be a metric space, let $S \subseteq A$, and let $a \in A$. Then a is an ***isolated point*** of S provided that $a \in S$ and there is a deleted neighborhood of a that contains no elements of S.

Example 4: In R^2, consider the set $H = \{(x, y):y > 0\}$. All points in H as well as all points (x, y) with $y = 0$ (the x-axis) are cluster points of H. H has no isolated points. The derived set is

$$H' = \{(x, y):y \geq 0\}$$

Example 5: Let A be any nonempty set with the discrete metric (see Example 4, Section 6.1). If $S \subseteq A$, every element of S is an isolated point of S; S has no cluster points, so $S' = \varnothing$.

Closure

Definition 6.3.5 Let (A, ρ) be a metric space and let $S \subseteq A$. Then the ***closure*** of S is the set $\overline{S} = S \cup S'$.

Note: A point belongs to $\overline{S}$ if and only if it is the limit of a sequence of points of S.

Example 6: Consider the following subset of R^2:

$$S = \{(0, 0)\} \cup \{(x, y):x^2 + y^2 > 1\}$$

The derived set is

$$S' = \{(x, y):x^2 + y^2 \geq 1\}$$

and the closure of S is

$$\overline{S} = S \cup S' = \{(0, 0)\} \cup \{(x, y):x^2 + y^2 \geq 1\}$$

Closed Sets

Definition 6.3.6 Let (A, ρ) be a metric space. A set $F \subseteq A$ is ***closed in*** A (briefly, ***closed***) provided that every cluster point of F belongs to F.

Clearly F is closed if and only if $\overline{F} = F$.

The relationship between "closed" and "open" involves their complements.

Definition 6.3.7 If S is a subset of A, then the ***complement*** of S in A is the set $A \dashv S$.

The next three theorems generalize Theorems 3.10.4, 3.10.5, and 3.10.6. The proofs are identical.

Theorem 6.3.3 *Let (A, ρ) be a metric space. Then a subset F of A is closed if and only if its complement $A \dashv F$ is open.*

Theorem 6.3.4 *In a metric space:*

(i) The intersection of any collection of closed sets is closed
(ii) The union of a finite collection of closed sets is closed

Theorem 6.3.5 *Let (A, ρ) be a metric space. A subset F of A is closed if and only if every convergent sequence in F converges to a point in F.*

Example 7: The set $\{(x, y): x^2 + y^2 \geq 1\}$ in R^2 is closed because its complement is open.

Example 8: The set of points $\{(x, 2x + 1): x \in R\}$ is a closed subset of R^2. For if $\{(x_n, 2x_n + 1)\}_{n=1}^{\infty}$ converges in R^2, then, by Theorem 6.2.3, both $\{x_n\}_{n=1}^{\infty}$ and $\{2x_n + 1\}_{n=1}^{\infty}$ converge in R, say, $x_n \to a$ and $2x_n + 1 \to b$. But if $x_n \to a$, then $2x_n + 1 \to 2a + 1$, so $b = 2a + 1$. Thus, the sequence $\{(x_n, 2x_{n+1})\}_{k=1}^{\infty}$ converges to the point $(a, 2a + 1)$, which belongs to the given set.

Separability

Definition 6.3.8 Let S and T be subsets of a metric space. Then S is ***dense*** in T provided that $\overline{S} = T$.

Example 9: In R, the open interval (a, b) where $a < b$ is dense in $[a, b]$.

Example 10: The rationals are dense in the reals.

Definition 6.3.9 A metric space (A, ρ) is ***separable*** provided there is a countable subset of A that is dense in A.

Example 11: R with the absolute value metric is separable because Q is countable and dense in R.

EXERCISES 6.3

1. Verify that if (x_0, y_0, z_0) is a point in the first octant of three-dimensional space, and if $\delta = \min\{x_0, y_0, z_0\}$, then $N_\delta((x_0, y_0, z_0))$ is contained in the first octant.
2. Show that in a metric space, an open set is a union of neighborhoods.
3. Show that in a metric space, a deleted neighborhood of a point a is an open set.
4. Let (A, ρ) be a metric space and let $S \subseteq A$. Show that $\overline{S}$ is closed and that $\overline{S}$ is contained in every closed set that contains S.
5. Let (A, ρ) be a metric space and let $a \in A$ and $S \subseteq A$. Then a is an *interior point* of S provided that some neighborhood of a is contained in S. The *interior* of S is the set S^0 of all interior points of S.
   (a) Show that the interior of S is open.
   (b) Show that every open subset of S is contained in S^0.
   (c) Show that S is open if and only if $S = S^0$.
   (d) Show that $\overline{S^c} = (S^0)^c$ where the bar denotes closure and c denotes complement in A.
6. Show that in a metric space (A, ρ), the derived set of a subset of A is closed.
7. Show that the metric space R^n is separable.
8. Prove or disprove: If (A, ρ) is a metric space, $a \in A$, and $\delta > 0$, then $\overline{N_\delta(a)} = \{x : \rho(x, a) \le \delta\}$.

6.4 Completeness, Compactness

Bounded Sets in Metric Spaces

Definition 6.4.1 Let (A, ρ) be a metric space and let B be a nonempty subset of A. Then B is ***bounded*** provided that there is a real number M such that $\rho(x, y) \le M$ for all $x, y \in B$. If B is bounded, then its ***diameter*** is $\sup\{\rho(x, y) : x, y \in B\}$. A set that is not bounded is ***unbounded***.

Example 1: Let A be the subset $[a, b]$ of R $(a < b)$, and let ρ be the absolute value metric. Then A is bounded and its diameter is $b - a$.

Example 2: The sphere $\{(x, y, z) : x^2 + y^2 + z^2 = 1\}$ is a bounded subset of R^3 with diameter 2.

Example 3: The y-axis $\{(x, y) : x = 0\}$ is an unbounded subset of R^2.

Cauchy Sequences

The Cauchy property can be adapted to metric spaces generally.

Definition 6.4.2 Let (A, ρ) be a metric space and let $\{x_n\}_{n=1}^{\infty}$ be a sequence of points in A. Then $\{x_n\}_{n=1}^{\infty}$ is ***Cauchy*** provided that for an arbitrary $\varepsilon > 0$, there exists a positive integer N such that $\rho(x_m, x_n) < \varepsilon$ whenever $m, n \ge N$.

We saw that for sequences in R, the Cauchy property is equivalent to convergence.

Definition 6.4.3 A metric space (A, ρ) is ***complete**** provided that every Cauchy sequence in A is convergent (to a point in A).

Example 4: The closed interval $[0, 1]$ in R is a metric space under the absolute value metric. It is complete because if a sequence $\{x_n\}_{n=1}^{\infty}$ in that space is Cauchy, it is also Cauchy in R, and by Theorem 2.7.2, it must converge in R. But since each x_n belongs to the closed interval $[0, 1]$, the limit must also belong to $[0, 1]$.

Example 5: The open interval $(0, 1)$ under the absolute value metric is not complete. The sequence $\{1/n\}_{n=1}^{\infty}$ is Cauchy but does not converge to a point of $(0, 1)$.

Example 6: The set Z under the absolute value metric is complete. A sequence of integers cannot be Cauchy unless it has a tail that is constant, and such a sequence converges to an integer.

Theorem 6.4.1 **Completeness of R^n** *The metric space R^n is complete under the Euclidean metric.*

PROOF: Let $\{X_k\}_{k=1}^{\infty} = \{(x_k^{(1)}, x_k^{(2)}, \ldots, x_k^{(n)})\}_{k=1}^{\infty}$ be a Cauchy sequence in R^n. Then it can be shown (work Exercise 6.4.5) that each of the coordinate sequences $\{x_k^{(j)}\}_{k=1}^{\infty}$ is Cauchy in R. Therefore, each $\{x_k^{(j)}\}_{k=1}^{\infty}$ is convergent in R, and we can define X as follows:

$$X = (x^{(1)}, x^{(2)}, \ldots, x^{(n)})$$

where $x^{(j)} = \lim_{k\to\infty} x_k^{(j)}$ for $j = 1, 2, \ldots, n$. By Theorem 6.2.3, $X_k \to X$. ∎

Definition 6.4.4 A sequence in a metric space is ***bounded*** provided that its range is bounded.

The following two theorems are consequences of this definition.

Theorem 6.4.2 *In a metric space (A, ρ), every Cauchy sequence is bounded.*

PROOF: You are asked to prove this theorem in Exercise 6.4.6. ∎

Theorem 6.4.3 *A sequence $\{X_k\}_{k=1}^{\infty}$ in R^n is bounded if and only if the sequence of norms $\{\|X_k\|\}_{k=1}^{\infty}$ is bounded in R.*

PROOF: You are asked to prove this theorem in Exercise 6.4.7. ∎

*This use of the word *complete* differs from the usage in Section 1.2.

Next we take up some properties of bounded sets in R^n.

Theorem 6.4.4 **Bolzano-Weierstrass Theorem for Sequences in R^n** *Every bounded sequence in R^n has a convergent subsequence.*

PROOF: In order to avoid notational acrobatics, we will prove this theorem for the case $n = 3$. It should then be clear how the proof works for larger values of n.

Let $\{P_k\}_{k=1}^{\infty} = \{(x_k, y_k, z_k)\}_{k=1}^{\infty}$ be bounded in R^3. Then the three coordinate sequences $\{x_k\}_{k=1}^{\infty}$, $\{y_k\}_{k=1}^{\infty}$, $\{z_k\}_{k=1}^{\infty}$ are bounded in R. By the Bolzano-Weierstrass Theorem for Sequences in R (Theorem 2.7.1), some subsequence of $\{x_k\}_{k=1}^{\infty}$ is convergent in R, say, $x_{k_l} \to x$. Now consider $\{y_{k_l}\}_{l=1}^{\infty}$, which is the corresponding subsequence of y's; it, too, is a bounded sequence in R and therefore has a convergent subsequence $\{y_{k_{l_m}}\}_{m=1}^{\infty}$, say, $y_{k_{l_m}} \to y$. Observe that $x_{k_{l_m}} \to x$ because $\{x_{k_{l_m}}\}_{m=1}^{\infty}$ is a subsequence of $\{x_{k_l}\}_{l=1}^{\infty}$. Next we consider the subsequence of $\{z_k\}_{k=1}^{\infty}$ corresponding to the subscripts k_{l_m}; that subsequence has a convergent subsequence, say, $z_{k_{l_{m_q}}} \to z$. But then $x_{k_{l_{m_q}}} \to x$ and $y_{k_{l_{m_q}}} \to y$, and finally $\{P_{k_{l_{m_q}}}\}_{q=1}^{\infty}$ (a subsequence of $\{P_k\}_{k=1}^{\infty}$) converges to (x, y, z) by Theorem 6.2.3. ■

Corollary 6.4.1 **Bolzano-Weierstrass Theorem for Sets in R^n** *Every bounded infinite subset of R^n has at least one cluster point in R^n.*

PROOF: You are asked to prove this corollary in Exercise 6.4.9. ■

Compactness

Definition 6.4.5 Let (A, ρ) be a metric space. A subset $C \subseteq A$ is ***compact*** provided that every sequence in C has a subsequence that converges to a point in C.

Since this definition is exactly the same as the one given for subsets of R in Section 3.11, we know that any closed bounded interval in R is compact; also, any finite set of real numbers is compact.

It is easy to characterize the compact subsets of R^n.

Theorem 6.4.5 *A subset of R^n is compact if and only if it is closed and bounded.*

PROOF: Let S be a nonempty compact subset of R^n. To show that S is closed, we let X be a cluster point of S and show that $X \in S$. Now X is the limit of some sequence of points of S. Since S is compact, this sequence has a subsequence converging to a point of S. But the only point to which the subsequence can converge is X, and so $X \in S$. To show that S is bounded, suppose otherwise and let X_0 be some fixed point in S. Since S is unbounded, it is then possible to choose a sequence $\{X_k\}_{k=1}^{\infty}$ with the property that $\rho(X_0, X_k) > k$, so that $\rho(X_0, X_k) \to \infty$. By the compactness of S, there is a subsequence $\{Y_l\}_{l=1}^{\infty}$ of $\{X_k\}_{k=1}^{\infty}$ with $Y_l = X_{k_l}$ that converges to a point of S, say, $Y_l \to Y$. Then for l sufficiently large, say, $l \geq L$, we should have

$$\rho(Y_l, Y) < 1$$

Thus, for terms X_{k_l} with $l \geq L$, it follows that

$$\begin{aligned}\rho(X_0, X_{k_l}) = \rho(X_0, Y_l) &\leq \rho(X_0, Y) + \rho(Y, Y_l) \\ &< \rho(X_0, Y) + 1\end{aligned}$$

This inequality contradicts the way we chose X_k, and so we conclude that S is bounded.

Conversely, let S be a closed and bounded subset of R^n. To prove that S is compact, let $\{X_k\}_{k=1}^{\infty}$ be an arbitrary sequence in S. Since S is bounded, the Bolzano-Weierstrass Theorem assures us that $\{X_k\}_{k=1}^{\infty}$ has a convergent subsequence; since S is closed, the limit of this convergent subsequence must belong to S (Theorem 6.3.5), and so S is compact. ■

In Section 3.11 we discussed open covers and the Heine-Borel property. We now summarize the relevant generalizations to metric spaces.

Definition 6.4.6 Let (A, ρ) be a metric space and let $S \subseteq A$. An ***open cover*** of S is a collection C of open sets such that $S \subseteq \cup C$.

Definition 6.4.7 Let (A, ρ) be a metric space and let $S \subseteq A$. Then S has the ***Heine-Borel property*** provided that if C is an arbitrary open cover of S, then there is a finite subcollection C_0 of C such that C_0 is an open cover of S.

For metric spaces, it can be shown that compactness and the Heine-Borel property are equivalent (see Simmons, Chapter Four*). You are asked to prove one part of this equivalence in Exercise 6.4.10.

Theorem 6.4.5 shows us that in R^n, compactness is equivalent to the condition of being closed and bounded, just as it is in R. However, there are metric spaces for which this equivalence does not hold; in Section 6.6 we will discuss a space in which there is a subset that is closed and bounded but not compact (Exercise 6.6.3).

*George F. Simmons, *Introduction to Topology and Modern Analysis* (New York: McGraw-Hill, 1963).

EXERCISES 6.4

1. Which of the following subsets of R^2 are bounded? In case of boundedness, find the diameter.

(a) $\{(x, y): x^2 + y^2 \leq 36\}$
(b) $\{(x, y): x^2 - 4y^2 = 36\}$
(c) $\{(x, y): y = 1/x, x \geq 1\}$
(d) $\{(x, y): |x| + |y| = 1\}$
(e) $\{(x, y): x^3 + y^3 = 1\}$
(f) $\{(x, y): x^2 + y^2 < 4\}$

2. (a) Show that if a subset S of a metric space has diameter d, then there exist sequences $\{x_n\}_{n=1}^{\infty}$ and $\{y_n\}_{n=1}^{\infty}$ of points of S such that $\rho(x_n, y_n) \to d$.
(b) Show further that if S is compact, then there exist points $a, b \in S$ such that $\rho(a, b) = d$.

3. Determine whether the following sequences are Cauchy in R^2.

(a) $\left\{\left(\frac{\pi}{n}, \sin\frac{\pi}{n}\right)\right\}_{n=1}^{\infty}$
(b) $\left\{\left(\frac{\pi}{n}, \cos\frac{\pi}{n}\right)\right\}_{n=1}^{\infty}$
(c) $\left\{\left(\cos\frac{\pi}{n}, \sin\frac{\pi}{n}\right)\right\}_{n=1}^{\infty}$
(d) $\left\{\left(\cos\frac{n\pi}{2}, \sin\frac{n\pi}{2}\right)\right\}_{n=1}^{\infty}$

4. (a) Is the metric space consisting of the set $\{r: r \in Q \text{ and } 0 \leq r \leq 1\}$ with the absolute value metric, complete?
(b) Let B be the set of all bounded sequences of real numbers. For $X, Y \in B$ with $X = \{x_n\}_{n=1}^{\infty}$, $Y = \{y_n\}_{n=1}^{\infty}$ define

$$\rho(X, Y) = \sup\{|x_n - y_n| : n \in Z^+\}$$

Show that ρ is a metric on B. Is B bounded with respect to this metric?
(c) Is the metric space (B, ρ) in part (b) complete?

5. Regarding the proof of Theorem 6.4.1, show that if $\{X_k\}_{k=1}^{\infty}$ is Cauchy in R^n with the Euclidean metric, then each coordinate sequence $\{x_k^{(j)}\}_{k=1}^{\infty}$ is Cauchy in R.

6. Prove Theorem 6.4.2. [*Hint:* Start by proving that if $\{x_n\}_{n=1}^{\infty}$ is Cauchy, then $\{\rho(x_1, x_n)\}_{n=1}^{\infty}$ is bounded.]

7. Prove Theorem 6.4.3.

8. Show that the sequence $\left\{\left(\cos\frac{n\pi}{2}, \sin\frac{n\pi}{2}\right)\right\}_{n=1}^{\infty}$ is bounded in R^2 and find a convergent subsequence.

9. Prove Corollary 6.4.1.

10. (a) Show that if S is a subset of a metric space and if $\{X_n\}_{n=1}^{\infty}$ is a sequence of points in S that has no subsequence convergent to an element of S, then every element of S has some neighborhood that contains $\{X_n\}_{n=1}^{\infty}$ for at most a finite number of values of n.
(b) Show that in a metric space, a set with the Heine-Borel property is compact.

11. Show that a subset S of a metric space is compact if and only if every infinite subset of S has a cluster point in S.

6.5 *Continuous Mappings of Metric Spaces*

Continuity

We adopt a sequential definition for continuity of mappings from one metric space to another.

Definition 6.5.1 Let (A, ρ) and (B, σ) be metric spaces and let $f: A \to B$. If $a \in A$, then f is ***continuous*** at a provided that for every sequence $\{x_n\}_{n=1}^{\infty}$ of points of A converging to a, it is the case that $\{f(x_n)\}_{n=1}^{\infty}$ converges to $f(a)$ in B. A mapping $f: A \to B$ that is continuous at all points of A is ***continuous***.

In the case of a real function with domain D, the definitions are equivalent to our earlier definitions, provided that we use the absolute value metric in the domain and range.

Example 1: The mapping $f:(0, \infty) \to (0, 1)$ defined by $f(x) = x/(1 + x)$ is continuous.

Example 2: The mapping $f:R^2 \to R$ defined by $f(x, y) = x^3 + xy^2$ is continuous. If $(a, b) \in R^2$, and if $(x_n, y_n) \to (a, b)$, then (by Example 8 in Section 6.1), $x_n \to a$ and $y_n \to b$. So

$$\lim_{n\to\infty} f(x_n, y_n)^* = \lim_{n\to\infty} (x_n^3 + x_n y_n^2) = a^3 + ab^2 = f(a, b)$$

Thus, $(x_n, y_n) \to (a, b)$ in $A \Rightarrow f(x_n, y_n) \to f(a, b)$ in B, and f is continuous at (a, b).

Example 3: The mapping $f:R^2 \to R$ defined by

$$f(x, y) = \begin{cases} \dfrac{xy}{x^2 + y^2} & \text{if } (x, y) \neq (0, 0) \\ 0 & \text{if } (x, y) = (0, 0) \end{cases}$$

is not continuous at $(0, 0)$ because the sequence $\{(1/n, 1/n)\}_{n=1}^{\infty}$ converges to $(0, 0)$ while $\{f(1/n, 1/n)\}_{n=1}^{\infty}$ converges to $\frac{1}{2} \neq f(0, 0)$. The sequence $\{(1/n, -1/n)\}_{n=1}^{\infty}$ also converges to $(0, 0)$, and in this case $\{f(1/n, -1/n)\}_{n=0}^{\infty}$ converges to $-\frac{1}{2}$. Thus, there is no way to redefine f at $(0, 0)$ so as to make the redefined function continuous there.

Example 4: The mapping $f:R^2 \dashv (0, 0) \to R^2$ defined by

$$f(x, y) = \left(\frac{x}{\sqrt{x^2 + y^2}}, \frac{y}{\sqrt{x^2 + y^2}}\right)$$

is continuous. There is no way, however, to extend this function continuously to all of R^2 (work Exercise 6.5.2).

General Theorems about Continuous Mappings

We first turn to a theorem characterizing continuity at a point in terms of ε, δ, and neighborhoods. Recall that if $f:A \to B$, the inverse image under f of a subset T of B is defined to be $f^{-1}(T) = \{x : x \in A \text{ and } f(x) \in T\}$.

Theorem 6.5.1 *Let (A, ρ) and (B, σ) be metric spaces, let $a \in A$, and let $f:A \to B$. Then the following statements are equivalent:*

(i) f is continuous at a

*Strictly speaking, we should write $f((x_n, y_n))$ here.

(ii) *For every $\varepsilon > 0$, there is a $\delta > 0$ such that $\sigma(f(x), f(a)) < \varepsilon$ whenever x is an element of A such that $\rho(x, a) < \delta$*

(iii) *For every neighborhood M of $f(a)$, there exists a neighborhood N of a such that $N \subseteq f^{-1}(M)$*

PROOF: We prove that (i) $\Rightarrow$ (ii). Assume that f is continuous at a and let $\varepsilon > 0$ be given. Suppose that there is no $\delta > 0$ satisfying the condition stated in (ii). Then for $\delta_n = 1/n$ $(n \in Z^+)$ there is some $x_n \in A$ with $\rho(x_n, a) < 1/n$ and $\sigma(f(x_n), f(a)) \geq \varepsilon$. Then $x_n \to a$ but $\{f(x_n)\}_{n=1}^{\infty}$ does not converge to $f(a)$, contrary to (i). Now property (iii) follows immediately from (ii). You are asked to prove that (iii) $\Rightarrow$ (i) in Exercise 6.5.1. ■

Theorem 6.5.2 *Let (A, ρ) and (B, σ) be metric spaces and let $f{:}A \to B$. Then f is continuous if and only if the inverse image of every open subset of B is an open subset of A.*

PROOF: This theorem follows directly from part (iii) of Theorem 6.5.1. You are asked to prove the theorem in Exercise 6.5.5a. ■

Theorem 6.5.3 **Composition** *Let (A, ρ), (B, σ), and (C, τ) be metric spaces, and let $f{:}A \to B$ and $g{:}B \to C$. If f is continuous at $a \in A$ and g is continuous at $f(a) \in B$, then the composite mapping $g \circ f$ is continous at a.*

PROOF: If $x_n \to a$ in A, then by the continuity of f, we have $f(x_n) \to f(a)$ in B; and by the continuity of g at $f(a)$, we have $g(f(x_n)) \to g(f(a))$. That is, $(g \circ f)(x_n) \to (g \circ f)(a)$. ■

Theorem 6.5.4 **Preservation of Compactness** *Let (A, ρ) and (B, σ) be metric spaces and let $f{:}A \to B$ be continuous. If S is a compact subset of A, then $f(S)$ is compact in B.*

PROOF: Let S be compact in A; we show that $f(S)$ is compact in B. Take any sequence $\{y_n\}_{n=1}^{\infty}$ of points in $f(S)$. Each y_n is the image under f of at least one element of S, say, $y_n = f(x_n)$. Then $\{x_n\}_{n=1}^{\infty}$ is a sequence in the compact set S and has a subsequence that converges to a point of S, say, $x_{n_k} \to x$. But then $f(x_{n_k}) \to f(x)$, and so the subsequence $\{y_{n_k}\}_{k=1}^{\infty}$ converges to a point of $f(S)$. ■

Theorems about R^n

Theorem 6.5.4 has an immediate consequence of great importance in the theory of functions of several variables:

Theorem 6.5.5 *If $f: R^n \to R$ is continuous and if S is a compact subset of R^n, then $f(S)$ is compact.*

Corollary 6.5.1 **Extreme Value Theorem** *If $f: R^n \to R$ is continuous and if S is a compact subset of R^n, then f assumes an absolute minimum and an absolute maximum on S.*

PROOF: A compact subset of R has a least element and a greatest element. ■

Theorem 6.5.6 *Let f and g be real functions, continuous on a subset $D \subseteq R$. Then the mapping $F:D \to R^2$ defined by $F(x) = (f(x), g(x))$ is continuous.*

PROOF: If $\{x_n\}$ is any sequence in D converging to a point $a \in D$, then by the continuity of f and g we have $f(x_n) \to f(a)$ and $g(x_n) \to g(a)$. Then, by Example 8 in Section 6.1, $(f(x_n), g(x_n)) \to (f(a), g(a))$. In other words, $F(x_n) \to F(a)$. ■

Example 5: The mapping $T:[0, 2\pi) \to R^2$ defined by $T(t) = (\cos t, \sin t)$ is continuous. Its range is the unit circle in the coordinate plane.

Example 6: The *graph* of a real function f with domain D is the subset of R^2 defined by $G_f = \{(x, f(x)):x \in D\}$. If f is continuous, then by Theorem 6.5.6, the mapping $h:D \to R^2$ defined by $h(x) = (x, f(x))$ is continuous, and its range is G_f. It follows then by Theorem 6.5.4 that the graph of a continuous real function on a closed bounded interval is a compact subset of R^2.

Distance Functions

If b is a fixed point in a metric space, then as x varies over the space, the distance of x from b varies continuously.

Theorem 6.5.7 *Let (A, ρ) be a metric space and let $b \in A$. Then the function f defined for $x \in A$ by $f(x) = \rho(x, b)$ is continuous.*

PROOF: Let $a \in A$ and let $\{x_n\}_{n=1}^{\infty}$ be a sequence in A such that $x_n \to a$. We are to prove that $f(x_n) \to f(a)$, that is, that $\rho(x_n, b) \to \rho(a, b)$. By the Triangle Inequality,

$$\rho(x_n, b) \leq \rho(x_n, a) + \rho(a, b)$$

Also by the Triangle Inequality,

$$\rho(a, b) \leq \rho(x_n, a) + \rho(x_n, b)$$

and it follows that

$$|\rho(x_n, b) - \rho(a, b)| \leq \rho(x_n, a)$$

In other words,

$$|f(x_n) - f(a)| \leq \rho(x_n, a)$$

Since $\{\rho(x_n, a\}_{n=1}^{\infty}$ is null, $f(x_n) \to f(a)$. ■

Definition 6.5.2 Let (A, ρ) be a metric space and let B be a nonempty subset of A. For $x \in A$, the ***distance*** of x from B is defined to be

$$\rho(x, B) = \inf_{y \in B} \rho(x, y)$$

Example 7: In R, the distance of the origin from the open interval $(3, 4)$ is 3; the distance of any element of R from Q is 0.

Example 8: In R^2, the distance of $(0, 0)$ from the hyperbola $xy = 1$ is $\sqrt{2}$. (Verify.)

Theorem 6.5.8 *If (A, ρ) is a metric space and B is a nonempty subset of A, then $\rho(x, B) = 0$ if and only if $x \in \overline{B}$.*

PROOF: If $\rho(x, B) = 0$, then, by definition, no positive number is a lower bound of the set $\{\rho(x, y): y \in B\}$. Thus, for every $n \in Z^+$, there is an element $y_n \in B$ such that $\rho(x, y_n) < 1/n$. Therefore, $y_n \to x$ and $x \in \overline{B}$. Conversely, if $x \in \overline{B}$, then a sequence $\{y_n\}_{n=1}^{\infty}$ of points of B can be found such that $y_n \to x$. Therefore, $\rho(x, y_n) \to 0$ and $\rho(x, B) = 0$. ■

Theorem 6.5.9 *If (A, ρ) is a metric space and B is a nonempty subset of A, then the function f defined for $x \in A$ by $f(x) = \rho(x, B)$ is continuous.*

PROOF: We use part (ii) of Theorem 6.5.1 to prove the continuity of f at an arbitrary point $a \in A$. We will show that if $\varepsilon > 0$ is given, then there is a $\delta > 0$ such that $|f(x) - f(a)| < \varepsilon$ whenever $\rho(x, a) < \delta$. In fact, for a given $\varepsilon > 0$, we can choose $\delta = \varepsilon/2$.

Let $\rho(x, a) < \delta = \varepsilon/2$. Since $\rho(a, B) = \inf_{y \in B} \rho(a, y)$, we can choose $y \in B$ such that $\rho(a, y) < \rho(a, B) + \varepsilon/2$. Then

$$\begin{aligned}\rho(x, B) &\leq \rho(x, y)\\ &\leq \rho(x, a) + \rho(a, y) < \rho(x, a) + \rho(a, B) + \frac{\varepsilon}{2} < \rho(a, B) + \varepsilon\end{aligned}$$

We see that $f(x) < f(a) + \varepsilon$. Similarly, we can show that $f(a) < f(x) + \varepsilon$, and so $|f(x) - f(a)| < \varepsilon$ whenever $\rho(x, a) < \varepsilon/2$. ■

Definition 6.5.3 Let (A, ρ) be a metric space and let S and T be subsets of A. Then the ***distance*** of S from T is defined to be

$$\rho(S, T) = \inf_{x \in S} \rho(x, T)$$

Example 9: In R, the distance between the open intervals $(0, 1)$ and $(1, \infty)$ is 0.

Example 10: In R^2, the distance between the hyperbola $xy = 1$ and the x-axis is 0. The distance between the two branches of this hyperbola is $2\sqrt{2}$.

EXERCISES 6.5

1. Complete the proof of Theorem 6.5.1.
2. Show that there is no way to extend the function f of Example 4 continuously to all of R^2.
3. The following real-valued functions are defined and continuous on all of R^2 except for $(0, 0)$. Determine in each case whether the function can be extended continuously to all of R^2, and if so, what its value should be at $(0, 0)$.

   (a) $f(x, y) = \dfrac{x^2 - y^2}{x^2 + y^2}$

   (b) $f(x, y) = \dfrac{x^3 + xy^2}{x^2 + y^2}$

   (c) $f(x, y) = \dfrac{x}{x^2 + y^2}$

   (d) $f(x, y) = \dfrac{x^4 - y^4}{x^2 + y^2}$

4. Let $[a, b]$ be a closed bounded interval in R. For $n = 2$ or 3, a continuous mapping $f:[a, b] \to R^n$ determines a curve in R^n. Show that the following mappings are continuous, and sketch the resulting curves.

   (a) $f:[0, 1] \to R^2$; $f(t) = (t, 1 - t)$
   (b) $f:[0, 1] \to R^2$; $f(t) = (t, t^2)$
   (c) $f:[-1, 1] \to R^2$; $f(t) = (e^t, e^{-t})$
   (d) $f:[0, \pi] \to R^2$; $f(t) = \begin{cases} (t, \sin t/t) & \text{if } t \neq 0 \\ (0, 1) & \text{if } t = 0 \end{cases}$
   (e) $f:[0, 2\pi] \to R^3$; $f(t) = (\cos t, \sin t, 1)$
   (f) $f:[0, 2\pi] \to R^3$; $f(t) = (\cos t, \sin t, t)$

5. (a) Prove Theorem 6.5.2.
   (b) Let (A, ρ) and (B, σ) be metric spaces. Show that $f:A \to B$ is continuous if and only if the inverse image under f of every closed subset of B is a closed subset of A.
6. Show that if (A, ρ) is a metric space and if $f:A \to R$ is continuous, then $\{x:f(x) > b\}$ is open; $\{x:f(x) \geq b\}$ is closed; $\{x:a < f(x) < b\}$ is open; and $\{x:f(x) = 0\}$ is closed.
7. Show that if (A, ρ) is a metric space, if $b \in A$, and if $0 \leq r < s$, then $\{x:x \in A \text{ and } r < \rho(x, b) < s\}$ is open.
8. Show that if (A, ρ) is a metric space, if $x_0 \in A$, and if B is a compact subset of A, then there exists a point y in B such that $\rho(x_0, y) = \rho(x_0, B)$.
9. Prove or disprove each of the following statements involving a metric space (A, ρ), subsets S and T of A, and a point $x_0 \in A$:

   (a) If $x_0 \in S$, then $\rho(x_0, S) = 0$.
   (b) If $\rho(x_0, S) = 0$, then $x_0 \in S$.
   (c) If $\rho(x_0, S) = 0$ and S is closed, then $x_0 \in S$.
   (d) If $\rho(S, T) = 0$, then $S \cap T \neq \emptyset$.
   (e) If $\rho(S, T) = 0$ and S, T are closed, then $S \cap T \neq \emptyset$.
   (f) If $\rho(S, T) = 0$, if S is compact, and T is closed, then $S \cap T \neq \emptyset$.

PROJECT: Uniform Continuity

Let (A, ρ) and (B, σ) be metric spaces. A mapping $f: A \to B$ is *uniformly continuous* provided that for every $\varepsilon > 0$ there exists a $\delta > 0$ such that

$$\sigma(f(x), f(y)) < \varepsilon \text{ whenever } \rho(x, y) < \varepsilon$$

(1) Show that the functions defined in Theorems 6.5.7 and 6.5.9 are both uniformly continuous.

(2) Show that if $f: A \to B$ is uniformly continuous and if $\{a_n\}_{n=1}^{\infty}$ is a Cauchy sequence in A, then $\{f(a_n)\}_{n=1}^{\infty}$ is a Cauchy sequence in B.

(3) Adapt the proof of Theorem 3.11.3 to show that a mapping that is continuous on a compact set is uniformly continuous. (You may assume the equivalence of the Heine-Borel property and compactness for this proof.)

(4) Prove the following theorem: If (A, ρ) and (B, σ) are metric spaces, if D is dense in A, if C is complete in B, and if $f: D \to C$ is uniformly continuous, then there exists a uniformly continuous extension of f mapping A into B.

PROJECT: Homeomorphism

The metric spaces (A, ρ) and (B, σ) are *homeomorphic* provided that there exists a bijection $f: A \to B$ such that f and f^{-1} are continuous. If such a bijection exists, it is called a *homeomorphism*.

(1) Show that being homeomorphic is an equivalence relation.

(2) Show that any two nonempty bounded open intervals in R are homeomorphic.

(3) Show that any two nonempty open intervals in R (bounded or not) are homeomorphic.

(4) Show that if $f: A \to B$ is a homeomorphism, then a subset G of A is open in A if and only if $f(G)$ is open in B. Conversely, a bijection with this property is a homeomorphism.

(5) Show that if ρ and σ are equivalent metrics on a set A (see Exercise 6.1.5), then (A, ρ) and (A, σ) are homeomorphic.

(6) Show that if we are given any metric space (A, ρ), it is possible to find a metric σ for A such that (A, ρ) and (A, σ) are homeomorphic and (A, σ) is bounded. [*Hint:* Show that the metric $\sigma = \rho/(1 + \rho)$ of Exercise 6.1.6 is equivalent to ρ.]

6.6 *Hilbert Space*

In this section we study another important metric space whose elements are functions; in this case, they are sequences.

We wish to go from the space R^n, whose elements are vectors with n coordinates, to a space in which each "vector" is in fact a sequence, having an infinite number of "coordinates". If we try to apply the obvious generalization of Euclidean distance, then the distance between two sequences

$$X = \{x_1, x_2, \ldots, x_n, \ldots\} = \{x_n\}_{n=1}^{\infty} \quad \text{and}$$
$$Y = \{y_1, y_2, \ldots, y_n, \ldots\} = \{y_n\}_{n=1}^{\infty}$$

would be given by

$$\sqrt{\sum_{n=1}^{\infty} (x_n - y_n)^2}$$

and so the existence of such a distance depends on the convergence of a series.

Suppose we take Y to be the zero sequence $O = \{0, 0, \ldots, 0, \ldots\}$. If the distance from X to O is to be defined as above, we must require the series $\sum_{n=1}^{\infty} x_n^2$ to converge. By making this requirement, we can go on to develop a beautiful metric space known as l^2. It is one of a family of metric spaces named after the great mathematician David Hilbert.

DAVID HILBERT

(b. January 23, 1862; d. February 14, 1943)

After achieving fame in several areas of mathematics (the theory of invariants, algebraic number fields, and foundations of geometry), Hilbert turned to research in functional analysis. He is well remembered for his famous address before the Second International Congress of Mathematicians held in Paris in 1900. Looking to the coming century, Hilbert challenged the mathematical community by proposing 23 problems. He believed correctly that the pursuit of solutions to these problems would lead to new concepts, new methods, new problems, and new solutions. He enjoyed the company of men and women of outstanding intellect, especially at his traditional birthday celebrations. However, his last years were darkened by the loss of comradeship resulting from the political scene in Nazi Germany.

Definition 6.6.1 l^2 is the set of all real sequences $\{x_n\}_{n=1}^{\infty}$ such that the series $\sum_{n=1}^{\infty} x_n^2$ converges.

Example 1: The harmonic sequence $\{1/n\}_{n=1}^{\infty}$ belongs to l^2; $\{(-1)^n/\sqrt{n}\}_{n=1}^{\infty}$ does not.

Definition 6.6.2 If $X = \{x_n\}_{n=1}^{\infty} \in l^2$, then the ***norm*** of X is

$$\|X\| = \sqrt{\sum_{n=1}^{\infty} x_n^2}$$

The norm of an element of l^2 is analogous to the magnitude, or length, of a vector in R^n.

Example 2: For $k \in Z^+$, let $E_k = \{0, \ldots, 0, 1, 0, \ldots\}$ be the sequence with 1 in the kth position, 0 elsewhere. The sequences $E_1, E_2, \ldots$ are elements of l^2 and are analogous to the unit vectors in finite-dimensional spaces. Their norms are 1.

Inner Products

The inner product of two elements of l^2 is also a useful tool, but before we can define it properly we need a theorem about convergence.

Theorem 6.6.1 **Cauchy's Inequality** *If* $\sum_{n=1}^{\infty} x_n^2$ *and* $\sum_{n=1}^{\infty} y_n^2$ *are convergent series of real numbers, then the series* $\sum_{n=1}^{\infty} x_n y_n$ *is absolutely convergent, and*

$$\left|\sum_{n=1}^{\infty} x_n y_n\right| \le \sqrt{\sum_{n=1}^{\infty} x_n^2}\sqrt{\sum_{n=1}^{\infty} y_n^2}$$

PROOF: Our goal is to prove that the series $\sum_{n=1}^{\infty} |x_n y_n|$ is convergent. We accomplish this by showing that the partial sums $\sum_{n=1}^{m} |x_n y_n|$ are bounded. Now from Inequality (2) of Section 6.2, it follows (after replacing x_n by $|x_n|$ and y_n by $|y_n|$ and squaring) that

$$\left(\sum_{n=1}^{m} |x_n||y_n|\right)^2 \le \left(\sum_{n=1}^{m} |x_n|^2\right)\left(\sum_{n=1}^{m} |y_n|^2\right) = \left(\sum_{n=1}^{m} x_n^2\right)\left(\sum_{n=1}^{m} y_n^2\right)$$

Clearly, $\sum_{n=1}^{m} x_n^2 \le \sum_{n=1}^{\infty} x_n^2$ and $\sum_{n=1}^{m} y_n^2 \le \sum_{n=1}^{\infty} y_n^2$, so that the partial sums in question are bounded:

$$\sum_{n=1}^{m} |x_n y_n| \le \sqrt{\sum_{n=1}^{\infty} x_n^2}\sqrt{\sum_{n=1}^{\infty} y_n^2} = \|X\|\|Y\|$$

By Theorem 2.8.4, $\sum_{n=1}^{\infty} |x_n y_n|$ is convergent, and since an absolutely convergent series is convergent, we conclude that $\sum_{n=1}^{\infty} x_n y_n$ converges. For any $m \in \mathsf{Z}^+$, we have

$$\left|\sum_{n=1}^{m} x_n y_n\right| \le \sum_{n=1}^{m} |x_n y_n| \le \|X\|\|Y\|$$

and we can take the limit as $m \to \infty$:

$$\left|\sum_{n=1}^{\infty} x_n y_n\right| = \left|\lim_{m\to\infty} \sum_{n=1}^{m} x_n y_n\right| = \lim_{m\to\infty} \left|\sum_{n=1}^{m} x_n y_n\right|$$
$$\le \|X\|\|Y\| = \sqrt{\sum_{n=1}^{\infty} x_n^2}\sqrt{\sum_{n=1}^{\infty} y_n^2} \qquad \blacksquare$$

Because of Theorem 6.6.1, the series in the following definition converges.

Definition 6.6.3 For $X, Y \in l^2$ with $X = \{x_n\}_{n=1}^{\infty}$ and $Y = \{y_n\}_{n=1}^{\infty}$, the ***inner product*** of X and Y is

$$X \cdot Y = \sum_{n=1}^{\infty} x_n y_n$$

In terms of inner products and norms, Cauchy's Inequality can be stated as follows:

$$|X \cdot Y| \leq \|X\|\|Y\|$$

The Metric ρ

The next theorem shows that the sum of two elements of l^2 also belongs to l^2.

Theorem 6.6.2 **Minkowski's Inequality** *If $X, Y \in l^2$, then $X + Y \in l^2$ and*

$$\|X + Y\| \leq \|X\| + \|Y\|$$

PROOF: The sum of the two sequences $X = \{x_n\}_{n=1}^{\infty}$ and $Y = \{y_n\}_{n=1}^{\infty}$ is the sequence $\{x_n + y_n\}_{n=1}^{\infty}$. Now by Inequality (4) from Section 6.2, we have, for $m \in Z^+$,

$$\sqrt{\sum_{n=1}^{m} (x_n + y_n)^2} \leq \sqrt{\sum_{n=1}^{m} x_n^2} + \sqrt{\sum_{n=1}^{m} y_n^2}$$

Since $X, Y \in l^2$, the series $\sum_{n=1}^{\infty} x_n^2$ and $\sum_{n=1}^{\infty} y_n^2$ both converge and so

$$\sqrt{\sum_{n=1}^{m} (x_n + y_n)^2} \leq \sqrt{\sum_{n=1}^{\infty} x_n^2} + \sqrt{\sum_{n=1}^{\infty} y_n^2} = \|X\| + \|Y\|$$

If we square these terms, we obtain

$$\sum_{n=1}^{m} (x_n + y_n)^2 \leq (\|X\| + \|Y\|)^2$$

Since the partial sums of the series $\sum_{n=1}^{\infty} (x_n + y_n)^2$ are bounded, that series converges (Theorem 2.8.4) and therefore $X + Y \in l^2$. Moreover,

$$\sum_{n=1}^{\infty} (x_n + y_n)^2 \leq (\|X\| + \|Y\|)^2$$

so that

$$\|X + Y\| \leq \|X\| + \|Y\| \qquad \blacksquare$$

Clearly if $Y \in l^2$ and c is a real number, then $cY \in l^2$. It follows that $X - Y = X + (-1)Y \in l^2$ whenever $X, Y \in l^2$. We are in a position to define a metric ρ on l^2.

Definition 6.6.4 If $X, Y \in l^2$, then $\rho(X, Y) = \|X - Y\|$.

The function ρ is easily seen to be a metric. Given that $X, Y \in l^2$, then $X - Y \in l^2$ by the remarks above, and certainly $\|X - Y\|$ is a nonnegative real number. The only way the series defining $\|X - Y\|$ can be zero is for each term to be zero, hence $\rho(X, Y) = 0$ only if $X = Y$. Obviously, $\rho(X, Y) = \rho(Y, X)$, and the Triangle Inequality follows from Theorem 6.6.2.

Example 3: The sequences $X = \{1/2^n\}_{n=1}^{\infty}$ and $Y = \{1/3^n\}_{n=1}^{\infty}$ belong to l^2. The distance between these elements of l^2 is

$$\rho(X, Y) = \sqrt{\sum_{n=1}^{\infty} \left(\frac{1}{2^n} - \frac{1}{3^n}\right)^2} = \sqrt{\frac{7}{120}}$$

Example 4: Consider again the sequences $E_k (k \in Z^+)$ introduced in Example 2. The distance between E_k and $E_j (j \neq k)$ is equal to $\sqrt{2}$. Thus, the set $\{E_k : k \in Z^+\}$ is bounded.

Completeness of l^2

Theorem 6.6.3 **Completeness** *The metric space (l^2, ρ) is complete.*

PROOF: Let $\{X_k\}_{k=1}^{\infty}$ be a Cauchy sequence in l^2. We are to show that this sequence converges to some element X of l^2. Obviously a first step is to obtain a candidate for X.

Now each X_k is itself a sequence, say, $X_k = \{x_k^{(1)}, x_k^{(2)}, \ldots, x_k^{(j)}, \ldots\}$, in which $x_k^{(j)}$ is the jth "coordinate" of X_k. Since $X_k \in l^2$, the series $\sum_{j=1}^{\infty} \left(x_k^{(j)}\right)^2$ is convergent. (For convenience we will omit a pair of parentheses and write such a series as $\sum_{j=1}^{\infty} x_k^{(j)2}$.) We can arrange the sequence $\{X_k\}_{k=1}^{\infty}$ vertically in an array:

$$\begin{aligned}
X_1 &= \{x_1^{(1)}, x_1^{(2)}, \ldots, x_1^{(j)}, \ldots\} \\
X_2 &= \{x_2^{(1)}, x_2^{(2)}, \ldots, x_2^{(j)}, \ldots\} \\
&\vdots \\
X_k &= \{x_k^{(1)}, x_k^{(2)}, \ldots, x_k^{(j)}, \ldots\} \\
&\vdots
\end{aligned}$$

Looking at the vertical columns on the right, we see that for each $j \in Z^+$, $\{x_k^{(j)}\}_{k=1}^{\infty}$ is a Cauchy sequence in R: If K is a positive integer such that

$$\rho(X_k, X_l) < \varepsilon \qquad k, l \geq K$$

then, for that same K, we have for each $j \in Z^+$ and $k, l \geq K$,

$$|x_k^{(j)} - x_l^{(j)}|^2 \leq \sum_{j=1}^{\infty} (x_k^{(j)} - x_l^{(j)})^2 = [\rho(X_k, X_l)]^2 < \varepsilon^2$$

so that

$$|x_k^{(j)} - x_l^{(j)}| < \varepsilon \qquad k, l \geq K$$

Since $\{x_k^{(j)}\}_{k=1}^{\infty}$ is Cauchy, it converges to a real number, which we can denote by $x^{(j)}$. The sequence $\{x^{(j)}\}_{j=1}^{\infty}$ is our candidate for X. We need to show two things: first that $X \in l^2$, and then that $X_k \to X$.

Now since $\{X_k\}_{k=1}^{\infty}$ is Cauchy, it is bounded (by Theorem 6.4.2) and so the sequence of norms $\{\|X_k\|\}_{k=1}^{\infty}$ is bounded in R, say, by M. Thus, for every $k \in Z^+$,

$$\|X_k\|^2 = \sum_{j=1}^{\infty} x_k^{(j)2} \leq M^2$$

and therefore, for the partial sums, we have

$$\sum_{j=1}^{m} x_k^{(j)2} \leq M^2$$

We can take a limit of this finite sum as $k \to \infty$:

$$\lim_{k \to \infty} \sum_{j=1}^{m} x_k^{(j)2} = \sum_{j=1}^{m} \lim_{k \to \infty} x_k^{(j)2} = \sum_{j=1}^{m} x^{(j)2} \leq M^2$$

Thus, the partial sums of the series $\sum_{j=1}^{\infty} x^{(j)2}$ are bounded and the series converges, showing that

$$X = \{x^{(1)}, x^{(2)}, \ldots, x^{(j)}, \ldots\} \in l^2$$

Finally we show that $X_k \to X$. Let $\varepsilon > 0$ be given and use the Cauchy property to choose $K \in Z^+$ such that $\rho(X_k, X_l) < \varepsilon/2$ whenever $k, l \geq K$. Thus, for $k, l \geq K$, we have

$$\sum_{j=1}^{\infty} (x_k^{(j)} - x_l^{(j)})^2 < \frac{\varepsilon^2}{4}$$

Now any partial sum of this series is also less than $\varepsilon^2/4$:

$$\sum_{j=1}^{m} (x_k^{(j)} - x_l^{(j)})^2 < \frac{\varepsilon^2}{4} \qquad k, l \geq K$$

By holding k fixed for the moment, but still greater than or equal to K, and letting $l \to \infty$, we get

$$\lim_{l\to\infty} \sum_{j=1}^{m} (x_k^{(j)} - x_l^{(j)})^2 = \sum_{j=1}^{m} (x_k^{(j)} - x^{(j)})^2 \le \frac{\varepsilon^2}{4} \qquad k \ge K$$

Then

$$\sum_{j=1}^{\infty} (x_k^{(j)} - x^{(j)})^2 \le \frac{\varepsilon^2}{4} \qquad k \ge K$$

and so for $k \ge K$ we have

$$\rho(X_k, X) \le \frac{\varepsilon}{2} < \varepsilon \qquad \blacksquare$$

EXERCISES 6.6

1. Determine whether the following sequences are in l^2.

(a) $\left\{\frac{(-1)^n}{\sqrt[3]{n}}\right\}_{n=1}^{\infty}$

(b) $\left\{\frac{1}{n^{2/3}}\right\}_{n=1}^{\infty}$

(c) $\left\{\frac{2^{n+1}}{3^n}\right\}_{n=1}^{\infty}$

(d) $\left\{\frac{1}{n} \cos \frac{n\pi}{2}\right\}_{n=1}^{\infty}$

2. Find $\rho(X, Y)$ for the following pairs of elements of l^2.

(a) $X = \{(2/3)^n\}_{n=1}^{\infty}$; $Y = \{1/3^n\}_{n=1}^{\infty}$

(b) $X = \{1/n\}_{n=1}^{\infty}$; $Y = \{-1/n\}_{n=1}^{\infty}$ (Recall that $\sum_{n=1}^{\infty} 1/n^2 = \pi^2/6$.)

(c) $X = \{1/n\}_{n=1}^{\infty}$; $Y = \{(-1)^n/n\}_{n=1}^{\infty}$ [*Hint:* $\sum_{k=1}^{\infty} 1/(2k-1)^2 = \sum_{n=1}^{\infty} 1/n^2 - \sum_{k=1}^{\infty} 1/(2k)^2$]

3. Let $S = \{E_k : k \in Z^+\}$, where E_k is defined as in Example 2.

(a) Find the diameter of S.

(b) Show that S consists only of isolated points.

(c) Show that the Bolzano-Weierstrass property (Corollary 6.4.1) does not hold in l^2.

(d) Show that there is a closed bounded subset of l^2 that is not compact.

4. Let us call a sequence $X = \{x_n\}_{n=1}^{\infty}$ *rational* provided x_n is rational for all $n \in Z^+$; also, let us call X *terminating* if $x_n = 0$ for all sufficiently large n. Now prove or disprove the following statements.

(a) The set of rational sequences belonging to l^2 is dense in l^2.

(b) The set of terminating rational sequences is dense in l^2.

(c) The set of rational sequences belonging to l^2 is countable.

(d) The set of terminating rational sequences is countable.

(e) l^2 is separable.

APPENDIX

THE ISOMORPHISM THEOREM FOR ORDERED FIELDS

The statement "R is a field" does not describe the real number system very adequately because there are many fields that are quite different from R. Neither does the statement "R is an ordered field" give an adequate description because there are also ordered fields that are different from R. However, there is essentially only one *complete* ordered field* in the sense that any two complete ordered fields are structurally identical (*order-isomorphic* is the technical term). It is the purpose of this appendix to explain the meaning of order-isomorphic and to prove the Isomorphism Theorem.

Subfields

Let F be an arbitrary field with operations addition and multiplication, which we will denote by the usual symbols ($+$ for addition and $\cdot$ or juxtaposition for multiplication).

Definition A.1 A subset S of a field F is ***closed*** under addition (multiplication) provided that

$$a, b \in S \Rightarrow a + b \in S \qquad (a, b \in S \Rightarrow ab \in S)$$

The set of rational numbers is closed under addition and multiplication in R. In fact, we know that the rationals form a field under these operations.

*Ordered fields and completeness are defined in Sections R.1 and 1.2.

Definition A.2 Let F be a field. A ***subfield*** of F is a nonempty subset S of F that forms a field under the operations of F.

Example 1: Q is a subfield of R.

Example 2: The set $Q[\sqrt{2}] = \{a + b\sqrt{2} : a, b \in Q\}$ is a field under ordinary addition and multiplication and therefore is a subfield of R. (To see this for yourself, work Exercise R.1.6.)

Among the subfields of R, Q is the "smallest" in the sense that it is contained in every subfield of R.

Theorem A.1 *Every subfield of R contains the rational numbers.*

PROOF: Let S be any subfield of R. Since S is a field, it has a zero and a unity, which we denote by z and u, respectively. The equation

$$z + u = u$$

must hold in S and therefore also in R. By subtracting u from both sides, we get $z = 0$, showing that the zero of the subfield S must be the real number 0. It can be shown similarly that the unity of S is the real number 1.

Because $1 \in S$ and S is closed under addition, we have $2 = 1 + 1 \in S$, $3 = 2 + 1 \in S$, and so on. By induction, every positive integer belongs to S. Since S contains the opposite of each of its elements, the negative integers also belong to S. Finally, since division (except by 0) is possible in every field, we conclude that $p/q \in S$ whenever $p, q \in Z$ and $q \neq 0$. Thus, $S \supseteq Q$. ∎

The following theorem is useful for determining whether a subset of a given field is a subfield.

Theorem A.2 *Let S be a subset of an arbitrary field F and suppose that S contains at least two elements. Then S is a subfield of F if and only if the following conditions hold:*

(i) *S is closed under the addition and multiplication of F*
(ii) *The opposite of each element of S is also an element of S*
(iii) *The reciprocal of each nonzero element of S is also an element of S.*

PROOF: The commutative, associative, and distributive laws hold in S because they hold in F and S is a subset of F. The field postulates that state the existence of certain elements follow immediately from (i), (ii), and (iii). Thus, (i), (ii), and (iii) imply that S is a subfield of F. Conversely, any subfield of F must satisfy (i), (ii), and (iii) by definition. ∎

Preservation of Operations and Order

Definition A.3 Let S be a subset of a field F_1 and let $\varphi:S \to F_2$ be a mapping of S into a field F_2.

(i) If S is closed under the addition of F_1, then φ ***preserves addition*** provided that for $x, y \in S$,

$$\varphi(x + y) = \varphi(x) + \varphi(y)$$

(ii) If S is closed under the multiplication of F_1, then φ ***preserves multiplication*** provided that for $x, y \in S$

$$\varphi(xy) = \varphi(x)\,\varphi(y)$$

(iii) If F_1 and F_2 are ordered fields, then φ ***preserves order*** provided that for $x, y \in S$,

$$x < y \Rightarrow \varphi(x) < \varphi(y)$$

Note: We have taken some notational liberties here. For instance, in the equation

$$\varphi(x + y) = \varphi(x) + \varphi(y)$$

the plus sign on the left side refers to addition in F_1 and the plus sign on the right side refers to the addition in F_2. Since there is no other reasonable interpretation, we allow ourselves to use the same symbol for both operations. For similar reasons we have used the same symbols for the two multiplication operations, and also for the two order relations.

To say that $\varphi:S \to F_2$ preserves addition means that if the equation

$$z = x + y$$

holds in S, then the equation

$$\varphi(z) = \varphi(x) + \varphi(y)$$

holds in F_2 (see Figure A.1). Similar remarks can be made concerning preservation of multiplication and preservation of order.

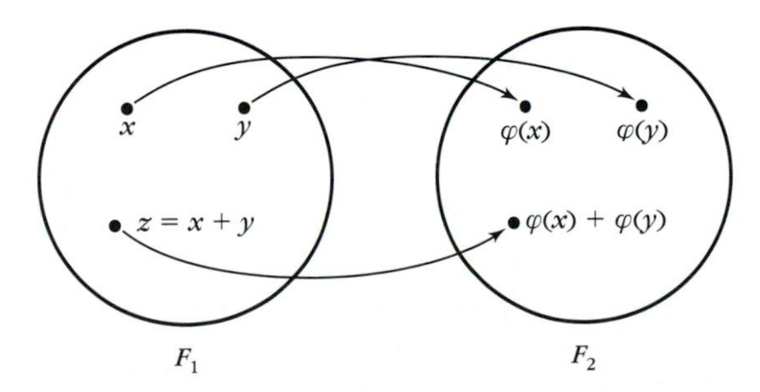

Figure A.1
Preservation of addition

Example 3: Let Q be the rationals and let $\varphi: Q \to Q$ be defined by $\varphi(r) = 2r$. Then φ preserves addition because $\varphi(r_1 + r_2) = 2(r_1 + r_2) = 2r_1 + 2r_2 = \varphi(r_1) + \varphi(r_2)$. It also preserves order because if $r_1 < r_2$, then $2r_1 < 2r_2$. However, φ does not preserve products.

Isomorphism

When the structures of two algebraic systems are similar in all details they are called *isomorphic* (which means literally "having the same form"). A simple example of isomorphism will be found by considering the finite fields F_1 and F_2 with addition and multiplication tables:

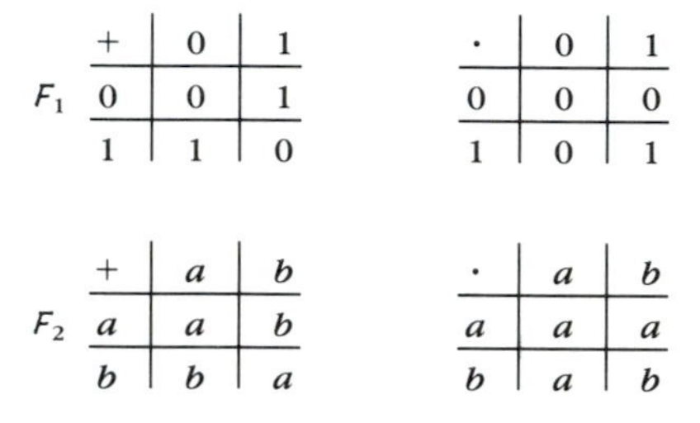

F_1

+	0	1
0	0	1
1	1	0

·	0	1
0	0	0
1	0	1

F_2

+	*a*	*b*
a	*a*	*b*
b	*b*	*a*

·	*a*	*b*
a	*a*	*a*
b	*a*	*b*

Figure A.2

Observe that F_1 and F_2 each have two elements, and if we replace the symbols "0" and "1" in the tables for F_1, respectively, by "*a*" and "*b*", we obtain the tables for F_2.

Definition A.4 Let F_1 and F_2 be fields. Then F_1 and F_2 are ***isomorphic*** provided there exists a bijection $\varphi: F_1 \to F_2$ that preserves addition and multiplication; such a bijection $\varphi: F_1 \to F_2$ is an ***isomorphism***.

An isomorphism preserves sums and products by definition; the following theorem shows that it preserves other algebraic properties as well.

Theorem A.3 *Let F_1 and F_2 be fields with zeros z_1, z_2 and unities u_1, u_2, respectively. Let $\varphi: F_1 \to F_2$ be an isomorphism. Then:*

(i) $\varphi(z_1) = z_2$

(ii) $\varphi(u_1) = u_2$

(iii) *For* $x, y \in F_1$, $\varphi(x - y) = \varphi(x) - \varphi(y)$

(iv) *For* $x, y \in F_1$ *and* $y \neq z_1$, $\varphi\left(\dfrac{x}{y}\right) = \dfrac{\varphi(x)}{\varphi(y)}$

PROOF: (i) For any fixed element a of F_1,

$$\begin{aligned}\varphi(z_1) + \varphi(a) &= \varphi(z_1 + a)\\ &= \varphi(a)\\ &= z_2 + \varphi(a)\end{aligned}$$

Subtracting $\varphi(a)$ gives

$$\varphi(z_1) = z_2$$

Thus, under any isomorphism, the zeros must correspond. The parallel statement (ii) about unities is proved similarly.

(iii) We have $x - y + y = x$, and because φ preserves sums,

$$\begin{aligned} \varphi(x - y) + \varphi(y) &= \varphi(x) \\ \varphi(x - y) &= \varphi(x) - \varphi(y) \end{aligned}$$

(iv) You are asked to prove property (iv) in Exercise A.4. ∎

Theorem A.4 *Let F_1, F_2 be fields. If $\varphi: F_1 \to F_2$ is an isomorphism, then $\varphi^{-1}: F_2 \to F_1$ is an isomorphism.*

PROOF: We learned in Section C.1 that $\varphi^{-1}: F_2 \to F_1$ is a bijection. Now let $u, v \in F_2$ and set $\varphi^{-1}(u) = x$ and $\varphi^{-1}(v) = y$. By hypothesis, φ is an isomorphism, so

$$\begin{aligned} \varphi(x + y) &= \varphi(x) + \varphi(y) \\ &= u + v \end{aligned}$$

Applying φ^{-1} to both sides gives

$$\begin{aligned} \varphi^{-1}[\varphi(x + y)] &= \varphi^{-1}(u + v) \\ x + y &= \varphi^{-1}(u + v) \end{aligned}$$

and, finally,

$$\varphi^{-1}(u + v) = \varphi^{-1}(u) + \varphi^{-1}(v)$$

Thus, φ^{-1} preserves sums, and a similar proof shows that it preserves products. ∎

Isomorphism of fields is an equivalence relation; that is:

(1) *Reflexive law:* Any field is isomorphic to itself (work Exercise A.5).
(2) *Symmetric law:* If F_1 and F_2 are fields such that F_1 is isomorphic to F_2, then F_2 is isomorphic to F_1 (by Theorem A.4).
(3) *Transitive law:* If F_1, F_2, and F_3 are fields such that F_1 is isomorphic to F_2 and F_2 is isomorphic to F_3, then F_1 is isomorphic to F_3 (work Exercise A.6).

Note: An injective (one-to-one) mapping $\varphi: F_1 \to F_2$ may not map F_1 onto all of F_2; however, the range of φ is a subfield of F_2, and this subfield is isomorphic to F_1. (Exercise A.7 asks you to verify these facts.)

Theorem A.1 says that every subfield of the reals contains the rationals. We now show that something like this is true for an arbitrary ordered field: namely, *every* ordered field has a subfield that is isomorphic to the rationals by means of an isomorphism that preserves order.

Definition A.5 Let F_1 and F_2 be ordered fields. Then F_2 is ***order-isomorphic*** to F_1 provided that there exists an isomorphism $\varphi: F_1 \to F_2$ that preserves order.

Order-isomorphism, like isomorphism, satisfies the properties of an equivalence relation: it is reflexive, symmetric, and transitive.

Theorem A.5 *Let F be an arbitrary ordered field. There exists a subfield Q_0 of F that is order-isomorphic to Q.*

PROOF: We construct Q_0 and the required isomorphism in stages. In the first stage we look for a subset of F that behaves like the positive integers. It is natural to associate the positive integer 1 with the unity u of F. Then we can associate 2 with $u + u$, 3 with $(u + u) + u$, and so on. To carry out this idea, consider the mapping $\varphi_0: Z^+ \to F$, defined recursively as follows:

(i) $\varphi_0(1) = u$
(ii) $\varphi_0(k + 1) = \varphi_0(k) + u \qquad k \in Z^+$

Let us denote the range of φ_0 by Z_0^+, so that $\varphi_0: Z_0 \to Z_0^+$ is surjective. By induction it is easy to verify that the elements of Z_0^+ are positive elements of F.

To prove that φ_0 preserves sums, we must show that the equation

$$\varphi_0(m + n) = \varphi_0(m) + \varphi_0(n) \tag{1}$$

holds for all $m, n \in Z^+$. We let m be an arbitrary but fixed positive integer, and then proceed by induction on n.

It is clear that Equation (1) holds for $n = 1$:

$$\begin{aligned} \varphi_0(m + 1) &= \varphi_0(m) + u && \text{by Equation (ii)} \\ &= \varphi_0(m) + \varphi_0(1) && \text{by Equation (i)} \end{aligned}$$

Next we assume that (1) holds for $n = k$ and show that it also holds for $n = k + 1$. The following chain of equations begins with the associative law in Z^+:

$$\begin{aligned} \varphi_0(m + (k + 1)) &= \varphi_0((m + k) + 1) \\ &= \varphi_0(m + k) + u && \text{by Equation (ii)} \\ &= [\varphi_0(m) + \varphi_0(k)] + u && \text{by the induction hypothesis} \\ &= \varphi_0(m) + [\varphi_0(k) + u] && \text{by the associative law for addition in } F \\ &= \varphi_0(m) + \varphi_0(k + 1) && \text{by Equation (ii)} \end{aligned}$$

Thus, Equation (1) holds for $n = k + 1$ whenever it holds for $n = k$, and by mathematical induction, (1) holds for all $n \in Z^+$. Since m is an arbitrary element of Z^+, (1) holds for all $m, n \in Z^+$. We conclude that φ_0 preserves sums. The fact that φ_0 also preserves products is left for you to prove by induction in Exercise A.8.

If $m < n$, then $n = m + p$ for some positive integer p, and by (1),

$$\varphi_0(n) = \varphi_0(m + p) = \varphi_0(m) + \varphi_0(p) > \varphi_0(m)$$

showing that φ_0 preserves order. Since any order preserving mapping is necessarily injective, we see that $\varphi_0 : Z^+ \to Z_0^+$ is a bijection.

In the next stage we extend φ_0 to the positive rationals. We define a mapping φ_1 from the positive rationals into F that agrees with φ_0 on the positive integers; that is, if $n \in Z^+$, then $\varphi_1(n) = \varphi_0(n)$. To do this, we let r be a positive rational and suppose that $r = p/q$, where $p, q \in Z^+$; then we define $\varphi_1(r)$ to be $\varphi_0(p)/\varphi_0(q)$. (Even though there are many ways of expressing r as a quotient, the definition of $\varphi_1(r)$ leads to a unique value; you are asked to verify this fact in Exercise 9a.) Now φ_1 does indeed agree with φ_0 on the positive integers because for $n \in Z^+$,

$$\varphi_1(n) = \varphi_1\left(\frac{n}{1}\right) = \frac{\varphi_0(n)}{\varphi_0(1)} = \frac{\varphi_0(n)}{u} = \varphi_0(n)$$

That φ_1 preserves addition is seen as follows. If $r = p_1/q_1$ and $s = p_2/q_2$, where p_1, p_2, q_1, q_2 are positive integers; then

$$\begin{aligned}
\varphi_1(r + s) &= \varphi_1\left(\frac{p_1q_2 + p_2q_1}{q_1q_2}\right) \\
&= \frac{\varphi_0(p_1q_2 + p_2q_1)}{\varphi_0(q_1q_2)} \\
&= \frac{\varphi_0(p_1)\varphi_0(q_2) + \varphi_0(p_2)\varphi_0(q_1)}{\varphi_0(q_1)\varphi_0(q_2)} \qquad \text{because } \varphi_0 \text{ preserves addition and multiplication} \\
&= \frac{\varphi_0(p_1)}{\varphi_0(q_1)} + \frac{\varphi_0(p_2)}{\varphi_0(q_2)} = \varphi_1(r) + \varphi_1(s)
\end{aligned}$$

Exercise 9 asks you to prove that φ_1 preserves multiplication and order, and that φ_1 is injective.

The final stage in our construction is to extend φ_1 to all rationals. We define a mapping $\varphi : Q \to F$ that extends φ_1 by making the opposite of r correspond to the opposite of $\varphi_1(r)$:

$$\varphi(r) = \begin{cases} \varphi_1(r) & \text{if } r > 0 \\ z & \text{if } r = 0 \\ -\varphi_1(-r) & \text{if } r < 0 \end{cases}$$

It can be shown that φ preserves sums and products (we omit the details). Because φ preserves sums and $\varphi_1(r)$ is positive when r is a positive rational number, it follows that φ preserves order; so φ is injective. If Q_0 is the range of φ, then $\varphi : Q \to Q_0$ is an order-preserving isomorphism, and so Q_0 is a subfield of F that is order-isomorphic to the rationals. ■

Isomorphism of Complete Ordered Fields

We now turn to the proof that any two complete ordered fields are order-isomorphic. We do this by showing that if F is a complete ordered field, then F is order-isomorphic to R.

As we have just seen, every ordered field F contains a subfield Q_0 that is order-isomorphic to the rationals. It is natural to call the elements of Q_0 the "rational elements of F". Likewise, the elements of Z_0^+ in the proof of Theorem A.5 will be called the "positive integers of F".

If the ordered field F is complete, then it is easy to show that F is Archimedean and that a "Rational Density Theorem" holds for F: that is, between any two distinct elements of F there is a rational element of F. The proofs of these facts are just like the proofs of the corresponding facts for R (Section 1.2).

The rational density property makes it possible to prove a useful lemma:

Lemma A.1 *Let F be a complete ordered field, let Q_0 be the set of rational elements of F, and let x be an arbitrary element of F. Then*

$$x = \sup\{r : r \in Q_0,\ r < x\}$$

PROOF: Clearly x is an upper bound of $S = \{r : r \in Q_0, r < x\}$. If $y < x$, there exists by the Rational Density Theorem a rational element r such that $y < r < x$, showing that y is not an upper bound of S. Therefore, x is the least upper bound of S. ∎

Theorem A.6 **Isomorphism Theorem** *Every complete ordered field is order-isomorphic to R*

PROOF: Let F be a complete ordered field. By Theorem A.5, there is an order-preserving isomorphism φ that maps the set Q of rationals in R onto the set Q_0 of rational elements of F. We now extend φ to an isomorphism Φ mapping R onto F. For each $x \in R$ we let $S_x = \{r : r \in Q \text{ and } r < x\}$. Then the set $\varphi(S_x) = \{\varphi(r) : r \in S_x\}$ is bounded above in F (why?) and by the completeness of F, $\sup \varphi(S_x)$ exists. We define $\Phi(x) = \sup \varphi(S_x)$.

(i) *Φ extends φ.* We must show that $\Phi(r_0) = \varphi(r_0)$ whenever $r_0 \in Q$. The set $S_{r_0} = \{r : r \in Q, r < r_0\}$ is mapped by φ one-to-one onto a subset of Q_0, namely, $\varphi(S_{r_0}) = \{s : s \in Q_0, s < \varphi(r_0)\}$, and so $\Phi(r_0) = \sup \varphi(S_{r_0}) = \varphi(r_0)$ by Lemma A.1.

(ii) *Φ preserves order* We must show that if $x_1 < x_2$ in R, then $\Phi(x_1) < \Phi(x_2)$ in F. By the Rational Density Theorem there exist $r_1, r_2 \in Q$ such that

$$x_1 < r_1 < r_2 < x_2$$

Therefore,

$$S_{x_1} \subseteq S_{r_1} \subseteq S_{r_2} \subseteq S_{x_2}$$

and

$$\varphi(S_{x_1}) \subseteq \varphi(S_{r_1}) \subseteq \varphi(S_{r_2}) \subseteq \varphi(S_{x_2})$$

Consequently,

$$\Phi(x_1) \leq \Phi(r_1) \leq \Phi(r_2) \leq \Phi(x_2)$$

Since $r_1, r_2 \in Q$, we know that $r_1 < r_2$ implies $\varphi(r_1) < \varphi(r_2)$, and since Φ extends φ, $\Phi(r_1) < \Phi(r_2)$, and so $\Phi(x_1) < \Phi(x_2)$.

(iii) Φ *is injective* An order preserving mapping is injective.

(iv) Φ *is surjective* We must show that if $y \in F$, then there is an $x \in R$ such that $\Phi(x) = y$. We let $T_y = \{s : s \in Q_0 \text{ and } s < y\}$. The set $\varphi^{-1}(T_y) = \{r : r \in Q \text{ and } \varphi(r) < y\}$ is nonempty, bounded above in R, and we can let $x = \sup \varphi^{-1}(T_y)$. We claim that $\Phi(x) = y$. First, $\varphi^{-1}(T_y) \subseteq S_x$ by the definition of x. To show the reverse inclusion, we let $v \in S_x$ so that $v < x$ and $v \in Q$. Since $x = \sup \varphi^{-1}(T_y)$, v is not an upper bound of $\varphi^{-1}(T_y)$, so there is an element $w \in \varphi^{-1}(T_y)$ such that $v < w < x$. Then $\varphi(w) \in T_y$ and $\varphi(w) < y$; hence, $\varphi(v) < \varphi(w) < y$, and so $v \in \varphi^{-1}(T_y)$. Therefore, $S_x = \varphi^{-1}(T_y)$ and $\Phi(x) = \sup \varphi(S_x) = \sup \varphi(\varphi^{-1}(T_y)) = \sup T_y = y$.

The proofs of the remaining steps are outlined, with details left for you to work out in Exercise A.10. The notation $S + T$, where S and T are subsets of a field, means $\{s + t : s \in S, t \in T\}$, and ST means $\{st : s \in S, t \in T\}$.

(v) Φ *preserves addition* For $a, b \in R$, we see that $S_{a+b} = S_a + S_b$. Now

$$\Phi(a + b) = \sup \varphi(S_{a+b}) = \sup \varphi(S_a + S_b)$$

Since φ preserves addition, it follows that

$$\begin{aligned} \sup \varphi(S_a + S_b) &= \sup(\varphi(S_a) + \varphi(S_b)) \\ &= \sup \varphi(S_a) + \sup \varphi(S_b) \\ &= \Phi(a) + \Phi(b) \end{aligned}$$

(vi) Φ *preserves multiplication* For positive $x \in R$, we define $P_x = \{r : r \in Q, 0 < r < x\}$ and observe that $\sup \varphi(P_x) = \sup \varphi(S_x) = \Phi(x)$ for $x > 0$. Now if $a, b > 0$, we have

$$P_{ab} = P_a P_b$$

and since φ preserves multiplication,

$$\varphi(P_{ab}) = \varphi(P_a)\varphi(P_b)$$

Then

$$\sup \varphi(P_{ab}) = \sup \varphi(P_a) \sup \varphi(P_b)$$

so that

$$\Phi(ab) = \Phi(a)\Phi(b)$$

We have proved that Φ preserves products of *positive* elements. But we saw above that Φ preserves addition, and so it also preserves opposites. On this basis we can show that Φ preserves products in general. For example, if $a < 0$, $b > 0$, we already know that

$$\Phi((-a)b) = \Phi(-a)\Phi(b) = [-\Phi(a)]\Phi(b) = -\Phi(a)\Phi(b)$$

while on the other hand,

$$\Phi((-a)b) = \Phi(-ab) = -\Phi(ab)$$

Thus,

$$-\Phi(ab) = -\Phi(a)\Phi(b)$$

and so, finally,

$$\Phi(ab) = \Phi(a)\Phi(b) \qquad \blacksquare$$

EXERCISES A

1. Let $Q[\sqrt{3}]$ consist of all real numbers of the form $a + b\sqrt{3}$, where $a, b \in Q$. Use Theorem A.1 to show that $Q[\sqrt{3}]$ is a subfield of R.

2. (a) Let $\varphi: Q \to Q$ be defined by $\varphi(x) = x^2$. Does φ preserve addition? Multiplication? Order?
   (b) Let $\varphi: Z^+ \to Z^+$ be defined by $\varphi(x) = x^2$. Does φ preserve addition? Multiplication? Order?

3. (a) Let $\varphi: Q[\sqrt{2}] \to Q[\sqrt{2}]$ be defined by $\varphi(a + b\sqrt{2}) = a - b\sqrt{2}$. Show that φ is an isomorphism.
   (b) Assuming $Q[\sqrt{3}]$ is a field (see Exercise 1), define $\varphi: Q[\sqrt{2}] \to Q[\sqrt{3}]$ by $\varphi(a + b\sqrt{2}) = a + b\sqrt{3}$. Determine whether φ is an isomorphism.

4. Prove part (iv) of Theorem A.3.

5. Show that any field is isomorphic to itself.

6. Show that isomorphism is transitive: If F_1, F_2, F_3 are fields and if F_1 is isomorphic to F_2 and F_2 is isomorphic to F_3, then F_1 is isomorphic to F_3.

7. Show that if F_1 and F_2 are fields and $\varphi: F_1 \to F_2$ is an injective mapping that preserves addition and multiplication, then the range of φ is a subfield of F_2 and is isomorphic to F_1.

8. Show that φ_0, defined in the proof of Theorem A.5, preserves multiplication.

9. The following exercises refer to the proof of Theorem A.5.
   (a) Show that φ_1, defined in the proof of Theorem A.5, is well defined: If $p_1/q_1 = p_2/q_2$, then $\varphi_0(p_1)/\varphi_0(q_1) = \varphi_0(p_2)/\varphi_0(q_2)$.
   (b) Show that φ_1 preserves multiplication and order.
   (c) Explain why the fact that φ_1 preserves order implies that φ_1 is injective.

10. The following exercises refer to parts (v) and (vi) of the proof of Theorem A.6.
   (a) Show that $S_{a+b} = S_a + S_b$.
   (b) Show that $\varphi(S_a + S_b) = \varphi(S_a) + \varphi(S_b)$.
   (c) Show that $\sup(\varphi(S_a) + \varphi(S_b)) = \sup \varphi(S_a) + \sup \varphi(S_b)$.
   (d) Show that, for $x > 0$, $\sup \varphi(P_x) = \sup \varphi(S_x)$.
   (e) Show that, for $a, b > 0$, $P_{ab} = P_a P_b$.
   (f) Show that $\Phi(ab) = \Phi(a)\Phi(b)$ where $a < 0$ and $b < 0$.

11. A construction of the field C of complex numbers is outlined in the project following Section R.1.
   (a) Show that R is isomorphic to the subset of C consisting of all ordered pairs of the form $(x, 0)$ for $x \in R$.
   (b) Show that the subset of C consisting of all ordered pairs (r_1, r_2), where $r_1, r_2 \in Q$, is a subfield of C.

LIST OF SPECIAL SYMBOLS

Below are listed symbols used in this book that have special meaning, and the page number where each symbol is introduced.

Limit Statements

$s_n \to L$ (or $\lim_{n\to\infty} s_n = L$) **(144)**

$s_n \to \infty$ **(157)** $s_n \to -\infty$ **(157)**

$s_n(x) \to s(x)$ $(x \in D)$ **(368)**

$s_n(x) \rightrightarrows s(x)$ $(x \in D)$ **(372)**

$f(x) \to L$ as $x \to a$ (or $\lim_{x\to a} f(x) = L$) **(201)**

$f(x) \to L$ as $x \to a^+$ (or $\lim_{x\to a^+} f(x) = L$) **(209)**

$f(x) \to L$ as $x \to a^-$ (or $\lim_{x\to a^-} f(x) = L$) **(209)**

$f(x) \to \infty$ as $x \to a$ (or $\lim_{x\to a} f(x) = \infty$) **(212)**

$f(x) \to -\infty$ as $x \to a$ (or $\lim_{x\to a} f(x) = -\infty$) **(212)**

$f(x) \to L$ as $x \to \infty$ (or $\lim_{x\to\infty} f(x) = L$) **(213)**

$f(x) \to \infty$ as $x \to \infty$ (or $\lim_{x\to\infty} f(x) = \infty$) **(214)**

Metric Space Notation

$x_n \to L$ (or $\lim_{n\to\infty} x_n = L$) **(415)**
$\|X\|$ **(418, 438)**
$X \cdot Y$ **(418, 440)**
$\rho_n(X, Y)$ **(420)**
$\rho(x, b)$ **(434)**
$\rho(x, B)$ **(435)**

Special Functions and Relations

$d|n$ **(21)**
xRy **(36)**
$a \equiv_m b$ **(39)**
$|a|$ **(69)**
$[x]$ **(79)**
$n!$ **(93)**
a^n **(93)**
$\binom{n}{k}$ **(99)**
$|A| = |B|$ **(111)**

Function Notation

$f(x)$ **(44)**
$f(S)$ **(45)**
$f: A \to B$ **(45)**
$g \circ f$ **(46)**
f^{-1} **(47)**
$f^{-1}(S)$ **(256)**

Summation Symbols

$\sum_{k=1}^{n} u_k$ (97) $\sum_{n=1}^{\infty} a_n$ (175) $\sum_{k=1}^{\infty} u_k(x)$ (384)

Calculus Symbols

$f', f'(x), \dfrac{dy}{dx}, D_x f(x), D_x y$ (260, 261)

$f'', f''', \ldots, f^{(n)}$ (272)

du (307)

$P[a, b]$ (280)

$U_\pi(f), L_\pi(f)$ (281)

$R(f; \pi; \xi_1, \ldots, \xi_n)$ (350)

$\underline{\int_a^b} f, \overline{\int_a^b} f, \int_a^b f$ (283)

$\int_a^\infty f, \int_{-\infty}^a f, \int_{-\infty}^\infty f$ (334, 336)

SUGGESTED REFERENCES

Analysis

Bartle, Robert G., *The Elements of Real Analysis,* 2nd ed., Wiley, New York, 1976.

Bartle, Robert G. and Donald R. Sherbert, *Introduction to Real Analysis,* 2nd ed., Wiley, New York, 1992.

Boas, Ralph P., Jr., *A Primer of Real Functions,* 3rd ed., Carus Monograph Number 13, Mathematical Association of America, 1981.

Clark, Colin W., *Elementary Mathematical Analysis,* 2nd ed., Wadsworth, Belmont, California, 1982.

Goffman, Casper, *Introduction to Real Analysis,* Harper and Row, New York, 1966.

Goldberg, Richard R., *Methods of Real Analysis,* 2nd ed., Wiley, New York, 1976.

Olmsted, John M.H., *Advanced Calculus,* Prentice Hall, Englewood Cliffs, New Jersey, 1961.

History, Biography, Foundations

Belhoste, Bruno *Augustin-Louis Cauchy A Biography,* Springer-Verlag, New York, 1991.

Boyer, Carl B., *History of Mathematics,* Wiley, New York, 1989.

Burton, David M., *Elementary Number Theory,* Wm. C. Brown, Dubuque, Iowa, 1994.

Burton, David M., *The History of Mathematics,* Wm. C. Brown, Dubuque, Iowa, 1991.

Cantor, Georg, *Contributions to the Founding of the Theory of Transfinite Numbers,* Dover Publications, New York, 1955.

Dunham, William, *Journey Through Genius,* Wiley, New York, 1990.

Eves, Howard, *An Introduction to the History of Mathematics,* Saunders, Philadelphia, 1990.

Grattan-Guinness, *The Development of the Foundations of Mathematical Analysis from Euler to Riemann,* MIT Press, Cambridge, 1970.

Hollingdale, Stuart, *Makers of Mathematics,* Penguin Books, London, 1989.

Reid, Constance, *Hilbert,* Springer-Verlag, New York, 1970.

Simmons, George F., *Calculus Gems,* McGraw Hill, New York, 1992.

Stillwell, John, *Mathematics and its History,* Springer-Verlag, New York, 1989.
Struik, Dirk J., *A Concise History of Mathematics,* Dover Publications, New York, 1987.
Wilder, R.L., *The Foundations of Mathematics,* Wiley, New York, 1965.

INDEX

D

E

F

G

H

I

L

M

N

O

P

Q

R

S

T